Lanchester Library
WITHDRAWN

Handbook of Functional Lipids

FUNCTIONAL FOODS AND NUTRACEUTICALS SERIES

Series Editor

G. Mazza, Ph.D.

Senior Research Scientist and Head
Food Research Program
Pacific Agri-Food Research Centre
Agriculture and Agri-Food Canada
Summerland, British Columbia

Functional Foods: Biochemical and Processing Aspects
Volume 1
Edited by G. Mazza, Ph.D.

Herbs, Botanicals, and Teas
Edited by G. Mazza, Ph.D. and B.D. Oomah, Ph.D.

Functional Foods: Biochemical and Processing Aspects
Volume 2
Edited by John Shi, Ph.D., G. Mazza, Ph.D., and Marc Le Maguer, Ph.D.

Methods of Analysis for Functional Foods and Nutraceuticals
Edited by W. Jeffrey Hurst, Ph.D.

Functional Foods: Biochemical and Processing Aspects
Volume 3
Edited by John Shi, Ph.D., G. Mazza, Ph.D., and Marc Le Maguer, Ph.D.

Handbook of Functional Dairy Products
Edited by Collete Short and John O'Brien

Handbook of Fermented Functional Foods
Edited by Edward R. Farnworth, Ph.D.

Handbook of Functional Lipids
Edited by Casimir C. Akoh, Ph.D.

Handbook of Functional Lipids

EDITED BY

Casimir C. Akoh

Boca Raton London New York Singapore

A CRC title, part of the Taylor & Francis imprint, a member of the Taylor & Francis Group, the academic division of T&F Informa plc.

Coventry University

Published in 2006 by
CRC Press
Taylor & Francis Group
6000 Broken Sound Parkway NW, Suite 300
Boca Raton, FL 33487-2742

© 2006 by Taylor & Francis Group, LLC
CRC Press is an imprint of Taylor & Francis Group

No claim to original U.S. Government works
Printed in the United States of America on acid-free paper
10 9 8 7 6 5 4 3 2 1

International Standard Book Number-10: 0-8493-2162-X (Hardcover)
International Standard Book Number-13: 978-0-8493-2162-7 (Hardcover)
Library of Congress Card Number 2005041864

This book contains information obtained from authentic and highly regarded sources. Reprinted material is quoted with permission, and sources are indicated. A wide variety of references are listed. Reasonable efforts have been made to publish reliable data and information, but the author and the publisher cannot assume responsibility for the validity of all materials or for the consequences of their use.

No part of this book may be reprinted, reproduced, transmitted, or utilized in any form by any electronic, mechanical, or other means, now known or hereafter invented, including photocopying, microfilming, and recording, or in any information storage or retrieval system, without written permission from the publishers.

For permission to photocopy or use material electronically from this work, please access www.copyright.com (http://www.copyright.com/) or contact the Copyright Clearance Center, Inc. (CCC) 222 Rosewood Drive, Danvers, MA 01923, 978-750-8400. CCC is a not-for-profit organization that provides licenses and registration for a variety of users. For organizations that have been granted a photocopy license by the CCC, a separate system of payment has been arranged.

Trademark Notice: Product or corporate names may be trademarks or registered trademarks, and are used only for identification and explanation without intent to infringe.

Library of Congress Cataloging-in-Publication Data

Handbook of functional lipids / edited by Casimir C. Akoh.
p. cm. -- (Functional foods and nutraceuticals series ; 7)
Includes bibliographical references and index.
ISBN 0-8493-2162-X (alk. : paper)
1 Lipids in human nutrition. 2. Functional foods. I. Akoh, Casimir C., 1955- II. Functional foods & nutraceuticals series ; 7.

QP751.H327 2005
612.3'97--dc22 2005041864

Taylor & Francis Group
is the Academic Division of T&F Informa plc.

Visit the Taylor & Francis Web site at
http://www.taylorandfrancis.com

and the CRC Press Web site at
http://www.crcpress.com

Series Editor's Preface

The Functional Foods and Nutraceuticals Book Series, launched in 1998, was developed to provide a timely and comprehensive treatment of the emerging science and technology of functional foods and nutraceuticals which are shown to play a role in preventing or delaying the onset of diseases, especially chronic diseases. The first five titles in the Series, *Functional Foods: Biochemical and Processing Aspects. Volumes 1 and 2; Herbs, Botanicals, and Teas; Methods of Analysis for Functional Foods and Nutraceuticals; Handbook of Fermented Functional Food*; and *Handbook of Functional Dairy Products*, have received broad acceptance by food, nutrition, and health professionals.

Functional Foods: Biochemical and Processing Aspects, Volume 1, the first volume of the series, is a best seller and is devoted to functional food products from oats, wheat, rice, flaxseed, mustard, fruits, vegetables, fish, and dairy products. In *Volume 2*, the focus is on the latest developments in the chemistry, biochemistry, pharmacology, epidemiology, and engineering of tocopherols and tocotrienols from oil and cereal grain, isoflavones from soybeans and soy foods, flavonoids from berries and grapes, lycopene from tomatoes, limonene from citrus, phenolic diterpenes from rosemary and sage, organosulfur constitutes from garlic, phytochemicals from Echinacea, pectin from fruit, and omega-3 fatty acids and docosahexaenoic acid (DHA) from flaxseed and fish products. *Volume 2* also covers solid–liquid extraction technologies for manufacturing nutraceuticals and dietary supplements.

The volume *Herbs, Botanicals, and Teas* provides an in-depth literature review of the scientific and technical information on the chemical, pharmacological, epidemiologic, and clinical aspects of garlic, ginseng, Echinacea, ginger, fenugreek, St. John's wort, Ginkgo biloba, kava kava, goldenseal, saw palmetto, valerian, evening primrose, licorice, bilberries and blueberries, and green and black teas. The book, which is superbly referenced, also contains chapters on international regulations and quality assurance and control for the herbal and tea industry.

The volume *Methods of Analysis for Functional Foods and Nutraceuticals* presents advanced methods of analysis for carotenoids, phytoestrogens, chlorophylls, anthocyanins, amino acids, fatty acids, flavonoids, water-soluble vitamins, and carbohydrates. The fifth volume of the Series, *Handbook of Fermented Functional Foods*, provides a comprehensive, state-of-the-art treatment of the scientific and technological information on the production of fermented foods, the microorganisms involved, the changes in composition that occur during fermentation, and, most importantly, the effect of these foods and their active ingredients on human health.

The volume *Handbook of Functional Dairy Products* addresses the latest developments in functional dairy ingredients and products, with a clear focus on the effect of these foods and their active ingredients on human health. This volume contains outstanding chapters dealing with probiotic lactobacilli and bifidobacteria, lactose hydrolyzed products, *trans*-galactooligosaccharides as prebiotics, conjugated linoleic acid (CLA) and its antiatherogenic potential and inhibitory effects on chemically induced tumors, immunoenhancing properties of milk components and probiotics, and calcium and iron fortification of dairy products.

The current volume, *Handbook of Functional Lipids*, edited by Professor Casimir C. Akoh, presents up-to-date information on all major scientific and technological aspects of functional lipids, including isolation, production, and concentration of functional lipids; lipids for food functionality; lipids with health and nutritional functionality; and the role of biotechnology for functional lipids. Some distinctive features of this book include in-depth treatments of structured lipids, γ-linolenic acid, marine lipids and omega-3 fatty acids, fortification of foods with eicosapentaenoic acid (EPA) and docosahexanoic acid (DHA), fat substitutes, *trans* fatty acids and *trans*-free lipids, the action of CLA in human subjects, extraction and purification of fatty acids, physical properties of lipids, and potential markets for functional lipids. The book contains 23 excellent chapters written by 34 international experts at the forefront of lipid science and technology. It is hoped that the effort will be beneficial to food, nutrition, and health practitioners as well as students, researchers, and entrepreneurs in industry, government, and university laboratories.

G. Mazza, Ph.D., FCIFST
Series Editor

Preface

Functional lipids can be broadly described as lipids that provide specific health benefits when consumed and/or that impact a specific functionality of a food product. The desired functionality may be a physical or chemical property. Other terms that can be used to describe functional lipids are lipids with physiological function, nutritional lipids, and medical and pharmaceutical lipids.

The food industry and consumer are interested in functional foods and hence the development of functional lipids to serve consumer needs. Obviously, this book is intended as a reference source for members of the food industry, nutritionists, product development scientists, and researchers who have an interest in providing the consumer with healthy and beneficial foods. In addition, academicians and students will find it helpful in their research to avoid reinventing the wheel. The book contains up-to-date references and emerging areas of industry and research interests. Experts in the field from different parts of the world were carefully selected to develop chapters in their areas of interest.

The functional food market is growing and will soon reach $20 billion. Functional lipids have general as well as niche market appeal. Asian countries have already embraced the concept of functional lipids by marketing various lipids with nutraceutical properties. The introduction of infant formula containing docosahexaenoic acid into the American market is a step in the right direction. Various sterol esters are now delivered in various forms to the consumer. Omega-3-containing oils, such as fish oil, are sought after worldwide because of their many health benefits in alleviating chronic human diseases.

Many books have addressed the issue of dietary lipids and health, but none are devoted to functional lipids. Many articles have been published on the isolation, synthesis, applications, and physical and chemical properties of lipids. This book also covers the market potential for functional lipids that was not addressed by previous books in the same general area. This book is divided into four sections of related topics for easy comprehension.

Casimir C. Akoh

About the Editor

Professor Casimir C. Akoh has a B.S. (1981) and an M.S. (1985) in biochemistry from the University of Nigeria and Washington State University, respectively, and a Ph.D. (1988) in food science (lipid chemistry and biochemistry) from Washington State University. Dr. Akoh is a distinguished research professor of food science at the University of Georgia, Athens, where he teaches courses on food biotechnology, food lipids, and food carbohydrates. Dr. Akoh's research work is in lipid biochemistry, chemistry, biotechnology, and nutraceuticals. He is an internationally recognized expert on low-calorie fat substitutes and structured lipids (e.g., olestra). He has obtained three patents on fat substitutes, structured lipids, and frying oil recovery.

Dr. Akoh has edited the books *Carbohydrate Polyesters as Fat Substitutes* and *Food Lipids* published by Marcel Dekker (1994; 1998 and 2002, respectively). He has produced more than 130 refereed publications and 27 book chapters and has given 120 presentations at professional meetings and 60 invited speeches nationally and internationally. He is associate editor of the *Journal of the American Oil Chemists' Society*, former associate editor of *Inform*, member of the editorial board of the *Journal of Food Lipids Biotechnology Letters, Journal of Food Science and Food Chemistry*, and past editorial board member of the *Journal of Food Quality*.

Dr. Akoh received the International Life Sciences Institute Future Leader Award in 1996 to 1997 for his work on structured lipids. He also received the 1998 Institute of Food Technologists Samuel Cate Prescott Award for Outstanding Ability in Research in Food Science and Technology, the University of Georgia 1999 Creative Research Medal Award, the 2000 Gamma Sigma Delta Distinguished Senior Faculty Research Award, the 2003 D.W. Brooks Research Award, the 2004 Distinguished Research Professor Award, and the 2004 American Oil Chemists' Society Stephen S. Chang Award. He is a governing board member of the American Oil Chemists' Society, and a 2005 Fellow of the Institute of Food Technologists.

Contributors

Robert G. Ackman
Canadian Institute of Fisheries Technology
Dalhousie University
Halifax, Nova Scotia, Canada

Casimir C. Akoh
University of Georgia
Athens, Georgia

William E. Artz
University of Illinois
Urbana, Illinois

Armand B. Christophe
University of Ghent
Ghent, Belgium

John N. Coupland
Pennsylvania State University
University Park, Pennsylvania

N.A. Michael Eskin
University of Manitoba
Winnipeg, Manitoba, Canada

Kelley C. Fitzpatrick
University of Manitoba
Winnipeg, Manitoba, Canada

Jeanet Gerritsen
Loders Croklaan
Wormerveer, Netherlands

Douglas G. Hayes
University of Tennessee
Knoxville, Tennessee

Jenifer Heydinger Galante
Stepan Company
Maywood, New Jersey

Oi-Ming Lai
Department of Bioprocess Technology
Faculty of Biotechnology and Biomolecular Sciences
Universiti Putra Malaysia
Serdang, Selangor, Malaysia

Jae Hwan Lee
Seoul National University of Technology
Seoul, South Korea

Jeung-Hee Lee
Chungnam National University
Taejeon, South Korea

Ki-Teak Lee
Chungnam National University
Taejeon, South Korea

Yong Li
Purdue University
West Lafayette, Indiana

Seong-Koon Lo
Department of Bioprocess Technology
Faculty of Biotechnology and Biomolecular Sciences
Universiti Putra Malaysia
Serdang, Selangor, Malaysia

Alejandro G. Marangoni
University of Guelph
Guelph, Ontario, Canada

David B. Min
Ohio State University
Columbus, Ohio

Inge Mohede
Loders Croklaan
Wormerveer, Netherlands

Reto Muggli
AdviServ Consulting
Hofstetten, Switzerland

Hannah T. Osborn-Barnes
Kemin Foods
Des Moines, Iowa

Marianne O'Shea
Loders Croklaan
Channahon, Illinois

Trine Porsgaard
Technical University
of Denmark
Lyngby, Denmark

Colin Ratledge
Department of
Biological Sciences
University of Hull
Hull, United Kingdom

Sérgio D. Segall
University of Illinois
Urbana, Illinois

Yugi Shimada
Osaka Municipal
Research Institute
Osaka, Japan

Vijai K.S. Shukla
International Food Science
Centre AS
Lystrup, Denmark

Richard R. Tenore
Stepan Company
Maywood, New Jersey

Phillip J. Wakelyn
National Cotton Council
Washington, D.C.

Peter J. Wan
USDA/ARS/SRRC
New Orleans, Louisiana

Udaya N. Wanasundara
POS Pilot Plant Corporation
Saskatoon, Saskatchewan, Canada

Janitha P.D. Wanasundara
Agriculture and Agri-Food
Canada
Saskatoon, Saskatchewan, Canada

Bruce A. Watkins
Purdue University
West Lafayette, Indiana

Vasuki Wijendra
Brandeis University
Waltham, Massachusetts

Amanda J. Wright
University of Guelph
Guelph, Ontario, Canada

Vivienne V. Yankah
Aventis Pasteur Ltd.
Toronto, Ontario, Canada

Suk-Hoo Yoon
Korea Food Research Institute
Gyeonggi-Do, South Korea

Contents

PART I Isolation, Production, and Concentration of Functional Lipids

PART II Lipids for Food Functionality

Part I

Isolation, Production, and Concentration of Functional Lipids

1 Functional Lipids within the Global Functional Food and Nutraceutical Sector

Kelley C. Fitzpatrick and N.A. Michael Eskin

CONTENTS

1.1 INTRODUCTION

In the late 1980s consumers began to recognize that certain foods and food supplements could have an impact on health. The first phase of this "health and wellness revolution" belongs to dietary supplements that, through the passage of the Dietary Supplement Health and Education Act (DSHEA) in the United States in 1994 [1], market dynamics changed significantly — initially to the positive and then to the negative.

For the food industry, it has been a somewhat different story that has yet to play itself out. The United States responded to industry pressure in 1990 by enacting the Nutrition Labeling and Education Act (NLEA) [2] to allow manufacturers to promote the benefits of their food products. This regulatory framework "kick" started the functional foods revolution in the North American marketplace, which welcomed new food product introductions, a revival of science, and consumer interest.

The 1980s were a period of rapid expansion in scientific knowledge about polyunsaturated fatty acids (PUFAs), in general, and omega-3 PUFA, in particular. Both omega-3 and omega-6 PUFAs are precursors of hormone-like compounds known as eicosanoids, which are involved in many important biological processes in the human body. Recently it has been suggested that the typical "Western" diet, which is relatively high in omega-6 PUFA and low in omega-3 PUFA, may not supply the appropriate balance of PUFAs for proper biological function. As such, a significant industry has developed to produce and market lipids for dietary supplements as well as ingredients for functional foods.

1.2 FUNCTIONAL FOODS

"Functional foods" is essentially a marketers' or an analysts' term and globally is not recognized in law or defined in the dictionary. For the purposes of this report, the definition of functional foods as proposed by Health Canada will be used. In 1998, it was proposed that a functional food was "similar in appearance to a conventional food, consumed as part of the usual diet, with demonstrated physiological benefits, and/or to reduce the risk of chronic disease beyond basic nutritional functions" [3].

As a category, functional foods include

- Conventional foods containing naturally occurring bioactive substances (such as dietary fiber in wheat bran to promote digestive regularity or β-glucan in oat bran to lower blood cholesterol)
- Foods that have been modified, by enrichment or other means, in terms of the amount, type, or nature of their bioactive substances, example, margarine that contains added phytosterol, an extract from plant sources that is known to interfere with cholesterol absorption, thereby lowering serum cholesterol levels
- Synthesized food ingredients, such as some specialized carbohydrates intended to feed microorganisms in the gut

1.2.1 United States

The most commonly referenced definition for a functional food in the United States is that used by the California-based *Nutrition Business Journal* (*NBJ*) [4]. *NBJ* defines a functional food as fortified with added or concentrated ingredients and/or marketed to emphasize "functionality" to improve health or performance. Unlike Health Canada's definition, *NBJ* includes "substantially fortified," "inherently functional," and "performance" foods within its definition.

1.2.2 Japan

Japan is the only country with a regulatory framework for functional foods. Foods for Specified Health Uses (FOSHU) are defined as those to which a functional ingredient has been added for a specific health effect, designed to promote or maintain good health. Under FOSHU such foods include those that contain functional substances that affect the physiological function and biological activities of the body; and those that claim if used in the daily diet, one can hope for a specified health benefit. Under the legislation, such foods must be evaluated individually and approved by the government. FOSHU can also be used for dietary supplements [5].

1.2.3 Europe

As in North America, several definitions of functional foods are used throughout the 15 countries of the European Union (EU). The most widely accepted definition is "foods that by virtue of physiologically active food components provide health benefits beyond basic nutrition" [6].

1.3 DIETARY SUPPLEMENTS

1.3.1 United States

In 1994, the DSHEA was passed in the United States as an amendment to the Federal Food Drug and Cosmetic Act [7]. DSHEA regulates dietary supplements, which are defined as "a product, other than tobacco, intended to supplement the diet that contains at least one or more of the following ingredients: a vitamin, a mineral, an herb or other botanical, an amino acid, or a dietary substance for use to supplement the diet by increasing the total dietary intake; or a concentrate, metabolite, constituent, or extract or combination of any of the previously mentioned ingredients."

1.3.2 Canada

Canada is the only global jurisdiction that has legislation related to "natural health products (NHPs)." Final regulations for this new category were published in *Canada Gazette II* on Wednesday, June 18, 2003 [8]. In Canada, NHPs include homeopathic preparations, substances used in traditional medicine, a mineral or trace element, a vitamin, an amino acid, an essential fatty acid, or other botanical, animal, or microorganism-derived substance. These products are generally sold in a medicinal or "dosage" form. Until publication of *Canada Gazette II*, the working definition for a nutraceutical in Canada has been "a product that has been isolated or purified from foods and generally sold in medicinal forms not usually associated with food. Nutraceuticals have been shown to exhibit a physiological benefit or provide protection against chronic disease." Health Canada decided that the product category of nutraceuticals would be encompassed within NHP regulations.

However, like functional foods, although the nutraceutical category is not recognized in law, it is used extensively as a marketing term for plant- and animal-based bioactives and ingredients that are sold in a medicinal form.

1.3.3 Europe

In the EU [9], dietary supplements are referred to as "food supplements," the purpose of which is to supplement the normal diet and to be concentrated sources of nutrients or other substances with a nutritional or physiological effect, alone or in combination, marketed in dose form. Such dosages are capsules, pastilles, tablets, pills and other similar forms, sachets of powder, ampoules of liquids, drop dispensing bottles, and other similar forms of liquids and powders designed to be taken in measured small unit quantities.

The EU adopted Directive 2002/46/EC on June 10, 2002, with the requirement that it be passed into national laws by all member states by July 31, 2003 [10]. It deals with laws relating to food supplements. The nutrients that are included are restricted to vitamins and minerals. Additional nutrients such as amino acids, essential fatty acids, and fiber may be added later. In countries where products currently exist that include ingredients not yet on the EU Directive List, these will be able to continue to be marketed, but not permitted, for EU-wide use.

1.3.4 Japan

Since 2001, 13 vitamins, 13 minerals, and 101 herbal supplements have been regulated within the food category, rather than as drugs. In April 2001, Japan implemented new regulations allowing dietary supplement labels to provide health and efficacy information for the first time. Dietary supplements may now carry claims

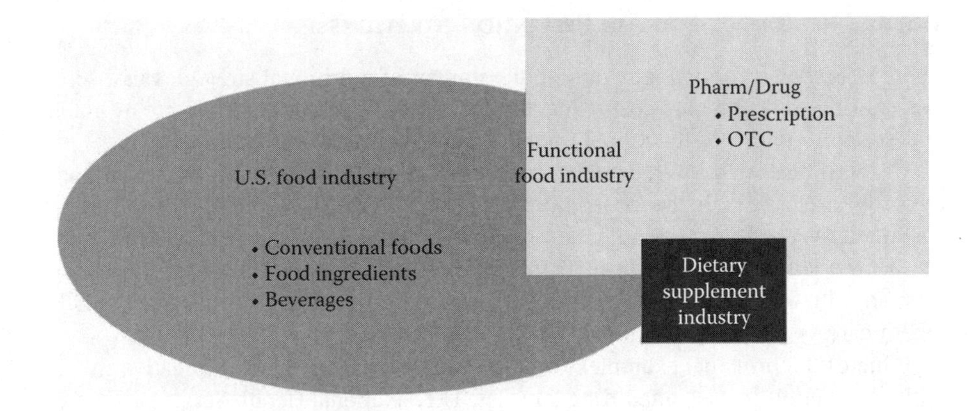

FIGURE 1.1 Context for functional foods and dietary supplements.

that are regulated under FOSHU or nutrient content claims. Herbals that carry health claims are considered within the more stringent FOSHU regulations. Currently, 12 vitamins and 2 minerals have been placed within the nutrient content category. This area represents 20% of the nutritional supplements market, while the remaining 80% is classified as either FOSHU (requiring individual approvals for health claims) or "other food" for which no health or efficacy claims are allowed.

With regard to market segregation, functional foods lie between the health-care continuum of foods and drugs as indicated in the Figure 1.1 [4].

1.4 THE GLOBAL MARKET

For 2003, the NBJ estimated current world consumption of NHPs, nutraceuticals, and functional foods to be approximately US$172 billion [11]. The primary markets for supplements and functional foods are the United States, Europe, Japan, and Asia [12].

1.5 MACRO TRENDS AFFECTING THE FUNCTIONAL FOOD INDUSTRY

Several macro trends are driving the growth of the global functional foods industry.

1.5.1 The Rising Cost of Health Care Related to Chronic Disease

In 2001 [13], chronic diseases contributed approximately 60% of all deaths worldwide and 46% of the total burden of disease. Almost half of these deaths are from cardiovascular diseases, obesity, and diabetes. In most developed countries health-care costs average between 9 and 14% of the gross national product.

1.5.2 The Role of Food in Prevention Strategies

There is limited information regarding the impact of nutritional strategies in disease prevention and health-care cost reduction. Some case studies, however, are available that support the use of functional foods to improve the health of populations.

According to Unilever, a reduction of low-density lipoprotein (LDL) cholesterol by 14% through the consumption of a cholesterol-lowering margarine, if sustained over a 5-year period, could result in a roughly 25% decrease in coronary risk in the U.K. population [14]. If this risk reduction were achieved in practice, this would reduce U.K. heart disease patient numbers by 250,000 and save the U.K. health-care system £433 million.

One of the principal examples is the introduction of margarine spreads fortified with plant sterols in the United Kingdom. The U.K. National Health Service estimated that these products have the potential to lower health-care costs for cardiovascular disease by £100 million per year. An added benefit to this cost savings is the fact that this reduction can be delivered at an annual cost of only £70 per patient per annum, which is borne at the expense of the patient and not the government [15].

In Canada, Holub suggests that there is a potential for the Canadian government to save an approximate Can$19 billion in health-care costs per annum [16] with the introduction of several functional foods and nutraceutical ingredients, including citrus pectins, guar gum, plant sterols, long-chain omega-3 PUFAs, cholestin (red rice yeast), and policosanol into the diet.

1.5.3 Safety and Toxicology Aspects Related to Functional Foods and Nutraceuticals

Functional or nutraceutical ingredients, although intended to produce a physiological effect, which can reduce chronic disease risk or otherwise optimize health, may also produce adverse effects under certain circumstances due to the fact that they encompass elements of drugs, nutrients, and food additives [17]. Thus, assurance of safety for functional/nutraceutical ingredients is critical, particularly because they are consumed by the general population in an unsupervised manner (e.g., without medical oversight). When evaluating the safety of functional/nutraceutical ingredients, clinical substantiation of safety is critical and, in particular, whether current or historical human exposure is associated with an adverse health outcome.

Although a functional food or nutraceutical ingredient is intended to produce optimized consumer health, because there is potential for lifetime exposure and consumption if unsupervised, the assurance of safety is critical. The primary safety must be derived from well-controlled randomized, double-blind human clinical intervention trials. Safety must encompass an understanding of the physiological activity of the functional/nutraceutical component, as it relates not only to a potential health benefit, but also to any potential toxicological effect that may result.

1.5.4 Globalization of the Food Industry

Many large multinational companies see the functional food industry as a way to escape the commoditization of their products, as the profit margins in functional foods tend to be higher. In North America, 5 companies control nearly 40% of the market, while the top 50 control nearly 80% of the market.

1.6 UNITED STATES

1.6.1 MARKET

In 2003, the sale of nutrition products consisting of dietary supplements and herbs, natural and organic foods, functional foods, and natural personal care products generated US$61.9 billion in consumer sales. The dietary supplement market generated $19.4 billion in the United States, growing only 1 to 3% from 2002 [18]. The functional food market grew 10% to sales of $21.9 billion in 2003 [18].

A key characteristic of the U.S. functional food market is a large focus on disease and its prevention. Products aimed at lowering blood cholesterol levels and cancer risk, as well as weight loss, have characterized the U.S. market. The use of botanicals in functional foods is much more popular than in Europe or Japan. Breads and grains, especially enriched breakfast cereals, followed by functional beverages (teas and energy drinks) are the primary functional food products in the U.S. market. Snack foods including nutrition bars are one of the fastest growing functional categories, experiencing a 23% rate of growth in 2003 [18].

1.6.2 CONSUMER TRENDS

In the United States, chronic disease represents about 70% of health-care costs. The U.S. Centers for Disease Control and Prevention (CDC) [19] predicted that obesity and overweight may contribute as much as $120 billion to health-care costs. Approximately 129.6 million Americans, or 64%, are overweight or obese. This has led to an increased interest in and purchase of weight loss products. The Valen Group [20] predicts that 59 million, or 28%, of Americans will purchase low-carbohydrate foods in the next year. *NBJ* [21] estimates sales of low-carb foods reached $1.4 billion in 2003 and could reach $3 billion in the next few years.

With regard to omega-3 fatty acids, a consumer survey quoted in *NBJ* found that in 2002, 58% of consumers were aware of omega-3, up from 46% in 1999 [22]. In addition, the notion of "mega" fish oil consumption has been championed by Barry Sears in *The Omega Rx Zone: The Miracle of the New High Dose Fish Oil*, and commercialized as OmegaRx high-dose fish oil and OmegaZone Bar with eicosapentaenoic acid (EPA) and docosahexaenoic acid (DHA). Sears' diet specifically recommends the inclusion of omega-3 fatty acids in low and moderate carbohydrate diets [23]. Other early entrants in the high-potency foods include salmon burgers by AQUACUISINE with 1000 mg of omega-3 per serving [24].

1.6.3 REGULATORY REVIEW

The United States has a wide variety of fortification policies and health claim labeling, which permit consumers to make informed dietary choices. Health claims that can be used on food and dietary supplement labels fall into three categories:

risk reduction claims, nutrient content claims, and structure/function claims. The responsibility for ensuring the validity of these claims rests with the manufacturer, Food and Drug Administration (FDA), or, in the case of advertising, with the Federal Trade Commission (FTC).

The NLEA of 1990 directed the FDA to change the way that food labels were regulated in order to make additional nutritional information available to consumers. As defined in this act, a health claim is a "statement that characterizes the relationship of a substance to a disease or a health-related condition, typically in the context that the regular dietary consumption of a substance may reduce the risk of a specific disease or health condition." Within the labeling claims, references that the food is intended to "cure," "mitigate," or "prevent" any disease are not permitted.

In 1994, the FDA reviewed several diet–disease relationships and established seven allowable risk-reduction health claims that can be made on conventional foods as long as they meet specific nutritional criteria related to fat, saturated fat, cholesterol, sodium, vitamin A, vitamin C, iron, calcium, protein, and fiber.

In January 1997, the FDA approved the first food-specific health claim under the NLEA, in response to a petition from the Quaker Oats Company. The authorized health claim describes the relationship between consumption of whole oat products and coronary heart disease risk reduction.

The FDA Modernization Act (FDAMA) became law in November 1997 [25]. It contains provisions to reduce the regulatory hurdles in the health claim approval process. Specifically, it directs the FDA to authorize health claims that are based on the published authoritative statements from U.S. government agencies such as the CDC, the National Academy of Sciences (NAS), or the National Institutes of Health (NIH). An authoritative statement is about the relationship between a nutrient and a disease or health-related condition. Within FDAMA, health claims can be made without going through the lengthy FDA review process, if they have already been published by these agencies. Premarket notification to the FDA of 120 days is required. In July 1999, General Mills received no objection for a claim linking whole grain foods to reduced risk of heart disease and cancer, based on statements from the NAS.

In December 2003, the FDA's Consumer Health Information for Better Nutrition Initiative was announced [26]. The new system allows the use of qualified health claims when there is emerging evidence for a relationship between a food, food component, or dietary supplement and reduced risk of a disease or health-related condition. In this case, the evidence is not well enough established to meet the significant scientific agreement standard required for the FDA to issue an authorizing regulation. Qualifying language is included as part of the claim to indicate that the evidence supporting the claim is limited. Both conventional foods and dietary supplements may use qualified health claims.

The claims currently permitted in the United States are listed below:

NLEA

- Calcium: osteoporosis
- Sodium: hypertension
- Dietary fat: cancer

- Saturated fatty acids and cholesterol: coronary heart disease
- Fiber-containing grain products, fruits, and vegetables: cancer
- Fiber-containing grain products, fruits, and vegetables: risk of coronary heart disease
- Fruits and vegetables: cancer
- Folate: neural tube defects
- Dietary sugar alcohols (sorbitol, xylitol, mannitol): dental caries
- Dietary soluble fiber (whole oats and psyllium): coronary heart disease
- Soy protein: coronary heart disease
- Sterols/stanols: risk of coronary heart disease

FDAMA

- Whole grains: coronary heart disease and certain cancers
- Potassium and sodium: hypertension

Qualified

- Omega-3 fatty acids (EPA and DHA): coronary heart disease
- Vitamin B_6, vitamin B_{12}, and folic acid: coronary heart disease
- Walnuts and other tree nuts: coronary heart disease
- Selenium: cancer
- Antioxidant vitamins (E and/or C): coronary heart disease
- Phosphatidylserine: cognitive dysfunction and dementia

1.7 EUROPEAN UNION

1.7.1 MARKET

For dietary supplements like functional foods, the EU represents a dichotomous market with regional differences apparent in market growth and individual product sales. According to EuroMonitor [27], the overall European market for dietary supplements experienced a growth rate of 2.2% in 2001 with sales at US$3.6 million. The European market for functional foods in 2002 was worth about €5.7 billion, and is forecast to grow by an average of 7.5% per year over the next 5 years to reach a value of €8.2 billion by 2007, at 2002 prices and exchange rates.

1.7.2 CONSUMER TRENDS

Several consumer reports have identified the EU consumer market for functional foods as very diverse and heterogeneous, with different populations concerned with, and suffering from, differing health concerns. However, as in North America, the need to maintain a healthy, active lifestyle, to look and feel good, and to enjoy tasteful food is paramount for the EU consumer.

1.7.3 REGULATORY REVIEW

Directives within the EU continue to pose problems for regulators and marketers alike. The EU adopted Directive 2002/46/EC on June 10, 2002, with the requirement that it be passed into national laws by all member states by July 31, 2003 [10]. It deals with laws relating to food supplements and includes only vitamins and minerals.

It should be emphasized that the obstacles to marketing functional foods are very high. There are currently EU-wide initiatives to modify health claims regulation, which are expected to be complete by 2005. Many industry experts believe that only large EU-based companies will be able to fulfill requirements for scientific support for health claims that will be required and to ensure accurate translation of claims into over ten different languages.

1.8 JAPAN

1.8.1 Market

The dietary supplement market in Japan remains undeveloped in comparison with those of the United States and Europe. The regulatory environment for nutritional supplements is restrictive and cumbersome [28]. Most dietary supplements are marketed through nonretail venues including the Internet and multilevel marketing firms. Japan is widely recognized as the most developed and established market for functional foods in the world and has the second largest functional food market behind the United States with a value of approximately US$12.5 billion.

1.8.2 Consumer Trends

In 2001, of the Japanese population 18% was 65 years or older in comparison to 13% in the United States [29]. By 2025, this percentage will increase to 28% and 18% in Japan and the United States, respectively. The Ministry of Heath and Welfare for Japan estimated that of deaths per 100,000 people, cancer claimed 239; heart diseases, 118; cardiovascular diseases, 105; pneumonia, 66; and for 1998, 10 deaths were related to diabetes [30]. The Japanese consumer spends US$166 per capita on nutraceuticals per year, which is higher than in any other country.

1.8.3 Regulatory Review

Planning a strategy for functional food marketing in Japan is not an easy task. Food and supplement companies are faced with a number of pathways in a maze of regulations. FOSHU is one of the five categories covered under "Foods for Special Dietary Uses." Since the passage of FOSHU, the public's interest in the relationship between nutrition, diet, and lifestyle-related diseases and disorders has flourished.

Since the inception of FOSHU in 1991, functional food sales have been growing at an average of 25% annually. The FOSHU market currently accounts for nearly US$4.1 billion of the total health food industry in Japan, which was estimated to be worth $12 billion in 2004. Currently, the "nonhealth claim" functional food market represents the largest sector of Japan's food industry at around $8 billion. This is followed by a newer category known as Foods with Nutrient Function Claims at an estimated $5 billion and including over 1000 products according to a recent Japanese market report [31].

Of the seven health benefit categories approved under FOSHU, 54% of products and 67% of sales are generated by the gastrointestinal health foods category. Functional pre- and probiotic drinks and yogurts dominate the market. Of the 396 FOSHU products currently approved, 150 use fiber sources, 70 use oligosaccharides, 70 use peptides and proteins, 55 use lactic acid bacteria, and 10 each use diacylglycerides and minerals such as calcium and iron. Other ingredients include noncarcinogenic sweeteners, polyphenols, chitosan, β-carotene, green tea catechins, and DHA.

1.9 OMEGA-3 FATTY ACIDS AS INGREDIENTS IN FUNCTIONAL FOODS AND NUTRACEUTICALS

Over the past 150 years, significant changes in the composition of the food supply of Western societies have occurred resulting in an increase in consumption of omega-6 PUFA and a corresponding decrease in intake of omega-3 PUFA. Today, the ratio of omega-6 to omega-3 in North America is estimated to be in the range of 10:1 to 25:1, in comparison with a ratio of 1:1 characteristic of diets of the Paleolithic era [32].

This alteration in the food supply and subsequent shift in the ratio of omega-6 to omega-3 PUFA is attributed to a number of factors. Modern food production methods decreased the omega-3 fatty acid content of many foods, including animal meats, eggs, and fish. The use of grain feeds, which are rich in omega-6 but poor in omega-3 PUFA, has altered the fatty acid composition of domestic livestock and thus of meats and eggs in the modern food supply. Today, domestic beef contains little to no detectable amounts of omega-3 PUFA. Aquaculture produces fish that contain less omega-3 PUFA than those that grow naturally and feed on plankton in the oceans, rivers, and lakes. The industrial revolution introduced vegetable oil technology and popularized the use of cooking oils from sunflower, peanut, and corn, all good sources of omega-6 PUFA.

1.10 THE FUNCTIONAL LIPID INDUSTRY

In 2002, fish oil surpassed other specialty supplements with growth of 32% for total consumer sales of US$131 million, of which $78 million was in the mass market and $35 million in the natural retail channel [25]. Fish oils comprised 6% of the specialty supplement segment in 2002. According to industry players such as Roche Vitamins (now DSM Nutritionals), strong, positive science is driving sales of omega-3 from fish oil. Growth of fish oil supplements has been averaging 30% for the last 2 years and is currently sustaining around 10 to 15% growth.

Plant oils grew at 17% in 2002 to consumer sales of $154 million and also comprising 6% of the specialty segment. Flaxseed oil is growing in popularity, as it is a vegetarian source of omega-3 fatty acids and has fewer taste and stability issues. Flax contains α-linolenic acid (ALA) in addition to linoleic acid (LA), which is omega-6. Bioriginal Food & Science Corp. (Saskatoon, Canada), a supplier of marine and plant bulk oils, concentrates, soft gels, and powders, saw flax sales grow at a greater rate than fish in 2002. Based on North American sales for fiscal year 2003 (ending September 30, 2003), the company anticipated 35% growth over 2002 and 66% growth in flax sales (all forms), ahead of fish oil sales.

In addition to their use in dietary supplements, the fats and oils industry is focusing upon adding these "good" oils to high-fat food products.

1.11 DELIVERY SYSTEMS FOR FUNCTIONAL LIPIDS

The addition of functional lipids to foods must not have a deleterious effect on flavor if it is to be accepted by the consumer. Unlike drugs, which are not designed for consumer acceptability, the primary concern is for the treatment of a particular disease condition. In sharp contrast, functional lipids must be presented to the consumer in a palatable food, as they are not perceived as drugs and are bought for the added health benefits they provide.

"Novel" refining techniques are available that produce plant and marine oils, which can be added to a range of foods without affecting the flavor profile of the product. Previously, fish oils have been used only in a "hardened" or hydrogenated form to prevent the occurrence of fishy off-tastes and smells. Currently, fish oils and dry powders, with microencapsulated oils, are available for food fortification use. The oily forms can be added to the lipid phase, with care taken to protect the readily oxidizable PUFA. The powdered forms are used mainly in dry goods such as bakery products and milk powders. Microencapsulated powders are dispersible in cold water and are exceptionally stable with a neutral taste, making such powders available for enrichment of foods such as reduced fat products, milk drinks, salad dressings, orange juice, drinks, and bread.

Omega-3 and -6 enriched products are currently being marketed in the United Kingdom, Korea, Taiwan, and Scandinavian countries. However, resistance from producers and consumers is still to be overcome and technical problems need to be resolved in some areas. Nutritionists and food developers realize that these foods need not only to be healthy but to taste as good as similar products. There is a degree of prejudice against fish and fish derivatives. For instance, the idea of yogurt or bread containing 1% fish oil is not particularly attractive.

Despite the technical issues of incorporating long-chain PUFAs into foods, there is no doubt that this concept offers food processors the opportunity to introduce a new range of foods associated with definite health benefits, which will enjoy the support of the scientific community as well as help in disease prevention.

Some functional lipids have limited bioavailability and require relatively large doses to ensure a certain amount is taken. Thus, the delivery system still remains the crucial step in ensuring efficacy of the particular functional lipid. This has considerable economic implications as enhanced absorption means a smaller amount is needed because of higher bioavailability. This book covers a wide range of functional lipids including the highly unsaturated fatty acids EPA and DHA, which, in addition, present challenges related to their oxidative stability.

Improved solubility and stability appear to be the key challenges for functional lipids. To this end, new developments in emulsion technology appear to hold considerable promise for their incorporation into foods. The development of micro- and double emulsions appear to be ways for improving the bioavailability, stability, and bioefficacy of functional lipids. For example, the development of a phospholipid-based microemulsion formulation for all-*trans*-retinoic acid, an active metabolite of retinol (vitamin A), by Hwang and co-workers [33] was shown to improve both its solubility and stability. This was achieved without affecting its pharmacokinetic and cancer properties. They further demonstrated that this technology could be used as an alternative parenteral formulation of all-*trans*-retinoic acid. Microemulsions are easy to form, have excellent thermodynamic stability, and are capable of solubilizing considerable quantities of oil-soluble compounds. In addition, they provide the ability for controlled and sustained drug or nutraceutical release [34]. Double emulsions, on the other hand, are unstable thermodynamically. Naturally occurring emulsifiers, such as protein macromolecules, bovine serum albumin, casein, gelatin, and vegetable proteins, were suggested to be added as possible stabilizers [35]. The improved stability of double emulsions afforded by these macromolecules permitted entrapping any type of nutraceutical, including functional lipids, with controlled release over a prolonged period of time. Further work in this area appears extremely promising for increasing the bioavailability and efficacy of functional lipids.

1.12 WORLD RECOMMENDATIONS FOR OMEGA-3 AND -6 FATTY ACIDS

Many countries and international organizations have made formal population-based dietary recommendations for omega-3 PUFA to ensure adequate intakes (AIs). Recommendations have typically been between 0.3 and 0.5 g/d of EPA + DHA and 0.8 to 1.1 g/d of ALA [36].

1. In 1999, the NIH sponsored an international workshop, which made recommendations for dietary intakes for omega-3 and omega-6 fatty acids. The NIH Working Group proposed AIs of 2 to 3% of total calories for LA, 1% of total calories for ALA, and 0.3% of total calories for EPA and DHA [37].
2. The United Kingdom recommends that 1% of energy be obtained from ALA and 0.5% from EPA and DHA combined [38].
3. The British Nutrition Task Force recommends a minimum of 0.5% of energy from ALA [39].
4. In September 2002, the NASs Institute of Medicine (IOM) set an AI for ALA of 1.6 and 1.1 g/d of ALA for men and women aged 19 to >70 years, respectively. The AI is based on the highest median daily intake of ALA by U.S. adults and represents an intake not likely to be associated with a deficiency of this nutrient. In conjunction with the Food and Nutrition Board and the National Academies, in collaboration with Health Canada, the IOM estimated an Acceptable Macronutrient Distribution Range (AMDR) for ALA to be 0.6 to 1.2% of energy, or 1.3 to 2.7 g/d, on the basis of a 2000 calorie diet [40].
5. The American Heart Association Scientific Advisory and Coordinating Committee has recommended eating a variety of fish at least twice a week to provide AI of omega-3 fatty acids, as well as flaxseed, canola oil, and other foods rich in ALA. As reviewed by Kris-Etherton et al. [36], total ALA intakes of about 1.5 to 3 g/d appear to be beneficial for individuals without coronary heart disease.

1.13 SAFETY

The two key elements that will determine the success of functional lipids, in the management of human health, are efficacy and safety. Functional lipids must have generally recognized as safe (GRAS) status. The safety of functional ingredients, including functional lipids, should follow the guiding principles described by Kruger and Mann [16] and summarized as follows:

1. The pharmacological effects of the particular functional lipid as well as its toxicological potential must be understood to predict the impact of exposure to different dose levels.
2. Each functional ingredient may have unique safety issues and must therefore be dealt with individually.
3. The safe level of ingestion or exposure must be determined as the margin of safety between the intended level of ingestion and its toxic level may be very small.
4. The potential for possible interaction with drugs must be determined.

Prior to human trials, evaluation of safety is based largely on animal studies. However, the most powerful evidence is that obtained from careful human trials. Since, for the most part, functional lipids occur naturally in food, the main concern is the possible toxicity of the isolated bioactive lipid ingredient, which may be concentration dependent.

Because nutraceutical such as functional lipids have the potential to enhance consumer health, decrease health care costs, and enhance economic development, Lachance [41] pointed out the reliance on diversified research that addresses safety and assures chemical and biological efficacy. Lachance and Saba [42] also noted that while microbiological and chemical assays can ensure the intended quality of functional ingredients themselves, following the terrorist attacks of September 11, 2001, there is also a need to prevent the introduction of counterfeit products and agents of terrorism. The latter can be assured through traceability by combining the following existing technologies: (1) global positioning, so that each ingredient or functional component has a chemical identification and authentication by collecting global positioning descriptors; (2) bar codes as a universal method to identify source, key dates, HACCP data, and the like; and (3) HACCP to assure the microbiological safety of the ingredient.

1.14 CONCLUSION

Functional ingredients including lipids have now become a high research priority by universities, industries, and governments around the world. This is evident by the establishment of institutes and centers that are totally focused on functional foods and nutraceuticals. The success of this investment will hopefully lead to new and exciting functional ingredients, including functional lipids, that will help to reduce the increasing health costs associated with an aging population by slowing down or preventing the onset of chronic diseases.

REFERENCES

1. United States Food and Drug Administration. Dietary Supplements. February 11, 2004. http://www.cfsan.fda.gov/~dms/supplmnt.html (accessed March 11, 2004).
2. U.S. Food and Drug Administration. Center for Food Safety and Applied Nutrition. September 2003. Claims that can be made for Conventional Foods and Dietary Supplements. http://www.cfsan.fda.gov/~dms/hclaims.html (accessed March 20, 2004).
3. Health Canada. Standards of Evidence for Evaluating Foods with Health Claims — Fact Sheet. November 2000. http://www.hc-sc.gc.ca/food-aliment/ns-sc/ne-en/health_claims-allegations_sante/e_soe_fact_sheet.htm (accessed January, 23, 2004).
4. Anon., Functional foods VI. *Nutr. Bus. J.* 8(2/3), 1, 2003.
5. Japan Health Food and Nutrition Food Association, 2004.
6. Fielding, B.A. and Frayn, K.N., An international vision of nutritional science. *Br. J. Nutr.* 80(1), 8, 1998.
7. United States Food and Drug Administration. Dietary Supplements. February 11, 2004. http://www.cfsan.fda.gov/~dms/supplmnt.html (accessed March 11, 2004).
8. Health Canada. Natural Health Products Regulations. Canada Gazette Part 2. June 2003. http://canadagazette.gc.ca/partII/2003/20030618/html/sor196-e.html (accessed June 18, 2003).
9. Directive 2002/46/EC of the European Parliament and of the Council. June 10, 2002. http://europa.eu.int/eur-lex/pri/en/oj/dat/2002/l_183/l_18320020712en00510057.pdf (accessed March 10, 2004).
10. Hanssen, M., Making sense of European Dietary Supplement Regulation. *Nut. Outlook* March 2004.
11. Ferrier, G., Nutrition Business Journal's Annual Industry Overview. Presented at Nutracon 2004, Anaheim, CA, 2004.
12. Anon., *Nutr. Bus. J.* 3(10/11), 1, 1998.
13. WHO, Technical Report Series 916. Diet, Nutrition and the Prevention of Chronic Diseases, 2003.
14. Unilever UK, As quoted in: Mellentin, J., Can functional foods/nutraceuticals make a major contribution to public health? *New Nutr. Bus.* 37, 2001.
15. U.K. National Health Service, 2000.
16. Holub, B.J., Report to Agriculture and Agri-Food Canada: Potential Benefits of Functional Foods and Nutraceuticals to Reduce the Risk and Costs of Diseases in Canada (May 2002).
17. Kruger, C.L. and Mann, S.W., Safety evaluation of functional ingredients. *Food Chem. Toxicol.* 41, 793, 2003.
18. Ferrier, G., 2004. Nutrition Business Journal's Annual Industry Overview. Presented at Nutracon 2004, Anaheim, CA, 2004.
19. Center for Disease Control and Prevention, Statistics. 2004. http://www.cdc.gov/nchs/fastats/ (accessed March 15, 2004).
20. The Valen Group. January 2004. Low Carb Consumers Speak: The Definitive Habits & Practices Study of U.S. Low Carb Consumers. http://www.valengroup.com/low_carb_consumers_speak.htm (accessed February 11, 2004).
21. Anon., Consumer research III. *Nutr. Bus. J*. 8(11), 2003.
22. Anon., Niche markets VI. *Nutr. Bus. J.* Speciality Supplements 8(7), 2003.
23. Sears, B., DrSears.com. The Zone Diet. http://www.drsears.com (accessed March 28, 2004).
24. Aquacuisine, http://www.aquacuisine.com (accessed March 28, 2004).

25. U.S. Food and Drug Administration. Center for Food Safety and Applied Nutrition. June 2001. Notification of a Health Claim or Nutrient Content Claim Based on an Authoritative Statement of a Scientific Body. http://www.cfsan.fda.gov/~dms/hclmguid.html (accessed March 20, 2004).
26. U.S. Food and Drug Administration. Center for Food Safety and Applied Nutrition. July 2003. Interim Procedures for Qualified Health Claims in the Labeling of Conventional Human Food and Human Dietary Supplements. http://www.cfsan.fda.gov/~dms/hclmgui3.html (accessed March 20, 2004).
27. Euromonitor, Supplements and functional beverages: A global overview, December 2002.
28. International Business Strategies, Dietary Supplements in Japan, December 2001.
29. Ministry of Health, Labor and Welfare, Japan, 2002.
30. Ministry of Health, Labor and Welfare. National Nutrition Survey in Japan, 2002.
31. The Condition of the Market — Report by the Japan Health Food and Nutrition Food Association HI-Japan Conference, Toyko, 2003.
32. O'Keefe, J.R. and Cordain, L., Cardiovascular disease resulting from a diet and lifestyle at odds with our Paleolithic genome: How to become a 21st-century hunter-gatherer. *Mayo Clin. Proc*. 79, 101, 2004.
33. Hwang, S.R., Lim, S.-J., Park, J.-S., and Kim, C.-K., Phospholipid-based microemulsion formulation of all *trans*-retinoic acid for parenteral administration. *Int. J. Pharm.* 276, 175, 2004.
34. Garti, N., Microemulsions as microreactors for food reactions, *Curr. Opin. Colloid Interface Sci.* 8, 197, 2003.
35. Garti, N., Progress in stabilization and transport phenomena of double emulsions in food applications. *Lebensm.-wiss, u-Technol.* 30, 222, 1997.
36. Kris-Etherton, P.M., Harris, W.S., Appel, L.J. et al., American Heart Association Scientific Statement: Fish consumption, fish oil, omega-3 fatty acids and cardiovascular disease. *Circulation* 106, 2747, 2002.
37. Simopoulos, A.P., Leaf, A., and Salem, N., Workshop on the essentiality of and recommended dietary intakes for n-6 and n-3 fatty acids, National Institutes of Health, Bethesda, MD, 1999.
38. COMA Cardiovascular Review Group. Report on health and social subjects: Nutritional aspects of cardiovascular disease, Department of Health, London, 1994, p. 46.
39. British Nutrition Foundation. In: *Unsaturated Fatty Acids: Nutritional and Physiological Significance,* Chapman & Hall, New York, 1992.
40. Anon., Dietary Reference Intakes for Energy, Carbohydrates, Fiber, Fat, Protein and Amino Acids (Macronutrients), National Academy of Sciences, Institute of Medicine, Health and Human Service's Office of Disease Prevention and Health Promotion, U.S., 2002.
41. Lachance, P.A., Nutraceutical/drug/anti-terrorism safety assurance through traceability. *Toxicol. Lett.* 150, 25, 2004.
42. Lachance, P.A. and Saba, R.S., Quality management of nutraceuticals, intelligent product delivery systems through traceability, in Ho, C.-T. and Zheng, Z., Eds., *Quality Management of Nutraceuticals* , American Chemical Society Series 803, 2002, pp. 2–9.

2 Microbial Production of γ-Linolenic Acid

Colin Ratledge

CONTENTS

2.1 INTRODUCTION

Although there has been interest in the possible production of oils and fats using various yeasts and filamentous fungi for well over 75 years, it has only been in the past 20 years that commercialization of these materials has occurred. The first process that was developed for a microbial oil was for the production of an oil rich in γ-linolenic acid (GLA, n-6), which forms the basis of this chapter. Although this process is no longer run, due to the availability of cheaper plant sources for GLA, the process nevertheless has provided an important milestone and benchmark for all future microbial oil productions.

Today, several large-scale commercial processes are operated for the production of oils rich in either arachidonic acid (20:4, n-6) or docosahexaenoic acid (22:6, n-3) using various microorganisms [1]. The presence of such oils is now a very important part of the nutraceutical business, especially for incorporation into infant formulae.

In this chapter, an account will be given of the first commercial microbial process for the production of a specific oil. Much of this information has not been available previously, as the details of the process itself were regarded as industrial know-how. No patents were ever taken out on this process because it was considered that the essential information about the occurrence of GLA in fungi was already in the public domain, as were details of the cultivation systems that had to be used to ensure maximum lipid production in an organism. With the advent of other microbial oil processes, and the divulgence of many of the details of these processes [1], there is no longer a need to restrict information about the GLA process. However, certain details concerning the origins of the organisms that were screened for GLA production, as well as the exact results of the production runs, have still remained restricted, as it is not impossible that the process could, one day, be resurrected and brought back into production.

The microbial GLA process has opened up the way in which other similar oils could be produced using bioreactors of up to 200 m^3 by using the type of technology that had been developed for the production of biomaterials such as antibiotics, amino acids, and even whole microbial cells for animal feed purposes. Furthermore, the GLA process demonstrated, for the first time, that extracting oil from the harvested microbial cells was not a major problem and could, in fact, be accomplished using conventional oil extraction technologies. In addition, many of the questions of safety had to be addressed with the first microbial oil, as clearly this oil was a novel product and could not be sold to the general public without extensive toxicological trials having been done. Again, establishing the protocols needed to show complete safety of use has since been extended to all current microbial oils being offered for sale. Like any product that is the first of its kind, there is considerable, and often intense, scrutiny to ensure that the product poses no threat to health and will be safe to consume even if taken at many times the suggested daily dose. As each subsequent product comes along, these fears are lessened and testing can eventually be simplified, the specifications quickly defined, and the product then swiftly cleared by the regulatory authorities. Therefore, we have a lot to be grateful for in the pioneering work that led to the world's first commercially viable single cell oil (SCO).

2.2 SINGLE CELL OILS

The term *Single Cell Oil* was coined [1] as an obvious parallel to the term *Single Cell Protein* (SCP), which was used in the 1960s to cover microbial biomass generated as a source of protein, principally for use as an animal feed material. The term *SCP* was meant to allay fears of the public about eating microbes, which, in most people's minds, are probably harmful and certainly noxious. Consequently, SCP became a widely used euphemism avoiding mention of bacteria, fungus, mold, or any other word that might have had negative connotations with the public. SCO was similarly meant to define edible oils being produced by unicellular (i.e., single-celled) organisms and, again, allowed their use and consumption without specifically mentioning that they were derived from microorganisms. After all, very few people would be attracted to the idea of consuming a fungal oil! Thus the term SCO was born [2] and has remained in use ever since.

2.2.1 Microbial Oils in General

The ability of certain microorganisms to accumulate up to 70%, and even greater, of their biomass as oil has long been known. Such species are known as oleaginous [3] and the phenomenon of lipid accumulation is therefore known as oleaginicity though the shorter neologism of *oleaginy* has been heard. This descriptor has then been used for microorganisms capable of accumulating 20% or more of their biomass as lipid or oil. This lower limit, though, is imprecise and the term *oleaginicity* was therefore originally intended to indicate a general propensity for lipid accumulation rather than defining, with any accuracy, what was the minimum amount of lipid that had to be accumulated for an organism to be classified in this way. Not all microorganisms, however, are oleaginous and, even when grown under the most appropriate conditions for lipid accumulation, they will not accumulate more than a few percentages of their biomass as oil.

For an oleaginous microorganism to produce its maximum content of lipid, it is necessary to grow it in a culture medium which is so formulated that the content of nitrogen (usually supplied as an ammonium salt) is exhausted after a day or so (see Figure 2.1 as a general example). Up until this point, growth of the organism has been in a balanced nutrient situation, known as the *tropophase*, where all the necessary nutrients for growth—C, H, N, O, S, P, K, Mg, etc.—are available. After nitrogen exhaustion, the cells enter an unbalanced phase, sometimes known as the *idiophase*, in which further cell proliferation is prevented by the absence of nitrogen, which, of course, is essential for both protein and nucleic acid synthesis. Provided a supply of carbon, usually glucose, remains available, the cells then continue to assimilate it but, because of their inability to generate new cells, this carbon is essentially surplus to growth requirements and is converted into a reserve storage material within the cells. If the cells should ever become starved of carbon, they are able to mobilize this reserve storage material as a source of carbon and energy. In the oleaginous yeasts, molds, and some algae, this reserve storage material takes the form of triacylglycerol (or triglyceride) oil. In other organisms, the storage material might be polysaccharides or materials such as poly-β-hydroxybutyrate and poly-β-hydroxyalkanoates, as are found in many bacteria.

2.2.2 The Oil Accumulation Process in Oleaginous Microorganisms

The biochemistry of the conversion of the surplus carbon into lipid has been studied in some detail in the author's laboratory [4]. In brief, the oleaginous organism

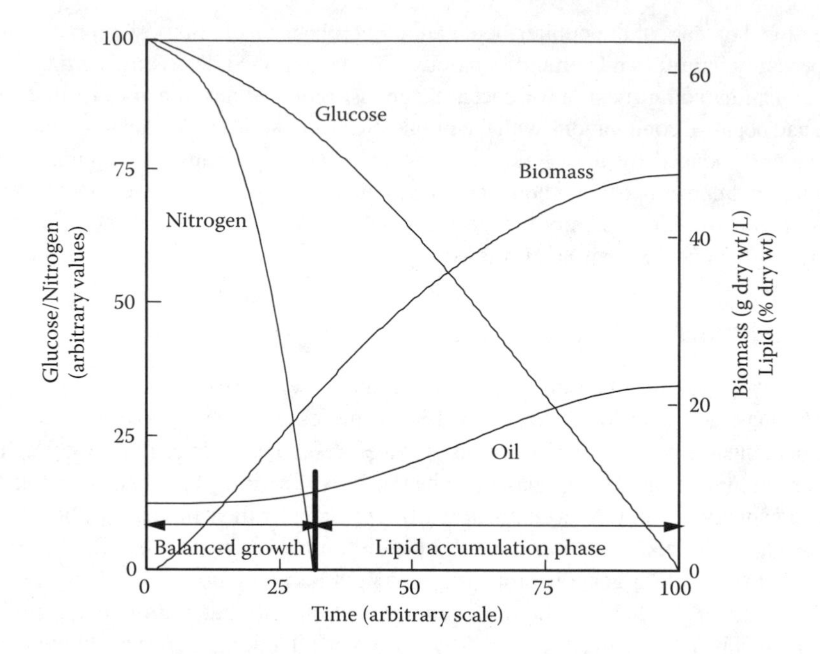

FIGURE 2.1 Idealized representation of lipid accumulation in an oleaginous organism. The organism grows initially with all nutrients in excess; the composition of the medium is so adjusted, however, that the supply of nitrogen (usually an ammonium salt) is quickly exhausted but a surfeit of carbon (usually glucose though other substrates can be used) remains. The excess carbon continues to be assimilated by the cells and, as it no longer can be converted to protein or nucleic acids because of the absence of nitrogen, the carbon is converted to triacylglycerols as a reserve storage material. The extent of lipid accumulation depends on the strain of the organism being used.

possesses a different mechanism of generating acetyl-coenzyme A (CoA) units, as the essential precursor of fatty acid biosynthesis, to the nonoleaginous species. The key to the process is that when nitrogen is exhausted from the growth medium, there is a change in the respiration of the cells so that citric acid, produced from glucose, is no longer oxidized by the citric acid cycle (Krebs cycle) and instead begins to accumulate. It is then cleaved by an enzyme called adenosine triphosphate (ATP): citrate lyase, which is found only in the oleaginous cell. This enzyme then generates acetyl-CoA and is probably physically linked to the fatty acid synthesizing system to ensure rapid conversion into fatty acids. In oleaginous species, but perhaps not in all, there is a second enzyme that generates the reducing power needed to reduce the acetyl units as they are elongated into saturated alkyl chains. This enzyme is malic enzyme. It uses malic acid and nicotinamide adenine dinucleotide phosphate (NADP) as substrates and generates pyruvic acid and NADPH (reduced form of NADP), which then acts as the requisite reductant for the fatty acid synthase. No other enzyme seems to be able to carry out this function and malic enzyme, like ATP:citrate lyase, is thought to be physically attached to the fatty acid synthase protein.

The biochemistry of oleaginicity has been described in detail elsewhere and the erudite reader wishing to know more about this should consult a recent review by the author and one of his colleagues [4]. The lists of oleaginous microorganisms, covering yeasts, molds, and algae [5–7], are not long ones. The attribute of accumulating large amounts of oil in a microbial cell probably rests with just about 100 or so species throughout all three of these groups, though more may be discovered as further exploration of this topic still continues. It is clear from an examination of these lists that the maximum amount of oil that any particular microorganism can accumulate is fixed and, moreover, can vary from species to species and even from strain to strain with a single species (see Table 2.2 for examples). Thus each organism appears to have its own ceiling of the maximum amount of lipid that it can accumulate; some species can have a limit of, say, 25% of their biomass as lipid; in others it may be 40 or even 50%, with the very highest levels being 70 to 80%.

This limitation of lipid accumulation therefore appears to be genetically controlled, as it is not possible to exceed it under any growth condition that might be tried. It is therefore part of the genetic makeup of an organism and is possibly due to the manner in which the gene regulating the synthesis of malic enzyme is controlled. Malic enzyme, as mentioned above, is the enzyme that generates NADPH for fatty acid synthase and it would appear that, in the oleaginous species that produce the lower amounts of lipid, the gene for malic enzyme synthesis is switched off shortly after nitrogen exhaustion. This then means that the supply of reducing power into fatty acid synthesis ceases and no more lipid is produced. The cells have the ability to limit their own obesity. Whether or not this is the general phenomenon, and whether it might be used to explain the variation in oil accumulation ability in plants, is still an open question.

2.3 COMMERCIAL ASPECTS OF SCO PRODUCTION

Early commercial attempts to produce SCOs focused mainly on yeasts [8], which were the most prolific producers of triacylglycerol oils. However, these oils were essentially the equivalent to those that could be obtained from plant seeds. The oils were high in contents of oleic acid (18:1), linoleic acid (18:2), and palmitic acid (16:0), but did not offer anything that could not be obtained much more cheaply from agriculture. Even attempts to produce a cocoa butter substitute fat using yeast technology [9,10] did not succeed in being economic even though cocoa butter itself commands a price of between US$2000 and $3000 per ton compared with, say, $400 to $600 per ton for soybean oil, corn oil, and the like.

The economics of SCO production dictate that only the most expensive of oils can be produced in this way. For the production of 1 ton of oil, a minimum of 5 tons of sugar are needed, but more sugar is needed to generate the cells within which the lipid is stored. A typical SCO process may therefore consume between 7 and 10 tons of sugar to produce 1 ton of oil. Sugar itself may not be expensive but even at about $250 to $300 per ton we have a total substrate cost of around $1800 to 3000 though this might be less if glucose syrup or molasses is used. Add to this the high cost of fermentation technology, which is probably not less than $1000 per ton of biomass generated, and we have an inherently expensive oil.

To produce microbial cells with an oil content of, say 50%, would probably cost a minimum of $2000 per ton of cells (i.e., $4000 for a ton of oil). For an organism with an oil content of 25%, where 4 tons of biomass would be needed, the fermentation costs per ton of oil now rises to $8000. Thus the minimum cost of producing 1 ton of SCO cannot be much less than $4000 and could be as much as double or even treble this if a slow-growing, low-yielding organism were to be used. To these costs, the costs of oil extraction and refining have to be added and, though this may not add much more than $500 per ton of oil, it still has to be taken into account.

From simple fermentation economics, only the most valuable of oils can thus be contemplated for production. Using microorganisms to try to produce substitutes for corn oil, groundnut oil, canola oil, or even cocoa butter is just not an economic proposition. Therefore, one has to aim for much higher valued oils.

2.4 MICROBIAL OILS AS A SOURCE OF γ-LINOLENIC ACID (GLA)

2.4.1 GLA: Sources and Applications

GLA (18:3, n-6) has a very long history of use, occurring as it occurs in the seed oil of the evening primrose (*Oenothera biennis*). Evening primrose oil (EPO) has been used as a quasi-medicinal oil for centuries; it is known as "King's Cure-All" and thus has been recommended for the alleviation of a wide number of illnesses [11–16], including the improved well-being of the elderly [17]. In the 1970s, it was suggested as a possible treatment for multiple sclerosis, a claim that has since been rescinded. Also at this time, EPO was regarded as highly useful for the relief of premenstrual tension and also for the treatment of eczema, particularly childhood and atopic eczema [15,16]. All these properties were attributed to the presence of GLA in the oil, even though the amount of GLA in the oil being sold at this time was only about 10%. GLA itself is a precursor of certain prostaglandins that fulfill essential physiological roles in the body, and thus a continual supply of GLA is needed to ensure continuous synthesis of the prostaglandins, as they have only a short half-life in the body. Normally, though, GLA is synthesized in the body from linoleic acid (18:2, n-6) by a specific Δ6 desaturase and supplementation of the diet should be unnecessary. However, under certain conditions, the activity of the Δ6 desaturase may decline, leaving the body with a deficiency in GLA and therefore in prostaglandins. Under such conditions, supplementation of the diet by oils containing GLA has been advocated [11–17]; however, it has to be stated that the evidence is not without its critics, and the efficacy or otherwise of EPO as a dietary supplement has been questioned, principally because 90% of it is not GLA and whatever effects it may bring about could be attributed to something other than GLA.

GLA, however, continues to be an attractive nutraceutical as a source of n-6 fatty acids. Oils rich in this fatty acid are still recommended for the treatment or alleviation of a number of clinical conditions: further information about the medical applications of GLA may be found in two recent monographs devoted to this topic [13,14].

The demand for EPO began to increase in the mid-1970s because of the many claims of its benefits. With this came a concomitant increase in price as demand began to exceed the supply. Top prices for EPO in the mid-1970s and early 1980s reached over $50 per kg (i.e., over $50,000 per ton). It was then one of the highest, if not the highest, valued triacylglycerol oils produced.

2.4.2 The Search for the "Best" GLA-Producing Fungus

The monopoly enjoyed by EPO led to searches for alternatives sources of GLA. The presence of GLA in *Phycomyces* fungi had been known since the 1940s [18], and it was subsequently found to be a common fatty acid in fungi classified as "the lower fungi," which included the genus of *Phycomyces* [19–21]. The lower fungi are sometimes known collectively as the Phycomycetes order but are more correctly classified into two subgroups: Mastigomycotina and Zygomycotina, with members of both groups producing GLA in their lipids. Interestingly, no other group of microorganism — bacteria, yeast, or higher fungi — produce GLA, though it is found in many marine and freshwater algae.

With the known occurrence of GLA in fungal microorganisms, it was therefore attractive to consider a biotechnological route for its production. Up to this time, no process had been developed on a commercial scale for the production of any microbial oil. Nevertheless, it seemed that if an appropriate organism could be identified, then production of a GLA-SCO should be feasible using the technology that was available for large-scale cultivation of microorganisms, including filamentous fungi.

Work began in the author's laboratory in 1976 to identify a possible fungal source of GLA. Over 300 species and strains were eventually screened over the next 6 years for GLA production. Table 2.1 lists the main genera that were examined. Within each genus, a diversity of species was examined to ensure that as wide a net as possible was cast to find the most promising species. All organisms were obtained

TABLE 2.1
Genera of the Lower Fungi (Phycomycetes) Screened for GLA Production

Absidia	*Mucor*
Basidiobolus	*Phlyctochytrium*
Choenephora	*Phycomyces*
Cunninghamella	*Pythium*
Delacroixia	*Rhizopus*
Entomophthora	*Zygorhynchus*
Mortierella	

In each case, a number of representative species were screened for growth performance, lipid production, and GLA content of the lipids. Those species that appeared to be the most promising (see text) were then examined in detail. See Table 2.2.

from major culture collections; no attempt was made to isolate possible organisms from the environment.

The criteria that were used to evaluate performance in this screening process were as follows:

- The organism should grow readily in submerged culture attaining at least 10 g/l in 3 d or less in simple stirred vessels. It should not pose problems for extensive filamentous growth or for pellet formation.
- It should have an extractable oil content of not less than 20% of the biomass.
- It should preferably have a GLA content of the total fatty acids of as near to 20% as possible and, preferably, over this value.
- The oil should be over 90% triacylglycerol.
- The organism should pose no hazard for large-scale cultivation. It should not be an animal pathogen or a plant pathogen. It also should have no history of causing any allergic reaction or toxicity.
- Ideally, the organism should grow at 30°C or slightly higher because if it had a lower growth temperature requirement, this could otherwise incur additional expenditure for cooling large-scale cultures.

After an initial cultivation in shake-flask cultures, all organisms were grown in vortex-aerated 1 l bottles [22]. In all cases, a glucose-based medium was used and, as the pH was not controlled in these vessels, diammonium tartrate was used as the N source with small amounts of yeast extract, malt extract, and peptone (totaling 0.5 g/l) added as sources of any vitamin that may be required. (Ammonium tartrate was used as the N source because this prevents acidification of the medium that occurs when ammonium chloride or ammonium sulfate is used, as the uptake of the ammonium ion leaves HCl or H_2SO_4 behind that quickly acidifies the medium and prevents microbial growth.) The medium composition was carefully balanced so that it would become N-limited after about 24 to 30 h (see Figure 2.1); a C:N ratio of 40:1 was therefore selected as able to induce lipid accumulation in an oleaginous microorganism.

After evaluating the performance of over 200 organisms, it seemed that the most promising species lay in the genera of *Cunninghamella* and *Mucor*, with *Rhizopus stolonifer* a further candidate. Accordingly, extensive examination then began of the key species in each of the two likely genera by looking at the performance of as many strains of each species as could be obtained from culture collections worldwide. Table 2.2 lists the results of 57 individual strains taken from just 8 species.

As can be seen, the lipid and GLA content of the fatty acids could vary enormously within a single species: for example, with *Mucor circinelloides* the content of lipid in the cells, all grown under identical conditions, ranged from 3 to 37% and the GLA ranged from 8 to 32%. This wide divergency indicates that screening can often miss the best organism, and therefore one should be prepared to examine the complete range of strains that are available in order to ensure that the best organism is not overlooked.

TABLE 2.2
Final Screening of Individual Strains for the Best GLA-Producing Organism

Organism	Lab. Strain No.	Lipid (% Dry Wt)	GLA (% Total Fatty Acids)
Cunninghamella blakesleeana	237	33	8
	239	28	13
	247	32	7
	248	36	8
	249	34	10
	281	23	10
C. echinulata	151	21	16
	244	22	18
	245	18	18
	246	20	17
	282	23	13
Mucor circinelloides	116	13	15
	119	37	8
	162	12	18
	232	6	13
	233	13	12
	240	8	17
	252	21	12
	253	9	20
	254	14	18
	255	17	8
	271	21	13
	272	6	11
	273	19	16
	274	3	22
M. genevensis	158	10	32
	241	2	11
	256	17	13
	269	27	20
	275	20	17
	158	10	32
	241	2	11
	256	17	13
	269	27	20
	275	20	17
M. mucedo	118	17	25
	223	28	1
	226	10	8
	242	14	22

(*continued*)

TABLE 2.2 (Continued)
Final Screening of Individual Strains for the Best GLA-Producing Organism

Organism	Lab. Strain No.	Lipid (% Dry Wt)	GLA (% Total Fatty Acids)
	278	9	24
M. racemosus	217	9	5
	218	14	7
	219	3	0
	231	12	0
	243	24	7
	259	23	14
	262	22	10
	265	28	11
	266	32	9
	267	30	13
	277	24	12
Rhizopus stolonifer	121	7	24
	227	6	8
	228	7	16
	229	11	20
	279	9	21
	280	11	28

Each organism was grown in a 1 l vortex-stirred bottle [32] with a medium having a high C:N ratio to encourage maximum lipid accumulation (see Figure 2.1 and text).

From the results given in Table 2.2, combined with observations on the growth patterns of the organisms and their ability to grow as dispersed cultures in 5 l, fully controlled, stirred bioreactors, the final list of three organisms was chosen. These were

- *M. circinelloides* strain 119
- *M. genevensis* strain 269
- *M. racemosus* strain 267

(The strain numbers given here and in Table 2.2 refer to the numbers allocated to the strains in the author's laboratory and do not correspond in any way to coding systems used by the various culture collections from which these organisms were obtained.) This then completed the laboratory work on these fungi.

Further work beyond the laboratory scale now had to be carried out in partnership with the company that intended to produce the GLA oil commercially, using its equipment and expertise to translate a laboratory process into an economically viable production. This company was J. & E. Sturge, Ltd. of Selby, North Yorkshire,

United Kingdom, which was major producer of citric acid using *Aspergillus niger* and had expressed a keen interest in this GLA process. The company had available four 220-m^3 stirred fermenters for citric acid production plus a number of smaller pilot-scale fermenters that were used to evaluate possible process improvements. Production of the GLA-SCO could then be contemplated by using one of the production-scale fermenters, as the technology used for citric acid production would be similar in most respects to that which would be needed for GLA production.

Each of the three fungi listed above were then grown in pilot-scale fermenters up to 10 to 15 m^3 to produce sufficient guidance as to their likely performance in the final production vessel of 220 m^3. The quality of the oil extracted from each organism was also evaluated and compared. From a consideration of all the various factors, the final choice of which organism to use was *M. circinelloides* strain 119. It was found, however, that the performance of the organism was not quite the same as that which had been initially found in the preliminary survey shown in Table 2.2. Instead of the cells reaching 37% oil content as they had done earlier, the ceiling now appeared to be about 25% oil, but, at the same time, the content of GLA in the fatty acids increased from 8 to 18%. This was then taken as the best overall yield for GLA that could be reasonably attained.

2.4.3 Production of GLA-SCO

The first large-scale production run for GLA was carried out in 1984 using one of the citric acid fermenters with full pH, temperature, and aeration controls. *Mucor. circinelloides* was grown at 30°C, which was just high enough to obviate the necessity for refrigeration to regulate the temperature, as would undoubtedly have been needed if the optimum growth temperature had been 28°C or less. The medium used, however, was different from that used for citric acid production and was formulated with a high C to N ratio so as to promote lipid accumulation and also to achieve as high a biomass as possible (see Figure 2.1). In the event, it was found advisable, because of the morphological characteristics of the organism, to restrict the biomass density to about 50 g dry wt/l. Above this density, there were problems with pumping the cells from the production vessel to the holding tank (see Figure 2.2). The final cell density was therefore controlled by limiting the initial glucose concentration to about 125 g/l.

2.4.3.1 Problems with the First Product

The first run produced sufficient biomass to evaluate methods for oil extraction and, when this had been solved, for there to be sufficient oil to carry out toxicological trials. The initial toxicological trials used brine shrimp as a very sensitive indicator organism. The first results with the brine shrimp were, however, a disaster. The oil was apparently toxic and most of the brine shrimp were killed; however, it was quickly realized that the culprit was the presence of free fatty acids in the oil (the reasons for their occurrence are given below). Once these had been removed, the oil was found to be extraordinarily safe: further and extensive animal feeding trials were carried out using mice, rats, and, of course, brine shrimp. The oil proved to be equal or superior to any other plant oil that was used in comparison.

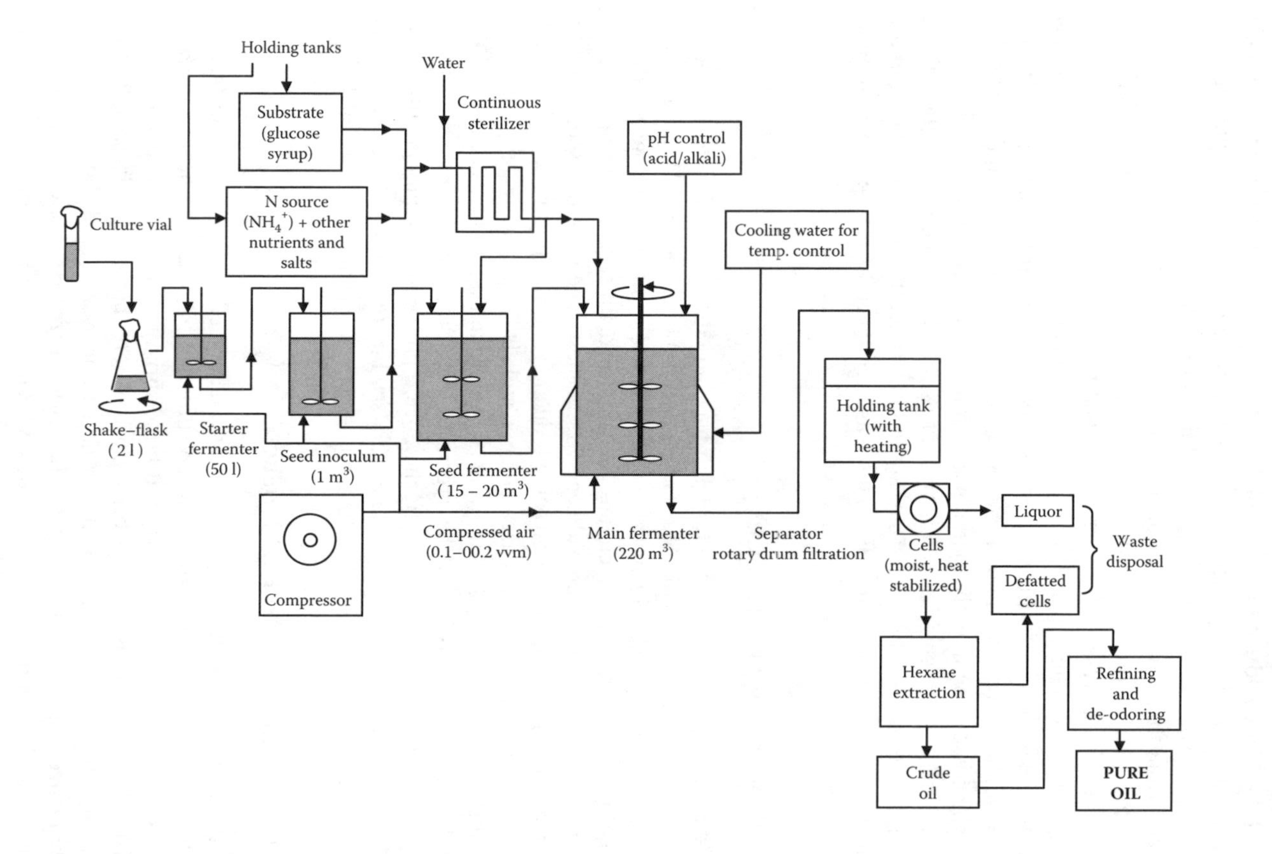

FIGURE 2.2 Diagrammatic representation of the process used for the production of GLA-SCO. The process is described in detail in Section 2.4.3.

The presence of the free fatty acids in the first batch of oil produced was something of a surprise, though we had known for many years that these materials were artifacts arising very readily in microbial oils if appropriate precautions to avoid their formation were not taken. As oils are produced as a reserve storage material in response to a nutritional deficiency other than carbon, it then follows that if the cells subsequently become devoid of a supply of external carbon, they will then commence to utilize the lipid that they have previously accumulated. After all, that is what it is there for. During the harvesting of cells from their culture medium, there will be only a little carbon remaining in the culture broth; and, even if some does remain, then this remnant is certainly removed during the harvesting of the cells. The harvested cells are thus starved of carbon. Accordingly, as they are still metabolically active, they immediately mobilize their lipid reserves and this entails the expression of a lipase to release fatty acids from the triacylglycerol, which are then oxidized to yield energy as well as acetyl-CoA to be used for synthesis of new cell material. The induction of lipase activity occurs instantly when the cells are starved of carbon. These events have been followed and reported for other oleaginous microorganisms [23]. Similarly in *M. circinelloides*, as it was harvested from the fermenter, lipase activity was induced in the cells; and then, during the oil extraction itself, the lipase continued to work, thereby releasing free fatty acids that subsequently contaminated the final oil.

Once the presence of free fatty acids was recognized, it was then important that the activity of the lipase was prevented at the earliest possible stage. To achieve this, the simplest way was to heat the culture broth prior to harvesting the cells. This heating could be applied to the entire fermenter itself, though it was found more expedient to evacuate the entire contents of the fermenter at the end of the lipid accumulation phase into a heated holding tank (see Figure 2.2). An alternative was to pass the culture broth through a heat exchanger in the transfer of the entire culture from final fermenter into the holding tank. Heating to about 60°C was found sufficient to inactivate all lipase activity; it also served to stabilize the cells from other hydrolytic enzyme activities.

2.4.3.2 Process Operations

The process of GLA-oil production is shown in Figure 2.2. In keeping with most other large-scale fermentation processes, the organism is grown through a series of fermenters of increasing size; each one is inoculated with the culture from the previous vessel after about 24 h to ensure that the cells are growing as rapidly as possible. The final production fermenter is inoculated with 10% of its volume with *M. circinelloides*; it is only in the final fermenter that there is a need to ensure that the medium has a surfeit of carbon and a deficiency of nitrogen. Prior to this, the medium is more or less balanced in terms of the C and N requirements of the cells, as this ensures the cells are growing at their maximum rate at the time they are inoculated into the next fermenter.

The pH, temperature, and aeration rates are carefully monitored and controlled throughout growth in all fermenters: it is important that the cells remain as dispersed as possible and that pellet formation, which is a common feature among filamentous fungi, is avoided as far as possible. *Mucor circinelloides*, fortunately, can be maintained as dispersed cells with very little clumping into pellets, which is then an advantage to ensure good aeration and oxygen transfer to the cells. These are all important characteristics which need to be considered before the final choice of organism is made. It is therefore vitally important that the biochemical engineers responsible for running the process and for the final harvesting of the biomass are consulted at the earliest stages of selecting the likely production organism. Simply judging the organism on its productivity, yield of oil and GLA content is not sufficient.

After completion of growth, the biomass is transferred into a holding tank and is heated to prevent lipid hydrolysis (see above). The use of a holding tank into which the entire final culture can be discharged is also expedient to give the minimum turnaround time with the main fermenter allowing it to be cleaned and recharged with fresh medium, and the next run then commenced with the minimum of delay. The holding tank also allows downstream processing of the heat-stabilized cells to take place at a rate that is commensurate with the capacity of the harvesting equipment, whether this be by centrifugation, rotary drum filtration, or any other means. In this case, the cells were harvested by rotary drum filtration with final pressing of the filtered cells to remove as much water from the cells as possible.

The final product was moist, packed, heat-stabilized cells. The cells did not need to be spray dried to remove all the water, as this was unnecessary for the final process of oil extraction (see next section). No deterioration of the oil apparently occurred in the cells once they had been heated. One production run from the 220 m^3 fermenter yielded over 10 tons dry wt (approximately 13 to 14 tons moist wt) of cells containing about 2.5 tons of oil. The duration of the final fermentation run was between 3 and 4 d.

2.4.3.3 Oil Extraction

Oil was extracted from the cells using a small-scale hexane extraction unit that was normally used for the extraction of essential oils from plant materials. Extraction units that deal with commercial quantities of plant oil seeds were simply too big to warrant use with the relatively small amount of fungal biomass that had to be dealt with. Such extraction units deal with tens of tons of material per hour and have a hold-up volume within the extraction plant of up to 100 tons — in other words, over 100 tons of new oil have to be extracted from any one source to displace the existing oil still in the system from the previous material used. The processing equipment used have could have handled the entire output of one of the production fermenters in less than a day of processing.

The extracted oil from *M. circinelloides* still required further processing (as do all oils, irrespective of their origins). This necessitated carrying out refinement and deodorization in order to remove nontriacylglycerol components of the oil, mainly phospholipids and some other polar lipids, as well as removing any volatile materials that could have adversely affected the smell or taste of the oil. This further processing was again carried out using small-scale equipment that was normally used for handling small batches (i.e., about 1 ton) of experimental oils coming from various exotic sources. These processes were exactly the same as those used in conjunction with other plant oils and did not need any major changes in their technology to handle the mold oil. The final product was a clear yellow oil with a triacylglycerols content of over 98% (see Table 2.3).

TABLE 2.3
Specifications of GLA-SCO

Production organism	*Mucor circinelloides* (= *javanicus*)
Production company	J. & E. Sturge Ltd, Selby, N. Yorks, United Kingdom
Cultivation system	220 m^3 fermenters; duration: ~ 4 d; biomass contains 25% extractable oil
Oil	
Trade name	Oil of Javanicus
Appearance	Pale yellow, clean and bright
Specific gravity	0.92 at 20°C
Peroxide value	3 (maximum)
Melting point	12–14°C
Free fatty acids	<<1%
Triacylglycerol content of oil	>97%
Added antioxidant	Vitamin E
Stability	No deterioration at 20°C over 12 months
Fatty Acyl Composition (rel.% w/w) of Oil	
14:0	1%
16:0	22–24%
16:1 (n-9)	1%
18:0	5–7%
18:1 (n-9)	38–40%
18:2 (n-6)	10–11%
18:3 (n-6)	18–19%
18:3 (n-3)	0.2%

2.4.3.4 Specifications of GLA-SCO, Safety Evaluation, and GRAS Status

Table 2.3 gives the specification of the final fungal oil that was offered for sale. As can be seen, a small amount of vitamin E was added to the oil to ensure long-term stability, though this was done merely as a precaution, as the oil itself proved remarkably stable. Preparations of the oil have been held in the author's laboratory, at room temperature, in air, and in sunlight for over 15 years without any sign of deterioration of the oil, with the content of GLA in the oil remaining completely unchanged. This indicates that the organism itself must produce effective antioxidants that are lipid soluble and so are coextracted with the oil and remain with it during the final clean-up processes. The nature of the antioxidant has not been investigated, though it is likely that it is related to known antioxidant materials such as the tocopherols.

The toxicological trials of the final oil preparation, including long-term feeding trials with experimental animals, confirmed the inherent safety of the oil. The oil was also able to claim a safe record of use, as the production organism, *M. circinelloides*, has a long history associated with fermented foods such as tempeh and tapé. Tempeh and tapé are traditional foods eaten throughout Indonesia and southeast Asia.

They are produced by a fermentation process that uses a mixture of fungal organisms, one of which is *M. circinelloides*. Thus, generations of people who have consumed tempeh as a staple component of their diet have, in fact, been eating *M. circinelloides* and this, of course, includes its lipid components. Therefore, as tempeh is a recognized food material with a long history of safe consumption, it was entirely reasonable to conclude that *M. circinelloides* was in itself safe to eat. Indeed, *M. circinelloides* is accorded generally recognized as safe (GRAS) status.

As the entire mold is safe to eat, it then follows that a component of it must also be safe. Further, as the oil itself could be defined in terms of its total composition, with all the components established as being present in a number of existing edible oils and food materials, there were no qualms about the safety of the *Mucor* oil. All the fatty acids that were present in the oil were, again, identifiable components of other edible oils. The production organism had no history of producing any unusual metabolites associated with any toxic condition; indeed the *Mucor* fungi are remarkable for their lack of producing "interesting" bioactive compounds. As far as the pharmaceutical industry is concerned, *Mucor* species are probably the most boring of all fungi with their complete inability to produce any biologically active compound.

Complete confidence in the safety of the oil could then be indicated, not only from a detailed toxicological examination of the oil, but also from its analysis showing that every component was already a component of other oils, and, finally, with the organism itself having a complete record of safe consumption over many centuries if not millennia.

2.4.3.5 Marketing

At this point, a GLA-rich oil had been produced, which exceeded the specification of EPO, against which it was designed to compete. It now had to be marketed and sold. The oil was given the name Oil of Javanicus, coming from the older name for the organism, *Mucor javanicus*. This name had arisen because the organism was originally isolated from Java, and, indeed, this is where its association with tempeh and tapé had first been recognized. Figure 2.3 shows some of the marketing material that was produced to help with the launch of the new oil.

Not without justification, a number of claims could be made for the oil to indicate its superiority over EPO. First, its specifications were superior to EPO, with almost twice the content of GLA (see Table 2.4). Then, the specifications of the oil itself could be guaranteed on a year-round basis, as clearly the fermentation process was not subject to the vagaries of the weather or climate. The quality of EPO could be somewhat variable and was dependent on where and when the plant had been grown, for how long the seed had been stored before extraction, and for how long the oil itself had been stored. The specifications of the EPO sold in the fall need not necessarily be the same as that sold the following summer. EPO from Eastern Europe need not have the same specifications as that produced in Western Europe; the quality of the oil this year might not be quite the same as last year's and may be different again next year.

Finally, the GLA-SCO could be claimed to be truly organic: agricultural crops are normally sprayed with various chemicals several times during their growth to keep down adventitious weeds and to prevent insect damage, and also fungicidal

FIGURE 2.3 Promotional material for the marketing of the world's first microbial oil — Oil of Javanicus.

agents may need to be applied to the seeds if these are to be stored for any length of time to prevent fungal growth and spoilage of the product. Products from fermenters do not have to be sprayed with any herbicide, pesticide, or fungicide and, consequently, Oil of Javanicus could be claimed to satisfy all the requirements to be labeled as "organically produced." It had the absolute minimum amounts of residues of these chemical spray materials. However, it was interesting to find that some minute residues of these materials could be found in the final mold oil, which were traceable to their presence in the original glucose feedstock used in the fermentation. The levels were, though, absolutely minute and were less than those found in EPO samples that were analyzed simultaneously; but, it should be emphasized that the level of chemical residues in EPO were well below permitted levels for such materials in plant seed oils and constituted no hazard to the consumer.

Not unnaturally, the arrival of this new oil, destined as a direct competitor to EPO, led to considerable opposition from the growers of evening primrose and from the suppliers of this oil to the general public. Their main response was to lower the price of EPO in an attempt to forestall the success of Oil of Javanicus. There were also major advertising campaigns to reinforce the benefits of EPO, not so much as a source of GLA, and for which Oil of Javanicus was demonstrably twice as good, but to keep the name of EPO at the fore as the oil of choice to alleviate various disorders, specifically premenstrual tension. Women who had found benefit from taking EPO were apparently reluctant to switch their allegiance to an "unproven" oil — after all, the success of EPO might have nothing to do with its GLA content. This was one of the major problems that had to be faced; the public wanted EPO because it was EPO — what GLA had to do with EPO's efficacy was not a concern to them. Indeed, at least 50% of all people buying EPO had no concept of its main active ingredient.

TABLE 2.4
Fatty Acid Profiles of Various Fungi and Plants Used, or Considered, for GLA Production

	Oil Content	Relative% (w/w) of Major Fatty Acids								
	(% w/w)	16:0	16:1	18:0	18:1	18:2 (n-6)	18:3 (n-6) GLA	18:3 (n-3)	20:1	22:1
Mucor circinelloides[a]	25	22	1	6	40	11	18	—	—	—
Mortierella isabellina[b]	~50?	27	1	6	44	12	8	—	0.4	—
M. ramanniana[b]	~40?	24	—	5	51	10	10	—	—	—
Syzygites megalolcarpus[c]	22	14	—	1	12	10	62	—	—	—
Evening primrose	16	6	—	2	8	75	8–10	0.2	0.2	—
Borage	30	10	—	4	16	40	22	0.5	4.5	2.5
Black currant	30	6	—	1	10	48	17	13	—	—

[a] Oil of Javanicus (see text).
[b] Production organisms used by Idemitsu Company. Ltd., Japan. Oil contents of cells uncertain.
[c] See Weete, J.D., Shewmaker, F. and Gandhi, S.R., *J. Am. Oil Chem. Soc.*, 75, 1367–1372, 1998.

EPO worked and that was sufficient: why change a winning preparation? Nevertheless, in spite of EPO's strong placement in the market, sales of Oil of Javanicus gradually began to take off and found greatest uptake by those people who considered a supplementary source of GLA in their diet beneficial.

Most Oil of Javanicus was sold under the separate trade name of GLA-Forte, which was distributed, mainly by mail order, by a small nutraceutical company based in Birmingham, United Kingdom. (Although this company no longer exists, the trade name of GLA-Forte continues to be used as a name for borage oil — see below.)

Some suggestions, though, were made that the GLA in Oil of Javanicus, and indeed in other oils containing GLA such as borage or black currant oils (see below and Table 2.4), might not be as nutritionally beneficial as the GLA in EPO. The inference was that the mold oil (and borage and black currant oils) was in some way inferior to EPO, even though EPO had the lowest of all contents of GLA [24]. This was based on the apparent differences in the rates of hydrolysis of the oils by pancreatic lipase [25], with the release of GLA governed by its positional distribution in the triacylglycerol and by the presence of neighboring long chain polyunsaturated fatty acids [25,26]. As Table 2.5 shows, there are differences in the distribution of the various fatty acyl residues in all four oils: EPO, Oil of Javanicus, borage, and black currant [27]. However, the positional distribution of GLA is not dramatically different at the *sn*-2 and *sn*-3 positions of all four oils; and, furthermore, it would be expected that the hydrolysis of all oils to their component fatty acids would be complete during the digestive process. The experiments that were done to show differences in digestibility of borage and EPO used an isolated lipase operating over 15 to 20 min [25], but this is probably not a sufficient indicator of what happens *in vivo* when there is longer contact between lipase and substrate to ensure complete hydrolysis of all the fatty acyl components of the oils, irrespective of their position on the original triacylglycerol molecule. Certainly, other workers have not reported on any difference between GLA availability from borage oil compared with EPO [28], whereas according to the work of Horrobin and colleagues [25], there were differences. As borage oil and Oil of Javanicus are somewhat similar in their distributions of GLA (see Table 2.5), the same conclusions reached by Raederstorff and Moser [28] might also then be taken to apply to the mold oil.

TABLE 2.5
Stereospecific Distribution (%mol/mol) of GLA in Triacylglycerol Oils from Four Sources

Position	Evening Primrose Oil	Borage Oil	Black Currant Oil	*Mucor Circinelloides* Oil
All	9.3	24.8	15.9	17.9
sn-1	3.6	4.0	4.1	13.3
sn-2	10.7	40.4	17.4	19.6
sn-3	13.5	30.1	25.8	19.6

Source: From Lawson, L.D. and Hughes, B.G., *Lipids* 23, 313–317, 1988. With permission.

It should perhaps be pointed out that the main evidence favoring EPO as a nutritional source of GLA, rather than Oil of Javanicus or borage oil, has come from the company Efamol Ltd., which was, and still is, one of the major suppliers of EPO in the United Kingdom and elsewhere. No declaration of interest was ever given in this key publication [25] or in others suggesting EPO was the preferential source of GLA.

2.4.3.6 The Demise of the GLA-SCO Process: Competitors and Economics

During the development of the fermentation process for GLA-SCO in the early and mid-1980s, there was increasing interest in finding other oils that contained GLA. These researches led to the discovery that GLA could also be obtained from the plant known as borage: *Borago officinalis*, which, like evening primrose, is often found growing as a wayside weed. GLA had also been identified as a component fatty acid in the oil extracted from blackcurrant seeds (*Ribes nigrum*) that are a waste product from the processing of this fruit [11]. Although the oil content of these seeds is quite high (30%), the oil also contains a high proportion of α-linolenic acid (ALA) along with the GLA (see Table 2.4), making it somewhat unattractive if one is trying to extol the virtues of an oil rich in n-6 fatty acids. This was therefore not regarded as a significant competitor of the mold oil. Borage oil, however, was a much more serious competitor. Interest in developing it as a cash crop was sharply stimulated when the company carrying out the initial work with borage approached J. & E. Sturge, the producers of the GLA-SCO, to enter into an agreement regarding GLA oils but were very firmly rebuffed. In retaliation, interest in developing borage oil as a source of GLA was then intensified.

The GLA content of borage oil is 22%, which is higher than that of the GLA-SCO (see Table 2.4), and could be produced more cheaply than either EPO or SCO. Borage, unlike evening primrose, is an annual plant: it can be grown and harvested in a single season. However, like evening primrose, borage produces minute seeds that pose some technical problems for harvesting, but, using the same procedures that were used with evening primrose, borage seeds could be harvested and their oil subsequently extracted. The oil, which has been given the name Starflower Oil after the appearance of the borage flowers, could also be produced more cheaply than the fungal oil, and, eventually, this price competition led to the fermentation process being discontinued in 1990.

A major additional reason for abandoning the GLA-SCO process was that the production company had been sold to another company, Rhone-Poulenc Ltd., which decided that the profitability of the *Mucor* process was too low and it would require substantial sums of money to be spent to develop the process and to market the oil to the level where sales could be substantially increased. The competitive price of borage oil, together with the falling price of EPO to now about $10 per kg, forced this decision, as clearly this was about the very lowest price that the mold oil could be produced, thereby leaving no margin for a reasonable profit.

To some extent, however, the GLA-SCO was not in a fair price competition with these plant oils. Both plant oils were classed, under European Union (EU) regulations, as nonfood crops and, as such, enjoyed subsidies from the EU to be grown in place of conventional food crops, which were in abundance with much of the produce surplus to requirements. The EU agricultural policy was, and still is, to diminish the amounts of food crops grown to the point where only a sensible margin of surplus is provided. Thus, the main agricultural competitors of Oil of Javanicus were subsidized materials and, if that was not enough, the costs of fermentation were penalized, again by EU agricultural rules.

As has been stated above, about 10 tons of sugar are need to generate 1 ton of oil using *M. circinelloides*, which has a oil content of only 25%. (The 75% of the biomass that is not oil still has to be produced from sucrose or glucose in the fermenter, and this pushes up the final conversion ratio of sugar to oil.) In Europe, the majority of sugar is produced from sugar beet but glucose can also be produced by hydrolysis of corn starch and this, as glucose syrup, is normally the favored feedstock to use in fermentations. Unfortunately, there is an EU tariff on sugar to protect the sugar beet growers; neither sucrose nor glucose can therefore be bought at world prices, but have to be purchased at the prevailing and artificially enhanced EU prices.

The GLA-SCO process thus suffered a double whammy: its production costs were forced higher than need be by the EU tariffs on the fermentation feedstock, and the final product then competed against plant oils whose growers received direct subsidies to lower their prices. Against such a doubly unfavorable pricing system, it is little wonder that the profitability of Oil of Javanicus was deemed insufficiently attractive for further investment and development. The process ceased in 1990, after 6 years of production, during which time about 50 tons of oil had been produced.

These artificial price restraints, of course, do not operate outside Europe and the opportunity therefore exists to reactivate this process in more economically favorable conditions. However, the demand for oils rich in GLA is probably more than satisfied by EPO and borage oil.

One of the main advantages of the fungal oil is its low content of linoleic acid (18:2) — see Tables 2.3 and 2.4 — which enables GLA to be fractionated from the oil in relatively high recoveries. This is more difficult, and costly, to achieve with either EPO or borage oil, where the content of 18:2 is much higher, and also with black currant oil with its high content of ALA. However, processes for the enrichment of GLA from these sources have been described [11,29,30]. For GLA enrichment, the entire oil is esterified to its methyl or ethyl esters and then can be subjected to the standard procedures of urea crystallization, low-temperature crystallization, and, if necessary, final purification by column chromatography. Purities of GLA from *M. circinelloides* of up to 98.5% have been obtained in this way (F.D. Gunstone, personal communication, 1980). If one were now to use preparative scale high-performance liquid chromatography as the final step, purifications of virtually 100% would be anticipated. While such a purification is possible from the fungal oil, it is by no means certain that pure GLA is required for any purpose.

2.5 OTHER SOURCES OF MICROBIAL GLA

The only other commercial process that was developed for the production of GLA using microorganisms was by Idemitsu Kosan Company Ltd, Tokyo, Japan, a major petroleum company, working (it is believed) in conjunction with scientists from the National Chemical Laboratory of Industry, Tsukuba, Ibaraki, Japan. This work began about 1980 [31]. Interest in the process appears to have been taken on by Meiji Seika Kaisha Ltd, Kanagawa, Japan, a food specialities company, though it is uncertain whether production still continues today.

In the initial phase of the work, which began with a survey of possible organisms, *Mortierella isabellina* and *M. ramanniana* were identified as the most promising species [31,32]. Unlike the screening process described above, the main emphasis of the Japanese work appeared to be to identify a high-oil producer with just a moderate level of GLA. Accordingly, both these organisms had oil contents of 40 to 50% of the biomass but with only 8 to 10% GLA in the oil (see Table 2.4). This was clearly a different philosophy of approach to our own where a 20% GLA content of the oil had been the target.

Further work with *M. ramanniana* did, though, succeed in pushing up the GLA content to over 18 to 20% of the oil [33], but this was by using a novel impeller (the Maxblend) in a small-scale (nonproduction) fermenter and by using a cultivation time of 9 d. Biomass yields were not exceptionally high at 63 g/l. The productivity of the process, expressed as grams of GLA produced per hour per liter of fermenter volume, was thus less than that achieved with the *Mucor circinelloides* process described above.

Interestingly, the GLA content of *Mortierella ramanniana* was also increased by isolating cold-tolerant mutants of it, which when regrown at 30°C, had GLA contents of the oil now at 13.5% instead of the original 7% [34]. When one such mutant was cultivated in a 600 l fermenter, the GLA content of the oil went up to over 18%; though, again, cultivation times needed to be rather lengthy at 8 to 10 d.

Further interest in developing a GLA process in the United Kingdom was initiated by Efamol in the 1980; Efamol was the major supplier of EPO and was somewhat anxious that the process developed at J. & E. Sturge (see Section 2.4.3) should not erode its market position. Overtures by Efamol to enter into an agreement with Sturge met with the same negative response as had been received by the company interested in developing borage oil (Section 2.4.3.6). A pilot-scale process was developed at the University of Dundee with the collaboration of Efamol with Professor Rod Herbert. Although no account of this work has been published, patents were taken out (see, e.g., [35]), which revealed that the organism to be used was also *Mucor circinelloides* (now deposited in a culture collection as IMI 307741). Oil contents of the organism were about 40% with a GLA content of about 14%. Work on the process, however, did not proceed beyond 3000 l; though clearly if large-scale production had been undertaken, it would have been a serious rival process to the Sturge Oil of Javanicus process.

Some potential interest in producing GLA using *M. circinelloides* grown on acetic acid as a feedstock was expressed by a South African group led by Professor Ludwig Kock working at the University of the Orange Free State in conjunction with Sasol Company Ltd., the large South African fuel oil company [36,37]. Acetic acid was chosen as a potential feedstock, as it was a waste by-product arising from the gasification of the coal process operated by Sasol. Improved oil production occurred using acetic acid rather than glucose, with over 30% oil contents of the biomass achieved [38–41]. A suggestion was also made that it might be possible to fractionate the final mold oil into a GLA-rich fraction with the remainder useful as a cocoa butter equivalent (CBE) fat [41]. However, although several patents for these processes were taken out (see, e.g., [36,37]), no subsequent production of a GLA oil, or of a CBE, appears to have taken place.

The results of a number of laboratory-based screening programs looking for GLA-producing microorganisms have been published over the years [42–47]. There have also been a relatively large number of publications reporting the production of GLA in a variety of fungi, but these are of insufficient value to warrant separate citations, as they merely continue to catalog well-established information. Not unexpectedly, almost all surveys of potentially useful microorganisms for GLA production have concentrated on various species of *Mucor*, *Mortierella*, *Cunninghamella*, and *Rhizopus*. There was, however, one exception.

The most significant finding of these surveys was the discovery of a hitherto unexamined fungus: *Syzgites megalocarpus* [47]. This fungus is a member of the order of Mucorales within the lower fungi division; one of its synonyms is *Mucor aspergillus*. It is a mycoparasitic fungus that lives on decaying mushrooms such as Boletus. In the initial survey, this organism produced only 10% lipid in its biomass but had an unparalleled level of 62% GLA in its oil (see Table 2.4). Subsequently, when it was grown on a medium with a high C to N ratio, the lipid content of the cells went up to 25%, though the GLA content of the total fatty acids was now between only 40 to 50% — the exact proportion depended on the growth period, which ranged from 8 to 14 d [47]. The biomass yield of this organism, however, was not very high at only 3.7 g/l, but little seems to have been done to optimize the overall growth performance using stirred tank bioreactors. Further studies on this organism as a potential source of GLA would appear to be strongly advisable if further interest in a microbial source of GLA is continued. *Syzgites megalocarpus* is, without exception, the highest GLA producer of any microorganism or plant.

The occurrence of GLA in marine and freshwater algae — both prokaryotic organisms such as *Spirulina* [5,48] and eukaryotic ones such as *Chlorella* [5] — has been known for many years. None of these sources, however, represent significant or even potential sources of the fatty acid: they are costly to cultivate because they are photosynthesizing organisms that require expensive photobioreactors. They grow very slowly, often needing several weeks to complete their growth, and final cell densities rarely exceed 2 to 3 g/l. Thus, although they get their carbon (CO_2) and energy (sunlight) for nothing, the cost of production is still an order of magnitude beyond that of a microbial SCO process.

2.6 A FUTURE FOR A GLA-SCO?

The possible future for a microbial source of GLA is unquestionably tied up with whether GLA itself has a future as a nutraceutical, dietary supplement, or medical agent. The jury seems to be out on this issue [21]. As the vast majority of published information has been done with EPO, which has a GLA content of 10%, this has made it difficult to be entirely sure of the efficacy of GLA itself, and the effects have not been brought about by the much higher content of linoleic acid (see Table 2.4) or even of vitamin E, which is included in the oil. The American Cancer Society has indicated [49] that GLA has had a beneficial effect on the neurological problems related to diabetes, especially in patients whose condition is already well under control [50]. However, neither EPO nor GLA has been shown effective in the treatment or prevention of any cancer in humans.

Whether oils containing GLA are useful in the treatment of disorders already mentioned above (see Section 2.4.1) is still subject to much research. Ultimately as oils, such as borage and evening primrose oils, are purchased as over-the-counter supplements, and their use is not governed by the same rules that apply to prescription medicines, their take-up by the public will depend very much on hearsay opinion as well as marketing strategies. Perception of effectiveness is a very powerful reason for purchase. If a GLA-SCO is to penetrate this market, it must either offer a clear improvement in efficacy for whatever ailment it is taken to alleviate or be much cheaper than any alternative. As the latter seems unlikely in view of the availability of evening primrose and borage oils, the case must therefore rest on some company prepared to carry out the necessary clinical and/or nutritional studies to establish a GLA-SCO as a main source of this fatty acid. This presupposes that a supply of the oil is created with which to carry out the necessary trials. As this also seems unlikely, the future for a GLA-SCO may therefore be over.

ACKNOWLEDGMENTS

I am very grateful for the information supplied to me over a number of years from various friends and colleagues associated with the various GLA-SCO processes that are mentioned in this chapter. In particular, I would like to acknowledge the help from a former colleague, Philip Milsom at J. & E. Sturge Ltd., Selby, North Yorkshire, United Kingdom, and Professors Rod Herbert (University of Dundee, Scotland), Frank Gunstone (formerly of the University of St. Andrews, Scotland), and Lodewyk Kock (University of the Orange Free State, Bloemfontein, South Africa) for their input and help both now and previously. I am particularly grateful to Professor Keith Coupland, formerly of Croda Universal Ltd., Hull, United Kingdom, and now an active colleague of the author's, for some invaluable behind-the-scenes information concerning the various machinations that went on in the early days of developing the *Mucor* process and the rival borage oil process. I would also like to take this opportunity to acknowledge the initial interest in beginning this work by the late John Williams of Bio-Oils Ltd., Nantwich, Cheshire, United Kingdom. It was John's original viewing of a short television clip on microbial oils made in 1976 for the television show "Tomorrow's World" in my laboratories that led to his contact with my group. It was due to his support and encouragement that the initial work on finding a good GLA-production organism was conducted. There are also a large number of technicians who carried out much of the hard work in my laboratories in developing the GLA process to whom I owe a collective debt of gratitude; in particular, it is a pleasure to acknowledge the outstanding input and help from David Grantham and Jennie Curson, who were stalwarts in carrying out much of the fermentation work with *Mucor* and the other molds over many years. There were also numerous research students who, again over many years, contributed to the biochemistry of oil accumulation in molds and yeasts: these individuals have been recognized elsewhere [4] for their work.

REFERENCES

1. Cohen, Z. and Ratledge, C., Eds., *Single Cell Oils,* American Oil Chemists' Society, Champaign, IL, 2005.
2. Ratledge, C., Microbial production of oils and fats, in *Food from Waste,* Birch, K.J., Palmer and Worgan, J.T., Eds., Applied Science Publishers, London, 1976, pp. 98–113.
3. Thorpe, R.F. and Ratledge, C., Fatty acid distribution in triglycerides of yeasts grown on glucose or n-alkanes, *J. Gen. Microbiol.,* 75, 151–163, 1972.
4. Ratledge, C. and Wynn, J.P., The biochemistry and molecular biology of lipid accumulation in oleaginous microorganisms, *Adv. Appl. Microbiol.,* 51, 1–51, 2002.
5. Ratledge, C., Microorganism as sources of polyunsaturated fatty acids, in *Structured and Modified Lipids,* Gunstone, F.D., Ed., Marcel Dekker, New York, 2001, pp. 351–399.
6. Ratledge, C., Microbial lipids, in *Biotechnology,* 2nd ed. Vol. 7 Products of Secondary Metabolism, Rehm, H.J. et al., Eds., VCH, Weinheim, Germany, 1997, pp. 133–197.
7. Ratledge, C., Yeasts, moulds, algae and bacteria as sources of lipids, in *Technological Advances in Improved and Alternative Sources of Lipids,* Kamel, B.S. and Kakuda, Y., Eds., Blackie, Glasgow, 1994, pp. 235–291.
8. Ratledge, C. and Evans, C.T., Lipids and their metabolism, in *The Yeasts,* 2nd ed., Vol. 3, Rose, A.H. and Harrison, J.S., Eds., Academic Press, New York 1989, pp. 367–455.
9. Davies, R.J., Scale up of yeast technology, in *Industrial Applications of Single Cell Oils,* Kyle, D.J. and Ratledge, C., Eds., American Oil Chemists' Society, Champaign, IL, 1992, pp. 196–218.
10. Smit, H. et al., Production of cocoa butter equivalents by yeast mutants, in *Industrial Applications of Single Cell Oils,* Kyle, D.J. and Ratledge, C., Eds., American Oil Chemists' Society, Champaign, IL, 1992, pp.185–195.
11. Clough, P.M., Specialty vegetable oils containing γ-linolenic acid and stearidonic acid, *Structured and Modified Lipids,* in Gunstone, F.D., Ed., Marcel Dekker, New York, 2001, pp.75–117.
12. Horrobin, D.F., Nutritional and medical importance of gamma-linolenic acid. *Prog. Lipid Res.,* 31, 163–194, 1992.
13. Huang, Y.S. and Mills D.E., Eds., *γ-Linolenic Acid: Metabolism and Its Roles in Nutrition and Medicine,* American Oil Chemists' Society, Champaign, IL, 1996.
14. Huang, Y.S. and Ziboh, A., Eds., *γ-Linolenic Acid: Recent Advances in Biotechnology and Clinical Applications*, American Oil Chemists' Society, Champaign, IL, 2001.
15. Horrobin, D.F. and Morse, P.F., Evening primrose oil and atopic eczema, *Lancet,* 345, 260–261, 1995.
16. Horrobin, D.F., Essential fatty acid metabolism and its modification in atopic eczema, *Am. J. Clin. Nutr.,* 71, 367S-372S, 2000.
17. Hornych, A. et al., The effect of gamma-linolenic acid on plasma and membrane lipids and renal prostaglandin synthesis in older subjects, *Bratisl. Lek. Listy,* 103, 101–107, 2002.
18. Bernard, K. and Albrecht, H., Die Lipide aus *Phycomyces blakeslaeeanus, Helv. Chim. Acta,* 31, 977–988, 1948.
19. Shaw, R., The occurrence of gamma-linolenic acid in fungi, *Biochim. Biophys. Acta,* 98, 230–237, 1965.
20. Shaw, R. The polyunsaturated fatty acids of microorganisms, *Adv. Lipid Res.,* 4, 107–174, 1966.

21. Shaw, R., The fatty acids of phycomycete fungi, and the significance of the γ-linolenic acid component, *Comp. Biochem. Physiol.,* 18, 325–331, 1966.
22. Marshall, B.J., Ratledge, C., and Norman, E., Improved design for a simple and inexpensive multi-place laboratory fermenter, *Lab. Pract.,* 22, 491–492, 1973.
23. Holdsworth, J.E. and Ratledge, C., Lipid turnover in oleaginous yeasts, *J. Gen. Microbiol.,* 134, 339–346, 1988.
24. Jenkins, D.K. et al., Effects of different sources of gamma-linolenic acid on the formation of essential fatty acid and prostanoid metabolites, *Med. Sci. Res.,* 16, 525–526, 1988.
25. Huang, Y.S. et al., In vitro hydrolysis of natural and synthetic γ-linolenic acid-containing triacylglycerols by pancreatic lipase, *J. Am. Oil Chem. Soc.,* 72, 625–631, 1995.
26. Phillips, J.C. and Huang, Y.S., Natural sources and biosynthesis of γ-linolenic acid: An overview, in *γ-Linolenic Acid: Metabolism and Its Roles in Nutrition and Medicine,* Huang, Y.S. and Mills, D.E., Eds., Americal Oil Chemists' Society, Champaign, IL, 1996, pp. 1–21.
27. Lawson, L.D. and Hughes, B.G., Triacylglycerol structure of plant and fungal oils containing γ-linolenic acid, *Lipids,* 23, 313–317, 1988.
28. Raederstorff, D. and Moser, U., Borage or primrose oil added to standardized diets are equivalent sources for γ-linolenic acid in rats, *Lipids,* 27, 1018–1023, 1992.
29. Gunstone, F.D., GLA — Occurrence and physical and chemical properties, *Prog. Lipid Res*., 31, 145–161, 1992.
30. Traitler, H., Wille, H.J., and Struder, A., Fractionation of blackcurrant seed oil, *J. Am. Oil Chem. Soc.,* 65, 755–760, 1988.
31. Suzuki, O., Yokochi, T., and Yamashina, T., Studies on production of lipids in fungi. II Lipid compositions of six species of Mucorales in Zygomycetes, *Yukagaku* (*J. Jpn. Oil Chem. Soc.*), 30, 863–868, 1981 (in Japanese).
32. Nakahara, T., Yokochi, T., Kamisaka, Y., and Suzuki, O., Gamma-linolenic acid from genus Mortierella, in *Industrial Applications of Single Cell Oils,* Kyle, D.J. and Ratledge, C., Eds., American Oil Chemists' Society, Champaign IL, 1992, pp. 61–97.
33. Hiruta, O. et al., Application of Maxblend fermentor® for microbial processes, *J. Ferment. Bioeng*, 83, 79–86, 1997.
34. Hiruta, O. et al., γ-Linolenic acid production by a low temperature-resistant mutant of *Mortierella ramanniana, J. Ferment. Bioeng*., 82, 119–123, 1996.
35. Herbert, R.A. and Keith, S.M., Fungal production of gamma-linolenic acid, European Patent 153,154, 1985.
36. Kock, J.F.L. and Botha, A., Biological treatment and cultivation of microorganisms, South African Patent 91,9749, 1993.
37. Kock, J.F.L. and Botha, A., Biological treatment and cultivation of microorganisms, U.S. Patent 07,987,958, 1994; and 5,429,942, 1995.
38. Kock, J.F.L. and Botha, A., Acetic acid — a novel source of cocoa butter equivalents and gamma-linolenic acid, *South African J. Sci.,* 89, 465–471, 1993.
39. Du Preez, J.C. et al., GLA production by *Mucor circinelloides* and *M. rouxii* in fed-batch culture with acetic acid as carbon substrate, *Biotechnol. Lett.,* 17, 933–938, 1995.
40. Botha, A. et al., Carbon source utilization and γ-linolenic acid production by mucoralean fungi, *Syst. Appl. Microbiol.,* 20, 165–170, 1997.
41. Roux, M.P. et al., *Mucor* — a source of cocoa butter and gamma-linolenic acid, *World J. Microbiol., Biotechnol.,* 10, 5417–5422, 1994.
42. Buranova, L., Rezanka, T., and Jandera, A., Screening for strains of the genus *Mortierella*, showing elevated production of highly unsaturated fatty acids, *Folia Microbiol.*, 35, 578–582, 1990.

43. Kristofikova, L. et al., Selection of *Rhizopus* strains for L(+)-lactic acid and γ-linolenic acid production, *Folia Microbiol.,* 36, 451–455, 1991.
44. Kennedy, M.J., Reader, S.L., and Davies, R.J., Fatty acid production and characteristics of fungi with particular emphasis on gamma linolenic acid production, *Biotechnol. Bioeng.,* 42, 625–634, 1993.
45. Kavadia, A. et al., Lipid and γ-linolenic acid accumulation in strains of Zygomycetes growing on glucose, *J. Am. Chem. Soc.,* 78, 341–346, 2001.
46. Certik, M., Balteszova, L., and Sajbidor, J., Lipid formation and γ-linolenic acid production by Mucorales fungi grown on sunflower oil, *Lett. Appl. Microbiol.,* 25, 101–105, 1997.
47. Weete, J.D., Shewmaker, F., and Gandhi, S.R., γ-Linolenic acid in Zygomycetous fungi: *Syzygites megalocarpus, J. Am. Oil Chem. Soc.,* 75, 1367–1372, 1998.
48. Murata, N., Deshnium, P., and Tasaka, Y., Biosynthesis of γ-linolenic acid in the cyanobacterium *Spirulina platensis,* in *γ-Linolenic Acid: Metabolism and Its Roles in Nutrition and Medicine,* Huang, Y.S. and Mills, D.E., Eds., American Oil Chemists' Society, Champaign, IL, 1996, pp. 23–32.
49. *American Cancer Society's Guide to Complementary and Alternative Cancer Methods,* 2000 (ISBN 0-944235-24-7).
50. Keen, H. et al., Treatment of diabetic neuropathy with gamma-linolenic acid. The gamma-linolenic acid multicenter trial group, *Diabetes Care,* 16, 8–15, 1993.

3 γ-Linolenic Acid: Purification and Functionality

Udaya N. Wanasundara and Janitha P.D. Wanasundara

CONTENTS

3.1 GENERAL ASPECTS AND OCCURRENCE

The very first scientific report on γ-linolenic acid (GLA) was from Heiduschka and Lueft in 1919 [1]. Their study reports on the seed oil of evening primrose (*Oenothera biennis* L.), a Native American herbal plant that was transplanted to Europe during the 17th century. An unusual form of linolenic acid present in the evening primrose seed oil, which yielded a hexabromide derivative of linolenic acid (LA) with very different physical characteristics from the other two common forms, α- and β-isomers present in linseed oil, has been described in their report. This unusual fatty acid was named as the γ-isomer or GLA [1]. Eibner and Schild [2] have characterized the exact chemical structure of GLA and Riley [3] verified it. GLA belongs to n-6 polyunsaturated fatty acid (PUFA) group and its three unsaturated double bonds are arranged in a methylene-interrupted fashion and are of *cis* configuration (Figure 3.1). GLA can be expressed as all-*cis*-6,9,12–octadecatrienoic acid or 18:3 (Δ6,9,12; Figure 3.1). Certain higher plants (in seeds and leaves), algae, fungi, and protozoans accumulate GLA in considerable amounts. Mammal tissues contain GLA in trace amounts and are not ubiquitously found.

3.1.1 SOURCES OF GLA

GLA-containing oil is legally approved as a single food ingredient and therefore not subjected to food additive regulation. As a result GLA-containing oils are available as nutrient supplements or nutraceuticals and are very popular in North America. The unreceding interest in this LA isomer has led to the exploration of GLA-rich sources

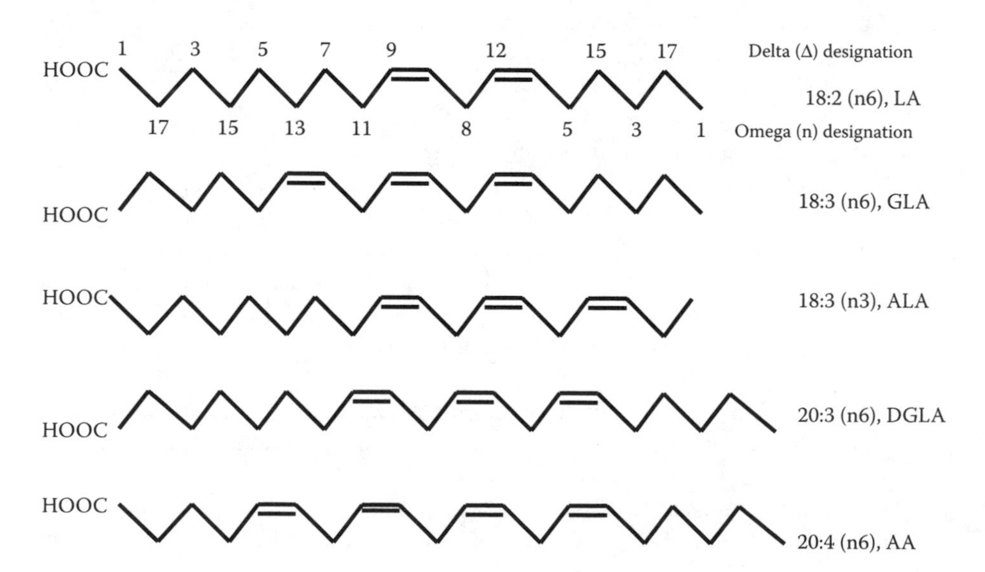

FIGURE 3.1 Chemical structures of LA, GLA, ALA, DGLA, and AA. Linoleic acid (LA), γ-linolenic acid (GLA), α-linolenic acid (ALA), dihomo-γ-linolenic acid (DGLA), and arachidonic acid (AA). Both omega (n) and delta (Δ) numbering system is indicated for LA. (Δ)-Numbering system starts carboxyl terminus and (n)-numbering starts at methyl terminus carbon.

as well as the ways to enrich GLA content in natural sources. The bioconversion of LA to arachidonic acid in animals results in very low levels of accumulation of the intermediate product, GLA, and makes animal lipids poor sources of GLA. A limited number of fungi and seed oils have been reported as sources of GLA. Table 3.1

TABLE 3.1
Contents of Oil and GLA in Selected Plants

Plant	Ref.	Seed Oil Content %	GLA Content % of Total Fatty Acids
Boraginaceae			
Adelocaryum coelestinum	[4]	22.0	12.4
Alkana froedinii	[4]	47.0	9.9
A. orientalis	[4]	23.0	12.4
Amsinckia intermedia	[4]	28.0	8.2
Borago officinalis	[5]	29.0–35.0	20.0–25.0
Brunnera orientalis	[4]	27.0	15.4
Echium plantagineum	[6]	23.7	9.2
E. aculeatum	[6]	17.1	22.3
E. giganteum	[6]	12.0	21.7
E. triste	[6]	19.9	17.2
E. fastuosum	[6]	13.7	23.8
E. sventenii	[6]	7.25	24.4
E. nervosum	[6]	20.5	24.5
E. acanthocarpum	[6]	15.1	24.5
E. callithyrsum	[6]	17.8	26.3
Lithospermum latifolium	[7]	16.3	23.9
Pectocarya platycarpa	[4]	15.0	15.2
Onagraceae			
Oenothera biennis	[4]	25	10.0
O. grandflora	[4]	4	9.3
O. lamarckiana	[4]	28	8.2
O. strigosa	[4]	29	7.0
Scrophulariaceae			
Scrophularia lanoceolata	[4]	26	8.0
S. marilandica	[4]	38	9.6
Saxifragaceae			
Ribes alpinum	[4]	19	8.9
R. grossularia (yellow gooseberry)	[8]	19.8–21.9	5.6–8.1
R. nigrum (black currant)	[8]	18.0–22.3	11.9–15.8
R. nigrum × *R. hirtellum* (jostaberries)	[8]	18.0–23.6	6.1–8.8
R. rubrum (red currant)	[8]	11.2–22.4	3.3–7.0
R. rubrum (white currant)	[8]	18.0–23.6	6.1–8.8
R. uva crispa (gooseberry)	[9]	18.3	10.0–12.0

provides a list of plants and the reported content of GLA in their seed oils. According to the literature, GLA is present in the oil of plant species from Boraginaceae, Onagraceae, Scrophulariaceae, and Saxifragaceae families. Only the Boraginaceae family includes species with GLA contents higher than 20% of total seed fatty acids. According to the review by Ucciani [10] 5 out of 87 species analyzed contain this. However, commercial production of seeds for GLA-rich oils is still limited to borage (*Borago officinalis* L.), black currant (*Ribes nigrum*), and evening primrose (*Oenothera biennis* L.). Oil fraction of hemp (*Cannabis sativa* L.) seed also contains GLA up to 4% of total fatty acids [11]. Among the microbiological sources, *Mucor javanicus* and *Spirulina platensis* are reported to contain 15 to 18 and 21% GLA, respectively [12–15]. The other reported unconventional sources of GLA are marine green algae *Chlorella spp*. NKG 042401 (10.5% of total fatty acids) [16] and safflower petals (2 to 3% of total fatty acids) [17]. A review compiled by Gunstone [5] provides more details on microbial sources of GLA.

3.1.2 Biosynthesis of GLA

3.1.2.1 In Plants

In the fatty acid synthesis pathway of higher plants, double bonds can be inserted at the Δ9, Δ12, and Δ15 positions (Figure 3.1). The presence of Δ9- and Δ12-desaturase results in formation of oleic (18:1[n-9]) and LA (18:2[n-6]) respectively, while the presence of Δ15-desaturase results in the occurrence of α-linolenic acid (ALA) (18:3, [n-3]) in plant tissues and seeds (Figure 3.2). In several plant species in the families of Onagraceae and Boraginaceae, the activity of Δ15-desaturase is absent; therefore ALA is not synthesized. However, the presence of Δ6-desaturase

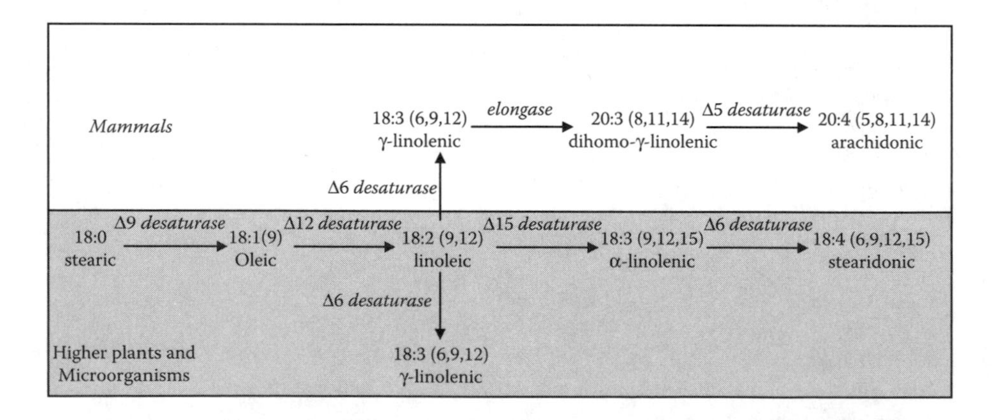

FIGURE 3.2 Simplified biosynthetic pathway of polyunsaturated fatty acids in plants and mammals. For each fatty acid numbers in parentheses indicate location of double bond according to Δ-numbering (refer to Figure 3.1). (Modified from Horrobin, D.F., *Prog. Lipid Res.*, 31, 163, 1992; Phillips, J.C. and Huang, Y.S., in *γ-Linolenic Acid: Metabolism and Its Roles in Nutrition and Medicine*, Huang, Y.-S. and Mills, D.E., Eds., 1996, p. 1. With permission.)

in these plant species allows the insertion of a double bond at the Δ6 position of LA and also to the accumulation of GLA in the tissues. In addition to Δ6-, Δ9-desaturases are active in *Ribes* spp. Accumulation of both GLA and stearidonic acid (SDA, 18:4[n-6], all-*cis*- 6, 9,12,15-) occurs. In contrast to animals, the synthesis of GLA in plants occurs in the linoleoyl moiety linked to phospholipids as the substrate for the Δ6-desaturase. The Δ6-desaturase is position specific and utilizes the linoleate only in position 2 of *sn*-phosphatidylcholine [18–20].

Lower plants such as algae and yeast use the phospatidylcholine-linked oleoyl (both *sn*-1 and *sn*-2) to form linoleoyl-phosphatidylcholine similar to higher plants. In GLA-rich microorganisms, the linoleoyl-phosphatidylcholine is desaturated by Δ6-desaturase to form GLA [21–23]. However, in certain strains of cyanobacteria (e.g., *Spirulina platensis, Synechocystis* PCC6714), Δ6-desaturation takes place with LA bound to glycerolipids as a mono- and digalactosyl diacylglycerol derivative [24–25].

Identification of genes involved in desaturation of PUFA at carbon 5 (Δ5-desaturase) and at carbon 6 (Δ6-desaturase) in lower organisms has led the way to generate transgenic microorganisms and plant cells [26–28]. This has been used to produce transgenic plants that are capable of producing the fatty acids arachidonic acid (AA), eicosapentaenoic acid (EPA, 20:5[n-3]), GLA, and/or SDA depending upon whether the nucleotide sequence encodes a Δ5- or Δ6-desaturase [29–31]. The development of oilseed crops and microbial sources that are designed to produce substantial quantities of GLA is a major goal for plant biotechnologists.

3.1.2.2 In Mammals

In mammals, GLA is produced constantly via desaturation (a further double bond is introduced) of LA, by Δ6-linoleate desaturase in the liver to dihomo-γ-linolenic acid (DGLA, 20:3 [n-6]; see Figure 3.2). Consequently, DGLA can be converted into AA [20:4, (n-6)], a critically important fatty acid. Chain elongation and desaturation of AA may continue in mammals and eventually produce docosahexaenoic acid (DHA, 22:6, [n-3], Figure 3.2).

In animal tissues, the enzyme Δ6-desaturase requires the coenzyme A (CoA) derivatives of LA as the substrate [34,35]. It is accepted that during the course of evolution, animals have lost the ability to insert double bonds into fatty acids beyond the Δ9 position (e.g., Δ12 and Δ15 positions) and, therefore, are not able to transform 18:1(n-9) to 18:2(n-6) and 18:2(n-6) to 18:3(n-3). However, the ability to insert double bonds between the existing bond and the carboxyl end of the acyl chain is retained. That is manifested as the ability to metabolize dietary 18:2(n-6) and 18:3(n-3) through Δ6-desaturation, elongation, and Δ5-desaturation to form long-chain (n-6) and (n-3) molecules, respectively [32].

The endogenous conversion or desaturation of LA to GLA is believed to be the rate-determining step in the formation of DGLA, hence AA and the rest of the crucial metabolites. In normal individuals, dietary provision of LA by virtue of its resulting conversion to GLA and AA satisfies the dietary need of GLA and AA. However, several disease states may impair Δ6-linoleic acid desaturase activity, thus resulting in reduced conversion of AA in the biosynthetic pathway. Extensive research has helped to identify n-6 essential fatty acids (EFAs) as fatty acids, which can reverse all the symptoms that result when all n-6 fatty acids are excluded from diet [18,36–38]. According to experiments, it has been demonstrated that LA, GLA, and AA supplementation can reverse the features of EFA deficiency [18,37,38].

3.2 PRODUCTION AND PURIFICATION

3.2.1 Extraction and Analysis

The predominant uses of GLA-rich oils are based on the pharmacological properties of GLA. Different methods of extraction are utilized for black currant, borage, and evening primrose seeds to recover their lipids. These include application of hot or cold pressure to express oils or extraction of lipids with hexane or supercritical CO_2 [39–42]. These oils are available either of virgin quality, which is obtained by cold pressing, or of refined quality that has gone through a specified purification process [39].

Determination of GLA in lipids is carried out the same any other fatty acid analysis by converting to respective methyl ester and separation on a suitable stationary phase for gas chromatographic analysis. Capillary columns of polar stationary phase material (e.g., Carbowax) are preferred as they give good separation of linoleic, γ-linolenic, and α-linolenic methyl esters that originate from fatty acids naturally present in GLA-rich sources [5].

High-performance liquid chromatography (HPLC) separation of GLA-containing triacylglycerols (TAGs) is possible with a silver ion or a reversed phase [43–45]. Although it is possible to separate in seed oil TAGs with identical acyl carbon numbers and degree of unsaturation (e.g., ALA or GLA) with reversed-phase HPLC alone [46,47], Christie [44] recommends using a combination of silver ion and reversed phase HPLC for the analysis of GLA-rich oils. Supercritical fluid chromatography (supercritical CO_2) may also be utilized to separate GLA containing TAGs. Blomberg and group [48] have utilized supercritical fluid as the mobile phase and a packed capillary column impregnated with silver nitrate to separate TAGs of borage oil. Using the stationary phase of 25% cyanopropyl–75% methylpolysiloxane in capillary supercritical chromatography, it is possible to separate GLA- and ALA-containing TAGs [49]. Manninen and co-workers [49] showed that by using two 10-m columns in tandem the resolution between 1,3-dioleoyl-2-γ-linolenoyl-*sn*-glycerol and 1,3-dioleoyl-2-α-linolenoyl-*sn*-glycerol can be enhanced.

The distribution of GLA in the TAG molecule is significant in determining the cleavage potential by pancreatic lipase, and thus the availability of GLA for absorption. Alternatively, the association of GLA with other fatty acids in the TAG species of the oils may modulate GLA potency in exerting its beneficial effects. There are several methods available for determining fatty acid stereospecific/positional distribution of TAGs based on thin-layer chromatography and chiral-phase or normal-phase HPLC [50]. The stereospecific distribution of the fatty acids in native evening primrose oil and borage oil has been determined by several research groups [50–53]. It is found that GLA in both borage and evening primrose oils is distributed asymmetrically and is preferentially located at *sn*-2 and *sn*-3 positions [50–53]. Positional fatty acid distribution of individual TAG fractions (Table 3.2) and intact oil (Table 3.3) of borage and evening primrose lipids shows that long-chain fatty acids (22:1[n-9] and 24:1[n-9]) are exclusively located at the *sn*-1 position [50].

3.2.2 Concentration and Purification

Because of the pharmaceutical value of GLA, a concentrated and pure form of GLA is essential for research studies as well as for commercial products. Natural sources contain GLA only up to 25 to 27% (w/w) of maximum (Table 3.1). In purification of polyunsaturated fatty acids that have pharmaceutical value, several methods have been described in literature. These include chromatography (high-performance liquid, silver-ion exchange, and supercritical fluid) and enzymatic methods, especially selective hydrolysis and selective esterification [54]. Low-temperature crystallization, urea complexation, and centrifugal partition chromatography (CPC) have been applied to obtain GLA in highly concentrated form.

Separation of fatty acids according to their solubility differences in organic solvents can be achieved at low temperatures in combination with the crystallization differences according to unsaturation. The low-temperature crystallization process is developed by exploiting this behavior and requires free fatty acids (FFAs) as the starting material. Use of low-temperature solvent crystallization to develop a GLA-concentrated product (as FFAs obtained from borage oil) has been described by Chen and Ju [55] and they were able to raise the GLA of borage oil (as FFA) from 23.4 to 88.9% with a yield of 62.0% under the optimum reaction conditions.

In urea fractionation, enrichment of PUFA in the liquid fraction occurs due to complex formation of saturated and monounsaturated fatty acids with urea. The hexagonal crystals (urea–fatty acid clathrates) can be easily separated from the supersaturated PUFA in the liquid (solvent) medium. Urea fractionation has been very successful in enriching GLA in a mixture of fatty acids obtained from natural seed oils (e.g., black current and borage) or microbial sources. Most GLA-rich oils will produce 70 to 90% concentrates in one to three crystallization steps [5]. Several patents [56–58] have been granted for the processes that use urea complexation for generating concentrates, including GLA (as free acid or methyl ester).

CPC may also be a useful tool in purifying and concentrating GLA. Borage oil FFAs (starting GLA content 21.8%) could be processed to obtain 98.3% GLA product by CPC using hexane/dichloromethane/acetonitrile (5:1:4, v/v) as the solvent [59]. CPC offers an alternative to other methods of chromatography in purifying PUFA, such as GLA, that have high commercial value and could scale up to produce large quantities.

Lipases are specific to the fatty acids in catalyzing hydrolysis reaction of TAGs, and thus discriminate between the fatty acids for their chain length and unsaturation. This opportunity has been used to develop concentrated forms of PUFA including GLA from source oils. The strategy is to enrich GLA in the FFA fraction by hydrolyzing GLA-containing oil with a lipase that selectively hydrolyze, the GLA–glycerol ester bond. However, a lipase that acts as such has not yet been discovered from natural sources or engineered. Therefore, a two-step approach is

TABLE 3.2
Positional Fatty Acids (Expressed as mol%) of TAG Fractions Isolated from Borage and Evening Primrose Oil

		Borage Oil TAG						Evening Primrose Oil TAG			
Fatty Acid[a]	TAG Species[b]	All	*sn*-1[c]	*sn*-2[d]	*sn*-3[e]	Fatty Acid[a]	TAG Species[b]	All	*sn*-1[c]	*sn*-2[d]	*sn*-3[e]
	LGG						LGG				
18:2		33.9	88.8	5.6	7.2	18:2		35.2	59.8	25.5	0.5
18:3		66.1	12.0	93.5	92.8	18:3		64.8	39.5	75.3	79.6
	LGL						LGL				
18:2		65.4	92.1	60.8	43.3	18:2		66.8	79.8	68.9	51.7
18:3		34.8	8.4	38.8	57.2	18:3		33.2	20.2	32.4	47.0
	OGG										
18:1[f]		27.5	65.4	2.8	14.2	18:1[f]					
18:2		12.2	14.5	47.1	79.7	18:2					
18:3		60.4	20.2	10.1	6.1	18:3					
	PGG										
16:0		30.4	57.4	38.4	–4.6	16:0					
18:2		3.3	3.9	9.3	–3.3	18:2					
18:3		66.3	38.5	53.0	107	18:3					
	OGL						OGL				
18:1[f]		33.0	20.2	23.4	55.4	18:1[f]		34.4	36.8	32.6	33.8
18:2		33.9	68.1	19.5	14.1	18:2		35.8	46.8	41.8	18.8
18:3		33.0	11.2	57.4	30.4	18:3		29.9	16.7	25.3	47.7
	PGL						PGL				
16:0		31.3	43.7	27.3	22.9	16:0		28.5	59.1	1.5	24.9
18:2		33.4	44.0	27.6	28.6	18:2		36.8	24.6	52.0	33.8
18:3		35.3	12.5	44.9	48.5	18:3		34.6	16.2	49.6	38.0

OLL					OLL				
18:1[f]	33.6	15.4	21.0	64.4	18:1[e]	33.6	27.2	41.4	32.2
18:2	66.4	84.6	30.0	35.5	18:2	66.4	72.8	58.6	67.8
					PLL				
					16:0	30.0	48.2	6.4	35.4
					18:1[f]	2.0	n.d.[g]	n.d.	n.d.
					18:2	67.0	42.8	103	55.6
					18:3	1.0	n.d.	n.d.	n.d.
SGL and PGO									
16:0	18.6	36.5	11.3	8.0					
18:0	13.5	2.4	22.2	15.6					
18:1[f]	19.1	8.5	24.3	24.5					
18:2	15.5	32.9	1.7	11.9					
18:3	33.4	19.8	40.6	39.8					

[a] All unsaturated fatty acids are n-6, unless otherwise indicated.

[b] G, γ-linolenic; L, linolenic; O, oleic; P, palmitic; S, stearic acids.

[c] 3 × [TAG} – 2 × [*sn*-2,3-DG urethane].

[d] 3 × [TG] – [*sn*-1] – [*sn*-3].

[e] 3 × [TG] – 2 × [*sn*-1,2-DG urethane].

[f] n-7 and n-9.

[g] n.d., not detected.

Source: Modified from Redden, P.R. et al., *J. Chromatogr. A*, 704, 99, 1995. With permission.

TABLE 3.3
Positional Fatty Acid Composition of Borage and Evening Primrose Oil (Expressed as mol%)

	Borage Oil				Evening Primrose Oil			
Fatty Acid	Total	*sn*-1[a]	*sn*-2[b]	*sn*-3[c]	Total	*sn*-1[a]	*sn*-2[b]	*sn*-3[c]
16:0	11.5	20.2	–3.1	17.5	6.0	11.3	–0.4	7.1
18:0	4.2	4.9	1.4	6.4	2.2	3.9	0.7	2.0
18:1	18.0	17.5	18.9	17.5	9.0	9.8	7.8	9.4
18:2n-6	42.0	38.5	53.4	34.2	74.4	70.0	81.5	71.6
18:3n-6	17.7	3.6	32.2	17.4	8.5	4.9	10.3	10.2
20:1n-9	3.3	5.9	0.0	4.0				
22:1n-9	2.0	6.1	–1.7	1.7				
24:1n-9	1.2	3.5	–0.9	0.9				

[a] 3 × [TAG} – 2 × [*sn*-2,3-DG urethane].
[b] 3 × [TG] – [*sn*-1] – [*sn*-3].
[c] 3 × [TG] – 2 × [*sn*-1,2-DG urethane].

Source: Modified from Redden, P.R. et al., *J. Chromatogr. A*, 704, 99, 1995. With permission.

often utilized in performing the selective esterification process. In the first step, generation of FFAs by hydrolyzing GLA-containing oil using a lipase that acts on PUFA/GLA as effective as on other constituent fatty acids is conducted. Then the enrichment of PUFA in the FFA fraction is achieved by esterification of FFAs other than GLA with an alcohol using a lipase that acts very weakly on GLA [54].

Shimada and group [54,60] have described a small-scale purification method for GLA starting with borage oil that contained 22% GLA at the start and obtained 70% GLA-containing concentrate. Lipase from *Pseudomonas* spp., which was found to be the most suitable for hydrolysis of borage acylglycerols, was used as the catalyst to obtain FFAs in the first step in this process. Several factors that affect hydrolysis reaction (e.g., oil-to-solvent ratio, enzyme-to-substrate ratio, reaction time, and temperature) have been studied. Under the specified conditions described by Shimada and group [60,61], hydrolysis of borage oil for 24 h resulted in the release of FFAs (92%), including 93% recovery of GLA in the FFA fraction. Free GLA obtained from this hydrolysis step was then selectively esterified with lauryl alcohol catalyzed by *Rhizopus delemer* lipase; single-step esterification resulted in a 74.2% recovery of GLA in the FFA fraction and concentration of GLA was raised to 93.7% by a second step of selective esterification under the same conditions [60]. This process has been adapted in large-scale preparation of GLA concentrate [61]. In this scaled-up process, a molecular distillation step has been included in removing the selectively esterified fatty acids. However, 15 to 20%

lauryl esters still remained in the concentrate and a urea adduct formation step was necessary to remove them completely. Starting from 45% (w/w) GLA-containing oil, Shimada and co-workers [54] were able to obtain concentrated GLA in FFA form (98.6% purity) with a recovery yield of 20% (w/w).

3.2.3 Structured Lipids with GLA

Several researchers have studied the possibility of developing structured TAGs with GLA and other selected fatty acids. The position of GLA in the TAG molecule is significant when it comes to the *in vivo* bioavailability of GLA. It is well proven that the ingestion of GLA-enriched oils results in the accumulation of DGLA in tissue phospholipids and TAGs; however, the absolute level of GLA in the oil may not be the sole determinant of biological efficacy of the oil. GLA bioavailability is influenced by the precise stereospecific composition of the TAG and the cellular kinetics of phospholipases and acyltransferases [62]. A good example is the formation of comparable amounts (per gram basis) of anti-inflammatory prostaglandin E_1 (PGE_1, Figure 3.3) by borage and primrose oil, although the GLA concentration of borage oil is twofold higher than that of primrose oil [63]. A previous section of this chapter describes the stereospecific distribution of GLA in borage and evening primrose oils.

A process based on esterification of GLA (obtained from natural sources) with glycerol in order to obtain 1-monoglyceride ester of GLA has been developed and patented by Jowett [64]. The GLA-monoacylglycerol so obtained may have the advantage of easy accessibility to GLA by digestive enzymes and also readily form emulsions with water, which are beneficial properties in medical and cosmetic applications. This patented process describes utilization of GLA obtained by chemical hydrolysis of natural sources and then reaction with glycerol to form monoacylglycerols (MAGs). Molecular distillation is suggested to recover the enriched product.

Kawashima and group [65] have reported preparation of TAG with medium-chain fatty acids (M) at the 1,3-position and the long-chain fatty acid (L) at the 2-position that resulted in a medium-chain/long-chain medium-chain fatty acid (MLM)-type TAG. The process is started with a GLA-enriched oil (45% w/w) and acidolysis reaction was performed with a GLA/caprylic acid (G/C) mixture using immobilized *Rhizopus oryzae* lipase as the catalyst. The resulting product contained 10.2 mol% MAG and 27.2 mol% diacylglycerol (DAG) and 44.5 mol% TAG (mainly 1,3-capryloy 1-2-γ-linolenoy glycerol [CGC]) based on total acylglycerols. Part of the reaction mixture contained tricaprylin (CCC) and partial acylglycerols, which could be removed by molecular distillation and would result in increase of CGC content up to 52.6 mol%. Using borage oil and *Candida rugosa* lipase, structured TAG containing 50 mol% CGC may be produced by this two-step process via selective hydrolysis.

The study carried out by Akoh and group [66] showed that EPA could be incorporated into the glycerol backbone of evening primrose oil by transesterification catalyzed by *Candida antarctica* lipase. Senanayake and Shahidi [67,68] have studied the possibility of esterifying EPA and DHA into GLA-containing borage and evening primrose oils via *Pseudomonas spp.* lipase catalyzed acidolysis in order to

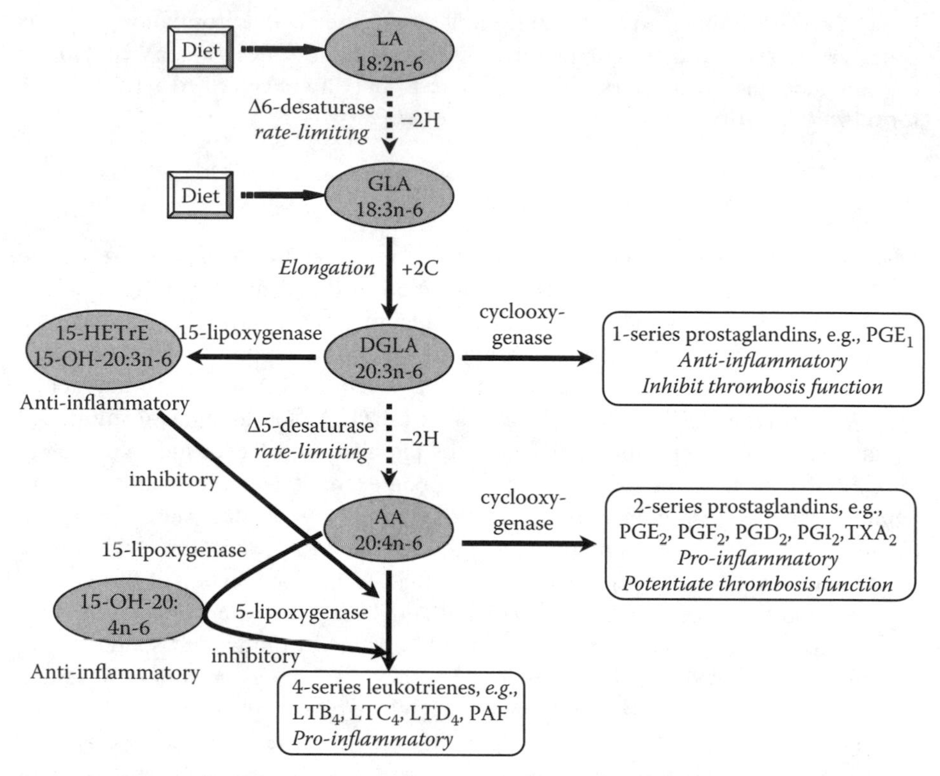

FIGURE 3.3 A simplified pathway to illustrate metabolic elongation of dietary linoleic and γ-linolenic acid. (Adapted from Fan, Y.-Y. and Chapkin, R.S., *J. Nutr.*, 128, 1411, 1998. With permission.)

obtain a TAG containing all these three PUFAs. According to their process, highest EPA incorporation of 39.9 and 37.4% in borage and evening primrose, respectively, occurred at the stoichiometric mole ratio of 1:3 for oil to EPA [67,68].

3.3 FUNCTIONS OF GLA IN MAMMALS

3.3.1 Metabolic Elongation of GLA

Distinct biological activities of individual PUFA are still debatable. It is generally accepted that mammals including humans require 1 to 2% of total dietary energy as LA to prevent EFA deficiency [69]. Changes in EFA intake may alter the fatty acid composition of cell membranes [70,71]. In many animal tissues, LA is converted to AA by an alternating sequence of Δ6-desaturation, chain elongation, and Δ5-desaturation, in which hydrogen atoms are selectively removed to create new double bonds and then two carbon atoms are added to lengthen the fatty acid chain (Figure 3.3). Dietary GLA bypasses the rate-limited Δ6-desaturation step and is quickly converted to DGLA by elongase, with only a very limited amount being desaturated to AA by Δ5-desaturase. However, limited activity of the Δ5-desaturase enzyme in rodents and humans results in conversion of only a small fraction of DGLA to AA. This was supported by the observation that in several cell types DGLA, the elongated product of GLA, is accumulated rather than AA after GLA supplementation [72,73]. DGLA can be converted to PGE_1 via the cyclooxygenase pathway and/or converted to 15-(*S*)-hydroxy-8,11,13-eicosatrienoic acid (15-HETrE) via the 15-lipoxygenase pathway [74]. The importance of 15-HETrE is that it is capable of inhibiting the formation of AA-derived 5-lipoxygenase (proinflammatory) metabolites. The oxygenated compounds, such as prostaglandin E_2 (PGE_2) and leukotriene B_4 (LTB_4), are also potent mediators of inflammation. As these eicosonoids are predominantly derived from EFAs, dietary manipulation or direct administration of precursor fatty acids may be employed to alter the eicosanoid profile.

The increase in the DGLA production relative to AA is capable of augmenting the bioconversion of AA metabolites, 2-series prostaglandins, 4-series leukotrienes, and platelet activating factor (PAF), which can exert anti-inflammatory effects on humans [73]. GLA supplementation helps to bypass a key regulatory rate-limiting step of Δ6-desaturase, which controls the formation of long-chain n-6 PUFA. Even if the intake of LA is adequate, substantial reduction of Δ6-desaturase activity due to various physiological/pathological conditions, including aging, impaired health (diabetes, viral infections, premenstrual syndrome, atopic dermatitis, rheumatoid arthritis, cancer, cardiovascular disease), lifestyle factors (alcohol, stress), certain dietary elements (high intake of cholesterol, saturated fats, and *trans* fatty acids), may result in a state of functional n-6 EFA deficiency [33,62,75]. Therefore, abundance of LA in the diet does not ensure the abundant generation of long-chain metabolites because the process of conversion is rate-limited by the activities of Δ6- and Δ5-desaturase, which is modulated by many nutritional and physiological factors. As the biosynthetic pathways of PUFA consist of several series of rate-determining desaturase and elongase steps (Figure 3.3), it is logical that dietary GLA and AA may have superior biopotency compared with LA. Much clinical evidence supports that dietary supplementation of GLA, the Δ6-desaturation product of LA, can alleviate this inadequacy.

3.3.2 Metabolic Oxidation of DGLA

The formed DGLA is cyclooxygenated (by Cox-1/2) to 1-series prostaglandins (PGE_1) and/or is metabolized (by 15-lipoxygenase) to 15-HETrE, depending on the cell type. These two oxidative metabolites of GLA are able to exert clinical efficacy in a variety of diseases, including suppression of chronic inflammation, vasodilation and lowering of blood pressure, and the inhibition of smooth muscle proliferation associated with atherosclerotic plaque development [72,76].

Several cell types, including neutrophiles, macrophages/monocytes, and epidermal cells, metabolize DGLA into 15-lipoxygenase product, 15-HETrE. It is evidenced that 15-lipoxygenase-derived hydroxy-fatty acids inhibit the synthesis of AA-derived 5-lipoxygenase metabolites [77,78] (Figure 3.3). AA-derived 5-lipoxygenase products (e.g., LTC_4 and LTB_4) are associated with several pathological inflammatory, hyperproliferative disorders [79]. It is also found that 15-HETrE can be incorporated into the membrane phospholipid, phosphatidylinositol 4,5-bisphosphate (Ptdlns 4,5-P_2), and released as 15-HETrE-containing DAG [80].

3.4 RELATIONSHIP TO DISEASE PREVENTION

The EFAs in human nutrition belong to both n-6 (LA) and n-3 (ALA) series. GLA is a derived EFA as it is produced via desaturation of a parent EFA, LA. As previously mentioned, the enzyme Δ6-desaturase is the rate-limiting factor in (n-6) EFA metabolism cascade, and it is influenced by changes in metabolic and endocrine regulation as well as by the progression of certain diseases.

3.4.1 Inflammatory Disorders and Related Conditions

Decreased levels of GLA are reported in subjects with inflammatory disorders. Abnormal EFA metabolism including that of eicosanoid production has been implicated in impairing the immune function and pathogeneses of inflammatory, autoimmune, and neoplastic diseases [81–83]. GLA is an intermediate precursor of both the 1- and 2-series prostaglandins (PGs) and 3- and 4-series leukotrienes (LTs) (Figure 3.3). Studies have shown that an increase in GLA intake enhances PGE_1 levels while decreasing PGE_2 levels [84–86]. PGE_2 for the most part has been responsible for exerting a suppressive effect on cell-mediated immunity, including lymphocyte proliferation and the production of T cell growth factor interleukin-2 (IL-2) [87,88]. Increased intake of GLA may help to influence immune function via these interconnected reactions.

3.4.1.1 Rheumatoid Arthritis

The clinical manifestation of inflammatory disease of the synovium is rheumatoid arthritis (RA), which is a chronic disease. RA results in pain, stiffness, swelling, deformity, and eventually loss of functions in the joints. Since GLA intake produces anti-inflammatory substances including PGE_1, that, in turn, helps to reduce clinical symptoms of RA. According to the proposed mechanism of how GLA helps to reduce RA-related inflammation, several studies have suggested the key contribution of DGLA-derived PGE_1 synthesis in macrophages. GLA and DGLA are able to suppress T cell activation by interfering with early events in the TcR/CD3 receptor-mediated single transduction cascade [89]. This, in turn, results in suppression of T cell proliferation after GLA administration [90].

Experimental findings of several research groups have indicated that PGs are solely mediators of inflammation; this has led to the investigation of PG-precursor fatty acids as potential therapeutic agents for RA and systemic lupus erythematosus, which are characterized by acute and chronic inflammation [91–93]. In several experimental animal models, suppression of acute and chronic inflammation as well as joint tissue injury has been observed when a GLA-enriched diet was provided [84,94,95]. Under such treatment, cells from inflammatory exudates are enriched with GLA and DGLA, while the PGE_2 and LTB_4 concentrations are reduced and the leukocyte effector function (chemotaxis, lysosomal enzyme release) is suppressed [94]. It has also been observed that the tissue DGLA to arachidonate ratio is increased substantially in animals fed a combination diet enriched in GLA as well as EPA. Callegari and Zurier [93] have explained that the increase is due to the inhibition of conversion of DGLA to AA by EPA makes more DGLA available to compete with AA for oxidation.

The study of Pullman-Moor and group [96] reported that administration of GLA (1.1 g/d) to patients having active RA resulted in increased proportion of total and phospholipid DGLA in circulating mononuclear cells. The ratio of DGLA to AA (DGLA:AA) and DGLA to stearic acid (SA) (DGLA:SA) showed a significant increase in these cells. Since SA is a major saturated fatty acid in the cell membrane and remains at a relatively fixed concentration, the changing ratio (DGLA:SA) reflects incorporation of DGLA into the cell membrane. Longer term administration (12 weeks) of GLA reduced production of stimulated peripheral blood monocytes (PBM) of PGE_2, LTB_4, and LTC_4, all of which are mediators of inflammation [96]. When the patients with RA and active synovitis were treated with GLA (1.4 to 2.8 g/d) for up to 1 year in a randomized, placebo-controlled trial [97,98], a progressive improvement was observed in the patients who continued the treatment for the entire year, suggesting that GLA functions as a slow-acting, disease-modifying antirheumatic drug [62]. A systematic review of the studies on RA patients given treatments prepared with GLA-containing oil indicates that there is moderate support for the idea that GLA has a medium to strong effect in reducing pain and tender joint count and a small effect in reducing stiffness [99].

3.4.1.2 Atopic Eczema

A series of studies published by Burr and Burr [100,101] reports about the "new deficiency disease produced by the rigid exclusion of fat from the diet" and also recognizes that the major defects associated with EFA deficiency in cutaneous biology are epidermal hyperproliferation and increased permeability of the skin to water. LA is the most abundant PUFA in human skin, and there is evidence to indicate that one functional role of LA is its involvement in the maintenance of the epidermal water barrier; disruption of this barrier is one of the major abnormalities in cutaneous EFA deficiency. AA is the second most prominent PUFA and makes up about 9% of the total fatty acids in the epidermal phospholipids of human skin [102–104]. The functional role of AA is due to generation of oxidative metabolites (PGE_2, $PGF_{2\alpha}$, and PGD_2) via the cyclooxygenase pathway and 15-HETrE via the lipoxygenase pathway. The amount of 5-lipoxygenase in the epidermis is negligible; therefore, most often 15-HETrE is found as the major lipoxygenase metabolite, not LTB_4, which is proinflammatory [105]. DGLA, the GLA elongated product, is metabolized to PGE_1 and 15-HETrE by epidermal cyclooxygenase and 15-lipoxygenase, respectively. The epidermis of both humans and guinea pigs contains active elongase enzyme that converts dietary GLA to DGLA [106].

In atopic eczema, a reduced rate of conversion of LA to GLA is observed; therefore lower levels of GLA metabolites, indicating impaired Δ6-desaturase activity, is present. Since Δ5-desaturase enzyme is deficient in the epidermis, dietary GLA is not metabolized in significant amounts to AA. Therefore, the beneficial effects of GLA-containing oils in clinical management of inflammatory hyperproliferative disorders of the skin [107,108] may be due, at least in part, to the epidermal generation of PGE_1 and 15-HETrE from elevated tissue DGLA concentrations [105].

Administration of GLA (2 to 6 g evening primrose oil/day) has produced a significant dose-related rise in plasma phospholipid DGLA in patients with atopic eczema and a less consistent rise in AA [109]. The rise in DGLA contributes to an anti-inflammatory and proliferation control effect in the skin in two ways. One way is that DGLA can be converted to PGE_1, which can stimulate cyclic AMP and inhibit phospholipase. This can reduce the selective hyperactivity of parts of the immune system [110]. The other possibility is 15-OH-DGLA (15-HETrE) inhibits lipoxygenases and, therefore, leukotriene formation is inhibited. Elevated levels of soluble IL-2 receptor concentrations have been observed in atopic eczema. Therefore, GLA induced suppression of IL-2 receptor concentration may have a significant effect on atopic eczema [111].

The first clinical study carried out on treating eczema with GLA (for 3 weeks) showed moderate beneficial effects [107]. Since then several randomized, placebo-controlled trials have been carried out and results show a mixed effect, but the balance of evidence indicates that moderate doses of GLA produce clinical improvement, particularly in itching. The review done by Horrobin [110] provides a comprehensive analysis of these clinical studies.

3.4.2 Cardiovascular Disease

3.4.2.1 Modulation of Atherogenesis

Smooth muscle cells (SMCs) and macrophages are two of the major reactive cell types involved in the progression of atherosclerosis. It is well known that macrophage-derived growth regulatory molecules (e.g., IL-1, nitric oxide, tumor necrosis factor-α, transforming growth factor-β, and macrophage derived growth factor [MDGF]) can influence SMC proliferation, and dietary lipids may influence the profile of these growth factors [112–114]. Atherosclerotic lesions result from an excessive inflammatory fibroproliferative response to various forms of disturbances to the endothelium and smooth muscle of arterial wall.

Dietary GLA can downregulate atherogenic potential by enhancing macrophage PGE_1 biosynthesis [63]. The antiproliferative cyclooxygenase product of PGE_1 elicits an array of biological responses by binding to select G protein-coupled surface receptors on SMCs, increasing intracellular 3′,5′-adenosine monophosphate (cAMP) levels [115]. cAMP is capable of stimulating the expression of numerous genes through the protein kinase A (PKA)-mediated phosphorylation of the nuclear cAMP response element binding (CREB) protein [116]. PGE_1 exhibits inhibition of vascular SMC proliferation through this mechanism [117]. It is known that agents that can reduce the migration and proliferation of vascular SMC can also retard the typical artherosclerotic plaque [118,119]. The macrophage-derived PGE_1 is capable of inducing SMC intercellular cAMP levels, resulting in the inhibition of vascular SMC proliferation, which is a major part of the atherogenic process. Also, it has been proved that dietary GLA can reduce the size of atherosclerotic lesions in ApoE genetic knockout mice, indicating that similar retardation may occur in humans [62,120].

3.4.2.2 Regulation of Blood Pressure and Hypertension

Current evidence indicates that dietary GLA is a potent blood pressure lowering nutrient, which may be useful in nutritional interventions in treating hypertension. In the studies carried out on hypertensive rats, significant alterations in the fatty acid composition of plasma, hepatic, and vascular tissues induced by dietary GLA have been found. GLA may alter lipid composition of vascular SMC membranes that affect receptor interaction and also it is a known fact that PUFAs influence regulation of membrane-bound receptors [121]. Compositional changes in fatty acids can affect cell membrane fluidity and, therefore, functional properties of the cell. Lowering of systolic blood pressure by GLA in adult spontaneously hypertensive rats (SHRs) may be mediated at least in part by interference with the rennin–angiotensin–aldesterone system at the level of the adrenal angiotensin II receptors [122]. GLA may inhibit receptor binding of angiotensin II and synthesis of aldosterone adrenal cells [122,123]. It is suggested that GLA may affect renal Na^+/K^+ adenosinetriphosphatase (ATPase) enzyme activity, which promotes natriuresis and results in blood volume reduction that lowers blood pressure [125,126]. A comparative study carried out by Engler et al. [127] showed that in relation to GLA, content of fungal oil is the most potent in lowering high blood pressure in SHRs, and also it is the most potent component in evening primrose oil. The active component GLA in evening primrose oil may reduce blood pressure through its simulative effects on the synthesis of 1-series PG, and also 2-series PG to a certain extent, which have vasoactive effects and might play a role in the control of renal function [125].

3.4.2.3 Hypocholestrolemic Effect

Studies by Takayasu and Yoshikawa [128] reported that addition of methyl-GLA can significantly lower plasma and liver cholesterol concentrations when animals were concomitantly fed a high-cholesterol diet. The studies carried out using the TAG form of GLA (as evening primrose oil) have found positive hypocholesterolemic effects on both animals and humans.

The change of PG profiles that results from GLA administration and subsequent influence on DGLA formation may influence cholesterol metabolism in several tissues. According to Horrobin [129,130], PGE_1 can inhibit cholesterol biosynthesis. In rat models, intraperitoneal administration of PGE_1 (1 mg/kg for 5 d) significantly decreased cholesterol, triacylglycerol, and phospholipid in serum and HDL [131]. Studies on isolated human mononuclear leukocytes showed that PGE_1 (and prostacyclin analogue, *iloprost*) is capable of suppressing sterol synthesis through an increase in cyclic AMP [132]. Cyclic AMP has the ability to inhibit 3-hydroxy-3-methylglutaryl coenzyme A (HMG-CoA) reductase and to suppress sterol synthesis [132,133] and also to decrease binding, uptake, and degradation of LDL by decreasing the number of LDL receptors in several cell lines [134,135]. Fox and colleagues [136] suggest that the hypocholesterolemic effect of GLA may be directly attributable to redistribution of plasma cholesterol to body tissues through an increase in tissue receptors, rather than a change in endogenous cholesterol synthesis or catabolism [136]. Huang and group [137] also hypothesized that the decrease in liver cholesterol levels by evening primrose oil in rats is attributed to HMG-CoA reductase activity, the rate-limiting enzyme of cholesterol biosynthesis; however, this hypothesis is not further confirmed.

3.4.3 Tumor and Cancer

One of the proposed mechanisms for the action of PUFAs in modulation of cancerous and tumorous cell growth is related to lipid oxidation. EFAs, especially GLA, AA, EPA, and DHA, are capable of inducing apoptotic cell death of tumor cells [138–142]. A review provided by Das [143] lists the following as the most relevant metabolic events in tumor cells that have relation to PUFA metabolism: (1) excess production of PGE_2 and $PGF_{2\alpha}$, which have immunosuppressive action; (2) decrease in free-radical generation coupled with a relative increase in antioxidative capacity; (3) a decrease in the content of PUFAs, which are necessary to trigger oxidative metabolism in human neutrophils and tumor cells; and (4) an increase in polyamines.

Impaired lipid oxidation has been observed in tumor cells [144–146]. EFAs are important structural components of cell membranes, and thus substrates for generation of lipid oxidation products, which have inhibitory activity on cell proliferation. Some PGs derived from *cis*-unsaturated fatty acids including GLA have antineoplastic properties [147,148]. Also, it has been noted that tumor cells (e.g., human macrophage-like cells, line U-937 and promyelocytic leukemia cells, HL-60) do not constitutively express phospholipase A_2 (PLA_2) activity, but do so when induced to differentiate *in vitro* [149,150]. The PLA_2 catalyzes the rate-limiting step in the release of *cis*-unsaturated fatty acids from the cell membrane lipid pool, which form precursors to various PGs. Therefore, it can be rationalized that the deficiency in PLA_2 activity in tumor cells prevents the formation of antineoplastic PGs. Tumor cells are also deficient in Δ6-desaturase enzyme and secrete excess of PGE_2, which is immunosuppressive and mutagenic [151–153].

GLA, AA, EPA, and DHA can selectively enhance both superoxide anion and hydrogen peroxide generation and the level of lipid peroxides in the tumor cells, but not in normal cells [141,154]. The observed high antioxidant activity and reduced levels of microsomal cytochrome P450 and cytochrome B5 [155,156] and increased levels of reduced glutathione content [156,157] in tumor cells is still puzzling. It has also been observed that decreased levels of superoxide dismutase during early stages of tumorrogenesis continue to remain depressed [158,159]. The reduction in lipid peroxidation in tumor cells has been attributed to the increased contents of lipophilic antioxidants in tumors, which is mainly due to α-tocopherol [160]. In normal cells, when DNA synthesis is at maximum, lipid peroxidation is suppressed and vice versa; therefore, cells have evolved a mechanism of changing their antioxidant content to protect the genetic material from free-radical damage [144,155,161].

DeVries and Van Noorden [159] suggest that the reduced lipid peroxidation in tumor cells may be due to the lack of adequate amounts of specific PUFA substrates and also to the loss of the lipid oxidation mechanism. The lack of adequate amounts of specific PUFA substrates may be due to reduced formation or uptake of PUFA by tumor cells. Several investigators have showed that tumor cells are deficient in PUFAs and contain higher amounts of oleic acid and decreased amounts of AA and C22 PUFAs [162]. Since the longer chain PUFAs stimulate higher rates of lipid oxidation than C18 PUFAs, this may be one of the rate-limiting factors in lipid peroxidation in tumor cells. Decomposed lipid peroxidation products (e.g., malonaldehyde, 4-hydroxynonadienal, 4-hydroxyhexenal) generated from PUFAs may have a regulating effect on DNA duplication enzymes, and thus function in regulating cell proliferation and uncontrolled tumor cell growth [159]. The reviews provided by Das [142] and De Vries and Van Noorden [159] extensively discussed how EFAs, including GLA, affect tumor cells via the lipid peroxidation pathway.

In the studies on growth and metastatis of rodent mammary tumor cells, attenuated levels of PGE_1 were found in opposition to PGE_2 in nonmetastatic cells [163–165]. AA-derived eicosanoids play an important role in down regulating growth and metastasis of tumor cells.

In mixed culture cells, GLA showed a more selective tumoricidal action than did AA or EPA [139] and *in vivo* studies showed antitumor [166] and antiangiogenic [167] activities and may increase the sensitivity of tumor cells to radiation and chemotherapeutic agents [168,169]. Das and co-workers [170] and Bakashi and group [171] described the use of GLA as a therapy for human gliomas, which can occur anywhere in the brain but usually affect the cerebral hemisphere. They have treated patients with grade-4 gliomas with 1 mg GLA for 7 d by intratumoral injection. It was concluded from this study that GLA is capable of enhancing the sensitivity of tumor cells to conventional anticancer and radiation treatments as found in studies of breast cancer [172,173] and also has the beneficial property of the lack of side effects.

Studies on GLA related to cancer therapy propose that GLA has a value as a new cancer therapeutic agent having selective antitumor properties with negligible systemic toxicity. The proposed mechanisms of activities include modulation of steroid hormone receptors (e.g., estrogen receptors in breast cancer). The clinical study carried out by Kenny et al. [172] concludes that GLA is a useful adjunct to primary tamoxifen treatment in endocrine-sensitive breast cancer. GLA may affect on estrogen receptor function and exert an additive or synergistic action with tamoxifen via enhanced down regulation of estrogen receptor-stimulated growth.

3.4.4 Diabetes and Related Conditions

It is found that desaturation and elongation of LA are decreased in subjects with type 1 diabetes (insulin-dependent, diabetes mellitus) [174]. Decreased levels of DGLA and AA in serum lipids in association with increased plasma levels of PGE_2 and $PGF_{2\alpha}$ have also been observed. In children with deficient levels of insulin and high levels of blood sugar supply, formation of DGLA from LA is lowered and decreased formation of PGE_1 has been observed [175]. Under conditions of diabetes mellitus, this restricted conversion of dietary LA to GLA has been attributed to the reduced $\Delta 6$-desaturase activity [176,177].

Platelet abnormalities are linked with vascular disease and diabetes has a high incidence of vascular complications. Studies on the fatty acid composition of platelet phospholipids showed that increased levels of AA in type 2 diabetes patients compared with the age-matched controls and the levels were disproportionately high compared with those in plasma total lipids [178]. Diabetic subjects with proliferative retinopathy showed significantly higher AA uptake activity than those with little or no background retinopathy; however, no difference was found for LA uptake activity in platelets between the two groups [179]. Positive results obtained from several clinical studies have evidenced that providing GLA to bypass blocked Δ6-desaturase activity can favorably alter platelet fatty acid composition and prostonoic metabolism [180–182]. Poisson and co-workers [183] provided a good review on this aspect of GLA.

Diabetic neuropathy is a characteristic of diabetes that affects sensory nerves and also of neuropathy affecting autonomic nerves. Lack of GLA metabolites is also a considered factor for normal neuronal structure, function, and microcirculation; thus impaired membrane function and nerve damage are observed in diabetic neuropathy [184,185]. Several observations made on diabetic animals suggest that the neuropathy may relate to changes in the phospholipid structure of the neuronal membrane and/or to abnormal neuronal microvascular function. Low levels of AA have been found in neuronal phospholipids (PLs) from diabetic animals with neuropathy [186].

TAGs containing GLA have been shown to normalize nerve conduction velocity and sciatic endoneurial blood flow [187,188]. GLA is able to correct nerve conduction abnormalities by enhancing the synthesis of the cyclooxygenase-derived vasodilator prostanoid, PGE_1, which is capable of increasing vasa nervorum perfusion [189]. Studies carried out on diabetic rats showed that ascorbate combined with GLA has therapeutic advantage over GLA alone; thus ascorbyl GLA may be suitable in clinical trials for diabetic neuropathy [62].

REFERENCES

1. Heiduschka, A. and Lueft, K., Fatty oil from the seed of the evening primrose (*Oenothera biennis*) and a new linolenic acid, *Arch. Pharm.*, 257, 33, 1919.
2. Eibner, A. and Schild, E., Quantitative analysis and technical evaluation of a Oenothera oil, *Chem. Umschau Fette, Oele, Wachse Harze,* 34, 339, 1927.
3. Riley, J.P., Seed fat of *Oenothera biennis* L., *J. Chem. Soc.,* 4, 2728, 1949.
4. Wolf, R.B., Kleiman, R., and England, R.E., New sources of γ-linolenic acid, *J. Am. Oil Chem. Soc.,* 60, 1858, 1983.
5. Gunstone, F.D., Gamma linolenic acid: occurrence and physical and chemical properties, *Prog. Lipid Res.,* 31, 145, 1992.
6. Guil-Guerrero, J.L. et al., Occurrence and characterization of oils rich in γ-linolenic acid Part I: *Echium* seeds from Macaronesia, *Phytochemistry,* 53, 451, 2000.
7. Guil-Guerrero, J.L. et al., γ-Linolenic acid from fourteen Boraginaceae species, *Ind. Crops Prod.,* 18, 85, 2003.
8. Goffman, F.D. and Galletti, S., Gamma-linolenic acid and tocopherol contents in the seed oil of 4 accessions from several *Ribes* species, *J. Agric. Food Chem.,* 49, 349, 2001.
9. Traitler, H. et al., Characterization of γ-linolenic acid in *Ribes* seed, *Lipids*, 19, 923, 1984.
10. Ucciani, E., Sources potentielles d'acide gamma-linolenique:une revue, *OC,* 2, 319, 1995.

11. Oomah, D.B. et al., Characteristics of hemp (*Cannabis sativa* L.) seed oil, *Food Chem.,* 76, 33, 2002.
12. Ratledge, C., Microbial oils and fats: an assessment of their commercial potential, *Prog. Ind. Microbiol.,* 16, 119, 1982.
13. Komisaka, Y. et al., Modulation of fatty acid incorporation and destauration by Trifluoperazine in fungi, *Lipids,* 25, 787, 1990.
14. Aggelis, G. et al., Possibility of gamma-linolenic acid production by culturing *Mucor circinelloides* CBS 172-27 on some vegetable oils, *Oleagineux,* 41, 208, 1991.
15. Aggelis, G. et al., Gamma-linolenic acid production by bioconversion of the linoleic acid of some oils, *Rev. Fr. Crops Gras,* 38, 95,1991.
16. Miura, Y. et al., Production of γ-linolenic acid from marine green alga *Chlorella* sp. NKG 042401, *FEMS Microbiol. Lett.,* 107, 163, 1993.
17. Srinivas, C.V.S., Praveena, B., and Nagaraj, G., Safflower petals: a source of gamma linolenic acid, *Plant Foods Human Nutr.,* 54, 89, 1999.
18. Stymme, S. and Stobart, A.K., Biosynthesis of γ-Linolenic acid in cotyledons and microsomal preparations from the developing seeds of common borage (*Borago officinalis*), *Biochem. J.,* 240, 385, 1986.
19. Griffith, G., Stobart, A.K., and Stymme, S., Δ6- and Δ12-Desaturase activities and phosphatidic acid formation in microsomal preparations from developing cotyledons of common borage (*Borago officinalis*), *Biochem. J.,* 252, 641, 1988.
20. Galle, A.M. et al., Biosynthesis of γ-linolenic acid in developing seeds of borage (*Borago officinalis* L.), *Biochim. Biophys. Acta,* 1158, 52, 1993.
21. Talamo, B., Chang, N., and Bloch, K., Desaturation of oleyl phospholipid to linoleyl phospholipid in *Torulopsis utilis, J. Biol. Chem.,* 248, 2738, 1973.
22. Pugh, E.L. and Kates, M., Desaturation of phosphatidylcholine and phosphatidylethanolamine by a microsomal enzyme from *Candida lipolytica, Biochim. Biophys. Acta,* 316, 305, 1973.
23. Pugh, E.L. and Kates, M., Characterization of a membrane bound phospholipid desaturase system of *Candida lipolytica, Biochim. Biophys. Acta,* 380, 442, 1975.
24. Sato, N., Seyama, Y., and Murata, N., Lipid linked desaturation of palmitic acid in monogalactosyl diacylglycerol in the blue-green algae (Cynobacterium) *Anabaena variabilis* studied *in vivo, Plant Cell Physiol.,* 27, 819, 1986.
25. Murata, N., Deshnium, P., and Tasaka, Y., Biosynthesis of γ-linolenic acid in the Cyanobacterium *Spirulina platensis,* in *γ-Linolenic acid: Metabolism and Its Roles in Nutrition and Medicine,* Huang, Y.-S. and Mills, D.E., Eds., AOCS Press, Champaign, IL1996, p. 22.
26. Huang, Y.-S. et al., Cloning of Δ12-and Δ6-desaturases from *Mortiella alpina* and recombinant production of γ-linolenic acid *Saccharomyces cerevisiae, Lipids,* 34, 649, 1991.
27. Sakuradani, E. et al., Δ6-Fatty acid desaturase from an arachidonic acid-producing *Mortierella* fungus gene cloning and its heterologous expression in a fungus, *Aspergillus, Gene,* 238, 445, 1999.
28. Mukerji, P. et al., Desaturase genes and uses thereof., U.S. Patent 6,635,451 B2, 2003.
29. Reddy, A.S. and Thomas, T.L., Expression of a cyanobacterial Δ6-desaturase gene results in gamma-linolenic acid production in transgenic plants, *Nat. Biotechnol.,* 14, 639, 1996.
30. Thomas, T.L., Production of gamma linolenic acid by a Δ6-desaturase, U.S. Patent 6,355,861 B1, 2002.
31. Knutzon, D., Polyunsaturated fatty acids in plants, U.S. Patent 6,459,018 B1, 2002.
32. Horrobin, D.F., Nrutitional and medical importance of gamma-linolenic acid, *Prog. Lipid Res.,* 31, 163, 1992.

33. Phillips, J.C. and Huang, Y.S., Natural sources and biosynthesis of γ-linolenic acid: an overview, in *γ-Linolenic Acid: Metabolism and Its Roles in Nutrition and Medicine,* Huang, Y.-S. and Mills, D.E., Eds., AOCS Press, Champaign, IL, 1996, p. 1.
34. Mead, J.F., Synthesis and metabolism of polyunsaturated acids, *Fed. Proc.,* 20, 952, 1961.
35. Okayasu, T. et al., Purification and partial characterization of linoleoyl-CoA desaturase from rat liver microsomes, *Arch. Biochem. Biophys.,* 206, 21, 1981.
36. Thomasson, H.J., Stearolic acid, an essential fatty acid?, *Nature,* 173, 452, 1954.
37. Rivers, J.P.W. and Frankel, T.L., Essential fatty acid deficiency, *Br. Med. Bull.* 37, 59, 1981.
38. Horrobin, D.F., Gamma-linolenic acid, *Rev. Contemp. Physiol.,* 1, 1, 1990.
39. Helme, J.P., Evening primrose, borage and blackcurrant seeds, in *Oils and Fats Manual Vol. I,* Karleskind, A., Ed., Intercept Ltd., 1996, p. 168.
40. Gawdzik, J. et al., Supercritical fluid extraction of oil from evening primrose (*Oenothera paradoxa* H.) seeds, *Chem. Anal.* 43, 695, 1998.
41. Gomez, A.M. and dela Ossa, E.M., Quality of borage seed oil extracted by liquid and supercritical carbon dioxide, *Chem. Eng. J.,* 88, 103, 2002.
42. Dauksas, E. Venskutonis, P.R., and Sivik, B., Supercritical fluid extraction of borage (*Borago officinalis* L.) seeds with pure CO_2 and its mixture with caprylic acid methyl ester, *J. Supercrit. Fluids,* 22, 211, 2002.
43. Ratnayake, W.M.N., Matthews, D.G., and Ackman, R.G., Triacylglycerols of evening primrose (*Oenothera biennis*) seed oil, *J. Am. Oil Chem. Soc.,* 66, 966, 1989.
44. Christie, W.W., Fractionation of the triacylglycerols of evening primrose oil by high-performance liquid chromatography in silver ion mode, *Fat Sci. Technol.,* 93, 65, 1993.
45. Tsevegsuren, N. and Aitzetmuller, K., γ-Linolenic acid in anemone spp. seed lipids, *Lipids,* 28, 841, 1993.
46. Perrin, J.-L. et al., Analysis of triglyceride species of blackcurrant seed oil by HPLC via a laser light scattering detector, *Rev. Fr. Crops Gras,* 34, 221, 1987.
47. Aitzetmuller, K. and Gronheim, M., Separation of highly unsaturated triacylglycerols by reversed phase HPLC with short wavelength UV detection, *J. High Resol. Chromatogr.,* 15, 219, 1992.
48. Blomberg, L.G., Bemirbuker, M., and Andersson, P.E., Argentation supercritical fluid chromatography for quantitative analysis of triacylglycerols, *J. Am. Oil Chem. Soc.,* 70, 939, 1993.
49. Manninen, P., Laakso, P., and Kallio, H., Separation of γ- and α-linolenic acid containing triacylglycerols by capillary supercritical fluid chromatography, *Lipids,* 30, 665, 1995.
50. Redden, P.R. et al., Stereospecific analysis of the major triacylglycerol species containing γ-linolenic acid in evening primrose oil and borage oil, *J. Chromatogr. A,* 704, 99, 1995.
51. Lawson, L.D. and Hughes, G.B., Triacylglycerol structure of plant and fungal oils containing γ-linolenic acid, *Lipids,* 23, 313, 1988.
52. Griffiths, G., Stobat, A.K., and Stymne, S., The Δ6- and Δ12-desaturase activities and phosphatidic acid formation in microsomal preparations from the developing cotyledons of common borage (*Borago officinalis*), *Biochem. J.,* 252, 641, 1988.
53. Laakso, P. and Christie, W.W., Chromatographic resolution of chiral diacylglycerol derivatives: potential in the stereospecific analysis of triacyl-sn-glycerols, *Lipids,* 25, 349, 1990.
54. Shimada, Y., Sugihara, A., and Tominaga, Y., Enzymatic purification of polyunsaturated fatty acids: a review, *J. Biosci. Bioeng.,* 91, 529, 2001.
55. Chen, T.-C. and Ju, Y.-H., Polyunsaturated fatty acid concentrates from borage and linseed oil fatty acids, *J. Am. Oil Chem. Soc.,* 78, 485, 2001.

56. Constantin, B. et al., Concentrate of polyunsaturated fatty acid ethyl esters and preparation thereof, U.S. Patent US 5,679,809, 1997.
57. Traitler, H., A process for enrichment with delta-6 fatty acids of a mixture of fatty acids, U.S. Patent 4,776,984, 1988.
58. Abril, J.R., Banzhaf, W., and Kohn, G., Production of human, animal or aquaculture organism diet supplement enriched with gamma linolenic acid and stearidonic acid by producing polar lipid rich fraction from plant seeds and microbes, International Patent, WO 200,292,073, 2003.
59. Wanasundara U.N. and Fedec, P., Centrifugal partition chromatography (CPC): emerging separation and purification technique for lipid related compounds, *Inform,* 13, 726, 2002.
60. Shimada, Y. et al., Purification of γ-linolenic acid from borage oil a two-step enzymatic method: hydrolysis and selective esterification, *J. Am. Oil Chem. Soc.,* 74, 1465, 1997.
61. Shimada, Y. et al., Large scale purification of γ-linolenic acid by selective esterification using *Rhizopus delemar* lipase, *J. Am. Oil Chem. Soc.,* 75, 1539, 1998.
62. Fan, Y.-Y. and Chapkin, R.S., Importance of dietary γ-linolenic acid in human health and nutrition, *J. Nutr.,* 128, 1411, 1998.
63. Fan, Y.-Y. and Chapkin, R.S., Mouse peritoneal macrophage prostaglandin E_1 synthesis is altered by dietary γ-linolenic acid, *J. Nutr.,* 122, 1600, 1992.
64. Jowett, P., 1-Mono-glyceride esters of gamma-linolenic acid, G.B. Patent 2 183 635 A, 1986.
65. Kawashima, A. et al., Production of structured TAG rich in 1,3-diacapryloyl-2-γ-linolenoyl glycerol from borage oil, *J. Am. Oil Chem. Soc.,* 79, 871, 2002.
66. Akoh, C.C. and Moussata, C.O., Lipase-catalyzed modification of borage oil: incorporation of capric and eicosapentaenoic acids to form structured lipids, *J. Am. Oil Chem. Soc.,* 75, 697, 1998.
67. Senanayake, S.P.J.N. and Shahidi, F., Enzyme-assisted acidolysis of borage (*Borago officinalis* L.) and evening primrose (*Oenothera biennis* L.) oils: incorporation of omega-3 polyunsaturated fatty acids, *J. Agric. Food Chem.,* 47, 3105, 1999.
68. Senanayake, S.P.J.N. and Shahidi, F., Structured lipid via lipase-catalyzed incorporatin of eicosapentaenoic acid to borage (*Borago officinalis* L.) and evening primrose (*Oenothera biennis* L.) oils, *J. Agric. Food Chem.,* 50, 477, 2002.
69. Chapkin, R.S., Reappraisal of essential fatty acids, in *Fatty Acids in Food and Their Health Implications,* 2nd ed., Chow, C.K., Ed., Marcel Dekker, New York, 2000, p. 557.
70. Warrington, J.R., Interleukin-2 abnormalities in systemic lupus erythematosus and rheumatoid arthritis: a role for overproduction of interleukin-2 in human autoimmunity, *J. Rheumatol.,* 15, 616, 1988.
71. Willis, A.L., Nutritional and pharmacological factors in eicosanoid biology, *Nutr. Rev.,* 39, 289, 1981.
72. Zurier, R.B. et al., Gamma-linolenic acid treatment of rheumatoid artheritis: a randomized, placebo-controlled trial, *Arthritis Rheum.,* 39, 1808, 1996.
73. Johnson, M.M. et al., Dietary supplementation of γ-linolenic acid alters fatty acid content and eicosanoid production of healthy humans, *J. Nutr.,* 127, 1435, 1997.
74. Borgeat, P., Hamberg, M., and Samuelsson, S., Transformation of arachidonic acid and homo-gamma-linolenic acid by rabbit polymorphonuclear leukocytes to monohydroxy acids from novel lipoxygenase, *J. Biol. Chem.,* 251, 7816, 1976.
75. Barre, D.E., Potential of evening primrose, borage, black currant, and fungal oils in human health, *Ann. Nutr. Metab.,* 45, 47, 2001.

76. Fan, Y.-Y., Ramos, K.S., and Chapkins, R.S., Dietary γ-linolenic acid modulates macrophage-derived soluble factor that down regulates DNA synthesis in smooth muscle cells, *Arterioscler. Thromb. Vasc. Biol.,* 15, 1397, 1995.
77. Chapkin, R.S. et al., Ability of monohydroxyeicosatrienoic acid (15-OH-20:3) to modulate macrophage arachidonic acid metabolism, *Biochem. Biophys. Res. Commun.,* 153, 799, 1988.
78. Miller, C.C. et al., Dietary supplementation with ethyl ester concentrates of fish oil (n-3) and borage oil (n-6) polyunsaturated fatty acids induce epidermal generation of local putative anti-inflammatory metabolites, *J. Invest. Dermatol,* 96, 98, 1991.
79. Goulet, J.L. et al., Altered inflammatory responses in leukotriene-deficient mice, *Proc. Natl. Acad. Sci. USA,* 91, 12852, 1994.
80. Cho, Y. and Ziboh, V.A., A novel 15-hydroxyeicosatrienoic acid substituted diacylglycerol (15-HETrE-DAG) selectively inhibits epidermal protein kinase Cβ, *Biochim. Biophys. Acta,* 1349, 67, 1997.
81. Brenner, R.R., De Tomas, M.E., and Peluffa, R.O., Effect of polyunsaturated fatty acids on the desaturation *in vitro* of linoleic to γ-linolenic acid, *Biochim. Biophys. Acta,* 106, 640, 1965.
82. Manku, M.S. et al., Essential fatty acids in plasma phospholipids of patients with atopic eczema, *Br. J. Dermatol.,* 110, 643, 1984.
83. Wu, D. and Meydani, S.N., γ-Linolenic acid and immune function, in *γ-Linolenic Acid: Metabolism and Its Roles in Nutrition and Medicine,* Huang, Y.-S. and Mills, D.E., Eds., AOAC Press, Champaign. IL, 1996, p. 106.
84. Tate, G. et al., Suppression of acute and chronic inflammation by dietary gamma linolenic acid, *J. Rheumatol.,* 16, 729, 1989.
85. Nerad, J.L., Meydani, S.N., and Dinarello, C.A., Dietary supplementation with gamma linolenic acid (GLA) and parameters of cell-mediated immunity, *Cytokine,* 3, 513, 1991.
86. Fan, Y.-Y., and Chapkin, R.S., Mouse peritoneal macrophage prostaglandin E synthesis is altered by dietary γ-linolenic acid, *J. Nutr.,* 122, 1600, 1996.
87. Goodwin, J.S. and Webb, D.R., Regulation of the immune response by prostaglandins, *Clin. Immunol. Immunopathol.,* 15, 106, 1977.
88. Rappaport R.S. and Dodge, G.R., Prostaglandin E inhibits the production of human interleukin 2, *J. Exp. Med.,* 155, 943, 1982.
89. Vassilopoulas, D. et al., Gamma linolenic acid and dihomo gamma linolenic acid suppress the CD3-mediated signal transduction pathway in human T cells, *Clin. Immunol. Immunopathol.,* 83, 237, 1997.
90. Rossetti, R.G. et al., Oral administration of unsaturated fatty acids: effects on human peripheral blood T lymphocyte proliferation, *J. Leukotr. Biol.,* 62, 438, 1997.
91. Krakauer, K.A., Torrey, S.B., and Zurier, R.B., Prostaglandin1 treatment of NZB/w mice III: preservation of spleen cell concentrations and mitogen induced proliferative responses, *Clin. Immunol. Immunopathol.,* 11, 256, 1978.
92. Dore-Duffy, P., Siok, C., and Zurier, R.B., Oral administration of prostaglandin E to humans: effects on peripheral blood leukocyte function, *J. Lab. Clin. Med.,* 104, 283, 1984.
93. Callegari, P.E. and Zurier, R.B., Botanical lipids: potential role in modulation of immunologic responses and inflammatory reactions, *Nutr. Rheum. Dis.,* 17, 415, 1991.
94. Kunkel, S.L. et al., Suppression of chronic inflammation by evening primrose oil, *Prog. Lipid Res.,* 20, 886, 1981.
95. Tate, G. et al., Suppression of monosodium urate crystal induced acute inflammation by diets enriched with gamma linolenic acid and eicosapentaenoic acid, *Arthritis Rheum.* 31, 1543, 1988.

96. Pullman-Moor, S. et al., Alteration of the cellular fatty acid profile and the production of eicosanoids in human monocytes by gamma-linolenic acid, *Arthritis Rheum.,* 33, 1526, 1990.
97. Leventhal, L.J., Boyce, E.G, and Zurier, R.B., Treatment of rheumatoid arthritis with gamma linolenic acid, *Ann. Intern. Med.,* 119, 867, 1993.
98. Zurier, R.B. et al., Gamma linolenic acid treatment of rheumatoid arthritis. A randomized placebo-controlled trial, *Arthritis Rheum.,* 39, 1808, 1996.
99. Soeken, K.I., Miller, S.A., and Ernst, E., Herbal medicines for the treatment of rheumatoid arthritis: a systemic review, *Rheumatology,* 42, 652, 2003.
100. Burr, G.O. and Burr, M.M., A new deficiency disease produced by the rigid exclusion of fat from diet, *J. Biol. Chem.,* 82, 3457, 1929.
101. Burr, G.O. and Burr, M.M., The nature and role of the fatty acids essential in nutrition, *J. Biol. Chem.,* 86, 587, 1930.
102. Vorman, H.E., Nemeck, R.A., and Hsia, S.L., Synthesis of lipids from acetate by human preputial and abnormal skin *in vitro, J. Lipid Res.,* 10, 507, 1969.
103. Ziboh, V.A., Biochemical abnormalities in essential fatty acid deficiency, *Models Dermatol.,* 3, 106, 1987.
104. Ziboh, V.A., The significance of polyunsaturated fatty acids in cutaneous biology, *Lipids,* 31, 249, 1996.
105. Ziboh, V.A., Miller, C.C., and Cho, Y., Metabolism of polyunsaturated fatty acids by skin epidermal enzymes: generation of anti-inflammatory and antiproliferative metabolites, *Am. J. Clin. Nutr.,* 7, 361S, 2000.
106. Miller, C.C. et al., Oxidative metabolism of dihomo-γ-linolenic acid by guinea pig epidermis: evidence of generation of anti-inflammatory products, *Prostaglandins,* 35, 917, 1988.
107. Lovell, C.R., Burton, J.L., and Horrobin, D.F., Treatment of atopic eczema with evening primrose oil, *Lancet,* 1, 278, 1981.
108. Wright, S. and Burton, J.L., Oral evening primrose-seed oil improves atopic eczema, *Lancet,* 99, 1120, 1982.
109. Manku, M.S. et al., Reduced levels of prostaglandin precursors in the blood of atopic patients: defective delta-6-desaturase function as a biochemical basis for atopy, *Prostag. Leukotr. Med.,* 9, 615, 1982.
110. Horrobin, D.F., Essential fatty acid metabolism and its modification in atopic eczema, *Am. J. Clin. Nutr.,* 7, 367S, 2000.
111. Ziboh, V.A., Implications of dietary oils and polyunsaturated fatty acids in the management of cutaneous disorders, *Arch. Dermatol.,* 125, 241, 1989.
112. Chapkin, R.S., Fan, Y. -Y., and Ramos, K.S., Impact of dietary γ-linolenic acid on macrophage-smooth muscle cell interactions: down-regulation of vascular smooth muscle cell DNA synthesis, in *γ-Linolenic Acid: Metabolism and Its Roles in Nutrition and Medicine,* Huang, Y.-S. and Mills, D.E., Eds., AOCS Press, Champaign, IL, 1996, p. 218.
113. Sperling, R., The effects of dietary *n*-3 polyunsaturated fatty acids on neutrophils, *Proc. Nutr. Soc.,* 57, 527, 1998.
114. Fan, Y.-Y., Ramos, K.S., and Chapkin, R.S., Modulation of atherogenesis by dietary gamma-linolenic acid, in *Eicosanoids and Other Bioactive Lipids in Cancer, Inflammation, and Radiation Injury, 4,* Honn, K.V., Marnett, L.J., Nigam, S., and Dennis, E.A., Eds., Kluwer Academic/Plenum, New York, 1999, p. 485.
115. Negishi, M., Sugimoto, Y., and Ichikawa, A., Prostaglandin receptors and their biological actions, *Prog. Lipid Res.,* 32, 417, 1993.
116. Foulkes, N.S. and Sasone-Corsi, P., Transcription factors coupled to the cAMP signalling pathway, *Biochim. Biophys. Acta,* 1288, F101, 1996.

117. Fan, Y.-Y., Ramos, K.S., and Chapkins, R.S., Cell-cycle dependant inhibition of DNA synthesis in vascular smooth muscle cells by prostaglandin E_1: relationship to intracellular cAMP levels, *Prostaglandins Leukotr. Essent. Fatty Acids,* 54,101, 1996.
118. Fan Y.-Y., Ramos, K.S., and Chapkin, R.S., Dietary lipid source alters murine macrophage/vascular smooth muscle cell interactions *in vitro, J. Nutr.,* 126, 2083, 1996.
119. Fan, Y.-Y., Ramos, K.S., and Chapkins, R.S., Dietary γ-linolenic acid enhances mouse macrophage-derived prostaglandin E_1 which inhibits vascular smooth muscle cell proliferation, *J. Nutr.,* 127, 1765, 1997.
120. Fan Y.-Y., Ramos, K.S., and Chapkin, R.S., Dietary gamma-linolenic acid suppresses aortic smooth muscle cell proliferation and modifies atherosclerotic lesions in apolipoprotein E knockout mice, *J. Nutr.,* 131, 1675, 2001.
121. Spector, A.A. and Yorek, M.A., Membrane lipid composition and cellular function, *J. Lipid Res.,* 26, 1015, 1985.
122. Engler, M.M. et al., Effects of dietary γ-linolenic acid on blood pressure and adrenal angiotensin receptors in hypertensive rats, *Proc. Soc. Exp. Biol. Med.,* 218, 234, 1998.
123. Goodfriend, T.L. et al., Fatty acids are potential regulators of aldosterone secretion, *Endocrinology,* 128, 2511, 1991.
124. Engler, M.M., γ-Linolenic acid: a potent blood pressure lowering nutrient, in *γ-Linolenic Acid: Metabolism and Its Roles in Nutrition and Medicine,* Huang, Y.-S. and Mills, D.E., Eds., AOCS Press, Champaign, IL, 1996, p. 200.
125. Leeds, A.R., Gray, I., and Ahmed, M., Effects of n-6 essential fatty acids as evening primrose oil in mild hypertension, in *Omega-6 Essential Fatty Acids: Pathophysiology and Roles in Clinical Medicine,* Horrobin, D.F., Ed., Alan Liss, New York, 1990, p. 56.
126. Deferne, J.-L. and Leeds, A.R., The antihypertensive effect of dietary supplementation with a 6-desaturated essential fatty acid concentrate as compared with sunflower seed oil, *J. Human Hypertens.,* 6, 1, 1992.
127. Engler, M.M., Comparative study of diets enriched with evening primrose, blackcurrent, borage, or fungal oils on blood pressure and pressor responses in spontaneously hypertensive rats, *Prostaglandins Leukotr. Essent. Fatty Acids,* 49, 809, 1993.
128. Takayasu, K. and Yoshikawa, I., The influence of exogenous cholesterol on the fatty acid composition of liver lipids in the rats given linoleate and γ-linolenate, *Lipids,* 6, 47, 1971.
129. Horrobin, D.F., A new concept of lifestyle-related cardiovascular disease: the importance of interactions between cholesterol, essential fatty acids, prostaglandin E_1 and thromboxane A_2, *Med. Hypothesis,* 6, 785, 1980.
130. Horrobin, D.F., The use of gamma-linolenic acid in diabetic neuropathy, *Agents Action Suppl.,* 37, 120, 1992.
131. Dionyssiou-Asteriou, A. et al., Influence of prostaglandin E_1 on high-density lipoprotein-fraction lipid levels in rats, *Biochem. Med. Metab. Biol.,* 36, 114, 1986.
132. Krone, W. et al., Prostacyclin analogue iloprost and prostaglandin E_1 suppress sterol synthesis in freshly isolated human mononuclear leukocytes, *Biochim. Biophys. Acta,* 835, 157, 1985.
133. Krone, W., Betteridge, D.J., and Galton, D.J., Mechanism of regulation of 3-hydroxy-3-methylglutaryl coenzyme: a reductase activity by low-density lipoprotein in human lymphocytes, *Eur. J. Clin. Invest.,* 9, 405, 1979.
134. Maziere, C., CyclicAMP decreases LDL catabolism and cholesterol synthesis in the human hepatoma cell line HepG2, *Biochem. Biophys. Res. Commun.,* 156, 424, 1988.

135. Stout, R.W. and Bierman, E.L., Dibutyryl cyclicAMP inhibits LDL binding in cultured fibroblasts and arterial smooth muscle cells, *Artherosclerosis,* 46, 13, 1983.
136. Fox, J.C., et al., *In vivo* regulation of hepatic LDL receptor mRNA in the baboon: differential effects of saturated and unsaturated fat, *J. Biol. Chem.,* 262, 7014, 1987.
137. Huang, Y.-S., Manku, M.S., and Horrobin, D.F., The effect of dietary cholesterol on blood and liver polyunsaturated fatty acid on plasma cholesterol in rats fed various types of fatty acid diet, *Lipids,* 19, 664, 1984.
138. Leary, W.P. et al., Some effect of gamma-linolenic acid on cultured human oesophageal carcinoma cells, *S. Afr. Med. J.,* 82, 681, 1984.
139. Begin, M.E. et al., Differential killing of human carcinoma cells supplemented with n-3 and n-6 polyunsaturated fatty acids, *J. Natl. Cancer Inst.,* 177, 1053, 1986.
140. Seigal, I. et al., Cytotoxic effects of free fatty acids on ascites tumor cells, *J. Natl. Cancer Inst.,* 78, 271, 1987.
141. Das, U.N., Selective enhancement of free radicals in tumor cells as a strategy to kill tumor cells both *in vitro* and *in vivo,* in *Biological Oxidation Systems,* Reddy, C.C. et al., Eds., Academic Press, New York, 1990, p. 607.
142. Das, U.N., Essential fatty acids, lipid peroxidation and apoptosis, *Prostaglandins Leukotr. Essent. Fatty Acids,* 61, 157, 1999.
143. Das, U.N., Gamma-linolenic acid, arachidonic acid and eicosapentaenoic acid as potential anticancer drugs, *Nutrition,* 6, 419, 1990.
144. Cheeseman, K.H. et al., Studies on lipid peroxidation in normal and tumor tissues, *Biochem. J.,* 250, 247, 1988.
145. Masotti, L., Cassali, E., and Galeotti, T., Lipid peroxidation in tumor cells, *Free Radic. Biol. Med.,* 4, 377, 1988.
146. Gerber, M. et al., Relationship between vitamin E and polyunsaturated fatty acids in breast cancer: nutritional and metabolic aspects, *Cancer Res.,* 64, 2347, 1989.
147. Tanaka, H. et al., The effect of PGD2 and 9-deoxy-9PGD2 on colony formation of murine osteosarcoma cells, *Prostaglandins,* 30, 167, 1985.
148. Sakai, T. et al., Prostaglandin D2 inhibits the proliferation of human malignant tumor cells, *Prostaglandins,* 27, 17, 1984.
149. Myers, R.L. and Siegal, M.I., The appearance of phospholipase activity in the human macrophage-like cell line U937 during dimethyl sulfoxide-induced differentiation, *Biochem. Biophys. Res. Commun.,* 30, 167, 1984.
150. Bonser, R.W. et al., The appearance of phospholipase and cyclo-oxygenase activities in the human promyelocytic leukemia cell line HL-60 during bimethyl sulfoxide-induced differentiation, *Biochem. Biophys. Res. Commun.,* 98, 614, 1981.
151. Dunbar, L.M. and Bailet, J.M., Enzyme deletion and essential fatty acid metabolism in cultured cells, *J. Biol. Chem.,* 250, 1152, 1975.
152. Das, U.N., Devi, G.R., and Res, K.P., Benzo(a)pyrene and gamma-radiation induced genetic damage in mice can be prevented by gamma-linolenic acid but not by arachidonic acid, *Nutr. Res.,* 5, 101, 1985.
153. Renner, H.W. and Declincee, H., Different anti-mutagenic acitons of linoleic and linolenic acid derivatives on Busulfan-induced chromosomal damage to human lymphocytes *in vitro, Nutr. Rep. Int.,* 36, 1276, 1987.
154. Das, U.N., Tumoricidal action of cis-unsaturated fatty acids and its relationship to free radicals and lipid peroxidation, *Cancer Lett.,* 50, 235, 1991.
155. Eriksson, L. et al., Distinctive biochemical pattern associated with resistance of hepatocytes in hepatocyte nodules during liver carcinogensis, *Environ. Health Perspect.,* 49, 171, 1983.

156. Faber, E., The biochemistry of generation of preneoplastic liver: a common metabolic pattern in hepatocyte nodules, *Can. J. Biochem Cell Biol.* 62, 486, 1984.
157. Sun, Y., Free radicals, antioxidant enzymes and carcinogenesis, *Free Radic. Biol. Med.,* 8, 583, 1990.
158. Oberlcy, L.W. and Buettner, G.R., Role of superoxide dismutase in cancer: a review, *Cancer Res.,* 39, 1141, 1979.
159. De Vries, C.E.E. and Van Noorden, C.J.F., Effects of fatty acid composition on tumor growth and metastatis, *Anticancer Res.,* 12, 1513, 1992.
160. Slater, T.F., Free radicals and tissue injury: fact and fiction, *Br. J. Cancer,* 55, Suppl., 8, 5, 1987.
161. Slater, T.F. et al., Studies on the hyperplasia (regeneration) of the rat liver, following partial hepatectomy: changes in lipid peroxidation and general biochemical aspects, *Biochem. J.,* 265, 51, 1990.
162. Calorini, L. et al., Lipid characteristics of RSV-transformed Balb/c 3T3 cell lines with different spontaneous metastatic potentials, *Lipids,* 24, 685, 1989.
163. Bennet, A., Prostanoids and cancer, *Ann. Clin. Res.,* 16, 314, 1984.
164. Dragao, J.R., Rohner, T.J., and Demers, L.M., The synthesis of prostaglandins by the NB rat prostate tumor, *Anticancer Res.,* 5, 393, 1985.
165. Fulton, A.M., Effects of indomethacine on the growth of cultured mammary tumors, *Cancer Res.,* 44, 2416, 1984.
166. El-Ela, S.H.A. et al., Effects of dietary primrose oil on mammary tumorgenesis induced by 7,12-dimethyl benz(a)anthracene, *Lipids,* 22, 1041, 1987.
167. Cai, J., Jiang, W.G., and Mansel, R.E., Inhibition of angiogenic factor- and tumor-induced angiogenesis by gamma linolenic acid, *Prostaglandins Leukotr. Essent. Fatty Acids,* 60, 21, 1999.
168. Sangeetha, P.S. and Das, U.N., Gamma linolenic acid and eicospentaenoic acid potentiate the cytotoxicity of anti-cancer drugs on cervical carcinoma (HeLa) cells *in vitro, Med. Sci. Res.,* 21, 457, 1993.
169. Vartak, S., Robbins, M.E., and Spector, A.A., Polyunsaturated fatty acids increase the sensitivity of 36B10 rat astrocytoma cells to radiation–induced cell kill, *Lipids,* 32, 283, 1997.
170. Das, U.N., Prakash, V.K.S., and Reddy, D.R., Local application of γ-linolenic acid in the treatment of human gliomas, *Cancer Lett.,* 94, 147, 1995.
171. Bakashi, A. et al., γ-Linolenic acid therapy of human gliomas, *Nutrition* 19, 305, 2003.
172. Kenny, F.S. et al., Gamma linolenic acid with tamoxifen as primary therapy in breast cancer, *Int. J. Cancer,* 85, 643, 2000.
173. Menendez, J.A. et al., Effects of gamma linolenic acid and oleic acid on paclitaxel cytotoxicity in human breast cancer cells, *Eur. J. Cancer,* 31, 402, 2001.
174. Tilvis, R.S. and Miettinen, T.A., Fatty acid composition of serum lipids, erythrocytes and platelets in insulin dependant diabetic women, *J. Clin. Endocrinol. Metab.,* 61, 741, 1985.
175. Arisaka, M. et al., Prostaglandin metabolism in children with diabetes mellitus, I. Plasma prostaglandin E_2, $E_{2\alpha}$, TBX_2 and serum fatty acid levels, *J. Pediatr. Gastroenterol. Nutr.,* 5, 878, 1986.
176. Mercuri, O., Peluffo, R.O., and Brenner, R.R., Depression of microsomal desaturation of linoleic to gamma-linolenic acid in the alloxan diabetic rat, *Biochim. Biophys. Acta,* 116, 409, 1966.
177. Poisson, J.P., Comparative *in vivo* and *in vitro* studies of the influence of experimental diabetes on rat liver linoleic acid delta-6- and delta-5-desaturation, *Enzyme,* 34, 1, 1985.

178. Morita, I. et al., Increased arachidonic acid content in platelet phospholipids from diabetic patients, *Prostaglandins Leukotr. Med.*, 11, 33, 1983.
179. Takahashi, R., Morse, N., and Horrobin, D.F., Plasma, platelet and arota fatty acids composition in response to dietary n-6 and n-3 fats supplementation in a rat model of non-insulin dependant diabetes, *J. Nutr. Sci. Vitaminol.*, 34, 413, 1988.
180. Van Doormaal, J.J., Idema, I.G., and Muskiet, F.A.J., Effects of short term high dose intake of evening primrose oil on plasma and cellular fatty acid compositions, alpha tocopherol levels and erythropoiesin in normal and type I (insulin-dependant) diabetic men, *Diabetologia*, 31, 576, 1988.
181. Uccella, R., Contini, A., and Sartorio, M., Action of evening primrose oil on cardiovascular risk factors in insulin-dependant diabetics, *Clin. Ter.*, 129, 381, 1989.
182. Arisaka, M., Arisaka, O., and Yamashiro, Y., Fatty acid and prostaglandin metabolism in children with diabetes mellitus II. The effect of evening primrose oil supplementation on serum fatty acid and plasma prostaglandin levels, *Prostaglandins Leukotr. Med.*, 43, 197, 1991.
183. Poisson, J.-P. et al., γ-Linolenic acid biosynthesis and chain elongation in fasting and diabetes mellitus, in *γ-Linolenic Acid: Metabolism and Its Roles in Nutrition and Medicine*, Huang, Y.-S. and Mills, D.E., Eds., AOCS Press, Champaign, IL, 1996, p. 252.
184. Horrobin, D.F., Gamma linolenic acid: an intermediate in essential fatty acid metabolism with potential as an ethical pharmaceutical and as a food, *Rev. Contemp. Pharmacother.*, 1, 1, 1990.
185. Horrobin, D.F., The use of gamma linolenic acid in diabetic neuropathy, *Agents Act. Suppl.*, 37, 120, 1992.
186. Zhu, X. and Eichberg, J., 1,2-diacylglycerol content and its arachidonyl-containing molecular species are reduced in sciatic nerve from streptoztozocin-induced diabetic rats, *J. Neurochem.*, 55, 1087, 1990.
187. Jamal, C.A. and Carmicheal, H., The effect of gamma linolenic acid on human diabetic peripheral neuropathy: a double-blind placebo-controlled trial, *Diabetic Med.*, 7, 319, 1990.
188. Dines, K.C., Cameron, N.E., and Cotter, M.A., Comparison of the effects of evening primrose oil and triglycerides containing GLA on nerve conduction and blood flow in diabetic rats, *J. Pharmacol. Exp. Ther.*, 273, 49, 1995.
189. Cameron, N.E. and Cotter, M.A., Comparison of the effects of ascorbyl gamma-linolenic acid and gamma-linolenic acid in the correction of neurovascular deficits in diabetic rats, *Diabetologia*, 39, 1047, 1996.

4 Purification of Free Fatty Acids via Urea Inclusion Compounds

Douglas G. Hayes

CONTENTS

4.1 INTRODUCTION

Urea inclusion compound (UIC)-based fractionation of free fatty acids (FFAs) has been employed for over 50 years on both analytical and preparative (kg) scales. This approach has potential value as a large-scale and continuous-mode process due to its mild operating conditions: low temperature and pressure, low cost, and its use of environmentally friendly and agriculturally derived materials (urea, ethanol, water). For these reasons, the method has recently received attention as a low-temperature, inexpensive, and environmentally favorable alternative or prefractionation step for molecular distillation in the purification of polyunsaturated FFAs (PUFAs). This chapter provides a physical description of UICs, their employment on a bench and preparative scale, and their potential use on a large scale for FFA fractionation and purification. A two-part review on UIC-based fatty acid fractionation has been recently published [1,2]; the goal of this review is to briefly summarize the review, provide updated information, and expand the discussion given therein on large-scale process development. In fact, Daniel Swern's review of UIC-based purification of lipids, written almost 50 years ago, remains a valuable tool to educate newcomers to the field [3]. In addition to serving as a means of downstream extractive purification, UICs also have potential applicability as a host system for chemical and free-radical polymerization reactions [4] and microscale chromatographic separations [5–7].

4.2 PHYSICAL DESCRIPTION OF UICS

Comprehensive reviews on the physical and chemical properties and behavior of UICs are available from the research groups of M.D. Hollingsworth at Kansas State University and K.D.M Harris at the University of Birmingham [4,8]. A short description will be given her.

At room temperature urea exists in tetragonal crystalline form; however, as discovered serendipitously by Bengen in the 1930s, in the presence of molecules with linear alkyl chains, urea forms long, needle-like clanthrates possessing hexagonal crystalline morphology [9,10] (Figure 4.1 [11]). The resultant UICs consist of a series of parallel, linear channels formed by a network of spiral, antiparallel, hydrogen-bonded strands of urea molecules with the linear alkyl chain inclusion compound guests residing in the channels' interior. Each oxygen atom of a urea molecule forms hydrogen bonds with four nitrogen atoms, and each nitrogen atom forms hydrogen bonds with two oxygen atoms. The inner diameter of the channels ranges between 0.55 and 0.58 nm. UICs are "incommensurate" (i.e., the vertical position of guest molecules in neighboring channels and the distance separating adjacent guests molecules in a given channel are variable). In other words, the channel walls can be considered "smooth," meaning that the "guests" will pack randomly with respect to the lengthwise direction of the channels. However, despite the incommensurate relationship, guest molecules in adjacent tunnels influence the chemical behavior of each other in a manner not fully understood [12]. Other physical and thermodynamic properties of UICs laden with FFA guests are listed in Table 4.1 [3,4,13–19].

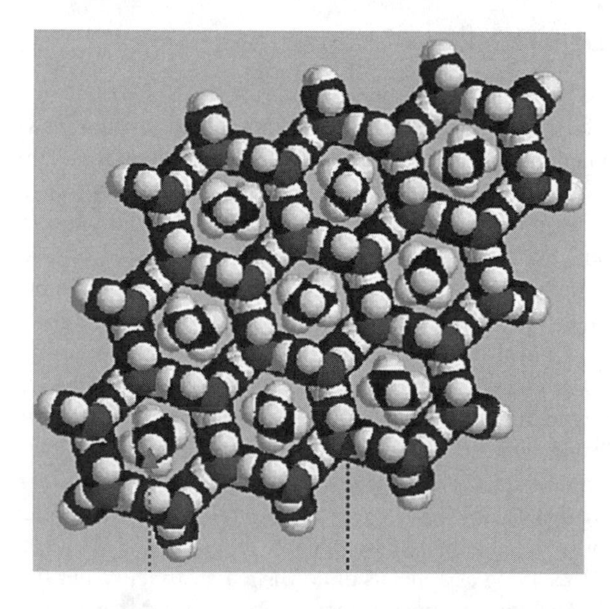

FIGURE 4.1 Front view of a UIC containing a hydrocarbon guest. Leftmost and rightmost arrows depict a UIC hydrocarbon guest and a urea molecule, respectively. Atoms depicted as white, black, red, and blue represent H, C, N, and O, respectively. (From Lee, S.-O. and Harris, K.D.M., *Chem. Phys. Lett.* 307(5,6), 327–332, 1999. With permission.)

TABLE 4.1
Physical and Thermodynamic Properties of Oleic Acid-Laden UICs

Property	Value for UICs	Value for Urea
Crystal type	Hexagonal [13]	Tetragonal [13]
Lattice parameters	$a = b = 0.823$ nm; $c = 1.1005$ nm [13]	$a = b = 0.567$ nm, $c = 0.4726$ nm [13]
Channel inner diameter	0.55–0.58 nm [4]	
Channel outside diameter	0.822 nm [4]	
Moles of urea per unit cell	6.0 [13]	4.0 [13]
Methylene groups of host per unit cell	3.5 [13]	
Moles of urea per mole of FFA[a]	12.2 [14]; 13.6 [15]	
Mass of urea per mass of fatty acid guest	3.0 ± 0.3 [4]	
Length of alkyl chain guests inside UIC channels	0.05 nm shorter than lengths of molecules in extended, linear conformations [4]	
Distance between adjacent UIC guests in the same channel	0.24 nm [16]	
Specific gravity	1.312 [13] 1.20 [17]	1.323 [17]
Enthalpy of formation[a]	114.7 kJ/mol of guest [14] 7.0 kJ/mol of urea [18]	14.4 kJ/mol of urea [18]
Change in enthalpy of formation per mole of CH_2	6.7 kJ/mol [4]	
Heat capacity of UICs at 23°C[b]	127.1 J/(mol urea* K) [19]	
Dissociation temperature[a]	110°C [3,15]	132.7°C [18]

[a] Ratio is proportional to alkyl chain length of guest.
[b] With hexadecane guest.

Inclusion compounds are also formed using homologues of urea. For instance, thiourea produces inclusion compounds with larger channel diameters than UICs (0.61 nm compared with 0.55 to 0.58 nm for urea), allowing for selective removal of branched and cyclic molecules from linear molecules, the latter of which are too narrow in width to serve as host [4,8]. Selenourea also forms inclusion compounds [4,8]. Furthermore, urea forms inclusion compounds of alternate structure with other small molecules, such as diacids and hydrogen peroxide [20–23].

4.3 BASIS OF SELECTIVITY FOR UIC-BASED FRACTIONATION

As introduced above, UIC formation occurs most readily when the UIC guest contains a long alkyl chain in an extended planar zigzag conformation; moreover, short-chain length, branching, cyclic groups, and multiple double bonds make UIC formation less energetically favorable. Branched molecules are more likely to form UICs when the branched group is located near one of the terminal positions of the guest's backbone and when a backbone carbon does not contain multiple branched groups. (Inclusion guests with bulky or hydrogen-bonding groups near the C_1 position, such as 2-undecanone, often form flat plate hexagonal crystals due to the hindrance of vertical crystal growth by the bulky terminal groups.) Guests with *trans* double bonds are more likely to form UICs than *cis*-containing guests due to the greater molecular linearity of the former. There usually exists a minimum chain length for a series of homologous organic molecules in order for UICs to occur. For instance, alkanes with chain length of C_5 or smaller do not form UICs, presumably because the chemical potential driving force for UIC formation is insufficiently small. A guest's stereochemistry may also be a factor, as suggested recently [24,25]. However, "poor" guests are often incorporated into UIC*s* in the presence of "good" guests, perhaps because bulk phase molecules replace guests in the UIC channels [5–7] or because poor guests become entrapped randomly in the clathrates during UIC formation.

Linear polymers such as poly(caprolactone), poly(ethylene glycol), and poly-L-lactic acid, as well as block copolymers containing linear polymer segments, also form UICs, some of which possess nonhexagonal crystalline structure [26]. Polymers recovered from UICs frequently exist in a highly crystalline morphological state.

4.4 METHODOLOGY FOR BENCH-SCALE UIC-BASED FRACTIONATION

The overall process consists of the following steps:

1. Cosolubilization of urea and the molecules to be fractionated (e.g., FFA) via use of a solvent and/or addition of heat
2. Heat removal to induce UIC formation
3. Sedimentation to separate the UICs, consisting solely of urea and linear and long-chain FFA (perhaps mixed with tetragonal urea crystals), from a solvent-rich phase containing urea, PUFA, and short-chain FFA
4. Isolation of the FFA product from the UICs and/or the solvent-rich phase

"Step zero," the prepurification of feed material, may be required. For instance, refining steps such as degumming are important for a crude FFA source since sulfur-containing compounds, peroxides, surfactants, and phospholipids inhibit UIC formation [27,28]. Perhaps the inhibitors function by increasing the solubility of the guest molecules in the bulk solvent-rich phase, reducing their propensity to partition to the UIC channels.

An important ingredient required for effective fractionation is the solvent, whose role is to cosolubilize urea and the molecules to be fractionated, for example, FFA, prior to UIC formation. The best solvents are polar and lack the ability to serve as UIC guest (e.g., methanol, ethanol, dioxane, and methylene chloride). The absence of solvent can result in the occurrence of complicated phase behavior [29,30].

The nature of the temperature programming to be employed during the cooling process is an important consideration. Most papers have employed a slow cooling process. Lee reports that a rapid temperature reduction leads to the coprecipitation of UICs and tetragonal urea, hence reducing extractive efficiency; however; as conceded by the author, a slow cooling process has disadvantages of long processing time and susceptibility to oxidative degradation for unsaturated FFA due to the long exposure at elevated temperatures [31]. To concur, Hayes and co-workers have performed preliminary investigations that demonstrate the amount of UICs formed for a given set of operating conditions decreases as the rate of cooling decreases (unpublished data). Lee recommends a moderate cooling rate of about 0.5 degrees per minute and the slow addition of FFA in small batch increments as a means of minimizing the disadvantages of both rapid and slow cooling processes [31]. Hayes, Van Alstine, and co-workers report that a rapid cooling process (cooling from ~60°C to room temperature within a minute) is sufficiently selective, efficient, and precise for FFA fractionation [27,32,33].

If desired, the occurrence of UICs can be verified by a variety of techniques, such as light microscopy, differential scanning calorimetry (DSC), x-ray diffraction, and Fourier transform infrared (FT-IR) spectroscopy. DSC can also be used to detect the presence of tetragonal urea and other molecules that do not reside in the UIC channels. Examples of the application of these techniques to analysis of UICs are given in [34–36] and discussed as an overview in [1].

To recover guest molecules from UICs, three different approaches can be employed, all of which lead to UIC decomposition:

1. An extractant is used in which the UIC guest is highly soluble; the extractant should not be capable of serving as UIC guest (e.g., isooctane).
2. A polar extractant is used that selectively removes urea (e.g., warm water, which should have a pH below 7 for use with FFA purification to prevent saponification).
3. Heat is applied to increase the UICs' temperature above their decomposition temperature, the latter of which is typically between 100 and 135°C (Table 4.1).

Usually, choice 2 is the simplest and most cost effective; however, a disadvantage of this approach is the difficulty and/or expense required for recovery of urea. Choice 1 is usually not practical due to the high cost of the solvent and the requirement of an additional process step to isolate the guest from the solvent. Choice 3 is rarely chosen due to the high energy cost and the possible thermal degradation of guest molecules and urea.

The recovery of guest molecules from the solvent-rich phase must also be carefully considered, since the solution also contains urea. The recommended procedure is to first remove solvent via evaporation, resulting in a physical solid-phase mixture of urea and guest molecules. Then warm water and/or guest-selective solvent (e.g., hexane for FFA) is applied to separate urea from the guest. For ethanol-rich solutions resulting from UIC-based fractionation of FFA, the direct addition of warm (~50°C), slightly acidic water (in the absence of solvent evaporation) resulted in the precipitation of ~75% of the FFA, which was recovered by sedimentation or extraction with hexane [32]. However, removal of the solvent via evaporation under reduced pressure prior to water addition increased the recovery of FFA to near 100% [32].

4.5 UIC-BASED FFA FRACTIONATION: APPLICATIONS

Molecular distillation, the most common method for fractionating FFA for decades [37], provides high selectivity, but, as a result of its high operating temperature (typically between 100 and 200°C) and ultralow pressure, it leads to high operating costs, possible thermal degradation of PUFA and oxygenated FFA species, and safety concerns. UIC-based fractionation does not possess the degree of selectivity of molecular distillation, but employs a much lower operating temperature and atmospheric pressure. Since the process involves relatively benign operating conditions and chemicals that are "generally regarded as safe" (GRAS) by the U.S. Food and Drug Administration (FFA, urea, water, ethanol), it can be labeled as "ecofriendly," and it can be safely employed in rural settings. Although its selectivity is lower than molecular distillation, it can be used as a preliminary fractionation step in conjunction with other separation methods, such as low-temperature crystallization, lipase-selective esterification, and chromatography, resulting in a high-purity FFA product. A multiple-stage UIC process greatly enhances FFA product purity.

UIC-based FFA fractionation has been employed for the following general purposes:

1. It is to isolate PUFA or ring-containing FFA of interest in the solvent-rich phase by forming saturated and monounsaturated FFA-rich UICs.
2. It is to improve the quality of food-related FFA by removing saturated FFA-rich UICs.
3. It is to improve the oxidative stability of food-related FFA by selectively forming a solvent-rich phase containing PUFA such as linoleic and α-linolenic acids. The FFA product resides in the UICs, resulting in the former's protection from auto-oxidation agents such as O_2 [38,39].

Specific examples of UIC-based FFA purification are given in Table 4.2 [27,32,40–55]. The most frequently encountered application is the isolation of PUFA from fish oils. Several of the citations employed a UIC-based fractionation in two stages to first remove saturated FFA, and then PUFA [31,49,53,54]. Note that UIC fractionation is also effective in fractionating fatty acid esters (e.g., methyl esters) and fatty alcohols [3].

TABLE 4.2
UIC-Based Fractionation of FFA

FFA Source[a]	Application
Fish oil	Isolation of AA, EPA, and DHA [27,40–45]
Black currant or borage oil	Isolation of GLA [27,46,47]
Conjugated linoleic acid reaction mixture	Isolation: $18{:}2^{9c,11t}$ and $18{:}2^{10t12c}$ [48,49]
Linseed oil	Isolation of ALA [27,50,51]
Bombax munguba (cotton) oil	Isolation of malvalic FFA [52]
Sterculia foetida (kapok) oil	Isolation of sterculic FFA [52]
Rapeseed oil	Removal of saturated FFA [32]
Meadowfoam oil	Fractionation: 20:1 [5] and 22:2 [5,13,27]
Milk fat	Separation of long-chain saturates from short-chain saturates and C_{14}–C_{18} unsaturates [53,54]
Shark liver oil 1-*O*-alkylglycerol ether lipids	Separation of saturated and unsaturated ether lipids [55]

[a] ALA, GLA, AA, EPA, DHA, malvalic, and sterculic acids refer to α-linolenic ($18{:}3^{9c,12c,15c}$), γ-linolenic ($18{:}3^{6c,9c,12c}$), arachidonic acid ($20{:}4^{5c,8c,11c,14c}$), eicosapentaenoic acid ($20{:}5^{5c,8c,11c,14c,17c}$), docosahexaenoic acid ($22{:}6^{4c,7c,10c,13c,16c,19c}$), *cis*-8,9-methylene heptadec-8-enoic, and *cis*-9,10-methylene octadec-9-enoic acid, respectively.

Source: From Hayes, D.G., *INFORM* 13 (Nov), 832–834, 2002. With permission.

4.6 CONSIDERATIONS FOR LARGE-SCALE UIC-BASED FATTY ACID FRACTIONATION

Bench-scale experiments should be performed first to help select operating conditions to be used for scale-up. With regard to selecting the amount of UICs that are formed, there exists a trade-off between product purity and yield. For instance, if one's product of interest is PUFA collected from the solvent-rich phase, as the overall weight fraction of UICs increases, the purity of the PUFA in the desired product increases at the expense of product yield. For recovery of product from the UICs, the purity increases and the recovery decreases as the weight fraction of UICs decreases [32]. The weight fraction of UICs is increased by increasing the initial weight fraction of urea, decreasing the initial weight fraction of solvent, decreasing the solvent's polarity (i.e., its solubility for urea), and decreasing the final temperature during the cooling process.

To predict the outcome of a UIC-based FFA fractionation, the author has developed a mathematical model based on lever rule equations derived from experimentally determined phase diagrams at several different temperatures, mass balances, and experimentally derived partition coefficients for the FFA species, which were dependent only upon the percentage uptake of total FFA by the UICs [33]. This approach was applied successfully to predict the resultant UIC and solvent-rich phase compositions for FFA derived from low erucic acid rapeseed oil/ethanol/water/urea system employing rapid cooling during the UIC formation process and water to ethanol ratios of 10:90 v/v or less [33]. Moreover, the bench-scale experiments provided data to create the phase diagrams and partition coefficients. In addition, phase equilibria were determined by measuring the UIC formation temperature as a function of the overall mixture's composition during slow cooling.

To date, UIC-based FFA fractionation has been employed at the kilogram scale [41] or smaller. A possible process plant design for large-scale continuous-mode UIC FFA fractionation is depicted in Figure 4.2. Low temperature and pressure would be employed throughout. The main unit operations of the process are a mixing vessel where the four components (FFA, urea, water, and ethanol) are cosolubilized at an elevated temperature (~50°C), followed by a crystallizer held at a reduced temperature (e.g., 25°C). Heat integration is possible in this design, where a water stream that removes heat in the crystallizer can be used to provide heat for the mixing vessel. The slurry that exits the crystallizer is separated into liquid and solid (UIC) phases by centrifugation. The FFA product will more frequently be in the former phase. The latter phase perhaps can be used directly as a fertilizer feedstock, or separated into FFA enriched in saturates and urea using warm water to extract the latter from the UICs. The FFA product is isolated by first evaporating away solvent (and a partial amount of urea since the later is slightly volatile); then, the remaining urea is extracted away from the FFA using water. The urea-rich water stream and the solvent-rich evaporate are then recycled back to the process. Fresh urea must be added to the system to compensate for the loss of urea in the UIC product. As an alternative process configuration, the FFA source can be purified using a pseudo-multistage scheme, where the FFA product/solvent-rich phase leaving the centrifuge is recycled back to the heated mixing vessel rather than subjected to evaporation (dotted line in Figure 4.2).

A key process parameter is the solvent type. The ideal solvent must be inexpensive, safe, and highly miscible with both urea and FFA and cannot serve as a UIC guest, as discussed above. Ethanol or ethanol/water mixtures rich in ethanol (such as the azeotropic 95% ethanol mixture) are recommended by this author. However, a concern for using ethanol (or methanol) is the formation of ethyl (or methyl) carbamate, a possible carcinogen. Carbamates form from a reaction between ethanol (or methanol) and urea at elevated temperatures [56,57]. It is possible that carbamate formation may occur in the mixing vessel and/or the evaporator in the proposed process scheme depicted in Figure 4.2, depending upon the temperature and the residence time. The presence of carbamate will be tested in the author's laboratory.

4.7 CONCLUSIONS

UIC-based fractionation is a valuable process step in the purification of FFA. Although currently employed at a bench-to-preparative scale, it can be readily applied to a large scale since it requires commonly used chemical processing equipment. The importance of UIC-based fractionation is expected to increase due to the increased need for energy efficiency, environmental friendliness, and increased development of novel FFA via enhanced capabilities in gene expression in plants and microorganisms.

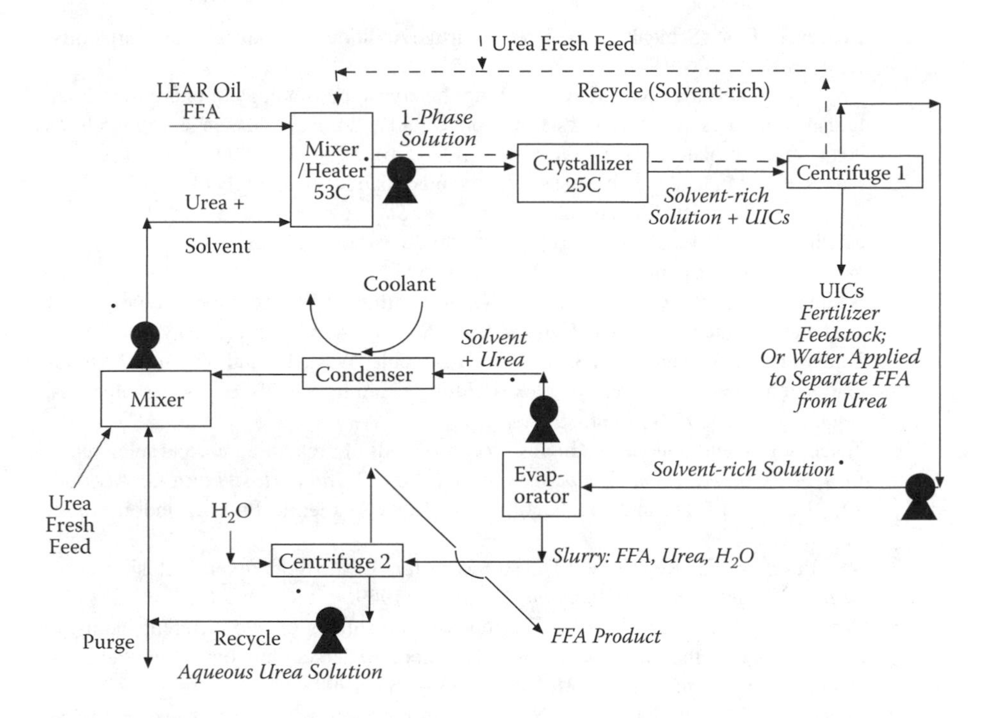

FIGURE 4.2 Proposed large-scale UIC-based FFA fractionation scheme.

REFERENCES

1. Hayes, D.G., Urea inclusion compound formation, *INFORM* 13 (Oct.), 781–783, 2002.
2. Hayes, D.G., Free fatty acid fractionation via urea inclusion compounds, *INFORM* 13 (Nov.), 832–834, 2002.
3. Swern, D., Urea and thiourea complexes in separating organic compounds, *Ind. Eng. Chem.* 47, 216–221, 1955.
4. Hollingsworth, M.D. and Harris, K.D.M., Urea, thiourea, and selenourea, in *Solid-State Supramolecular Chemistry: Crystal Engineering,* MacNicol, D.D., Toda, F., and Bishop, R., Eds., Pergamon Press, Oxford, 1996, pp. 177–237.
5. Mahdyafar, A. and Harris, K.D.M., Evidence for migration of molecules into the tunnel structure of urea inclusion compounds, *J. Chem. Soc., Chem. Commun.* 51–53, 1993.
6. Khan, A.A., Bramwell, S.T., Harris, K.D.M., Bariuki, B.M., and Truter, M.R., The design of a molecularly selective capillary based on an incommensurate intergrowth structure, *Chem. Phys. Lett.* 307, 320–326, 1999.
7. Deyhimi, F., Mermoud, F., Gülacar, F.O., and Buchs, A., Separation of hydrocarbon mixtures in the microgram range by inclusion in a urea-packed minicolumn, *Int. J. Environ. Anal. Chem.* 18, 75–85, 1984.
8. Harris, K.D.M., Understanding the properties of urea and thiourea inclusion compounds, *Chem. Soc. Rev.* 26, 279–289, 1997.
9. Bengen, M.F., Mein Weg zu den neuen Harnstoff-Einschlub-Berbindungen, *Angew. Chem.* 63, 207–208, 1951.

10. Bengen, M.F. and Schlenk, W., Jr., Uber neuartige Additionsverbindungen des Harnstoffs, *Experientia* 5, 200, 1949.
11. Lee, S.-O. and Harris, K.D.M., Controlling the crystal morphology of one-dimensional tunnel structures: induced crystallization of alkane/urea inclusion compounds as hexagonal flat plates, *Chem. Phys. Lett.* 307 (5,6), 327–332, 1999.
12. Harris, K.D.M., Investigating the structure and dynamics of a family of organic solids: the alkane/urea inclusion compounds, *J. Solid State Chem.* 106, 83–98, 1993.
13. Smith, A.E., The crystal structure of the urea-hydrocarbon complexes, *Acta Crystallogr., Sect. A: Found. Crystallogr.* 5, 224–235, 1952.
14. Redlich, O., Gable, C.M., Dunlop, A.K., and Millar, R.W., Addition compounds of urea and organic substances, *J. Am. Chem. Soc.* 72, 4153–60, 1950.
15. Knight, H.B., Witnauer, L.P., Coleman, J.E., Noble, W.R., Jr., and Swern, D., Dissociation temperatures of urea complexes of long-chain fatty acids, esters, and alcohols, *Anal. Chem.* 24, 1331–1334, 1952.
16. Takemoto, K. and Sonoda, N., Inclusion compounds of urea, thiourea, and selenourea, in *Aspects of Inclusion Compounds Formed by Organic Host Lattices,* Atwood, J.L., Davies, J.E.D., and MacNicol, D.D., Eds., Academic Press, London, 1984, pp. 44–67.
17. Van Bekkum, H., Remijnse, J.D., and Wepster, B.M., Selenourea inclusion compounds, *Chem. Commun. (London)* (2), 67–8, 1969.
18. White, M.A. and Harnish, R.S., Thermodynamic studies of two different inclusion compounds with the same guest: toward a general understanding of melting behavior in binary compounds, *Chem. Mater.* 10, 833–839, 1998.
19. Penberton, R.C. and Parsonage, N.G., Thermodynamic properties of urea + hydrocarbon adducts. Part 1. Heat capacities of the adducts of n-$C_{10}H_{22}$, n-$C_{12}H_{26}$, n-$C_{16}H_{34}$, and n-$C_{20}H_{42}$ up to 300 K, *J. Chem. Soc., Faraday Trans.* 61, 2112–2121, 1963.
20. Li, Q. and Mak, T.C.M., Hydrogen-bonded urea-anion host lattices. Part 4. Comparative study of inclusion compounds of urea with tetraethylammonium and tetraethylphosphonium chlorides, *J. Inclusion Phenom.* 28, 151–161, 1997.
21. MacNicol, D.D., Toda, F., and Bishop, R., *Solid-State Supramolecular Chemistry: Crystal Engineering,* Pergamon Press, Oxford, UK, 1996.
22. Dobado, J.A., Molina, J., and Portal, D., Theoretical study on the urea-hydrogen peroxide 1:1 complexes, *J. Phys. Chem. A* 102, 778–784, 1998.
23. Smith, G., Kennard, C.H.L., and Byriel, K.A., The preparation and crystal structures of a series of urea adducts: with fumaric acid (2:1), with itaconic acid (1:1), and with cyanuric acid (1:1), *Aust. J. Chem.* 50, 1021–1025, 1997.
24. Yeo, L. and Harris, K.D.M., Chiral recognition in incommensurate one-dimensional inclusion compounds: a computational investigation, *Tetrahedron Asym.* 7 (7), 1891–1894, 1996.
25. Yeo, L. and Harris, K.D.M., Temperature-dependent structural properties of a solid urea inclusion compound containing chiral guest molecules: 2-bromotetradecane/urea, *Can. J. Chem.* 77, 2105–2118, 1999.
26. Huang, L. and Tonelli, A.E., Polymer inclusion compounds, *J. Macromol. Sci, Rev.* C38 (4), 781–837, 1998.
27. Hayes, D.G., Van Alstine, J.M., and Setterwall, F.N., Urea-based fractionation of fatty acids and glycerides of polyunsaturated and hydroxy fatty acid seed oils, *J. Am. Oil Chem. Soc.* 77, 207–215, 2000.
28. Zimmerschied, W.J., Dinerstein, R.A., Weitkamp, A.W., and Marschner, R.F., Crystalline adducts of urea with linear aliphatic compounds: a new separation process, *Ind. Eng. Chem.* 42, 1300–1306, 1950.

29. Kuhnert-Brandstätter, M. and Burger, A., Phase diagrams of urea inclusion compounds. Part 1. Palmitic acid and urea, *Pharmazie* 51, 288–292, 1996.
30. Kuhnert-Brandstätter, M. and Burger, A., Phase diagrams of urea inclusion compounds. 2. Stearic acid and urea, *J. Therm. Anal.* 50, 559–567, 1997.
31. Lee, S.K., UD 6,664,405, 2003, Method for isolating high-purified unsaturated fatty acids using crystallization. November 28, 2001.
32. Hayes, D.G., Bengtsson, Y.C., Van Alstine, J.M., and Setterwall, F.N., Urea complexation for the rapid, cost effective, ecologically responsible fractionation of fatty acid from seed oil, *J. Am. Oil Chem. Soc.* 75, 1403–1408, 1998.
33. Hayes, D.G., Van Alstine, J.M., and Asplund, A.L., Triangular phase diagrams to predict the fractionation of free fatty acid mixtures via urea complex formation, *Sep. Sci. Technol.* 36, 45–58, 2001.
34. Vasanthan, N., Shin, I.D., Huang, L., Nojima, S., and Tonelli, A.E., Formation, characterization, and segmental mobilities of block copolymers in their urea inclusion compound crystals, *Macromolecules* 30, 3014–3025, 1997.
35. Vasanthan, N., Shin, I.D., and Tonelli, A.E., Conformational and motional characterization of isolated poly(ε-caprolactone) chains in their inclusion compound formed with urea, *Macromolecules* 27, 6515–6519, 1994.
36. Hayes, D.G., Asplund, A.L., Van Alstine, J.M., and Zhang, X., Molecular weight-based fractionation of poly-L and poly-D,L lactic acid via a simple inclusion compound based process, *Sep. Sci. Technol.* 37, 769–782, 2002.
37. Perry, E.S., Molecular Distillation of Lipids and Higher Fatty Acids, in *Methods of Enzymology, Vol. 3,* Colowick, S.P. and Kaplan, N.O, Academic Press, New York, 1957, pp. 383–391.
38. Schlenk, H. and Holman, R.T., Separation and stabilization of fatty acids by urea complexes, *J. Am. Chem. Soc.* 72, 5001–5004, 1950.
39. Schlenk, H. and Holman, R.T., The urea complexes of unsaturated fatty acids, *Science (Washington, DC, 1883-)* 112, 19–20, 1950.
40. Ju, Y.H., Huang, F.C., and Fang, C.H., The incorporation of n-3 polyunsaturated fatty acids into acylglycerols of borage oil via lipase-catalyzed reactions, *J. Am. Oil Chem. Soc.* 75, 961–965, 1998.
41. Ratnayake, W.M.N., Olsson, B., Matthews, D., and Ackman, R.G., Preparation of omega-3-PUFA concentrates from fish oils via urea complexation, *Fat Sci. Technol.* 90, 381–386, 1988.
42. Haagsma, N., van Gent, C.M., Luten, J.B., de Jong, R.W., and Van Doorn, E., Preparation of an ω3 fatty acid concentrate from cod liver oil, *J. Am. Oil Chem. Soc.* 59, 117–118, 1982.
43. Cartens, M., Molina Grima, E., Robles Medina, A., Giménez Giménez, A., and Ibáñez González, J., Eicosapentaenoic acid (20:5n-3) from the marine microalga *Phaeodactylum tricornutum*, *J. Am. Oil Chem. Soc.* 73, 1025–1031, 1996.
44. Ackman, R.G., Ratnayake, W.M.N., and Olsson, B., The "basic" fatty acid composition of Atlantic fish oils: potential similarities useful for enrichment of polyunsaturated fatty acids by urea complexation, *J. Am. Oil Chem. Soc.* 65, 136–138, 1988.
45. Gimenez, A.G., González, M.J.I., Medina, A.R., Grima, E.M., Salas, S.G., and Cerdán, L.E., Downstream processing and purification of eicosapentaenoic (20:5n-3) and arachidonic acids (20:4n-6) from the microalga *Porphyridium cruentum*, *Bioseparation* 7 (2), 89–99, 1998.
46. Shimada, Y., Sakai, N., Sugihara, A., Fujita, H., Honda, Y., and Tominaga, Y., Large-scale purification of γ-linolenic acid by selective esterification using *Rhizopus delemar* lipase, *J. Am. Oil Chem. Soc.* 75, 1539–1544, 1998.

47. Traitler, H., Willie, H.J., and Studer, A., Fractionation of blackcurrant seed oil, *J. Am. Oil Chem. Soc.* 65, 755–760, 1988.
48. Ma, D.W.L., Wierzbicki, A.A., Field, C.J., and Clandinin, M.T., Preparation of conjugated linolenic acid from safflower oil, *J. Am. Oil Chem. Soc.* 76 (6), 729–730, 1999.
49. Kim, Y.J., Lee, K.W., Lee, S., Kim, H., and Lee, H.J., The production of high-purity conjugated linoleic acid (CLA) using two-step urea-inclusion crystallization and hydrophilic arginine-CLA complex, *J. Food Sci.* 68 (6), 1948–1951, 2003.
50. Newey, H.A., Shokal, E.C., Mueller, A.C., Bradley, T.F., and Fetterly, L.C., Drying oils and resins. Segregation of fatty acids and their derivatives by extractive crystallization with urea, *Ind. Eng. Chem.* 42, 2538–2541, 1950.
51. Swern, D. and Parker, W.E., Application of urea complexes in the purification of fatty acids, esters, and alcohols. III. Concentrates of natural linoleic and linolenic acids, *J. Am. Oil Chem. Soc.* 30, 5–7, 1953.
52. Fehling, E., Schonwiese, S., Klein, E., Mukherjee, K.D., and Weber, N., Preparation of malvinic and sterculic acid methyl esters from *Bombax munguba* and *Serculia foetida* seed oils, *J. Am. Oil Chem. Soc.* 75 (12), 1757–1760, 1998.
53. Kim, Y.J. and Liu, R.H., Selective increase in conjugated linoleic acid in milk fat by crystallization, *J. Food Sci.* 64 (5), 792–795, 1999.
54. Kim, Y.J., Lee, K.W. and Lee, H.J., Increase of conjugated linoleic acid level in milk fat by bovine feeding regimen and urea fractionation, *J. Microbiol. Biotechnol.* 13 (1), 22–28, 2003.
55. Bordier, C.G., Sellier, N., Foucault, A.P., and Le Goffic, F., Purification and characterization of deep sea shark *Centrophorus squamosus* liver oil 1-O-alkylglycerol ether lipids, *Lipids* 31 (5), 521–528, 1996.
56. Adams, P. and Baron, F.A., Esters of carbamic acid, *Chem. Rev.* 65, 567–602, 1965.
57. Canas, B.J. and Yurawecz, M.P., Ethyl carbamate formation during urea complexation for fractionation of fatty acids, *J. Am. Oil Chem. Soc.* 76, 537, 1999.

5 Solvent Extraction to Obtain Edible Oil Products

Phillip J. Wakelyn and Peter J. Wan

CONTENTS

5.1 INTRODUCTION

Oils and fats are recovered from diverse biological sources by mechanical separation, solvent extraction, or combination of the two methods [1,2]. These materials include animal tissues (e.g., beef, chicken, and pork); crops specifically produced for oil or protein (e.g., soy, sunflower, safflower, rape/canola, palm, and olive); by-products of crops grown for fiber (e.g., cottonseed and flax); crops for food and their coproducts (e.g., corn germ, wheat germ, rice bran, coconut, peanuts, sesame, walnuts, and almonds); nonedible oils and fats (castor, tung, and jojoba); and other oil sources (oils and fats from microbial products, algae, and seaweed). There are many physical and chemical differences among these diverse biological materials. However, the similarities are that oils (edible and industrial) and other useful materials (e.g., vitamins, nutraceuticals, fatty acids, and phytosterols) can be extracted from these materials by mechanical pressing, solvent extracting, or combination of pressing and solvent extraction. The preparation of the various materials to be extracted varies. Some need extensive cleaning, drying (optional), fiber removal (cottonseed), dehulling, flaking, extruding, and the like, all of which affect the solvent–substrate interaction and, therefore, the yield, composition, and quality of the oils and other materials obtained.

Historically, the advancement of processing technology for recovering oils and other useful materials has been primarily driven by economics. Each extraction process was optimized through trial and error with the available technology to produce maximum yield of high-quality products at the lowest cost. For thousands of years, stone mills, and for several centuries, simple hydraulic or lever presses, were used as batch systems. The continuous mechanical presses became reality only during the early 1900s. It was not until the 1930s that extraction solvents were used more widely, which greatly enhanced the recovery of oil from oilseeds or other oil-bearing materials. In recent years, safety, health, and environmental regulations for the solvents used for extracting oil from oilseeds have prompted research efforts to find solvents to replace commercial hexane. These solvents, including ethanol, isopropanol, water, and supercritical carbon dioxide, are technically feasible as oil extraction solvents, but at present are economically unacceptable [1]. Recent research to use commercial isohexane in two separate cottonseed oil mills [3,4] has demonstrated energy savings and throughput increases. In the near term, it appears that commercial isohexane can be used as an alternative to commercial hexane. Research continues on acetone, since it is not regulated by the United States Environmental Protection Agency (U.S. EPA) as a hazardous air pollutant (HAP) or as a volatile organic compound (VOC) that is a precursor of ozone [5].

While oil extraction has a long development history, the oil refining methods were largely introduced during the 20th century. Until the recent past, crude oil production and oil refining were two separate industries. However, during the last quarter century, shear economics and product synergy have caused both horizontal (merge of similar operations) and vertical integration (combination of different but related operations) of these businesses to occur. Now many companies do both crushing and various degrees of oil refining, such as water degumming and caustic or physical refining. To ensure consistent oil quality, most of the cottonseed crushers in the United States incorporate miscella (oil–solvent mixture) refining, and some cottonseed companies further process the oil to final marketable refined, bleached, deodorized (RBD) oil.

Economic forces and governmental regulations will likely continue to prompt many changes in both oil extraction and oil refining industry. A general overview of the critical steps of the oil extraction and oil refining processes using soybean and cottonseed as examples are presented, and regulatory concerns and toxicity of extraction solvents (commercial hexane and other potential alternative solvents) are also discussed.

5.2 OIL EXTRACTION PROCESS

Four types of processing systems are used to extract oil from oil-bearing materials: hydraulic press, expeller or screw press, prepress solvent extraction, and direct solvent extraction. Selection of the process scheme is usually dictated by the oil content of the starting materials. High oil-containing materials such as canola or rapeseed, sunflower seed, corn germ, and peanut usually use the prepress solvent extraction process to produce oil and meal. Cottonseed kernel contains close to 34% oil, which is high enough for prepress solvent extraction and is now mostly processed with direct solvent extraction in the United States. Seed, such as soybean, contains less than 30% oil, and is routinely extracted by direct solvent extraction. Prior to extraction, oil-bearing materials have to be prepared for extraction to separate the crude oil from the meal [1,2] (see Figure 5.1). Careful control of moisture and temperature during processing must be exercised to maintain the quality of the protein in the meal, to minimize the damage to the oil, and to maximize oil extraction. Crude oils are refined by conditioning with phosphoric acid to make phospholipids more hydratable for removal and treating with sodium hydroxide (alkali or caustic

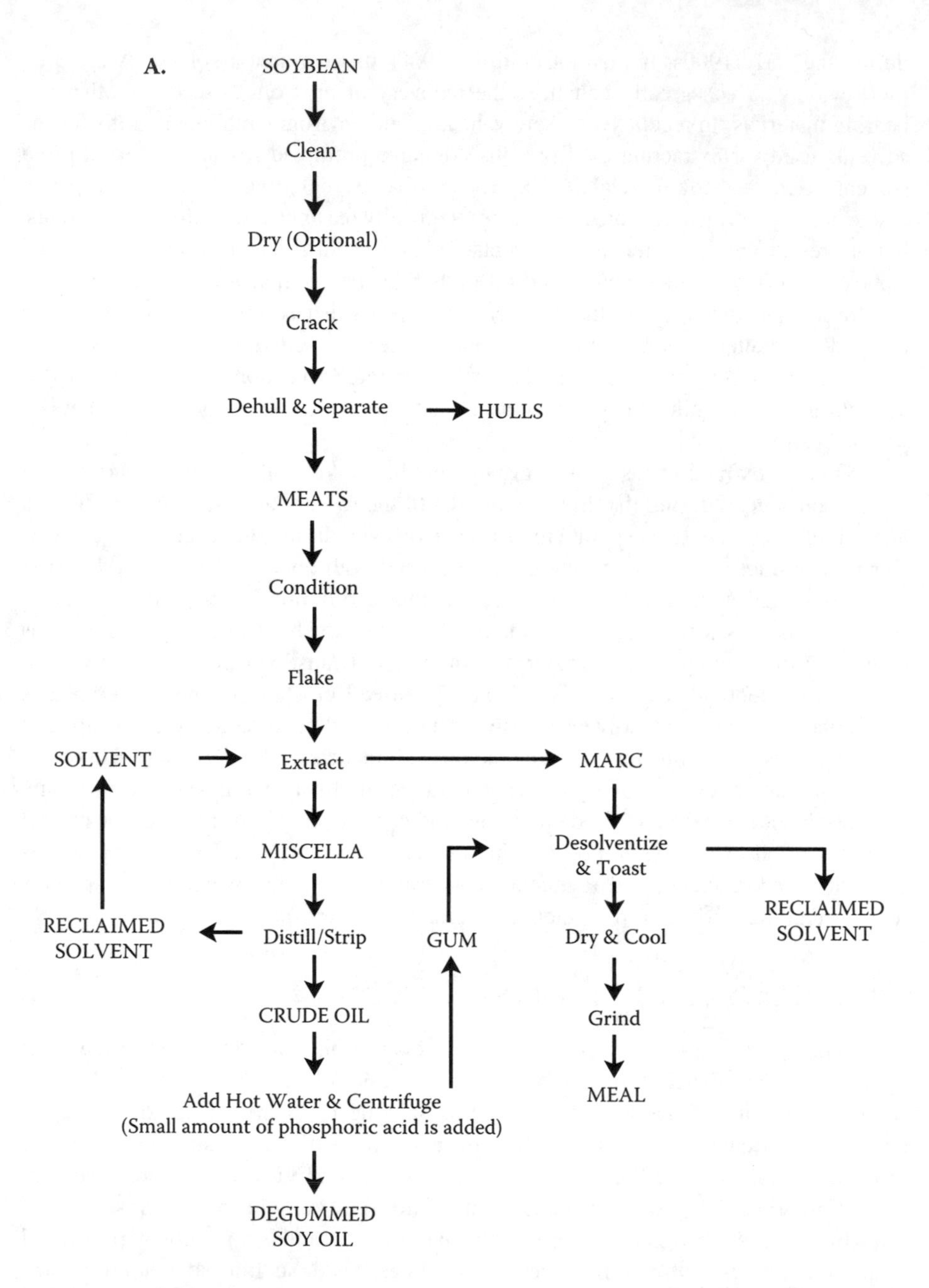

FIGURE 5.1 Flow diagrams of oilseed extraction process. (a) Soybean. (b) Cottonseed.

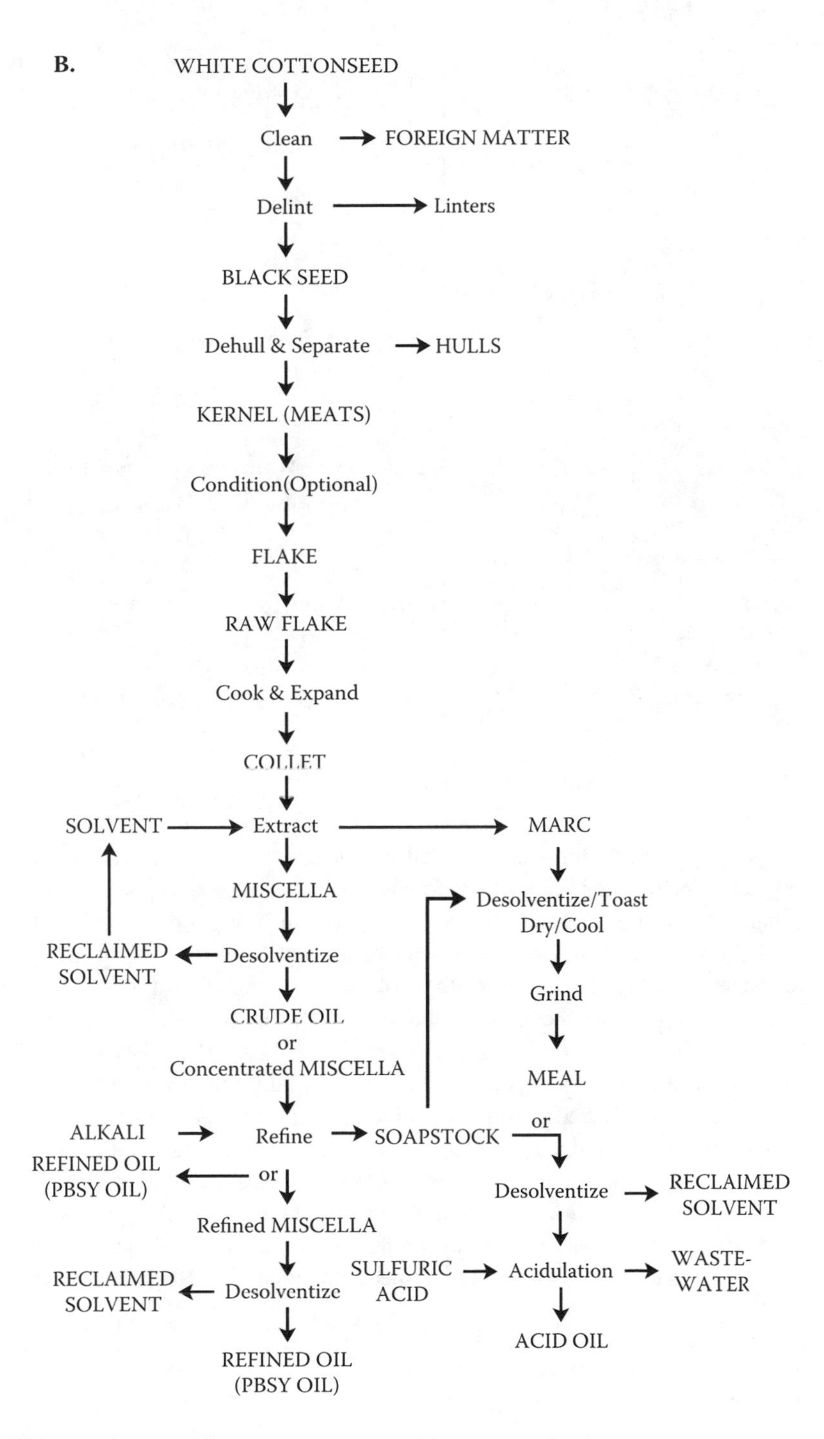

FIGURE 5.1 (Continued).

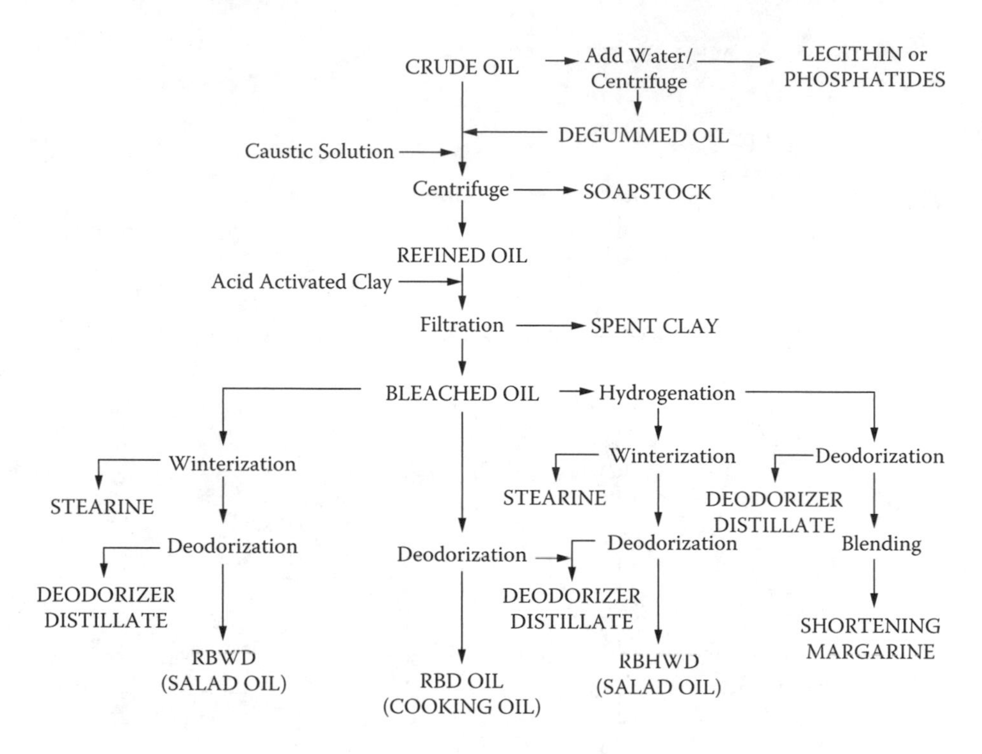

FIGURE 5.2 Flow diagram of edible oil processing.

refining) (see Figure 5.2). Refined oil is dried and bleached with activated clay to remove color pigments and residual soap. Bleached oils are then deodorized by steam distillation to eliminate odorous components. When the soybean oil is of good quality and the free fatty acid (FFA) content is low, the oil can be physically refined. Physical refining involves water washing to recover lecithin followed by phosphoric acid pretreatment and bleaching to remove the nonhydratable gum (primarily phospholipids). After bleaching, the oil is steam deodorized to remove the FFA. The RBD oil is used to produce various finished products (e.g., salad and cooking oils, shortenings, and margarine). Some of the finished products also require the oil to be hydrogenated, which changes the consistency and solid content of the oil and increases stability to oxidation, which extends the shelf life of the finished products. Also, some of the oils (e.g., cottonseed and sunflower) are winterized to remove the higher melting constituents or wax. The solid fraction removed from winterization can be used in confectionery products; the winterized oil is less likely to become cloudy in refrigerated storage.

5.2.1 Preparation for Extraction

5.2.1.1 Storage

For optimum extraction and quality of oil, the oil-bearing material should be stored at dry and relatively low temperature. If it is wet, it should be processed as soon as possible after harvest. Oils in the presence of water can deteriorate rapidly due to lipid hydrolysis catalyzed by lipases, forming FFAs and causing greater refining loss of usable triglycerides.

5.2.1.2 Seed Cleaning

The first step in the commercial processing of oilseeds is cleaning to remove foreign materials, such as sticks, stems, leaves, other seeds, sand, and dirt using dry screeners and a combination of screens and aspiration. Permanent electromagnets are also used for the removal of trash iron objects. Final cleaning of the seed usually is done at the extraction plant just prior to processing.

5.2.1.3 Dehulling

This process may include the removal of excess moisture by drying, cracking the seed, and removing the outer seed coat (hull) of the seed. The hull contains little or no oil, so its inclusion makes the extraction less efficient and dilutes the protein content of the meal. Also, the hull will reduce the total yield of oil by absorbing and retaining oil in the press cake. An acceptable level of hull removal must be determined, depending on the desired protein level of the final meal. Hulls are usually broken loose from kernels with impact hullers and removed by aspirators and undehulled seeds are removed from the kernels by screening and returned to the hullers. Some meats still adhere to the hulls, which are beaten, and then screened again to obtain the meat. In the case of high oil content seed for direct solvent extraction, such as cottonseed and sunflower seed, a certain quantity of hull material is added to the kernels to provide the structural strength and matrix needed for the solvent extraction process.

5.2.1.4 Grinding, Rolling, or Flaking

After dehulling, the meats are reduced in size and flaked to increase surface area and to facilitate oil removal. Proper moisture content of the seeds is essential for proper flaking, and, if the moisture level is too low, the seeds are "conditioned" with water or steam to raise the moisture to about 11%. In the case of soybean, heat will be applied to soften the meats prior to flaking. For solvent extraction, flakes are commonly not less than 0.203 to 0.254 mm (0.008 to 0.010 in.), which can be solvent extracted efficiently with less than 1% residual oil. Thinner flakes tend to disintegrate during the solvent extraction process and reduce the miscella percolation rate.

5.2.1.5 Cooking

Prior to extraction, the flakes are heated. The purpose of cooking the flakes is to (1) break down cell walls to allow the oil to escape; (2) reduce oil viscosity; (3) control moisture content (to about 7% for expanding operation); (4) coagulate protein; (5) inactivate enzymes and kill microorganisms; and (6) fix certain phosphatides in the cake, which helps to minimize subsequent refining losses. Flakes are cooked in stack cookers to over 87.8°C (190°F) in the upper kettle. Flakes with high phosphatide content may benefit from being cooked at slightly lower temperatures to avoid elevating refining losses. The temperature of the flakes is raised to 110 to 132.2°C (230 to 270°F) in the lower kettles. The seeds are cooked for up to 120 min. Overcooking lowers the nutritional quality of the meal and can darken both the oil and meal. Poor-quality seeds with high levels of FFAs cannot be cooked for as long a period as high-quality seeds because of darkening. Darker oil requires additional refining to achieve a certain bleachable color. For soybeans, this heat treatment is often done prior to flaking process.

5.2.1.6 Expanding

Sometimes low shear extruders called expanders are used. This equipment has the capability to process both low- and high-oil content materials. The meats are fed into an extruder after dehulling, flaking, and cooking and are heated as they are conveyed by a screw press through the extruder barrel. The meats are under considerable shear, pressure, and temperature when they reach the exit of the extruder. The change in pressure as the material leaves the extruder causes it to expand and most of the oil cells are ruptured, releasing the oil, which is rapidly reabsorbed to the porous "collets" or pellets. The expanded collets produced are then cooled and extracted with solvent. Because of the excellent efficiency of expanders, almost all cottonseed in the United States is direct solvent extracted from the expanded collets. Large amounts of soybean are also expanded and direct solvent extracted.

5.2.2 Oil Extraction

5.2.2.1 Mechanical Extraction

Olive oil is still routinely obtained from olives by using a low-temperature hydraulic press process, referred to as "cold press." This is done to minimize the heat-related degradation to olive oil. Palm fruit and some cottonseed are extracted with an expeller or screw press; both are continuous processes. To achieve a higher yield of oil, often a higher heat treatment of the cottonseed flakes is carried out prior to expelling. This process can extract up to 90% of the available oil from the cottonseed kernels and leaves about 3 to 5% residual oil in the pressed cake. The cake is ground or palletized as feed protein ingredients for the livestock.

5.2.2.2 Prepress Solvent Extraction

In this process, the oil-bearing material is first mildly pressed mechanically by means of a continuous screw press operation to reduce the oil by half to two thirds of its original level, before solvent extraction to remove the remaining oil in the prepressed cake. Pressing followed by solvent extraction is more commonly used when high-oil-content materials (e.g., canola/rapeseed, sunflower seed, flaxseed, corn germ, and cottonseed) are processed. This process reduces the amount of oil to be extracted by solvent and, therefore, requires a smaller extractor and less solvent than a direct solvent extraction facility of the same throughput.

5.2.2.3 Direct Solvent Extraction

This process involves the use of a nonpolar solvent, usually commercial hexane, which consists of 50 to 80% *n*-hexane and other isomers of six carbon paraffins, to dissolve the oil from oilseed flakes or collets without removing proteins and other non-oil-soluble compounds. Solvent extraction yields about 11.5% more oil than does the screw press method, and less oil remains in the meal. The cooked flakes or collets are mixed with solvent in a batch or continuous countercurrent extraction operation. The high vapor pressure of hexane limits the practical operating temperature of the extraction and its contents to about 50 to 55°C. The resulting miscella (oil–solvent mixture) and the marc (solvent-laden collets or flakes) are heated to evaporate the solvent, which is collected and reused. The oil is freed from the miscella by using a series of stills, stripping columns, and associated condensers, which were designed to maximize the recovery of solvent while minimizing the thermal damage of the extracted crude oil. The essentially hexane-free oil (i.e., crude oil) is cooled and filtered before leaving the solvent extraction plant for storage or further treatment. This is the crude oil normally traded in the commodity market. To minimize the settling problem during storage and shipping, crude soybean oil is often degummed before it is traded. Occasional overheating of the crude oil will cause irreversible color changes in the oil. The crude oil should be stored at the lowest temperature possible (<60°C, 140°F) without solidification and maintaining an easy flowing property.

Due to economic factors and product quality concerns, most of the cottonseed mills in the United States further integrate oil refining as part of the routine operation. The majority of the cottonseed mills conduct miscella refining with sodium hydroxide to produce a once refined or Prime Bleachable Summer Yellow (PBSY) cottonseed oil [6]. The primary benefit of this additional refining operation is to achieve a more consistent oil quality in terms of its color and reduced refining loss. Several U.S. cottonseed oil mills further process the cottonseed oil to finished RBD cottonseed oil, which is sold directly to the market or food processors.

The extraction solvent used in the extractor is normally recovered from the miscella and solvent-laden flakes or collets and reused. The small amount of solvent loss that occurs through vents, crude oil, and desolventized flakes is unavoidable (i.e., fugitive loss). Estimated solvent loss from each of these emission points is given in Figure 5.3. Management and control of emission loss during solvent extraction is an important issue for worker safety and protection of our environment.

5.2.2.3.1 Factors Affecting Extraction

There is little theoretical basis to be followed for the extraction of oilseeds [7–10]. The study of the extraction of oilseeds is complicated by the fact that the total extractable material is variable in quantity and composition [7,8]. Composition of the early extracted material is nearly pure triglycerides. As the extraction progresses, an increasing amount of nonglyceride material will be extracted [7,8]. It is believed that the majority of the oil from oilseed flakes is easily and readily extracted [7,9]. While the thickness of flakes affects extraction rate, the concentration of miscella below 20% does not greatly increase the amount of time to reduce the residual oil

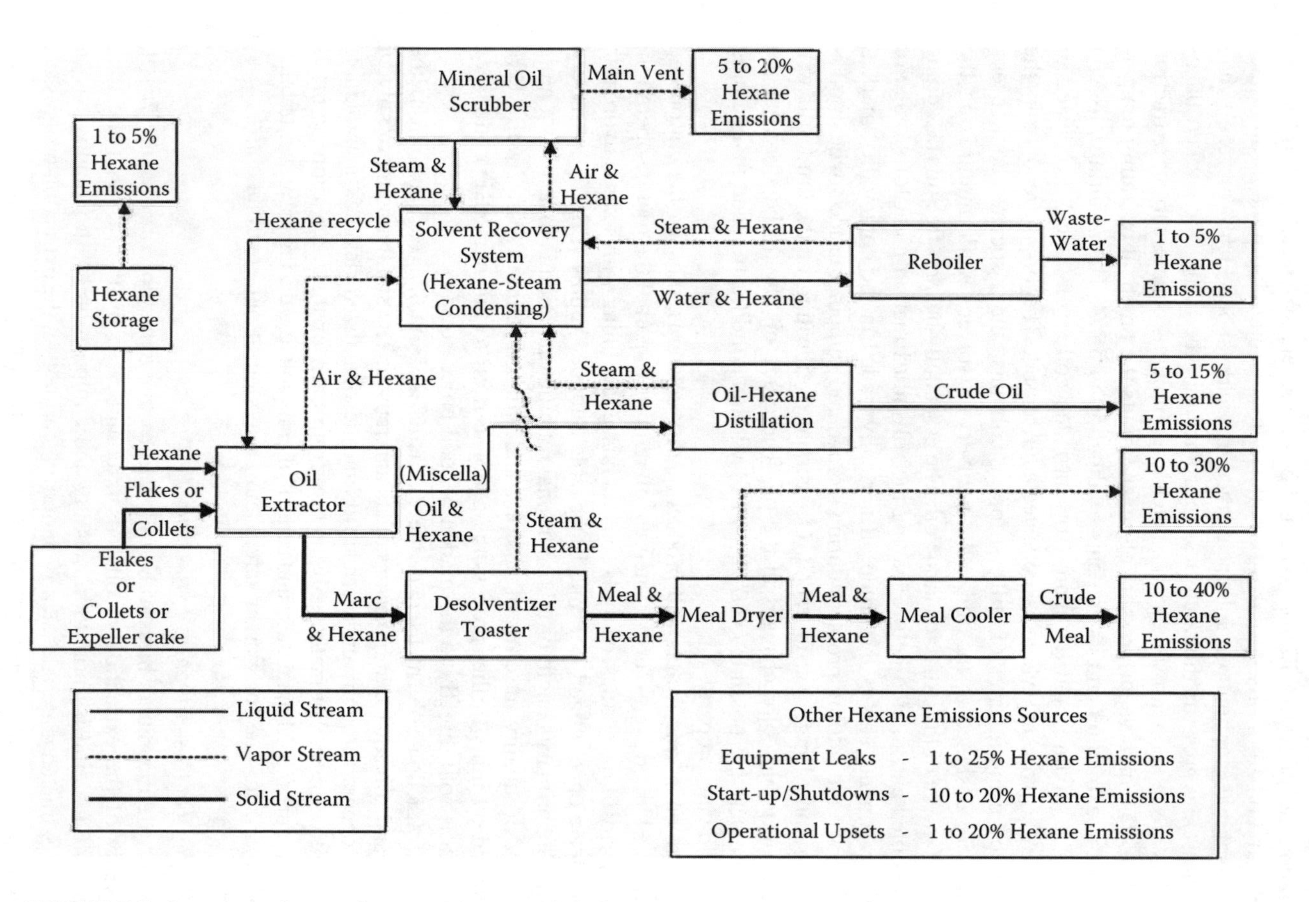

FIGURE 5.3 Overview of extraction operation and identification of emission sources.

in flakes to 1% [7]. Good [10] summarized much of the early effort in soybean extraction: (1) the first oil extracted is superior in quality to the last small fraction; (2) while other solvents have been used in the past, hexane has become the primary solvent due to a combination of properties; (3) flake thickness is the most important factor in achieving good extraction results; (4) higher extractor temperatures up to nearly the boiling point, improve extraction results; (5) moisture control is important throughout the extraction process; (6) heat treatment affects the total extractables; and (7) the soaking theory of extraction indicates that weak miscellas are very effective in helping to achieve good extraction results.

Particle size, which relates to the surface area available for extraction, is obviously one of the most important factors for extraction study. Coats and Wingard [11] noticed that particle size was more influential when the seed grit was being extracted. When oilseed flakes were being extracted, the flake thickness would be a more important factor than size of the flakes. Moisture content in oilseed can affect the extraction results [12–14]. Optimum moisture content of cottonseed meats for extraction was first reported by Reuther et al. [13] to be from 9 to 10%. Work by Arnold and Patel [14] indicated 7 to 10% to be the optimum moisture for cottonseed flakes and very little variation in extraction rate for soybean with moisture content between 8 and 12%.

Wingard and Phillips [15] developed a mathematical model to describe the effect of temperature on extraction rate using a percolation extractor as follows:

$$\begin{gathered}\text{Log (time, in minutes)} = n \log (\text{temp, in } ^\circ\text{F}) + \log k \\ \text{or time} = k\,(\text{Temp})^n\end{gathered} \tag{5.1}$$

where time is defined as the number of minutes required to reach 1% residual oil in the oilseed flakes. For all practical purposes, they concluded that the time in minutes required to reduce the oilseed to 1% residual oil content on a dry basis varied inversely with the square of the extraction temperature in degrees Fahrenheit.

5.2.2.3.2 Evaluation Methods

Except for the pilot plant batch or countercurrent extraction described by various laboratories [16–18], most of the solvent extraction evaluation work found in the literature was done in one or several of the lab-scale devices. The percolation batch extraction apparatus of the Soxhlet type has often been used to evaluate the rate of extraction of hydrocarbon solvents, such as the one described by Bull and Hopper [8]. Wingard and Shand [19] described a percolation-type extractor and a cocurrent batch extractor. They claimed that these extractors to be useful to study the factors influencing equipment design and plant operation as well as fundamental studies contributing to a general understanding of extraction. Wan et al. modified the design of the percolation-type extractor to closely simulate single-stage countercurrent miscella extraction conditions as practiced in the factory [20]. The cocurrent batch extractor with numerous variations was also frequently applied for evaluating the extraction properties of selected solvents. The extractions were often operated at room temperature [19,21]. Soxhlet extraction [20,22] and the Soxtec System HT6 (Perstorp Analytical, Herndon, Virginia) were also frequently used to evaluate solvents [23]. The Soxhlet extractor allows vaporized and condensed pure solvent to percolate through an oilseed sample. The temperature of the condensed solvent is normally lower than its boiling point. Depending upon the cooling efficiency of the condenser and the room temperature, the temperature of the condensed solvent and the temperature of the extracting solvent in the extractor largely varied from lab to lab. This extraction temperature variability was minimized with the Soxtec method by refluxing the oilseed sample in the boiling solvent for 15 min and was followed by a Soxhlet type of rinsing for 35 min. In theory, the Soxtec method is more efficient and better reproduced. However, the Soxtec method utilized only a 3-g oilseed sample. The heterogeneity of an oilseed sample could be a significant source of variation.

Flakes of oilseeds were most frequently used for the solvent extraction studies. Sometimes, ground oilseed kernels through a specified sieve size were used [23]. Residual oil content in the extracted flakes after a certain specified extraction condition or oil content in miscella was examined and the percentage of total oil extracted determined [8–15,19]. The total extractable oil of the flakes was determined by 4 h of Soxhlet extraction with petroleum ether or hexane. Wan et al. [20] used a precision densitometer to determine the miscella concentration (percentage of oil in miscella by weight) after a given time of extraction from which the percentage of oil extracted from cottonseed flakes was calculated. From these data, Wan et al. [20] were also able to estimate the initial rate of extraction and final extraction capacity for each solvent as fresh and at selected initial miscella concentrations up to 30%.

Bull and Hopper [8] conducted extraction of soybean flakes in a stainless steel batch extraction apparatus of the Soxhlet type with petroleum solvents, Skellysolve F (boiling range, 35 to 58°C) at 28°C and Skellysolve B (boiling range, 63 to 70°C) at 40°C. The extraction was carried out to permit the miscella obtained by each flooding of the flakes with solvent to be recovered separately. Their results showed that the iodine number decreased and the refractive index increased slightly with the extraction time, which implied that more saturated fat was extracted during the later stages of the extraction. Oils extracted during the later stages of the extraction were found to contain greater amounts of unsaponifiable matter and were rich in phosphatides, as high as 18% of the last fraction. Skellysolve B, which is a hexane-rich solvent, demonstrated a much faster initial rate of extraction than that of Skellysolve F, which is a pentane-rich solvent; therefore, it took longer to complete the extraction for Skellysolve F. The fatty acid profile of each fraction showed a slight increase in saturated fatty acid and a slight decrease in unsaturated fatty acid in the later fractions.

Arnold and Choudhury [24] reported results derived from a lab-scale extraction of soybean and cottonseed flakes in a tubular percolation extractor at 135 to 140°F with pure, high-purity, and commercial hexane and with reagent grade benzene. They claimed that pure hexane extracted soybean slower than high-purity and commercial hexane. During the first 60 min of extraction, benzene extracted more oil than the hexanes. However, at the end of 80 min, benzene extracted only slightly more than pure hexane but definitely less than commercial hexane. Similar results were obtained for the four solvents when cottonseed flakes were extracted.

A laboratory extraction study of cottonseed flakes using various hydrocarbon solvents was reported by Ayers and Dooley [22]. A Soxhlet extractor and Waring blender were used for these experiments. Among the petroleum hydrocarbon solvents tested were branched, normal, and cycloparaffins, as well as aromatic hydrocarbons with various degrees of purity. They were pure grade (99 mol% purity) *n*-pentane, isopentane, cyclohexane, benzene, and *n*-heptane; technical grade (95 mol% purity) neohexane, diisopropyl, 2-methylpentane, 3-methylpentane, *n*-hexane, and methylcyclopentane; technical grade (90 mol% purity) cyclopentane; and commercial grade *n*-heptane, isohexanes, *n*-hexane, isoheptane, and *n*-heptane. To assess the performance of these solvents, they used the following empirical formula:

$$\text{Quality-efficiency rating} = 0.4\ (\text{oil yield factor}) + 0.4\ (\text{refining loss factor}) + 0.2\ (\text{refined and bleached oil color factor}) \quad (5.2)$$

When comparing the oil yield factor alone, 3-methylpentane was rated the best. When comparing the solvents based on the empirical quality-efficiency rating formula, they concluded that methylpentanes (3- and 2-methylpentane) were superior extraction solvents for cottonseed oil. The normal paraffins, highly branched isohexanes, cycloparaffins, and aromatics were progressively rated as less efficient than methylpentanes. Therefore, they recommended that a tailor-made solvent for the extraction of cottonseed should exclude aromatic hydrocarbons, have low limits on cycloparaffin content, and consist largely of normal and isoparaffin hydrocarbons.

A more recent study by Wan et al. [20] using a laboratory-scale dynamic percolation-type extractor (Figure 5.4) operated at the following conditions: temperature (5°C below the boiling point of each solvent) and miscella flow rate (9 gal/min/ft^2), similar to those applied in the oil mill practice. Commercial grade hexane, heptane, isohexane, neohexane, cyclohexane, and cyclopentane were used to extract cottonseed flakes, which had 5.8% moisture and 31.4% oil. When these solvents were tested near their boiling points, hexane apparently extracted cottonseed oil at a higher initial rate (>94% oil extracted after 2 min) than all other solvents. Both heptane and hexane were also able to extract more oil at the end of 10 min of extraction. Isohexane demonstrated an adequate initial extraction rate (80% oil extracted after 2 min) and extraction capacity (93% oil extracted after 10 min of extraction), but is noticeably less effective than hexane. Similar to findings by Ayers and Dooley [22], results from the study by Wan et al. [20] also demonstrated that neohexane, cyclohexane, and cyclopentane performed distinctly less efficiently than hexane, heptane, and isohexane. Conkerton et al. [23] tested commercial heptane vs. hexane in a Soxtec extractor. Under this extraction condition, heptane actually extracted more oil than hexane from ground cottonseed kernel passed through a 20-mesh screen. The oil and meal quality were not appreciably affected by the higher temperature extraction of heptane.

5.2.2.3.3 Plant Scale Results

Although hydrocarbon solvents have been used for oilseed extraction since the 1930s, very little plant operating data are available. During the spring of 1994, Wan et al. [3] conducted plant trials with commercial heptane and isohexane at a 300 ton/d cottonseed crushing plant, which was constructed more than 90 years ago as an expeller plant and routinely used hexane as the extraction solvent at the time of the plant trials. Test results indicated that heptane performed well as an extraction solvent. However, it required extra energy and time to recover and consequently reduced the

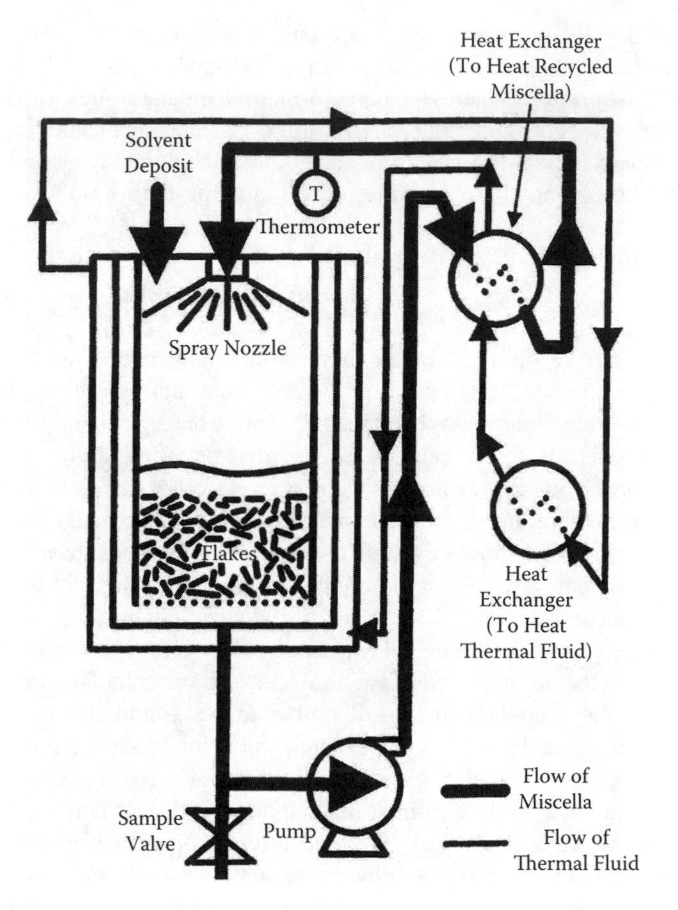

FIGURE 5.4 Schematic of bench-scale dynamic percolation extractor.

throughput rate of cottonseed processed. Isohexane, on the other hand, was termed by the plant engineers as an "easier" solvent to operate than hexane. The plant also experienced a 40% steam savings and better than 20% throughput increase when it was operating with isohexane [3]. This encouraging result prompted a second plant trial with commercial isohexane [4]. The second plant trial was carried out at a cottonseed oil mill with a relatively new extraction and miscella refining facility, which was constructed in 1988 with a designed capacity of 500 ton/d but operated at only 270 ton/d due to limited delinting capacity. After week-long testing with commercial isohexane, this plant experienced more than 20% natural gas usage and easily increased the throughput rate by close to 10% when compared with commercial hexane. This energy savings with commercial isohexane over commercial hexane may be largely attributed to the difference in the amount of water present in their corresponding azeotropes. Isohexane requires an additional step — isomerization — in manufacture and will always be priced higher than hexane. But based on the two cottonseed oil mill trials, isohexane can be a cost-efficient solvent [25]. One additional benefit is that the shorter residence time of the extracted cottonseed marc in the desolventizer/toaster because of the lower boiling range of isohexane will likely preserve more vegetable protein in the final meals, which has been observed by both plants during the tests [3,4]. The benefit in improved quality of oils was not obvious in both plant trials but might be realized with extended trials. These plant trials have motivated others to evaluate commercial isohexane in their extraction facilities. Currently, it is estimated that more than 20% of the oilseed extraction capacity in the United States uses commercial isohexane.

5.2.3 Solvent for Extraction

5.2.3.1 Solvents

Research in solvents for extraction has been carried out for more than 150 years. The effort intensified since the first patent was issued to Deiss of France in 1855 [1,26,27]. This cumulative effort has yielded the partial list of the various solvents being tested in Table 5.1. In the early effort of selecting the extraction solvent, the availability, operation safety, extraction efficiency, product quality, and cost were the major concerns. In recent decades, toxicity, biorenewability, and environmental friendliness have been added to the solvent selection criteria. Among this score of solvents tested, the majority of the candidate solvents was excluded on the grounds of toxicity and safety and only a handful of solvents are used in various degrees. These are acetone, alcohol, hexanes, heptane, and water [1,27–31]. Water is used in the rendering of fat from animal tissues and fish and in coconut processing [31], alcohol for spice and flavorants extraction [27,29], and acetone for lecithin separation and purification [28]. For commodity oils derived from vegetable sources, only hydrocarbon solvents have been used since the 1930s. Acetone was used by an Italian cottonseed oil mill during the 1970s [28]. Aqueous acetone and acetone–hexane–water azeotropes were studied by the scientists at the Southern Regional Research Center of Agricultural Research Service, U.S. Department of Agriculture (USDA) during the 1960s and 1970s [28]. The effort was stopped due to the cost of retrofitting the existing extraction facility required for the industry, the difficulties in managing the mixture of solvents with the presence of water, and the product quality concerns — a strong undesirable odor associated with the acetone extracted meals [28]. Ethanol and isopropanol were studied in the 1980s as a potential replacement for hexane in oil extraction. Both were proved technically feasible but economically unacceptable [29,30]. Therefore, the only cost-effective solvents for commodity extraction are still petroleum-based paraffins. Composition and properties of several commercial grade hydrocarbons are given in Tables 5.2 and 5.3 [3,20].

5.2.3.2 Toxicity of Extraction Solvents

Many halogenated and aromatic solvents have been examined in the past and are effective in extracting edible oils. However, these solvents have various degrees of toxicity and, therefore, are not likely to be used as alternates or replacements for commercial hexane for edible oil extraction. Several hydrophilic solvents have also been studied as oil extraction solvents. They are not toxic but their overall performance

TABLE 5.1
Chemical and Physical Properties of Extraction Solvents[a]

Solvent	Boiling Point (°C)	Latent Heat of Vapori-zation (cal/g)	Specific Heat (cal/g°C)	Specific Gravity at 20°C (g/cm³)	Liquid Viscosity at 20°C (cp)	Surface Tension (dyne/cm)	Dielectric Constant	Flash Point (°C)	Explosive Limits (% vol. in Air)	Water Solubility at 25°C (g/l)	Water Azeotrope (% wt)	Boiling Point (°C)	TLV–TWC (ppm)
Water	100.0	540.1	1.018	1000	1005	72.5	80.36	NF	NF	—	—	—	—
Petroleum fractions													
Gasoline	39.204			0.72–0.76				–45	1.3–6.0				300[b]
Petroleum ether	35–60			0.62–0.66				–40	1.4–8.0				
Pentanes	32–37			0.62–0.63				–54					600
Methyl pentanes	56–62			0.65–0.67				–32					500
Hexanes	65–69			0.67–0.72	0.32			–22	1.1–7.5				50
Heptanes	88–99			0.72–0.74	0.47			–5	1.2–6.7		10.3	80–85	400
Cyclohexane	65–85												300
Alkanes													
n-Propane	–41.7	101.6	0.520[bp]						2.4–9.5				
n-Butane	0.5	92.2	0.550[bp]						1.9–8.5				800
n-Pentane	36.2	85.3	0.342	0.626	0.234	16.05	1.84	–49	1.4–8.0	0.36			600
Isopentane	27.8	81.7	0.535	0.620	0.224	15.00	1.84	–5	1.4–8.3	0.14			
n-Hexane	69.0	79.9	0.533	0.659	0.312	18.40	1.89	–23	1.2–7.7	0.05	5.0		50
Isohexane	60.3	77.4	0.533	0.652	0.299	15.84		–7	1.2–7.0	0.00			500
n-Heptane	98.4	75.5	0.528	0.684	0.417	20.14	1.92	–4	1.0–7.0	0.05	13.0	79.4	400
Cycloparaffins													
Cyclopentane	49.5	92.9	0.422	0.751	0.438	22.42	1.97	–37	1.4-	0.00			600
Cyclohexane	80.7	77.2	0.433	0.778	0.977	24.98	2.02	–18	1.3–8.4	0.00			300

Aromatic hydrocarbons													
Benzene	80.1	94.0	0.410	0.879	0.647	28.88	2.28	10	1.3–7.9	0.82			10[b]
Toluene	110.6	86.7	0.402	0.867	0.585	28.53	2.38	6	1.2–7.1	0.47			50
Mixed xylenes;	138.7	81.1	0.408		0.65		28.31	2.4	29	1.0–5.3	0.00		100
Halogenated hydrocarbons													
Dichloromethane	39.8	78.7	0.276	1.326	0.513	28.00	9.08	NF	NF	20.00	1.5	38.1	50[b]
Ethyl chloride	12.3	92.5	0.370[bp]	0.923[bp]			6.29	–50	3.6–14.8				1000
Chloroform	61.2	59.0	0.234	1.477	0,562	27.10	4.81	NE	NF	0.00	2.8	56.1	10[b]
Carbon tetrachloride	76.7	46.5	0.200	1.585	0,971	26.84	2.24	NE	NF	0.80	4.1	66.0	5[b]
1,2-Dichloroethane	83.6	77.3	0.310	1.255	0.78		37.50	10.50	15	6.2–15.9	8.69	8.2	70.5200
1,2-Dichloroethylene	60.3	73.0	0.270	1.282	0.467	28.00	9.20	14		0.00			10[b]
1,1,1-Trichloroethane	74.1	68.7	0.240	1.324	0.858	25.12	7.50	NF	NF	0.00	4.3	65	350
1,2,-Trichloroethylene	86.7	57.3	0.300	1.456	0.550	32.00	3.27	NF	NF	1.00	5.4	73	50[b]
1,1,2,2-Tetra (chloroethylene	121.0	50.1	0.210	1.618	0.90	29.3	2.20	NF	NF	0.00	15.9	87.8	50[b]
1,2-Dichloropropane	95.9	72.2	0.310	1.157	0.865	31.40	8.93	21	3.1–14.5	2.7			75[b]
n-Butylbromide	101.4			1.274			7.07	24	2.6–6.6	0.00			
1,1,2-Trichloro-1,2,2-trifluomethane	47.6	35.1	0.213	1.374	0.694	19.0	2.44	NF	NF	0.27			1000

(continued)

TABLE 5.1 (Continued)
Chemical and Physical Properties of Extraction Solvents[a]

Solvent	Boiling Point (°C)	Latent Heat of vapori-zation (cal/g)	Specific Heat (cal/g°C)	Specific Gravity at 20°C (g/cm³)	Liquid Viscosity at 20°C (cp)	Surface Tension (dyne/cm)	Dielectric Constant	Flash Point (°C)	Explosive Limits (% vol. in Air)	Water Solubility at 25°C (g/l)	Water Azeotrope (% wt)	Boiling Point (°C)	TLV–TWC (ppm)
Alcohols													
Methanol	64.7	473	0.599	0.792	0.59	22.53	31.2	1.1	6.0–36.5	Inf	None	None	200
Ethanol	78.3	204	0.61	0.785	1.22	22.1	25.7	1.2	3.3–19.0	Inf	4	78.2	1000
n-Propanol	97.2	162.6	0.586	0.805	2.256	23.8	20.1	22	2.6–13.5	Inf	29.1	87.7	200
Isopropanol	82.5	159.3	0.596	0.786	2.4	20.8	18.6	11.7	2.5–12.0	Inf	12.3	80.4	400
n-Butanol	117.7	141.3	0.563	0.811	2.948	24.6	16.1	37	1.5–11.3	79	42.5	92.7	50
Isobutanol	107.9	138.0	0.716	0.803	6.68	22.8	17.7	28	1.7–10.9	95	33.0	89.8	100
Allyl alcohol	96.9	163.8	0.665	0.850	1.072	25.7	21.6	21	2.5–18.0	Inf	27.7	88.9	2
Furfuryl alcohol	1.71							79	1.8–16.3	Inf	43	94	5
Aldehydes													
Furfural	161.7	107.5	0.401	1.161	1.49	40.7	41.9	60	2.1-	83	35	97.9	2[b]
Ketones													
Acetone	56.1	124.5	0.51	0.791	0.316	23.7	21.5	–16	2.2–13.0	Inf	None	None	750
Methyl ethyl ketone	79.6	106.0	0.55	0.806	0.423	24.6	18.51	–4	2.0–10.2	353	11.3	73.4	200
Esters													
Methyl acetate	56.9	104.4		0.958	0.381	24.76	7.3	10	3.1–16		3.5	56.4	200
Ethyl acetate	77.0	102.9		0.897	0.473	23.75	6.02	0	2.3–11.4		8.5	70.4	400
Ethers													
Ethyl ether	34.5	84.0	0.548	0.715	0.233	17.0	3.88	–40	2.3–6.2	75	1.1	34.1	400
Isopropyl ether	68.4	68.2	0.526	0.722	0.379	32.0	2.21	–28	1.4–21	2	4.5	62.2	250

Dioxane	101.1	97.0		1.033	1.439	34.45	16.3	5	2.0–22.0	Inf	18.4	87.8	5
Ethylene glycol monomethyl ether	124.2	129.6	0.534	0.966	1.53	30.8		43		Inf	77.8	99.9	
Ethylene glycol monoethyl ether	134.7		0.530	0.931	1.84	28.2		40	2.6–15.7	Inf	71.2	99.4	
Amines													
Ethanolamine	172.2	197.1		1.108	24.1	48.3	5.3	90	1.7–9.8	Inf			3
Butylamine	77.0	104.8		0.741	0.68	24.0			1.2–8.0	Inf			5
Triethylamine	89.5			0.733		20.66	12.3		1.8–12.4	15	10	7.5	W
Pyridine	115.3	59.6		0.982	1.038	37.25		23		Inf			5
Other solvents													
Carbon disulfide	46.5	84.1	0.24	1.263	0.363	32.25	2.64	–30	1.0–50.0	2.2	2.8	42.6	10
Carbon dioxide	–78.5	83.1	0.20					NF	NF				5000
Solvent mixtures													
Hexane/acetic acid (96:4)	68												
Hexane/methanol (46:25)	51												
Hexane/ethanol (79:21)	58.6												
Hexane/isopropanol (77:23)	62.7												
Hexane/allyl alcohol (95:5)	65.5												
Aromatic hydrocarbon/ethanol (90:10)													

(continued)

TABLE 5.1 (Continued)
Chemical and Physical Properties of Extraction Solvents[a]

Solvent	Boiling Point (°C)	Latent Heat of Vapori-zation (cal/g)	Specific Heat (cal/g°C)	Specific Gravity at 20°C (g/cm³)	Liquid Viscosity at 20°C (cp)	Surface Tension (dyne/cm)	Dielectric Constant	Flash Point (°C)	Explosive Limits (% vol. in Air)	Water Solubility at 25°C (g/l)	Water Azeotrope (% wt)	Boiling Point (°C)	TLV–TWC (ppm)
Ethanol/water (96:4)	78.2		0.598	0.805				18					
Isopropanol/water (87.7:12.3)	80.4	204.1	0.770	0.818	2.106	21.1	27.0	17					
Methanol/trichloro-ethylene (75:25)													
Ethanol/trichloroeth ylene (75:25)													
Acetone/water (90:10)	59												
Acetone/hexane/ water (54:44:2)	49												

Abbreviations: bp, value at boiling point; NF, nonflammable; Inf, infinitely soluble; s, sublimes.

[a] Hron, R.J. in *Technology and Solvents for Extracting Oilseeds and Non-Petroleum Oils*, Wan, P.J. and Wakelyn, P.J., Eds., AOCS Press, Champaign, IL, 1997, pp. 186–191.

[b] Suspected of being carcinogenic.

TABLE 5.2
Composition of Commercial Grade Hydrocarbon Solvents[a]

	Heptane			Hexane		
Component	**Company A Lv%[b]**	**Company B Wt%**	**Company C Lv%[b]**	**Company A Lv%[b]**	**Company B Wt%**	**Isohexane Wt%**
Hydrocarbon with 5-C or less						0.9
2,2-Dimethylbutane		0.1				14
2,3-Dimethylbutane		0.1				15.9
2-Methylpentane		0.2		0.2		46.3
3-Methylpentane		0.1		4		20.1
n-Hexane	0.1–0.99			86.2	80.00–94.99	2.6
Methylcyclopentane				9.6	11.00–19.99	0.2
Cyclopentane derivatives		7.1				
Cyclohexane derivatives	1.00–2.99	0.8				
2-Methylhexane		21.6				
3-Methylhexane		28.2				
n-Heptane	80.00–94.99[c]	23.6	26.5			
Other isoheptanes		12.6				
Toluene	0.10–0.99	3.5				
Octane isomers	3.00–10.99					
Other unsaturated hydrocarbons		2.1				
Benzene, ppm	0.01–0.09					
Other paraffins			62.5			
Other cycloparaffins			11			

[a] Data were provided by the supplier.
[b] Lv% = Liquid volume %.
[c] *n*-Heptane and heptane isomers.

TABLE 5.3
Physical Properties of Commercial Grade Hydrocarbon Solvents

	Types of Solvent		
Properties	**Hexane**	**Heptane**	**Isohexane**
Boiling range, °F	152–156	195–212	131–142
Heat of vaporization, btu/lb	143.9	136	139
Liq. spec. heat, cal/g/°C	0.533	0.528	0.52
Vap. spec. heat, cal/g/°C	0.386	0.385	0.39
Specific gr., (16 °C/60 °F)	0.679	0.694	0.66
Density of liquid, lb/gal (at 60°F)	5.63	5.8	5.49
Vapor pressure, psia (100°F)	5	2.3	
Flash point, °F	–15	15	

as an oil extraction solvent is not acceptable. This is why hydrocarbon-based solvents are likely to be used for oil extraction for the foreseeable future.

5.2.3.2.1 Commercial Hexane

Commercial hexane has been used for decades as the solvent to extract oils from biological sources. It is a mixture of six carbon saturated compounds, with *n*-hexane as the predominant component. Pure *n*-hexane causes peripheral nerve damage in rats and humans when inhalation exposures are maintained for several months at 500 ppm in rats or 125 ppm in humans [32,33]. However, commercial hexane, which contains 52% *n*-hexane and a mixture of hexane isomers (see composition below), does not cause peripheral nerve damage in animals.

The composition of the commercial hexane tested [34] is as follows:

52% *n*-hexane
16% methylcyclopentane
16% 3-methylpentane
13% 2-methylpentane
3% cyclohexane

This was shown by extensive animal inhalation studies, which were mandated by the U.S. EPA under Section 4 of the Toxic Substances Control Act (TSCA) [35]. The test results, summarized by Galvin [34], showed that this commercial hexane blend was not a neurotoxin. In addition, the following tests also were negative: acute toxicology, subchronic neurotoxicity, mutagenicity (both *in vitro* and *in vivo* studies), oncogenicity, and development and reproductive studies (with the above described commercial hexane) at a vapor concentration as high as 9000 ppm for 6 h/d, 5 d/week up to 13 weeks. The EPA Integrated Risk Information System (IRIS) file [46] contains detailed toxicity information on *n*-hexane.

5.2.3.2.2 Commercial Isohexane and Hexane Isomers

One alternative to commercial hexane, which would require minimum retrofit of existing extraction facilities, is commercial isohexane. This solvent, a blend of hexane isomers (2-methlypentane or isohexane, 3-methylpentane, 2,3-dimethylbutane, 2,2-dimethylbutane, and <1% *n*-hexane), has not been tested as extensively as commercial hexane. The individual components, however, have been tested in various toxicological assays. Based on the available information commercial isohexane is not a neurotoxicant [34].

5.2.3.2.3 Other Solvents

Many other solvents have been examined by various research teams as potential alternatives to commercial hexane for the extraction of edible oils. Toxicity information for the most common ones is summarized in Table 5.4. More discussion on these solvents can be found in the summary by Wakelyn and Adair [36].

Most of the solvents mentioned in this section, with the exception of acetone, are regulated by the U.S. EPA as VOCs (40 Code of Federal Regulation (CFR)

TABLE 5.4
Some Toxicity Information for Potential Alternate Solvents

Solvent (CAS Number)	LD_{50}	Other Toxic Concerns
Acetone (67-64-1)	5.8 g/kg in rats	A central nervous system depressant in animals and humans
	20 g/kg in rabbits	
2-Butanone (methyl ethyl ketone) (78-93-3)	2.74 g/kg in rats	Eye and skin irritation and cause of narcosis
	13 g/kg in rabbits	
Cyclohexane (110-82-7)	12.7 g/kg in rats	Moderate irritation to eyes and mucous membrane
Cyclopentane (287-92-3)		Narcotic
Ethyl acetate (141-78-6)	5.6 g/kg in rats	Irritation to the eyes, mucous membranes, respiratory tract
	3.0 g/kg in cats	
Ethyl alcohol (ethanol) (64-17-5)	7.06 g/kg in rats	Irritation to the eyes, mucous membranes
Heptane (142-82-5)	2.22 g/kg in mice	Central nervous system depressant
Isopropyl acetate(108-27-4)	3.0 g/kg in rats	Irritation to the eyes, mucous membranes
Isopropyl alcohol (isopropanol) (67-63-0)	5.05 g/kg in rats	Irritation to the eyes, nose, throat
	12.8 g/kg in rabbits	
Methyl alcohol (methanol) (67-56-1)	5.628 g/kg in rats	Headaches and visual impairment
	15.8 g/kg in rabbits	
Methylcyclohexane (108-87-2)		Similar to that of heptane
Methylene chloride (dichloromethane) (75-09-2)		Decreased visual and auditory function; headaches, dizziness, nausea, and memory loss; a B2 probable human carcinogen
n-Propyl acetate (109-60-4)	9.37 g/kg in rats	Irritation to the eyes, respiratory system

51.100(s)) because they can undergo photochemical oxidation in the atmosphere in the presence of sunlight and nitrogen oxides (NOx) to form ozone at a greater rate than ethane (C_2H_6). They are also regulated by the U.S. EPA as HAPs. The main component of the hydrocarbon solvents used for edible oil extraction, *n*-hexane, is a neurotoxin and considered a HAP by the EPA. Thus, *n*-hexane-containing solvents are more stringently regulated than commercial isohexane (see Section 5.4).

5.3 THE OIL REFINING PROCESS

Since integration in oil refining processes has become a common practice in the oil extraction industry, the various steps of oil refining operation as outlined in Figure 5.2 are briefly described in the following [37].

5.3.1 Processing Crude Oil

Most crude edible oils, obtained from oil-bearing materials, consist primarily of triglycerides (triacylglycerols). The triglycerides (approximately 95% of the crude oil) are the constituents recovered for use as neutral oils in the manufacture of finished products. The remaining nontriglyceride portion contains variable amounts of other lipophilic compounds, such as FFAs, nonfatty materials generally classified as "gums," phospholipids (phosphatides), tocopherols, color pigments, trace metals, sterols, meal particles, oxidized materials, waxes, moisture, and dirt. Most of these minor lipid components are detrimental to the finished product color, flavor, and smoking stability and so must be removed from the neutral oil by a purification/separation process. The object of the purification/separation steps is to remove the objectionable impurities while minimizing possible damage to the neutral oil and tocopherols and loss of oil during such processing.

Lecithin and cephalin are the principal common phosphatides found in edible oils. Soybean, canola/rapeseed, corn, and cottonseed are the major oils that contain significant quantities of phosphatides. Alkaline treatment used for FFA reduction is also capable of removing most of the phosphatides from these crude oils. Tocopherols are important minor constituents of vegetable oils, which are natural antioxidants that retard the development of rancidity.

Refining, bleaching, and deodorization are necessary if the oil is to be used in food applications. Oil that has gone through these three critical processing steps is called RBD oil. These oils are most frequently found in the salad and cooking oil applications. Figure 5.2 illustrates the processing pathways.

5.3.1.1 Refining

Refining involves the removal of nonglyceride materials (phospholipids, color, and trace materials) and FFAs. The goal is to produce high-quality refined oil with the highest yield of purified triglycerides. Refining is by far the most important step in processing. Improperly refined oil will present problems in bleaching and deodorization and reduce quality.

Some solvent-extracted crude oils, including soybean and canola/rapeseed, contain approximately 2 to 3% gums, which are mainly phosphatides (lecithin and cephalin) and require degumming. Gums can cause problems through higher than necessary refining losses or by settling out in storage tanks. The degumming operation exploits the affinity of most phosphatides for water, converting them to hydrated gums that are insoluble in oil and readily separated by centrifugal action. Lecithin can be recovered and concentrated from the gums in a separate solvent extraction process (referred to as deoiling), usually with acetone.

Either water-degummed oil or crude oil can be treated with sodium hydroxide solution to saponify impurities that are subsequently removed as soap stock by a primary refining centrifuge. Conventional alkali refining is by far the most widespread method of edible oil refining. The success of the alkali refining operation is the coordination of five prime factors: (1) use of the proper amount of reagent (sodium hydroxide), (2) proper mixing, (3) proper temperature control, (4) proper residual contact time, and (5) efficient separation.

Oil is alkali refined by the addition of sodium hydroxide solution at a level sufficient to neutralize the FFA content of the oil. An excess of sodium hydroxide is required to reduce the color of the refined oil, to ensure the completion of the saponification reaction, and to remove other trace elements. The amount and strength of the sodium hydroxide solution needed to neutralize the FFAs is dependent on the amount of both FFAs and phosphatides present in the crude oil. Water-soluble soaps are formed in the primary reaction between the sodium hydroxide and FFAs. The hydratable phosphatides react with the caustic, forming oil-insoluble hydrates. The caustic used in alkali refining is normally diluted to about 8 to 14% NaOH, although higher concentrations are occasionally used to reduce color. The proper amount of NaOH solution added to the oil will produce adequately refined oil with a minimum of triglyceride oil loss. The amount of NaOH solution (neutralizing dose plus excess) is determined by experience and adjusted according to laboratory results.

After the NaOH solution is injected, it is mixed for 6 to 10 min to ensure thorough contact. The treated oil is then heated to assist in breaking of the emulsion prior to separation of the soap stock from oil in continuous centrifuges.

Any soap remaining after the primary soap stock separation is removed through continuous hot water washings. In this step, water is added at 10 to 15% at a temperature sufficient to prevent emulsification, generally 82 to 90.5°C (180 to 195°F). The oil is again separated from the soapy phase in water wash separators and dried under partial vacuum conditions prior to bleaching.

5.3.1.2 Physical Refining

This process uses steam distillation to remove the FFAs in the oil and is, therefore, called steam refining. The technology was first applied to high-FFA-containing oils in 1930. The palm oil industry adopted the process in the 1950s. Physical refining of soybean oil, which contains phospholipids and low FFAs, became a reality only recently. In physical refining, the crude oil is first water degummed and this may be followed by phosphoric acid treatment, which may be required to remove nonhydratable phospholipids prior to the bleaching step. The bleached oil is then steam distilled to remove FFAs. This is a normal operation of a deodorizer. The physical refining process has numerous advantages; it is a simpler and less costly process and has a reduced wastewater load.

5.3.1.3 Miscella Refining (Caustic Refining in the Presence of Solvent)

The final oil concentration of the miscella at the end of extraction is usually around 20 to 25% by weight. After going through the initial evaporator, the concentration of oil in the miscella is increased to close to 40 to 50%. In miscella refining, the concentrated miscella is then refined with caustic soda (16 to 24 degree Be°). The FFAs in the oil react with the sodium hydroxide in a continuous tubular mixer and reactor at 130 to 135°F (54 to 57°C). The refined oil in the solvent is then separated from the soap by centrifugation prior to entering the second stage evaporator and mineral oil scrubber. The once-refined oil should then have less than 80 ppm residual solvent and be ready for bleaching and deodorization or other further processing. In recent years, oilseed extraction mills have been motivated to use miscella refining because of improved neutral oil yield, more consistent oil color, and elimination of water washing when this process is used.

5.3.2 Bleaching

The oil is further purified by bleaching, which removes color bodies and trace metals, as well as entrained soaps and some oxidized products that are adsorbed into the surface of bleaching agents or adsorbents. This improves the appearance, flavor, taste, and stability of the final product. Types of adsorbents most commonly used include neutral clay, acid activated clay, and activated carbon. The choice of adsorbent will depend on a balance between activity of the adsorbent, oil retention loss, and adsorbent cost.

The process is generally carried out via batch or continuous bleaching. Adsorbent (e.g., bentonite clay) is mixed with the refined oil creating a slurry that is agitated to enhance contact between the oil and the adsorbent. This is generally carried out under a vacuum at 90 to 95°C (194 to 203°F) for 15 to 30 min. Vacuum bleaching offers the advantages of an oil with improved oxidative and flavor stability. Finally, the adsorbent is filtered from the oil using pressure leaf filters precoated with diatomaceous earth. Spent clay is steamed for efficient oil recovery.

5.3.3 Winterizing

The term "winterization" derives from the observation that refined cottonseed oil, stored in outside tanks during the winter months, will form a solid fraction. A batch process to mimic the winterization process has been used for decades to fractionate the solid portion of an oil (i.e., separate the higher melting fraction from the lower melting fraction). This process is applied to ensure that the finished oil will not easily become cloudy in cool storage conditions. The winterization process is commonly done in a chilled room held at 42°F (5.6°C) with the oil in deep, narrow, rectangular tanks. After 2 to 3 d, the solid portion of the oil will grow to desired crystal size, which is separated by a filtration step.

5.3.4 Deodorization

Deodorization, which removes the volatile compounds along with residual FFAs, is a critical step in ensuring the purity of any vegetable oil and improves flavor, odor, color, and oxidative stability. Many of the volatile compounds removed are formed by the auto-oxidation of fat, which produces aldehydes, ketones, alcohols, and hydrocarbons that are associated with undesirable rancid flavors and odors. The process also is effective in removing any remaining residues of pesticides or pesticide metabolites that may be in the oil.

Deodorization, which can be conducted as a batch operation in smaller plants or as a continuous or semicontinuous process in larger deodorizing facilities, consists of a steam distillation process in which the oil is heated to 230°C (446°F) under a vacuum of 2 to 10 mmHg. Steam is sparged through the oil to carry away the volatiles and provide agitation. The odor and flavor compounds, which are more volatile than the triglycerides, are preferentially removed. After deodorization and during the cooling stage, 0.005 to 0.01% citric acid is generally added to chelate trace metals, which can promote oxidation. Deodorized oils preferably are stored in an inert atmosphere of nitrogen to prevent oxidation. Tocopherols and sterols are also partially removed in the deodorization process. Tocopherols can be recovered from the deodorizer distillate in a separate operation.

The deodorization process is usually carried out as the last step prior to packaging or bulk shipping. If the oil is to be further processed, as described in the following section, the oil is normally processed through bleaching, the further processing step or steps (e.g., hydrogenation, esterification, fractionation, and formulation) are performed, and then the product is deodorized as the final step.

5.3.5 Further Processes of Fats and Oils

RBD oils are the largest and still growing segment in the U.S. fats and oils market. Fats and oils used for other major applications, such as shortenings for frying and baking, margarines, high-stability oil, and food emulsifiers, require additional processing steps to achieve the desired functional characteristics. In this section, only hydrogenation and blending, esterification and interesterification, and fractionation are briefly described [37].

5.3.5.1 Hydrogenation and Blending

The primary objective of hydrogenation is to modify the melting properties of oils. This is accomplished by converting a certain number of the unsaturated double bonds in the liquid oil to saturated bonds by reaction with hydrogen using a metal catalyst. Nickel is the most frequently used catalyst for this process. The targeted melting profile or solid content of the end product is controlled by the concentration of catalyst (0.01 to 1%), amount of hydrogen (hydrogen pressure 10 to 60 psig), temperature

(380 to 500°F), and reaction time (30 to 90 min). Most hydrogenation processes are carried out in a batch operation. Since the hydrogenation process is not entirely selective, several side reactions can also occur during the process, which create both geometric and positional isomers. Naturally occurring double bonds in the liquid oil are normally in the *cis* form, where the hydrogen atoms are on the same sides of the double bonds. Instead of saturating the double bond, the hydrogenation process can create a new double bond with the two hydrogen atoms on the opposite side of the double bond, which is the *trans* isomer. The formation of the *trans* double bonds is often accompanied with a migration of the double bonds from their naturally occurring positions. These are positional isomers. All three events occur simultaneously during the hydrogenation reaction but the degree of hydrogenation can be controlled in a repeatable manner. A series of hydrogenated oils with various melting profiles can be generated and serve as the base stocks. These hydrogenated base fats will then be used to formulate or blend into the final margarine, baking shortening, stable frying shortening, and the like. Hydrogenation and blending have many advantages in meeting the market demands: they are flexible, easy to manage, and cost effective. During the last 50 to 60 years, hydrogenation and blending gradually have replaced most of the traditional applications where animal fats were used.

The formation of *trans* fats by the hydrogenation process has become an important health-related issue. As a result, the U.S. Food and Drug Administration (FDA) published a final rule in 2003, which starting on January 1, 2006, requires manufacturers to list *trans* fats on the nutrition facts panel on food packaging and some dietary supplements sold at retail (see section 5.4.3.2). This regulation has caused the vegetable oil producing and extracting industries to reformulate many products. Instead of partially hydrogenating soy oil, some manufacturers are totally hydrogenating some soy oil and adding that to unhydrogenated soy oil to get the desired consistency and functionality. Other methods are being considered to lower or eliminate *trans* fats in human edible products.

5.3.5.2 Esterification and Interesterification

The esterification reaction between hydroxyl-rich moieties (glycerol, propylene glycol, polypropylene glycol, and sucrose) and fats or fatty acids are frequently applied to produce food emulsifiers, such as monoglycerides, propylene glycol monoglycerides, polyglycol esters, and sucrose esters. The interchange of fatty acids between two different fats and oils is called interesterification. Both esterification and interesterification reactions are commonly catalyzed by sodium methoxide. Interesterification can also be carried out using lipase. Chemically catalyzed interesterification is not selective like lipase interesterification; therefore, it is sometime called randomization or random rearrangement. Even though randomization is a mature technology, randomized interesterification was never popular due to the cost, limited choices of base fat, difficulties in carrying out the process, and product quality control. Therefore, lipase interesterification is favored by the industry as the means to produce various base fats for formulation and blending. Recently, nutritional concerns about *trans* fatty acids generated during the hydrogenation process have prompted scientists to reactivate the interesterification technology as the means to deliver functional fats.

5.3.5.3 Fractionation

Winterization is the most common type of fractionation used in the fats and oils industry. This technology is often used to fractionate cottonseed and palm oil into olein (liquid and oleic acid-rich) and stearin (solid at room temperature and stearic acid-rich) fractions. Fractionation using solvent, combined with temperature control, is used in deoiled lecithin production and for very high stability liquid oil from cottonseed for coating applications. Acetone is the primary solvent used in these fractionation applications.

5.4 REGULATORY ISSUES

Safety, health, and environmental issues for edible oil extraction facilities vary depending on the extraction method used. For example, aqueous extraction, supercritical fluid extraction, and mechanical extraction facilities have different requirements than those of organic solvent extraction facilities. Many workplace, environmental, food safety, and other regulations apply to oilseed processors and oil refiners [38]. Some of the regulations required in the United States are discussed here. Many other countries have similar requirements, but if they do not, it would be prudent for oilseed solvent extraction operations and vegetable oil refiners to consider meeting these regulations and for these industries to have environmental, health and safety, and quality management programs [39–41]. A glossary of terms for the abbreviations used in this section is included at the end of this chapter.

5.4.1 Workplace Regulations (OSHA)

Workplace regulations are promulgated and enforced in the United States by the Occupational Safety and Health Administration (OSHA), which is part of the U.S. Department of Labor, under the Occupational Safety and Health Act (OSH Act; PL 91-596 as amended by PL 101 552; 29 U.S. Code 651 etseq.). OSHA general industry health and safety standards (29 CFR 1910) apply to oilseed extraction and oil refining. In addition, even if there is not a specific standard, OSHA can site a facility under the "general duty clause" (Sec. 5(a)(1) of the OSH Act), since the OSH Act requires the employer to maintain a safe and healthful workplace. Some health and safety standards that affect oil extraction and refining are listed below.

5.4.1.1 OSHA Health Standards

- Air Contaminants Rule, 29 CFR 1910.1000 (the permissible exposure limit [PEL] for *n*-hexane is 500 ppm [1800 mg/m^3], 8 h time-weighted average; PELs for sodium hydroxide, sulfuric acid, vegetable oil mist, nuisance dust, etc.)
- Hazard Communication Standard, 29 CFR 1910.1200
- Cotton Dust Standard, 29 CFR 1910.1043 (cottonseed oil mills having medical surveillance, record keeping, and reporting requirements)
- Blood-Borne Pathogens, 29 CFR 1910.1030
- Occupational Exposure to Hazardous Chemicals in Laboratories, 29 CFR 1910.1450

5.4.1.2 OSHA Safety Standards

- Process Safety Management, 29 CFR 1910.119 (for *n*-hexane)
- Emergency Action Plan, 29 CFR 1910.38(a)(1)
- Fire Prevention Plan, 29 CFR 1910.38(b)(1)
- Fire Brigades, 29 CFR 1910.156
- Permit-Required Confined Space, 29 CFR 1910.146
- Lockout-Tagout, 29 CFR 1910.147
- Occupational Noise Exposure, 29 CFR 1910.95 and
  - Hearing Conservation Program, 29 CFR 1910.95(c)
- Ergonomics — no standard but can be regulated under the OSHA "general duty clause"
- Personal Protection Equipment
  - General Requirements, 29 CFR 1910.132
  - Eye and Face Protection, 29 CFR 1910.133
  - Respiratory Protection, 29 CFR 1910.134
  - Head Protection, 29 CFR 1910.135
  - Foot Protection, 29 CFR 1910.136

5.4.2 Environmental Regulations (EPA)

The U.S. EPA administers all regulations affecting the environment and chemicals in commerce. EPA regulations are intended to protect human health and welfare and the environment. The individual states and state environmental regulatory control boards implement and enforce most of the regulations.

The legislation that serves as the basis for the regulations can be divided into

1. Statutes that are media-specific (Clean Air Act [CAA] and Clean Water Act [CWA])
2. Statutes that manage solid and hazardous waste (Resources Conservation and Recovery Act [RCRA] and Comprehensive Environmental Response, Compensation and Liability Act [CERCLA; "Superfund"])
3. Statutes that directly limit the production rather than the release of chemical substance (Toxic Substances Control Act [TSCA] and Federal Insecticide, Fungicide and Rodenticide Act [FIFRA])

Some of the more important environmental regulations that affect oilseed extraction and refining are given below.

5.4.2.1 Clean Air Act (CAA; 42 U.S. Code 7401 et seq.)

States and state air control boards are required to implement regulations and develop state implementation plans (SIP) [42]. HAPs, such as *n*-hexane, are regulated with National Emissions Standards for Hazardous Air Pollutants (NESHAP) and criteria pollutants (e.g., ozone [O_3], particulate matter [PM], nitrogen oxides [NO_x], sulfur oxides [SO_x], carbon monoxide [CO], and lead [Pb]) are regulated with National Ambient Air Quality Standards (NAAQS).

n-Hexane is a regulated HAP, but isohexane and acetone are not. Regulated criteria pollutants such as O_3, PM, CO, and NO_x, can also be emitted during the extraction and refining of cottonseed oil. Industrial boilers emit some of these criteria pollutants as well as HCl and other regulated pollutants. In exhaust gases from stationary reciprocating internal combustion engines (RICE), about 27 HAPs have been measured, mainly formaldehyde, acrolein, methanol, and acetaldehyde.

5.4.2.1.1 HAPs or Air Toxics (40 CFR 61)

If a facility is a major emitter of *n*-hexane, the EPA requires sources to meet national emissions standards [43,44]. The air toxic control measures for source categories are technology-based emission standards (not health-based) established for major sources (10 ton/year of one HAP or 25 ton/year of total HAP) that require the maximum degree of reduction emissions, taking costs, other health and environmental impacts, and energy requirements into account. Compliance with an NESHAP involves the installation of maximum achievable control technology (MACT); MACT essentially is maximum achievable emission reduction. The NESHAP for Solvent Extraction for Vegetable Oil Production (4/12/01, 65 FR 34252; 40 CFR 63 subpart GGGG) requires all existing and new solvent extraction processes that are major sources to meet HAP emission standards, as a 12-month rolling average based on a 64% *n*-hexane content. Covered solvent extraction facilities are those that produce crude vegetable oil and meal products by removing crude oil from listed oilseeds (corn germ, cottonseed, flax, peanuts, rapeseed, safflower, soybeans, and sunflower) through direct contact with a solvent. HAP emission standards (solvent loss factor) for cottonseed oil production are as follows: for cottonseed, large (process >120,000 ton/year) 0.5 gal/ton; and for cottonseed, small (<120,000 ton/year) 0.7 gal/ton. Compliance with the vegetable oil NESHAP was required April 12, 2004. Since the emission loss factor values are 12-month rolling averages, the first compliance report was due no sooner than 48 months after the standard was promulgated (i.e., April 12, 2005).

If a solvent extraction facility, as of April 12, 2004, was not using commercial hexane, and the new solvent was addressed in their Title V federal operating permit, then they are not covered by the vegetable oil NESHAP. However, if a facility switched from commercial hexane to another extraction solvent that was low-HAP or non-HAP after 4/12/04, they would have been covered by all the burdensome recordkeeping and reporting requirements of the vegetable oil NESHAP, except for a direct final rule amendment (40 CFR 63.2840) to the vegetable oil NESHAP in 2004 (9/01/04; 69 FR 53338) for pollution prevention. This amendment eliminates the recordkeeping and reporting requirements that are unnecessary for determining compliance at these vegetable oil production facilities that exclusively use a qualifying low-HAP (e.g., isohexane containing less than 1% *n*-hexane) or non-HAP (e.g., acetone) extraction solvent. The new requirements are minimal-vegetable oil producers only need to document and report that they are using a solvent that meets the low-emitting criteria.

If a solvent extraction facility is not covered by the vegetable oil NESHAP but is using a volatile organic liquid (VOL; [40 CFR 60.11b(k)], any organic liquid that can emit volatile organic compounds into the atmosphere) with storage vessels ≥75 m^3, it would be covered by the Standards of Performance for Volatile Organic Liquid Storage Vessels (Including Petroleum Liquid Storage Vessels) (40 CFR 60 subpart Kb). There are control and record-keeping requirements. However the VOL requirement does not apply to vegetable oil processing facilities that are using commercial hexane and are in compliance with the solvent loss rate in the vegetable oil processing NESHAP (i.e., vessels subject to 40 CFR 63 subpart GGGG).

Vegetable oil processing plants also may be covered by the Industrial Boiler MACT (NESHAP for Industrial, Commercial, and Institutional Boilers and Process Heaters; 69 FR 55218, 9/13/04; 40 CFR 63 Subpart DDDDD) and the RICE MACT [NESHAP for stationary Reciprocating Internal Combustion Engines; 69 FR 33474, 6/15/04; 40 CFR 63 Subpart ZZZZ). The RICE MACT targets engines with a site rating of more than 500 brake hp; The U.S. EPA is also intending to promulgate a rule for small engines (<500 hp) located at major sources by December 2007.

5.4.2.1.2 NAAQS

The NAAQS are set at levels sufficient to protect public health, including the health of sensitive populations (primary air quality standards) and public welfare (secondary air quality standards) from any known or anticipated adverse effect of the pollutant with an adequate (appropriate) margin of safety.

VOCs are essentially considered the same as the criteria for pollutant ozone [41,42]. *n*-Hexane and hexane isomers are VOCs. Most U.S. cottonseed oil extracting facilities would be major sources of VOCs and would be covered by the requirements for ozone emissions and attainment, unless they used a solvent that was not classified as a VOC (e.g., acetone).

Most vegetable oil production facilities are major sources of particulate matter (PM). (see the Glossary of Terms). Depending on the oilseed processed, PM emissions can be 0.1 to 0.3 lb of total suspended particulate (TSP), which is about 50% PM 10 and less than 2% PM 2.5, per ton of seed processed. PM controls would also have to be part of a facility's federal and state permits. Cottonseed oil production facilities probably also have to include NO_x, SO_x, and CO emissions in their federal and state permits.

Any new or significantly modified facility would have to comply with the new source review (NSR) requirements. NSR is a preconstruction permitting program. If new construction or making a major modification will increase emissions by an amount large enough to trigger NSR requirements, then the source must obtain a permit before it can begin construction.

5.4.2.1.3 Odor

There are no specific federal regulations for odor. However, since odor is a nuisance, states can regulate odor if they choose to and some states do (e.g., Colorado and Idaho).

5.4.2.1.4 Federal Permits (40 CFR 70)

All major sources of regulated solvents are required to have federally enforceable operating permits (FOP) [42], also referred to as Title V permits. If a facility is a major source for any pollutant, e.g., hexane, and thus requires a Title V federal permit, then all emissions of HAPs have to be included in their federal operating permit.

5.4.2.1.5 State Permits

Most states require state permits for facilities that emit listed air pollutants [43,44]. In some states federal permits and state permits are combined, while in other states facilities are required to have both a state or county (air district) permit and a federal permit. As part of annual emission inventory reporting requirements, many states already require reporting of HAP and VOC because of their state implementation plan (SIP).

5.4.2.2 Clean Water Act (CWA; 33 U.S. Code 1251 et seq.)

Under the CWA, the U.S. EPA establishes water quality criteria used to develop water quality standards, technology-based effluent limitation guidelines, and pretreatment standards and has established a national permit program (National Pollution Discharge Elimination System [NPDES] permits; 40 CFR 122) to regulate the discharge of pollutants. The states have the responsibility to develop water quality management programs and enforce most of the regulations promulgated pursuant to the CWA.

Oilseed processing and oil refining may be covered by the following [45]:

- Basic discharge effluent limitations require NPDES permits (40 CFR 122) for pollutant discharges.
- Storm water regulations require a storm water permit (40 CFR 122 and 123); Phase I covers industrial activities and construction sites (land disturbing activities) of >5 acres; Phase II addresses construction sites of 1 to 5 acres (64 FR 68722, December 8, 1999; coverage due March 10, 2003). The storm water permit for runoff from industrial activities is now part of the NPDES permit; for construction a construction general permit is required.
- Oil spill prevention and response plans (40 CFR 112) were amended in 2002 to have an aggregate threshold of 1320 gal for oil spill prevention plan (spill prevention, control, and countermeasures [SPCC] plan). The response plan threshold is 42,000 gal and requires a federal response plan.
- Requirements applicable to cooling water intake structures (33 U.S. Code Section 316[b]) (40 CFR 125) cover new manufacturing facilities (65 FR 65256, December 12, 2001; 68 FR 36749, June 19, 2003) and are being developed for existing manufacturing facilities. Regulations are to ensure that the location, design, construction, and capacity of cooling water intake structures reflect the best available technology available for minimizing adverse environmental impact (applies to point sources that have a design intake flow >2 million gal/d).

5.4.2.3 Resource Conservation and Recovery Act (RCRA; 42 U.S. Code 6901 et seq.)

RCRA Subtitle D covers nonhazardous wastes. Subtitle C (40 CFR 261) is a federal "cradle-to-grave" system to manage hazardous waste (including provisions for cleaning up releases and setting statutory and regulatory requirements). Materials or items are hazardous wastes if and when they are discarded or intended to be discarded. Hazardous wastes are either listed wastes (40 CFR 261.30–.33) or characteristic wastes (40 CFR 261.21–.24). The U.S. EPA defines four characteristics for hazardous waste: ignitability (40 CFR 260.21); corrosivity (40 CFR 260.22); reactivity (40 CFR 260.23); and toxicity (40 CFR 260.24).

5.4.2.3.1 *Spent Bleaching Clay*

This is not an RCRA hazardous waste (40 CFR 302). It is usually disposed of by taking it to a regular landfill. There sometimes can be a spontaneous combustion problem (oxidation of unsaturated fatty acids in the retained oil, causing self-heating and leading to combustion) when it is taken to the landfill. The potential for spontaneous combustion in bleaching earth depends on the type and amount of oil retained and rises with increasing unsaturation of the fatty acids in the retained oil. The U.S. Department of Transportation classifies materials liable to spontaneous combustion as Class 4.2 hazardous materials (49 CFR 173.124 [b] and Appendix E 3). Spent bleaching clay can be finely ground and put in small quantities into animal meal in operations that do oil extraction. With the increase concern about dioxin in food and feed product by the FDA and EPA, this is discouraged.

5.4.2.3.2 *Spent Nickel Catalyst*

This is not considered an RCRA hazardous waste (40 CFR Table 302.4). No reporting of release of this substance is required if the diameter of the solid metal released is equal to or exceeds 100 μm. The reportable quantity (RQ) for particles greater than 100 μm is 100 lb. Most, if not all, spent nickel catalyst is recycled.

5.4.2.4 Emergency Planning and Community Right-to-Know Act (EPCRA; 42 U.S. Code 11001 et seq.)

EPCRA requires states to establish emergency planning districts with local committees to devise plans for preventing and responding to chemical spills and releases.

5.4.2.4.1 *Section 313 (40 CFR 372), Toxic Release Inventory (TRI)*

Businesses are required to file annual reports with federal and state authorities of releases to air, water, and land above a certain threshold for chemicals on the TRI/Section 313 list (40 CFR 372.65) by July 1 each year for the previous year's releases [46a]. TRI requirements are triggered if a facility is involved in manufacturing with 10 or more full-time employees, manufactures, processes, or other uses with one or more listed substances in a quantity above the statutory reporting threshold of 25,000 lb/year (manufactured or processed) or 10,000 lb/year (other use). Beginning with the 1991 reporting year, such facilities also must report pollution prevention and recycling data for such chemicals pursuant to Section 6607 of the Pollution Prevention Act (42 U.S. Code 13106). *n*-Hexane was added to the TRI list in 1994 with reporting for 1995 emissions [40,41]. Isohexane is not on the TRI list.

5.4.2.5 Toxic Substances Control Act (TSCA; 15 U.S. Code 2600 et seq.)

If a chemical's manufacture, processing, distribution, use, or disposal would create unreasonable risks, the U.S. EPA, under the TSCA (40 CFR section 700, et seq.), can regulate it, ban it, or require additional testing.

5.4.2.5.1 Inventory Update Rule (IUR) (40 CFR 710)

The IUR was established in 1986 to require manufacturers and importers of chemicals listed on the master TCSA inventory to report current data every 4 years on the production volume of chemicals imported or produced. The IUR was amended in 2003 (IURA) to add more reporting and use requirements for chemicals with a production volume greater than 300,000 lb. Food and feed products produced from natural agricultural products, such as oilseeds, are not required to be reported, but all oil and meal products obtained by solvent extraction that are sold for other than food or feed use (e.g., oils as chemical raw materials and meal as fertilizer) are. Cottonseed oil and other vegetable oils, soap stocks, acidulated soap stocks, deodorized distillates, and hydrogenated vegetable oils are some of the substances reported by extraction and refining operations under the IUR.

5.4.3 Food Safety (FDA)

In the United States, the U.S. FDA regulates all aspects of food under the Federal Food, Drug, and Cosmetic Act (FFDCA; 21 U.S. Code 321 et seq.) including food ingredients and labeling.

5.4.3.1 Premarket Approval of Solvents/Other Processing Aids

An extraction solvent or other processing substances that can be in food are subject to premarket approval by the U.S. FDA unless its use is generally recognized as safe (GRAS) by qualified experts [47]. Oilseed extraction solvents and food processing substances, to be legally used in the United States, must have been subject to an approval by the FDA or the U.S. Department of Agriculture (USDA) from 1938 through 1958 for this use (prior sanction); must be GRAS for this use by the FDA "GRAS affirmation" (21 CFR 170.35) (substances do not have to be specifically listed and there are several ways to determine GRAS); or must be used in accordance with food additive regulations promulgated by the FDA (21 CFR 170.3[h][I]). Food additives generally fall into two broad categories: (1) those added directly to food (21 CFR 172), and (2) those added indirectly to food through contact with packaging materials, processing equipment, or other food-contact materials (21 CFR 174–178).

Many prior sanctions and early GRAS determinations are not codified in the U.S. FDA regulations. Extracting solvents used in food manufacturing, such as *n*-hexane or isohexane, have been labeled as a food additive or incidental additives (i.e., "additives that are present in a food at insignificant levels and do not have any technical or functional effect in that food"). Incidental additives can be "processing aids" (i.e., "substances that are added to a food during processing but removed from the food before it is packaged"). Most food-processing substances, including solvents, can be regarded as "incidental additives" and thus are exempt from label declaration in the finished food product.

Since vegetable oil and other human food-grade oils undergo refining, bleaching, deodorization (steam distillation), and sometimes other purification processes as part of the manufacturing process prior to being used as a food product, they should not contain any (0 to <100 ppb) of the extraction solvent, if proper manufacturing practices are followed.

Commercial hexane, containing about 50 to 85% *n*-hexane (the rest are the hexane isomers, which make up commercial isohexane), has been in major use since the 1940s as an oilseed extraction solvent on the determination that it is GRAS and it may also be subject to a prior sanction. Like many other food-processing substances, there is no U.S. FDA regulation specifically listing *n*-hexane as GRAS or prior sanctioned. However, under FDA regulations hexane has been cleared as a solvent (residue not more than 5 to 25 ppm) for use in many products [47,48]. Isohexane/hexane isomers also are not specifically listed as GRAS or prior sanctioned.

In Europe, the maximum residue limit (MRL) for *n*-hexane in vegetable oils has been established as 5 ppm *n*-hexane (European Union [EU], 1988, Community Directive 88/344/EECC of June 13, 1988; *Official J. Eur. Commun.* L 157, 24 June 24, 1988, pp. 0028–0033). There is no MRL for isohexane/hexane isomers.

In summary, extraction solvents can be considered incidental additives/processing aids that are exempt from label declaration. GRAS status may be determined by a company or an industry ("GRAS self-determination" [49] or "GRAS notification" [April 17, 1997; 62 FR 18938]), an independent scientific organization (e.g., "FEMA GRAS" [50]), or the U.S. FDA (GRAS affirmation [21 CFR 170.35]). The FFDCA (21 U.S. Code 321 et seq.) does not provide for the U.S. FDA to approve all ingredients used in food, and the U.S. FDA explicitly recognizes that its published GRAS list is not meant to be a complete listing of all substances that are in fact GRAS food substances. Although there is no requirement to inform the U.S. FDA of a GRAS self-determination or to request FDA review or approval on the matter, the U.S. FDA has established a voluntary GRAS affirmation program under which such advice will be provided by the agency. Solvents that do not have prior sanction, a GRAS determination of some kind, or a tolerance set probably should be evaluated for compliance under food safety requirements, if a facility is considering changing its extracting solvent.

5.4.3.2 *trans* Fat Labeling

In Europe *trans* fatty acid labeling of retail foods is required. In July 2003, the U.S. FDA published a final rule requiring manufacturers to list *trans* fats on the nutrition facts panel of food and some dietary supplements sold at retail stores (68 FR 41434; July 11, 2003; 21 CFR 101). This rule, which goes into effect on January 1, 2006, requires all foods containing at least 0.5 g per serving of *trans* fat to declare the amount of *trans* fat per serving on the nutrition facts panel; products containing <0.5 g per serving are required to express *trans* fat content as "0."

5.5 FUTURE TRENDS

In the future, there most likely will be new demands for highly specialized extraction solvents as newly domesticated species that make useful novel oils [51] and other products. New or altered biological products with enhanced nutritional and industrial properties will be developed through conventional breeding and genetic engineering for use as "functional foods" [52] (e.g., phytosterols to achieve cholesterol lowering); as oils with altered lipid profiles [53] (e.g., for lower saturated fat) or with more vitamin E; as new drugs/nutraceuticals and industrial chemicals (e.g., fatty acids for lubricants, and for cosmetics, coatings, detergents, surfactants, flavors, and polymers); as sources for specialty chemicals; and as value-added products [54–59]. There will be demands for solvent systems for the simultaneous removal of undesirable meal components (e.g., mycotoxins, gossypol, flavors, and odors) that offer the potential for upgrading meal for use as higher value animal feeds and human foods. Solvents that offer energy savings and that pose lower health, environmental, and fire hazards will also be sought.

Oil refining process innovation will continue to be focused in the area of more efficient operation, lower energy demand, least or no waste generated, and minimum undesirable chemical modifications of the oils. Interesterification and fractionation may become more important choices of operation than hydrogenation to minimize the formation of geometric (*trans*- v. *cis*-double bonds) and positional (repositioned double bonds) isomers of unsaturated fatty acids in the final fats and shortenings.

Genetically engineered/biotech crops make up a growing share of the agricultural output [60,61]. In the United States in 2003, about 78% of the cotton acreage (21% globally; about 33% world production) [62], about 85% of the soy acreage (55% globally), about 40% of the corn acreage (11% globally), and about 16% of the canola acreage globally were biotech crops [61]. Biotechnology is a powerful tool in the hands of agricultural scientists. The ability to breed desirable traits or eliminate problematic ones can yield potentially spectacular benefits, such as various chemicals of importance, including improved fats and oils (e.g., high oleic acid oils), vaccines and medicine, improved nutrition (e.g., casaba, oilseeds, rice, and sweet potatoes), and improved yields with the use of fewer agricultural chemicals. Biotech crops could be increasingly developed as "biofactories" for a wide range of products, including nutrients, pharmaceuticals, and plastics. There is much promise for being able to produce products that would protect millions from disease, starvation, and death.

However, biotechnology and biotech crops have become very controversial and have run into serious problems in the European Union (EU), particularly in the United Kingdom [63]. Currently, traceability and labeling for biotech products for whole food products and animal feeds are required in some places in the world (e.g., the EU, Japan, Mexico, Australia, and New Zealand). Highly refined foods (e.g., vegetable oil, food thickeners, and starches) are explicitly excluded, except in the EU. In the EU, as of April 18, 2004, all foods, feeds, and additives are covered [64,65]. Vegetable oilseed meal, vegetable oil, and other food products made from cottonseed and other vegetable oils are covered or will be covered by the EU traceability and labeling regulations. For example, genetically modified (GM) feed is covered at present, and food additives produced from genetically modified organisms (GMOs) (e.g., refined cottonseed oil), feed additives produced from a GMO, and feed produced from a GMO are not covered at present but will be in the future. Even if DNA or protein from the transgenic plant cannot be detected in the refined vegetable oil, documentation is required to show that the product did not come from a biotech plant to be able to label a product "GM free." "Adventitious or unavoidable" presence of a biotech product has a threshold of 0.9% for EU-approved GMOs and 0.5% for EU-unapproved GMOs that have been through the regulatory process of the exporting country—after 3 years the threshold will be 0%. Thus, even though this technology has great promise for increased use of new and existing solvents for extraction of products from diverse biological materials, there are also many potential problems because of misperceptions and misinformation.

The *trans* fat labeling regulation (See section 5.4.3.2) is resulting in reformulation of many products by the vegetable oil producing and extracting industries, such as totally hydrogenating some soy oil and adding that to unhydrogenated soy oil to get the desired consistency and functionality. Other research is underway to develop other methods to lower or eliminate *trans* fats edible products.

It is clear that the future has much uncertainty at the same time that it offers much promise. As our knowledge about nutritional requirements and quality management of fats and oils is enhanced, more cost-efficient and innovative processing technologies will be needed. Continued advancement in software and hardware will further promote more accurate analytical methods. The new generation of professionals that enter into the fats and oils business with their improved knowledge and tools will likely face many exciting and more complicated challenges.

GLOSSARY OF TERMS

ACGIH American Conference of Industrial Hygienists, an independent standards setting organization

BACM Best Available Control Measures

BACT Best Available Control Technology

CAA Clean Air Act, 42 U.S. Code 1251 et seq.

CERCLA (Superfund) Comprehensive Environmental Response, Compensation, and Liability Act, 42 U.S. Code 9601 et seq.

CFR Code of Federal Regulations, where the U.S. federal regulations after promulgation are codified, The preceding number, the Title, the succeeding number (after CFR); is the Part of Section (e.g., 29 CFR 1910, Title 29 Code of Federal Regulations at Part 1910)

CWA Clean Water Act (Federal Water Pollution Control Act), 33 U.S. Code 1251 et seq.

EPA Environmental Protection Agency, 42 U.S. Code 4321 et seq.

EPCRA Emergency Planning and Community Right-to-know Act, part of CERLA/Superfund, Title III of SARA, the 1986 amended Superfund

FR Federal Register; where regulatory announcements and new rules and their justification are published; preceding number, the volume, the succeeding number (after FR); the page, usually followed by the date when it appeared (e.g., 51 FR 27956, Volume 51 Federal Register, page 27956)

GRAS Generally Recognized As Safe

HAP Hazardous Air Pollutant, 40 CFR 61

HCS Hazard Communication Standard, 29 CFR 1910.1200

IUR Inventory Update Rule, 40 CFR 710
LAER Lowest Achievable Emission Rate
MACT Maximum Achievable Control Technology
MSDS Material safety data sheet, required under OSHA HCS
NAAQS National Ambient Air Quality Standard, 40 CFR 50
NESHAP National Emission Standard for Hazardous Air Pollutants under the CAA
Nonattainment Areas that are not meeting NAAQS, 40 CFR 51.100 et seq.
NPDES National Pollution Discharge Elimination System, the national permit program under the CWA, 40 CFR 122
NSR New Source Review
OPA-90 Oil Pollution Act of 1990
OSHA Occupational Safety and Health Administration (part of the U.S. Dept. of Labor), 29 U.S. Code 651 et seq.
Ozone (O_3) One of the compounds on the NAAQS list that is formed through chemical reaction in the atmosphere involving VOC, NO_x, and sunlight; also a primary constituent of smog
NO_x Nitrogen Oxides
PEL Permissible Exposure Limit for an air contaminant under OSHA standards
PM Particulate Matter; one of the NAAQS; denotes the amount of solid or liquid matter suspended in the atmosphere; the EPA regulating PM as PM_{10} ("coarse" particulate 10 μm and less) and $PM_{2.5}$ ("fine" particulate 2.5 μm or less)
POTW Publicly Owned Treatment Works, for indirect wastewater discharge
PSD Prevention of Significant Deterioration, a requirement of NSR
PSM Process Safety Management standard
RACM Reasonably Achievable Control Measures
RACT Reasonably Available Control Technology
RCRA Resource Conservation and Recovery Act, 42 U.S. Code 6901 et seq.
RCRA-Characteristic Wastes Hazardous wastes that are ignitable, corrosive, reactive, or toxic, 40 CFR 260.64
RQ Reportable quantity
RCRA-Listed Wastes Specially listed hazardous wastes in 40 CFR 261.30–33
SARA Superfund Amendments and Reauthorization Act
TCLP Toxic characteristic leaching potential under RCRA, 40 CFR 261.24
Title V The part of the Clean Air Act that deals with federal permits, 40 CFR 70
TLV Threshold Limit Value for an air contaminant under ACGIH regulations
TRI Toxic Release Inventory, under Section 313 of EPCRA
TSCA Toxic Substances Control Act
TWA Time-weighted average
U.S. Code The United States Code where legislation, including health, safety, and environmental legislation, is codified once it is passed by Congress (e.g., 42 U.S. Code 7401 is Title 42 U.S. Code at paragraph 7401)
VOC Volatile Organic Compounds; a group of chemicals that react in the atmosphere with nitrogen oxides (NO_x) in the presence of heat and sunlight to form ozone; does not include compounds determined by EPA to have negligible photochemical reactivity

REFERENCES

1. Wan, P.J. and Wakelyn, P.J. *Technology and Solvents for Extracting Oilseeds and Non-Petroleum Oils*, AOCS Press, Champaigne, IL, 1997, pp. 1–353.
2. Williams, M.A. and Hron, R.J. Obtaining oils and fats from source materials, in *Bailey's Industrial Oils and Fat Products,* 5th ed, Vol. 4: Edible Oil and Fat Products: Processing Technology, Hui, Y.H., Ed., John Wiley Sons, New York, 1996, pp. 61–154.
3. Wan, P.J. et al. Alternate hydrocarbon solvents for cottonseed extraction: plant trials, *J. Am. Oil Chem. Soc.,* 72, 661, 1995.
4. Horsman, M. Using isohexane to extract cottonseed, *Oil Mill Gaz*., 105(8), 20, 2000.
5. Trading Rules, National Cottonseed Products Association. Memphis, TN, 61. 2001, P.
6. Wakelyn, P.J. et al. Acetone: an environmentally preferable choice? *INFORM,*12, 887, 2001.
7. Karnofsky, G. Theory of solvent extraction, *J. Am. Oil Chem. Soc*. 26, 564, 1949.
8. Bull, W.C. and Hopper, T.H. The composition and yield of crude lipids obtained from soybeans by successive solvent extractions, *Oil Soap*, 18, 219, 1941.
9. Coats, H.B. and Karnofsky, G. Solvent extraction II. The soaking theory of extraction, *J. Am. Oil Chem. Soc.,* 27, 51, 1950.
10. Good, R.D. Theory of soybean extraction, *Oil Mill Gaz.,* 75, 14, 1970.
11. Coats, H.B. and Wingard, M.R. Solvent extraction. III. The effect of particle size on extraction rate, *J. Am. Oil Chem. Soc.,* 27, 93, 1950.
12. Bull, W.C. Some observations on the effect of moisture on the quantitative extraction of lipids from soybeans, *Oil Soap,* 20, 94, 1943.
13. Reuther, Jr., C.G. et al. Solvent extraction of cottonseed and peanut oils. VIII. Of moisture on the preparation and flaking of cottonseed, *J. Am. Oil Chem. Soc*, 28, 146, 1951.
14. Arnold, L.K. and Patel, D.J. Effect of moisture on the rate of solvent extraction of soybeans and cottonseed meats, *J. Am. Oil Chem. Soc*, 30, 216, 1953.
15. Wingard, M.R. and Phillips, R.C. Solvent extraction IV. The effect of temperature on extraction rate, *J. Am. Oil Chem. Soc*, 28, 149, 1951.
16. Pominski, J. et al. Solvent extraction of cottonseed and peanut oils. IV. Pilot olant batch extraction, *Oil Mill Gaz*., 51(12), 33, 1947.
17. Sweeney, O.R. and Arnold, L.K. A new solvent extraction process for soybean oil, *J. Am. Oil Chem. Soc*., 26, 697, 1949.
18. Rao, F.K. and Arnold, L.K. Alcoholic extraction of vegetable oils. V. Pilot plant extraction of cottonseed by aqueous ethanol, *J. Am. Oil Chem. Soc*., 35, 277, 1958.
19. Wingard, M.R. and Shand, W.C. The determination of the rate of extraction of crude lipids from oil seeds with solvents, *J. Am. Oil Chem. Soc*., 26, 422, 1949.
20. Wan, P.J. et al. Alternative hydrocarbon solvents for cottonseed extraction, *J. Am. Oil Chem. Soc*., 72, 653, 1995.
21. Beckel, A.C., Belter, P.A., and Smith, A.K. Solvent effects on the products of soybean oil extraction, *J. Am. Oil Chem. Soc*., 25, 7, 1948.
22. Ayers, A.L. and Dooley, J.J. Laboratory extraction of cottonseed with various petroleum hydrocarbons, *J. Am. Oil Chem. Soc*., 25, 372, 1948.
23. Conkerton, E.J., Wan, P.J., and Richard, O.A. Hexane and heptane as extraction solvents for cottonseed: a laboratory study, *J. Am. Oil Chem. Soc*., 963, 1995.
24. Arnold, L.K. and Choudhury, B.R. Extraction of soybean and cottonseed oil by four solvents, *J. Am. Oil Chem. Soc*., 37, 458, 1960.
25. Wan, P.J. Isohexane as a solvent for extraction, *INFORM,* 7, 624, 1996.

26. Wan, P.J. Hydrocarbon solvents, in *Technology and Solvents for Extracting Oilseeds and Non-Petroleum Oils,* Wan, P.J. and Wakelyn, P.J., Eds., AOCS Press, Champaign, IL, 1997, pp. 170–185.
27. Johnson, L.A. Theoretical, comparative, and historical analyses of alternative technologies for oilseed extraction, in *Technology and Solvents for Extracting Oilseeds and Non-Petroleum Oils,* Wan, P.J. and Wakelyn, P.J., Eds., AOCS Press, Champaign, IL, 1997, pp. 4–47.
28. Hron, R.J. Acetone, in *Technology and Solvents for Extracting Oilseeds and Non-Petroleum Oils,* Wan, P.J. and Wakelyn, P.J., Eds., AOCS Press, Champaign, IL, 1997, pp. 186–191.
29. Hron, R.J. Ethanol, in *Technology and Solvents for Extracting Oilseeds and Non-Petroleum Oils,* Wan, P.J. and Wakelyn, P.J., Eds., AOCS Press, Champaign, IL, 1997, pp. 192–197.
30. Lucas, E.W. and Hernandez, E. Isopropyl alcohol, in *Technology and Solvents for Extracting Oilseeds and Non-Petroleum Oils,* Wan, P.J. and Wakelyn, P.J., Eds., AOCS Press, Champaign, IL, 1997, pp. 199–266.
31. Hagenmaier, R.D. Aqueous processing, in *Technology and Solvents for Extracting Oilseeds and Non-Petroleum Oils,* Wan, P.J. and Wakelyn, P.J., Eds., AOCS Press, Champaign, IL, 1997, pp. 311–322.
32. American Petroleum Institute (API). Neurotoxicity: n-Hexane and Hexane Isomers. Medical Research Pub. 30-32846 and 30-32858. Washington, DC, 1983.
33. Effects of n-Hexane in Man and Animals. Research Report 174-2, Berichte Deutsche Gesellschaft fur Mineralolwissenschaft und Kohlechemie e. V. Hamburg, Germany. 1982. (Distributed in English by the American Petroleum Institute, 2101 L St., N.W., Washington, DC.)
34. Galvin, J.B. Toxicity data for commercial hexane and hexane isomers, in *Technology and Solvents for Extracting Oilseeds and Non-Petroleum Oils,* Wan, P.J. and Wakelyn, P.J., Eds., AOCS Press, Champaign, IL, 1997, pp. 75–85.
35. U.S. Environmental Protection Agency. Commercial Hexane Final Test Rule. (53 Federal Register 3382; February 5, 1988).
36. Wakelyn, P.J. and Adair, P.K. Toxicity data for extraction solvents other than hexane and isohexane/hexane isomers, in *Technology and Solvents for Extracting Oilseeds and Non-Petroleum Oils,* Wan, P.J. and Wakelyn, P.J., Eds., AOCS Press, Champaign, IL, 1997, pp.86–100.
37. O'Brien, R.D., Farr, W.E., and Wan, P.J., Eds. *Introduction to Fats and Oils Technology,* AOCS Press, Champaign, IL, 2000, 618 pp.
38. Wakelyn, P.J. and Wan, P.J. Solvent extraction: safety, health, and environmental issues, in *Extraction Optimization in Food Engineering,* Tzia, C. and Liadakis, Eds., Marcel Dekker, New York, 2003, pp. 391–427.
39. Wakelyn, P.J. Regulatory considerations for extraction solvents for oilseeds and other nonpetroleum oils, in *Technology and Solvents for Extracting Oilseeds and Non-Petroleum Oils,* Wan, P.J. and Wakelyn, P.J., Eds., AOCS Press, Champaign, IL, 1997, pp. 48–74.
40. Wakelyn, P.J. and Adair, P.K. Assessment of risk and environmental management, in *Emerging Technologies, Current Practices, Quality Control, Technology Transfer, and Environmental Issues,* Vol. 1, Proc. of the World Conference on Oilseed and Edible Oil Processing, Koseoglu, S.S., Rhee, K.C. and Wilson, R.F., Eds., AOCS Press, Champaign, IL, 1998, pp. 305–312.
41. Wakelyn, P.J., Adair, P.K., and Gregory, S.R. Management of quality and quality control in cottonseed extraction mills, *Oil Mill Gaz.,* 103(6), 23, 1997.

42. Wakelyn, P.J. Overview of the Clean Air Act of 1990: cottonseed oil mills, *Cotton Gin Oil Mills Press,* 92(17), 12, 1991.
43. Wakelyn, P.J. and Wan, P.J. Regulatory considerations of VOC, HAP, *INFORM,* 9, 1155, 1998.
44. Wan, P.J. and Wakelyn, P.J. Regulatory considerations of VOC and HAP, from oilseed extraction plants. *Oil Mill Gaz.*, 104(6), 15, 1998.
45. Wakelyn, P.J. and Forster, L.A. Jr. Regulatory issues regarding air, water, solid waste and salmonella. *Oil Mill Gaz.*, 99(12), 21, 1994.
46. n-Hexane. U.S. EPA Integrated Risk Information System (IRIS) Substance File. U.S. EPA, 1999 (www.epa.gov/ngispgm3/IRIS/subst/0486.htm).

46a. US EPA. The Toxic Release Inventory (TRI) and Factors to Consider Using TRI Data. http://www.epa.gov/tri/2002_tri_brochure.pdf

47. Wakelyn, P.J. and Wan, P.J. Edible oil extraction solvents: FDA regulatory considerations, *INFORM,* 15, 22, 2004.
48. Wakelyn, P.J. Regulatory considerations for oilseed processors and oil refiners, in *Introduction to Fats and Oils Technology,* 2nd ed, Wan, P.J., O'Brien, R., and Farr, W., Eds., AOCS Press, Champaign, IL, 2000, pp. 302–325.
49. Allen, A.H. GRAS self-determination: staying out of the regulatory soup, *Food Product Design* (April 1996 supplement to Food Product Design, 6 pages), 1996.
50. Oser, B.L. and Ford, R.A. Recent progress in the consideration of flavoring ingredients under the Food Additive Amendments 6. GRAS Substances, *Food Technol.*, 27(1), 64, 1973.
51. Murphy, D.J. Development of new oil crops in the 21st century, *INFORM,* 11, 112, 2000.
52. Ryan, M.A. Functional foods hit supermarket shelves, *Today's Chemist at Work,* 8(9), 59, 1999.
53. Haumann, B.F. Structured lipids allow fat tailoring, *INFORM,* 8, 1004, 1997.
54. Flickinger, B. and Hines, E. A look into the future of food, *Food Quality,* 6(7), 18, 1999.
55. Hou, C.T. Value added products from oils and fats through bio-processes, in *Int. Symp. on New Approaches to Functional Cereals and Oils,* Beijing, China, Nov. 9–14, 1997, Chinese Cereals and Oils Association, Beijing, China, 1997, p. 669.
56. Kyle, D.J. New specialty oils: development of a DHA-rich nutraceutical product, in *Int. Symp. on New Approaches to Functional Cereals and Oils,* Beijing, China, Nov. 9–14, 1997, Chinese Cereals and Oils Association, Beijing, China, 1997, p. 681.
57. Daum, J.K. Modified fatty acid profiles in Canadian oilseeds, in *Int. Symp. on New Approaches to Functional Cereals and Oils,* Beijing, China, Nov. 9–14, 1997, Chinese Cereals and Oils Association, Beijing, China, 1997, pp. 659–668.
58. Henry, C.M. Nutraceuticals: fad or trend?, *Chem. Eng. News,* 77(48), 42, 1999.
59. Ohlson, R. Bonn symposia explore industrial uses, *INFORM,* 10, 722, 1999.
60. Persley, G.J. and Siedow, J.N. Applications of Biotechnology to Crops: Benefits and Risks, CAST Issue Paper no. 12, Council for Agricultural Science and Technology, Dec., 1999.
61. James, C. 2003. Global Status of Commercialized Transgenic Crops: 2003. International Service for the Acquition of Agri-biotech Applications (ISAAA). ISAAA Briefs no. 30-2003 (http://www.ISAAA.org).
62. Wakelyn, P.J., May,O.L., and Menchey, E.K. Cotton and biotechnology, in *Handbook of Plant Biotechnology,* Christou, P. and Klee, H., Eds., John Wiley & Sons, Chichester, UK, 2004, pp. 1117–1131.
63. Heylin, M. Ag biotech's promise clouded by consumer fear, *Chem. Eng. News,* 77(49), 73, 1999.

64. EU Traceability & Labeling Regulations: Regulation (EC) no. 1829/2003 of the European Parliament and of the Council of 22 Sept. 2003 on genetically modified food and feed. *Official Journal of the European Union* pp. L268/1–L268/23 [18 Oct. 2003]; Regulation (EC) no. 1830/2003 of the European Parliament and of the Council of 22 Sept. 2003 concerning the traceability and labeling of genetically modified organisms and the traceability of food and feed produces produced from genetically modified organisms and amending Directive 2001/18/EC. *Official Journal of the European Union* pp. L268/24–L268/28 [18 Oct. 2003].
65. EU Traceability & Labeling Implementation Guidance: Commission Regulation (EC) No. 641/2004 of 6 April 2004 on detailed rules for the implementation of Regulation no. 1829/2003 of the European Parliament and of the Council as regards the application for the authorization of new genetically modified food and feed, the notification of existing products and adventitious or technically unavoidable presence of genetically modified material, which has benefited from a favorable risk evaluation. *Official Journal of the European Union* pp. L102/14–L102/25 [7 April 2004].

Part II

Lipids for Food Functionality

6 Physical Properties of Fats and Oils

Amanda J. Wright and Alejandro G. Marangoni

CONTENTS

6.1 INTRODUCTION

Understanding the physical properties of lipids is critical to the formulation and manipulation of fats, oils, and fat-structured food products. Rheological properties, and ultimately texture, are closely related to the structures of underlying fat crystal networks (FCNs) in foods. Figure 6.1 shows the hierarchy of parameters that determine the physical properties of lipids and lipid-based materials. Composition and processing parameters determine the crystalline structure and rheological properties of FCNs. Differences at the nano- and microstructural levels in fats translate into

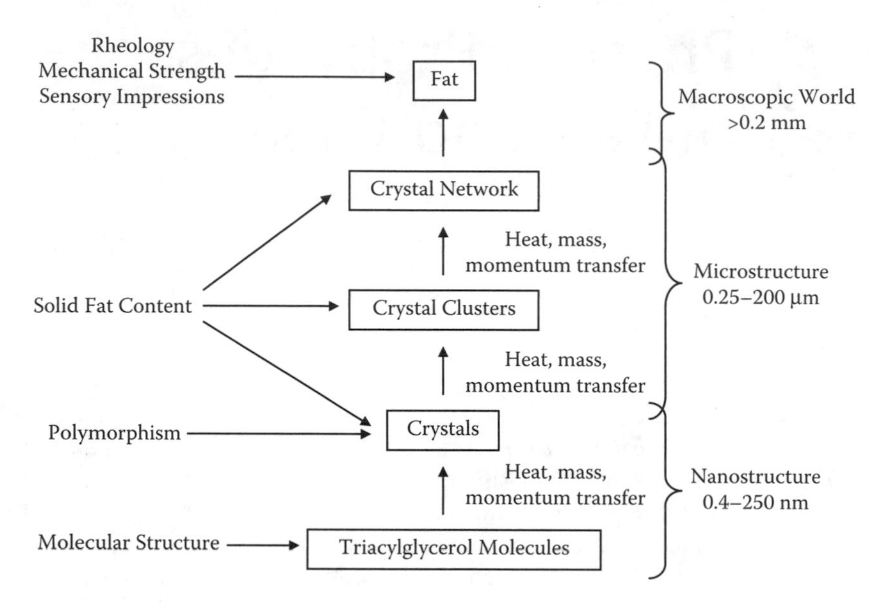

FIGURE 6.1 Levels of lipid structure that determine the rheological properties and texture of fats and fat-based foods.

differences in terms of functionality and sensory properties. Other networks, including protein, polysaccharide, and ice, may also be present in foods. Each network and the interactions between ingredients all contribute to the overall sensory impressions of foods. This chapter will review the physical properties of lipids, including melting, crystallization, phase behavior, polymorphism, microstructure, and rheology, and the ways in which the physical properties of fats can be modified.

6.2 THE PHYSICAL PROPERTIES OF LIPIDS

6.2.1 The Effects of Composition

Molecular composition lays the foundation for a fat's physical character. Triacylglycerols (TAGs) constitute the bulk of the lipid mass in natural fats and oils, although minor lipids and other lipid-soluble components may also be present and can impact on the behavior of fats and oils [1–3]. Considering the large number of fatty acids (FAs) that exist in nature and the potential for positional variability along the glycerol backbone, a wide range of TAGs is possible. For example, over 400 different FAs have been identified in milk fat [4]. Fatty acids may be saturated or unsaturated, branched or linear, and short- or long-chained [5]. Saturated FA chains tend to adopt a straight-chain configuration, with the long carbon chains adopting an in-plane zig-zag pattern. In the case of unsaturated FAs, a bend occurs at the position of the double bond and leads to a decrease in packing density. Because of their closer packing, saturated FAs have higher melting temperatures than their corresponding unsaturated FAs in the *cis* configuration. Geometric isomerism at sites of unsaturation allows for double bonds to exist in either the *cis* or *trans* configuration. *Trans* fatty acids (TFAs) have higher melting points than their corresponding *cis* isomers. For example, the melting point of oleic acid (18:1 *cis*) is 16°C, while its geometric isomer, elaidic acid (18:1 *trans*), has a melting temperature of 44°C.

The types and positioning of FAs within a TAG determine its melting behavior. For instance, long-chain unsaturated FAs in vegetable oils are responsible for their low melting points and liquid state of oils at room temperature. In comparison, palm stearin is solid at room temperature because it contains a much higher proportion of high-melting-point TAGs. Natural fats generally contain many different TAGs, each with its own unique melting temperature. Therefore, food lipids possess melting ranges rather than melting points. For example, the TAGs in milk fat have melting points between −40 and 40°C [6]. This results in a fairly wide range of plasticity and gradual melting behavior. In contrast, cocoa butter has a much sharper melting profile. The solid TAGs cocoa butter contains at room temperature melt between 32 and 35°C [7]. Cocoa butter is composed of mainly symmetrical TAGs with oleic acid (O) at the *sn*-2 position and saturated FAs at *sn*-1 and *sn*-3 positions. The most common TAGs in cocoa butter are POP, SOS, and POS, where P and S are palmitic acid and stearic acid, respectively. This relatively narrow distribution of TAG species leads to a steep melting curve, although melting behavior is also influenced by polymorphism. The positional distribution of FAs can also influence a fat's physical properties. For example, when lard is interesterified to decrease the proportion of palmitic acid at the *sn*-2 position, a less grainy consistency is obtained and the baking performance of the fat is improved. As stated before, molecular composition lays the foundation for a fat's physical properties. Inarguably, it has an enormous influence on melting behavior, crystallization behavior, polymorphism, and solid fat content (SFC).

6.2.2 Lipid Crystallization

Crystallization behavior determines the SFC, polymorphism, microstructure, and ultimately the rheological properties of a fat crystal network. At a temperature below its melting point, a molecule experiences supercooling. Sometimes termed undercooling, this is the equivalent of supersaturation. It represents the thermodynamic driving force toward crystallization (i.e., the transformation from the liquid to the solid state). The factors influencing crystallization are both kinetic and thermodynamic in nature [8]. Kinetic parameters include molecular clustering, solvation/desolvation, adsorption, surface/volume diffusion, and conformational rearrangements during fat crystallization. Thermodynamic considerations include the surface-melt interfacial free energy and the crystallization temperature, which is the chemical driving force to nucleation based on the supersaturation ratio [8].

Fat crystallization is an exothermic process. It includes both nucleation and crystal growth events. Nuclei must be formed before crystal growth can occur, but these events are not mutually exclusive. Nucleation can continue to take place while crystals grow on the surfaces of existing nuclei [9]. Nucleation begins when the melt deviates from thermodynamic equilibrium [8]. TAGs aggregate into tiny clusters called embryos. Embryos continue to form and redissolve until some critical size is reached. At this point, a stable nucleus is formed. This will occur only when the energy related to the heat of crystallization exceeds the energy required to overcome the increase in surface area [10]. The Gibbs free energy of a spherical crystal embryo (ΔG_n) involves contributions from both the surface and volume changes and is defined according to Equation 6.1:

$$\Delta G_n = 4\pi r^2\sigma - \frac{4\pi r^3 \Delta G_m}{3V_m} \tag{6.1}$$

where r is the nuclei radius, σ is the surface free energy per unit surface area, ΔG_m is the molar free energy change associated with the liquid–solid-phase change, and V_m is the molar volume [11]. ΔG_n will reach a maximum at some critical embryo radius (r_c) as defined by Equation 6.2.

$$r_c = \frac{2\sigma V_m}{\Delta G_m} \tag{6.2}$$

To minimize the free energy, clusters smaller than this critical size will dissolve and those larger than r_c will continue to grow [11]. At higher degrees of supercooling (i.e., at lower temperatures), the critical radius required for stability is decreased because of decreased crystal solubility [12]. Also, at higher degrees of supercooling the critical radius is reduced because ΔG_m is larger. Figure 6.2 shows the nucleation free energy as a function of nuclei radius. ΔG was calculated using Equation 6.1 and values of σ, ΔG_m, and V_m as indicated. The critical radius (r_c) occurs when ΔG_n reaches a maximum and can be calculated using Equation 6.2. For TAG crystallization, the concept of radius may be misleading. Dimers of TAG molecule probably align laterally and then arrange into bilayers [13].

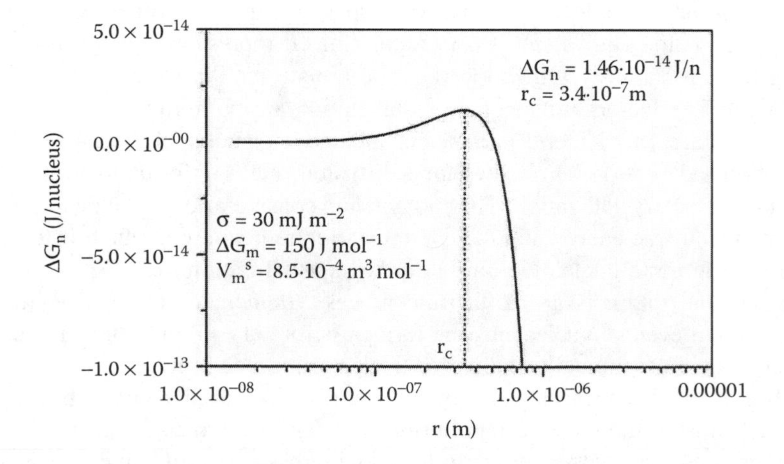

FIGURE 6.2 Free energy of nucleation (G_n) as a function of nuclei radius (r) for a lipid melt.

Primary homogeneous, primary heterogeneous, and secondary nucleation can occur. With significant undercooling (i.e., at a temperature much lower than the melting temperature of a fat), primary homogeneous nucleation can occur. This type of nucleation is, however, rare in fats. It can occur only in pure solutions in the absence of foreign interfaces. In reality, nucleation is usually catalyzed by the presence of foreign particles (existing crystals or other food ingredients) or by interfaces like the walls of a container. Primary heterogeneous nucleation needs significantly less supersaturation than homogeneous nucleation [11]. Foreign surfaces decrease the surface energy term in the crystallization free energy expression (Equation 6.1) [11,14]. Surfaces and foreign particles introduce small regions of order that exceed that of the bulk liquid [15]. In a mixture of TAGs, the highest melting species crystallize first. Secondary nucleation can take place after primary nucleation has occurred [13]. Secondary nucleation refers to the formation of new crystal nuclei upon contact with existing crystals. This can be very significant, particularly in bulk fats where the number of nuclei needed to induce crystallization is very low (1 per cubic millimeter or less). In an emulsion, the likelihood of every fat globule containing a nucleus or catalytic impurity to induce crystallization is low. As a result, emulsified fats require more supercooling (i.e., lower temperatures) than bulk fats [16].

The driving force required for crystal growth is relatively small compared with that for nucleation [8]. TAG crystal growth involves both the diffusion of molecules from the bulk solution across a boundary layer and the incorporation of TAGs into the crystal lattice of an existing nucleus or crystal [17]. It depends on a number of factors, including the degree of supersaturation, the rate of molecular diffusion to a crystal surface, and the time required for molecules to fit into a growing crystal lattice [18]. The entropic losses associated with a molecule incorporated into a crystal lattice are enormous. Therefore, the activation energy for this event is high. This tends to be the rate-determining step in fat crystallization [19–21]. In a multicomponent fat (i.e., containing many TAG species), crystallization is especially slow because the supersaturation for each TAG molecule is very small [21]. There is also competition between similar molecules for the same sites in a crystal lattice [22].

Crystallization and physical properties are influenced by both internal factors (diffusion, molecular compatibility, crystal size, TAG species, nuclei composition, and quantity) and external conditions (crystallization, temperature, tempering, cooling rate, shear, and storage conditions) [11,22]. Melt viscosity, for example, has a significant effect on crystal growth rate. It limits both molecular diffusion and the dissipation of the latent heat of crystallization away from a growing crystal face [16]. As a result, growth rate is inversely proportional to melt viscosity. Melt viscosity tends to increase with decreasing temperature [12] and is also affected by shear forces in a system.

Crystallization temperature and cooling rate have a large influence on solidification, SFC, and rheological properties. Temperature dictates how many TAGs are undercooled and to what extent. Lower temperatures and higher levels of supersaturation result in higher rates of nucleation and crystallization [23]. Faster cooling is associated with extensive nucleation, while more gradual cooling is generally correlated with increased crystal growth [14]. Samples cooled slowly tend to have fewer, but larger, crystals that cluster into aggregates. They also tend to be softer in nature. Figure 6.3 shows polarized light micrographs of lard cooled at 0.1°C/min (Figure 6.3A) and 5.0°C/min (Figure 6.3B). Although the SFCs of these samples were similar (30.3 ± 0.2% and 33.2 ± 0.3% for Figure 6.3A and 6.3B, respectively), the slowly cooled sample was much softer. Penetration depths for the rapidly and slowly cooled lard were $15.1 \pm 2.9 \times 10^{-1}$ mm and $40.2 \pm 3.5 \times 10^{-1}$ mm, respectively, as determined by cone penetrometry [24].

Fast cooling also favors the formation of more metastable polymorphs [25]. The presence of shear during and after crystallization can also have a huge impact on fats [26]. Shear increases the overall mixing, leading to more effective molecular interactions, better heat transfer, and increased nucleation rate. Agitation serves to break up crystallites, creating seed crystals, which will act as surfaces for secondary nucleation. Mazzanti et al. [27] demonstrated increases in crystalline orientation with increasing shear.

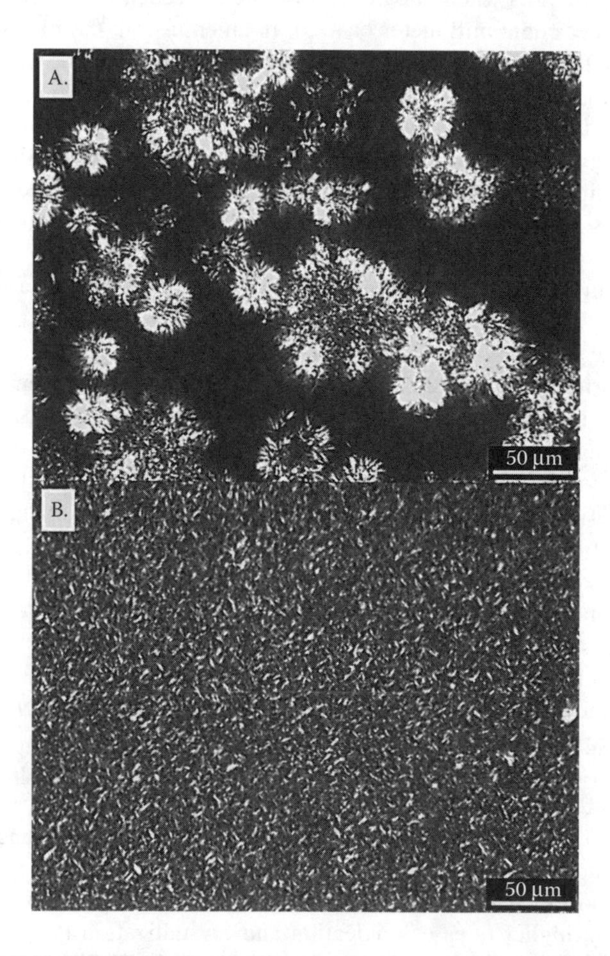

FIGURE 6.3 Polarized light micrographs of lard cooled slowly (0.1°C/min) and rapidly (5.0°C/min).

At present, there is debate concerning the nature of the liquid to solid-phase transitions in fats. There is some evidence that an intermediate state of matter exists between the amorphous liquid melt and crystalline solid states. The liquid crystalline state was observed in organic materials over a century ago and in synthetic polymers during the 1950s [28]. Such phases were called liquid crystals because they have properties that lie between those of completely amorphous substances and crystalline materials. They exhibit liquid-like flow, but show varying degrees of molecular ordering. In fats, however, the presence and nature of a true liquid crystalline state is still a matter of active debate. Fat and oil processors are familiar with a concept referred to as crystal memory, where remnants of the crystalline state can remain after a fat is melted. If crystal memory is not destroyed by sufficient heating, a fat seems to "remember" its previous crystalline history and is influenced toward that conformation. Some researchers would argue that this is evidence of liquid crystalline order in fats.

Larsson [13,29] was the first to propose that TAG molecules associate into bilayer regions in the liquid state. Support came from x-ray diffraction (XRD) studies showing the occurrence of a diffuse peak in the long spacing range of fats upon melting. The peak's position corresponded to an arrangement of TAG bilayers in which glycerol backbones are aligned together and the hydrocarbon chains are in a liquid-like conformation. Short-range lamellar ordering has also been observed above the melting temperature in liquid TAGs using very fast cooling freeze-etch electron microscopy [16]. Additional evidence of ordering in the melt [30] and liquid oils [31] has been reported. The debate continues surrounding the exact nature of the ordering in the liquid state of TAGs [32].

6.2.3 The Phase Behavior of Lipids

In the liquid state, TAGs are nearly ideally miscible. No changes in heat or volume occur when liquid fats are mixed [16]. However, in the solid state, incomplete miscibility can result in multiple distinct solid phases. Mixed or compound crystals contain more than one molecular species [18,33]. They form in natural fats with complex TAG compositions and fractions of TAGs [16,34–38]. The likelihood of compound crystallization increases when the molecular species are more similar in shape, size, and properties [17]. Also, mixed crystals form more extensively at lower temperatures because more TAGs are supersaturated and to a greater extent so crystallization proceeds more rapidly [16]. Compound crystals have lower densities and enthalpies of fusion than pure crystals of the same modification and tend to slowly rearrange into purer crystals [17].

A consequence of complex phase behavior is that rapidly cooled milk fat has a higher SFC than the same milk fat cooled in a slow and stepwise fashion [18,39]. Also, the higher melting TAGs in a fat tend to dissolve in the lower melting species. As a result, the melting curve of a fat does not equal the sum of its component TAGs [16]. Thermal eutectics occur when the melting point of a fat mixture is less than that of the individual fats present in the blend. Eutectics are a consequence of incompatibility of the solid phases present. When the solid phases present are compatible, solid solutions form. In contrast, eutectics are evidenced by softening in mixtures of fats. Eutectics are observed between cocoa butter and some cocoa butter equivalents, for example. In Figure 6.4, the melting temperatures of cocoa butter mixed with the middle- and high-melting fractions of milk fat (MMF and HMF, respectively) are shown. The melting point depression for blends of cocoa butter with MMF (Figure 6.4A) indicates a eutectic. When these fats are mixed, solid solutions do not form at all concentrations. Instead, softening occurs. Softening is not observed when HMF is mixed with cocoa butter (Figure 6.4B), but rather a solid solution forms.

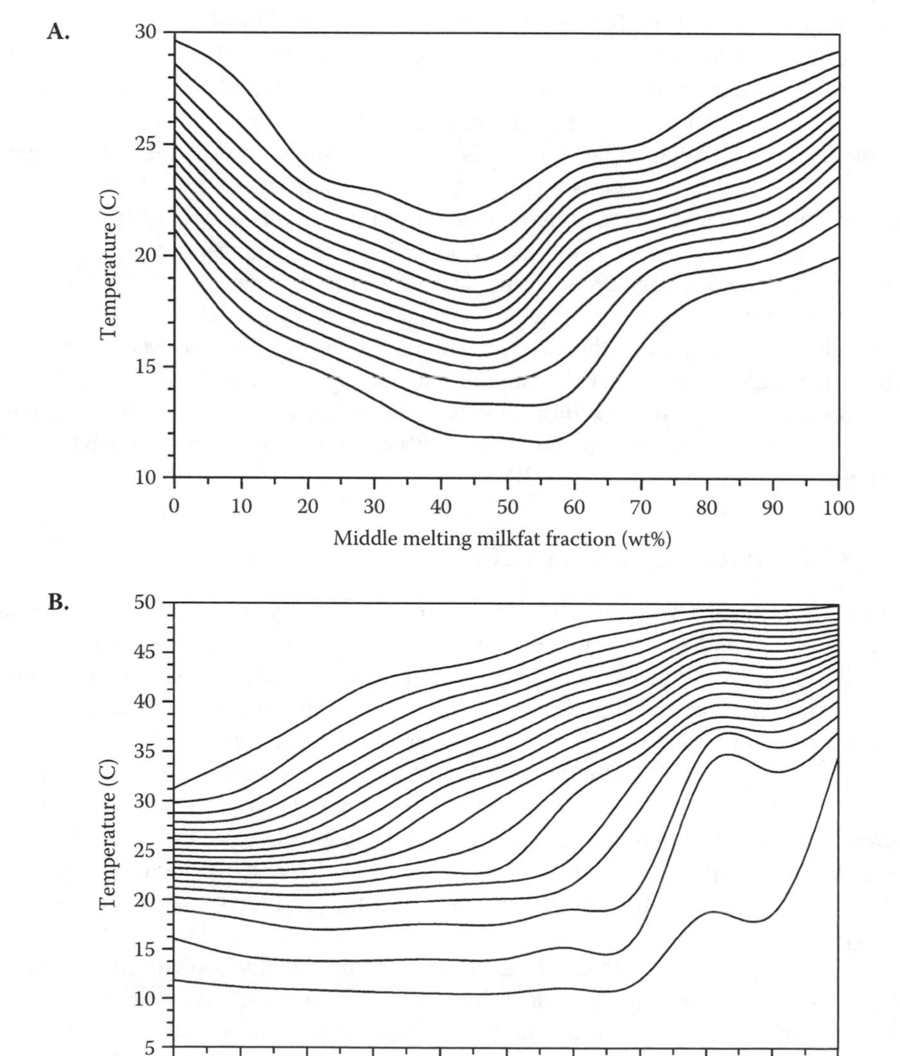

FIGURE 6.4 Blends of cocoa butter with middle- and high-melting fractions of milk fat (MMF and HMF, respectively) showing the formation of a eutectic with MMF (A) and compatibility in the solid phase with HMF (B).

6.2.4 Fat Crystal Polytypism and Polymorphism

In the solid state, lipids form lamella-type structures. Both the tilt of the FAs present and the packing of atoms between the FAs can vary. These differences lead to the existence of polytypism and polymorphism in fats. Polytypism refers to differences in the stacking of TAGs in crystal lamellae. It is described by the long Bragg spacings in XRD studies. This d-spacing or lamellae layer thickness depends on the length of the TAG molecules and also on the angle of tile between the chain axis and the basal lamellar plane [12]. Polytypism is indicated with a -2 or -3 designation following the polymorph type for a bilayer or trilayer arrangement, respectively (see Figure 6.5). The bilayer arrangement of FA chains is the most common packing structure for natural fats, although trilayer structures have also been identified. Trilayer structures are commonly observed in TAGs in which the FA at the *sn*-2 position differs significantly from those at the *sn*-1 and -3 positions [12].

Polymorphism has been well established and studied extensively in both pure TAG and natural fat systems. Polymorphism exists when a material of set chemical composition possesses different subcell packings in the solid state [5,40,41]. These materials are identical in the liquid state, but can arrange into more than one type of solid lattice. Polymorphism refers to differences within the unit cell structure of crystals [8]. It arises from variations in the tilt of the TAG molecules in a bilayer and also from variations in how the hydrocarbon chains are packed [13].

As early as 1849 (as reported by Chapman [42]), multiple melting events were observed in pure fats. Polymorphism explains this phenomenon. In 1934, Clarkson and Malkin [43] confirmed the existence of different crystalline forms in tristearin by XRD. Today, it is commonly accepted that three main crystal polymorphs exist in the solid state of TAGs: alpha (α), beta prime (β'), and beta (β), in order of increasing stability and melting temperature. Within these polymorphs, subforms are possible [44–46]. The α subcell has a hexagonal arrangement in which the FA chains show disorder and rotational mobility along the hydrocarbon chain axes [13].

FIGURE 6.5 Schematic representation of polytypism in triacylglycerols. Triacylglycerols aligned in bilayer and trilayer arrangements.

It appears as a close packing of oscillating FAs [47,48]. In contrast, the β′ polymorph has a more fixed subcell arrangement referred to as orthorhombic perpendicular. In this polymorph, every second zigzag plane (created by the zigzag pattern of the hydrocarbon chains) is perpendicular to the other planes. The β′ polymorph displays intermediate stability. The most stable β crystal has a triclinic subcell with the hydrocarbons arranged parallel to each other [5,49] and very tight stacking of the bilayers at the methyl end group planes [50]. Other metastable polymorphs are also observed in mixed TAG systems [48].

Differences in crystal packing give rise to differences in density, melting temperature, and heats of fusion between the polymorphs. Therefore, fat polymorphism can be explored using melting methods [41]. Vibrational spectroscopy [51,52] and nuclear magnetic resonance [53] are also used to study polymorphism, although XRD is the most definitive method. The α, β′, and β polymorphs have characteristic XRD short spacings of 0.415, 0.38 and 0.42, and 0.46 nm, respectively [5,54]. Instruments allowing temperature controlled, simultaneous time-resolved XRD at both small and wide angles and differential scanning calorimetry (DSC) measurements are valuable tools for studying polymorphic transitions [55,56].

Fat polymorphism is related to physical character and functionality. For example, α crystals tend to be very thin and small crystals, while the β′-polymorph morphology has a longer needle-like shape, typically less than 5 μm. In contrast, β crystals have been described as much larger and having a plate-like morphology [57]. In palm oil, α crystals appear as spherulites in contrast to the needle-like β′ crystals [58]. Polymorphism has implications on the functionality of fats and fat-containing foods. For example, shortenings, which contain β crystals, typically have poor aeration in cakes [59]. Instead, β′ crystals are preferred in shortenings and margarines. Similarly, in whipped toppings, the β′ polymorph is associated with desirable functionality and shine [60]. In chocolate, the defect known as bloom can occur when improperly tempered cocoa butter undergoes a polymorphic transformation. Phase separations and oil migration with subsequent recrystallization may also lead to the development of fat bloom [61,62]. Tempering is conducted to ensure that the desired crystalline form (form V or β in cocoa butter depending on naming scheme) is obtained in cocoa butter. When chocolate is tempered properly, it exhibits desirable snap and melting properties, high gloss, and stability against bloom [63]. Cocoa butter and chocolate polymorphism have been studied extensively [56,64–66].

The polymorphic form that develops during fat crystallization is influenced by several factors, including fat purity, TAG compatibility, temperature, supercooling, cooling rate, shear, catalytic impurities, solvents, and seed crystals [40]. For example, at higher cooling rates, the α polymorph predominates in milk fat [46]. Transitions from one polymorphic form to the next can occur during processing and storage depending on the internal crystal/solvent conditions and external processing conditions [40,41,67]. Polymorphic transitions in fats are accelerated by the presence of shear [67,68].

Because the β form has the highest melting point, and hence the lowest solubility, it is the most thermodynamically stable of the three main polymorphs. However, nucleation in the α polymorph is favored. Although the β′ and β crystals are relatively less soluble and more supersaturated, but the α crystal has a lower surface free energy and lower heat of crystallization [10]. Therefore, although the β polymorph

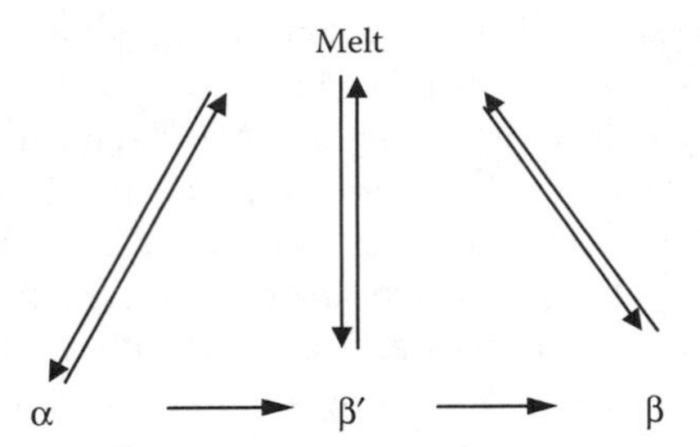

FIGURE 6.6 Dynamics of polymorphic transformations in fats.

is thermodynamically the most stable form, nucleation and crystal growth rates are higher for the β′ polymorph, and especially for the α polymorph [8]. Figure 6.7 shows that the activation free energy of nucleation ($\Delta G^{\#}$) for the α polymorph is lower than for the more stable polymorphic forms.

Polymorphic stability depends on van der Waals interactions and packing density [1]. When multicomponent fats do transform, they exhibit monotropic rather than enantiotropic polymorphism. Transformations proceed only from the less to the more thermodynamically stable forms [47]. To return to a less stable polymorph, the crystal structure must be completely destroyed through the melt as shown in Figure 6.6. TAG polymorphic transitions to more stable polymorphs can occur through either the melt or the solid state [69]. In monoacid short-chain TAGs, the α form can transform to the β′ form during heating either via melting and then recrystallization or directly without being melted. The direct transformation is preferred in short-chain TAGs (C12) while it occurs via melting for longer chain TAGs (C18) because of the relatively high stability of the α form [1]. At least some order remains during transitions, as evidenced by persistent short spacings in continuous XRD measurements [50].

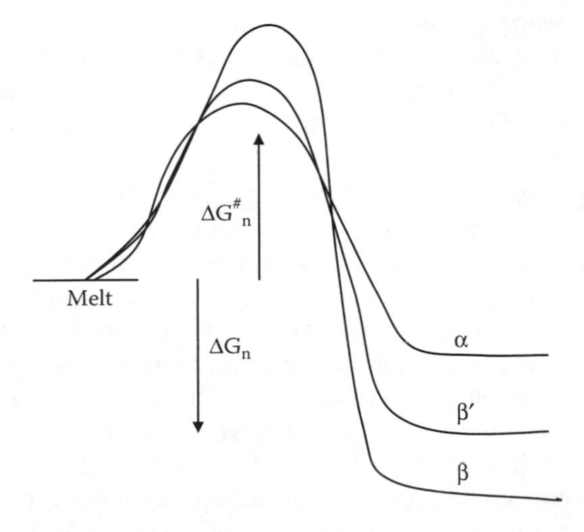

FIGURE 6.7 Free energies of nucleation for the α, β, and β′ polymorphs in fats.

Although the α crystal tends to form first in many fats, it typically transforms to a more stable polymorph. This is related to, among other things, chemical composition. The diversity of FA chain length, TAG carbon number, and amount of liquid oil present determine which polymorph is favored in a fat [13,41]. Fats tend to prefer either the β′ or β modification. For example, milk fat, beef tallow, and palm fats are β′-tending. The majority of their solids remain in the β′ modification even after prolonged storage. In contrast, lard and fully hydrogenated soybean oil are examples of β fats. Multicomponent fats tend to be more stable in metastable states than those fats with narrow and symmetrical distributions of FAs [47,70]. For example, when palm oil is introduced into hydrogenated canola oil, the increased FA diversity stabilizes the β′ polymorph [71]. Surfactants also delay polymorphic transitions by interfering with the transformation kinetics [1,72].

Fat polymorphism is related to and complicated by phase behavior [73]. In blends of fully hydrogenated canola oil and soybean oil, intersolubility and polymorphic changes occurred simultaneously [74]. In milk fat, the α crystal has a relatively long lifetime in milk fat, perhaps due to stabilization related to compound crystal formation [75].

6.2.5 Fat Crystal Networks

Fats and fat-based foods consist of partially crystalline fat suspended in liquid oil. Once crystals form, Brownian motion and van der Waals forces cause them to aggregate into a three-dimensional network [76]. With as little as 10% crystalline material, a semisolid network in which the oil is immobilized can be observed. The strength of FCNs can increase for an extended period of time because of continued crystallization, crystal rearrangement, and crystal aggregation [77–79]. Sintering refers to the development of solid bridges in the narrow gaps of fat crystal networks after crystallization [17,80,81]. Butter hardness can increase for months after manufacturing, depending on temperature [82–84]. Indeed, thermodynamic equilibrium may never be attained in a complex fat system. Segregation leading to crystals of higher purity and rearrangement into more stable polymorphs can continue indefinitely depending on several factors, including composition and storage temperature.

Although the ratio of solid to liquid fat is a big determinant of fat consistency, the type of solids formed, the way in which these solids are arranged, and the interactions between them are very important [78,85–88]. Fat microstructure refers to the level of structure between roughly 0.25 and 200 μm (Figure 6.1). The importance of fat microstructure on physical properties has been demonstrated [85,86,89]. At this level, crystal number, shape, density, clustering, and distribution all affect the properties of a fat. These parameters are themselves determined by chemical composition and processing conditions [80,90,91]. For example, when milk fat fractions are processed at higher agitation rates, nucleation is enhanced. The average crystal size is reduced and a more viscous nature and lower elastic modulus was observed [92]. When blends of a high melting milk fat fraction and sunflower oil were crystallized slowly, the crystals were larger and more densely arranged, and had more regular boundaries than when fast cooling was employed [93]. In turn, crystal properties can dramatically affect texture [39,41,96]. For example, smaller crystals are generally associated with a harder fat. In contrast, platelet-like crystals greater than 30 μm give foods a grainy and sandy mouthfeel.

Microscopy is a valuable tool for studying the relationships among processing conditions, microstructure, and physical properties in fats. Fat crystal networks can be visualized by various techniques including electron microscopy [86,95–98], confocal scanning light microscopy [86,92], multiple photon microscopy [99,100], and polarized light microscopy [57,88,101]. By visualizing the microstructure of FCNs, we can quantify them in terms of crystal, shape, density, and arrangement. Recent advances in quantifying the arrangement of fat crystals have strengthened the link between processing and physical properties [102–104]. Fractal dimensions (D_f), which can be determined from microscopy or rheology, indicate both the degree of occupancy in a crystal network and also the heterogeneity in the distribution of that mass. This concept strengthens the ability to quantify fat microstructure and the ability to explain rheological properties [24,89,105,106]. Milk fat and lard composition and processing conditions were related to microstructure and ultimately to rheology [24]. Compared with gradually cooled fats, samples cooled quickly had higher SFCs and smaller crystalline particles that formed in metastable states. Higher cooling rates were also associated with more homogeneous distributions of mass (i.e., lower D_f) and firmer consistencies as measured by cone penetrometry [24].

6.2.6 Rheological Properties

6.2.6.1 Small Deformation Testing

A sample's response to very low levels of stress or strain says something about its structure. Microstructure can therefore be probed using dynamic or oscillatory testing in which the amount of stress or strain applied is varied in a sinusoidal fashion. Such tests must be performed at such low levels of strain that sample integrity is maintained ($\ll 1\%$ deformation). Under these conditions, fat crystal networks demonstrate viscoelastic behavior. For viscoelastic materials, the phase shift between the applied strain and resulting stress (or vice versa) is determined by the contributions from the elastic and viscous components and ranges between 0 and 90° [107]. The storage modulus (G′) corresponds to the elastic (i.e., in phase) component of the material, whereas the loss modulus (G″) corresponds to the viscous (i.e., out of phase) component of the material. Based on these values, several other parameters, including the phase angle (tan δ), complex modulus, complex viscosity, dynamic viscosity, and complex compliance can be determined [108]. To characterize these parameters, experiments should be conducted within the linear viscoelastic region (LVR) where the relationship between stress and strain is linear. Figure 6.8 shows the results of an oscillatory stress sweep for a sample of cocoa butter. In the initial LVR, values of $\sim 5 \times 10^7$ Pa and 0.4×10^7 Pa, for G′ and G″, respectively, are observed. The sample "broke" at around 0.05% strain.

Rheological parameters determined by small deformation testing provide information about fat crystal networks and have been correlated with hardness and spreadability. This testing is a valuable tool for studying the physical properties of fats and for relating microstructure to rheological properties [103,105,109–112]. It can be used to characterize behavior such as melting [113] and to compare the effect of composition and processing conditions on structural integrity [39,89,111,114].

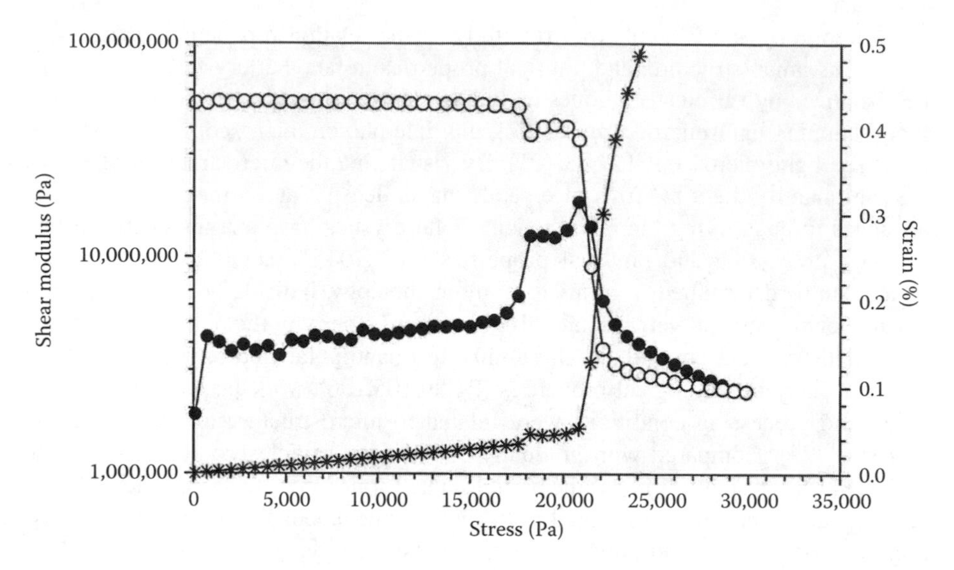

FIGURE 6.8 Results of small deformation oscillatory testing of cocoa butter. Storage modulus (open circles), loss modulus (closed circles), and strain (asterisks) as a function of applied shear stress.

For example, Toro-Vazquez et al. [115] used oscillatory rheometry to follow the polymorphic changes in cocoa butter during static crystallization.

6.2.6.2 Large Deformation Testing

Texture and the macroscopic properties of fats can be analyzed by sensory evaluation or by instrument. In both cases, large deformations are applied and the nature of the structure is explored by destroying it. Rheometers are typically used to measure a particular mechanical attribute, such as penetration resistance, yield force, cutting force, or resistance to flow. Rheometry does have the advantages of efficiency, reliability, and reproducibility, although the values obtained are arbitrary. They have little practical meaning unless they are correlated with findings from sensory evaluation or with a particular processing parameter. To achieve this, rheological tests and probes have been developed to imitate sensory panels [116,117]. The real power for a rheological test comes when the results correlate with sensory analysis [118–121].

Plastic fats consist of both solid crystals and liquid oil at room temperature. Because of their partially solid consistency, some measurement of "hardness" is commonly used to study the texture of these materials [122]. When a force is applied, plastic fats have some structural integrity with which to resist that force. However, above a certain point (i.e., at a yield value), they begin to flow [85,123]. A typical load-deformation curve for a plastic fat is shown in Figure 6.9A. There is an initial linear region in which elastic behavior is demonstrated and an elastic constant can be obtained (Figure 6.9B), but at some point (i.e., at the yield force) the material is broken.

Cone penetrometry is perhaps the most common method used to investigate the textural properties of fats [85]. In constant force cone penetrometry, the distance to

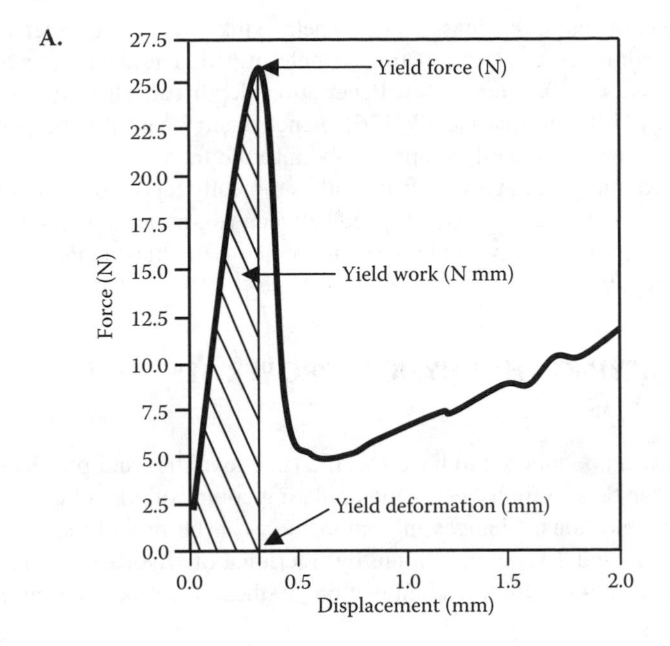

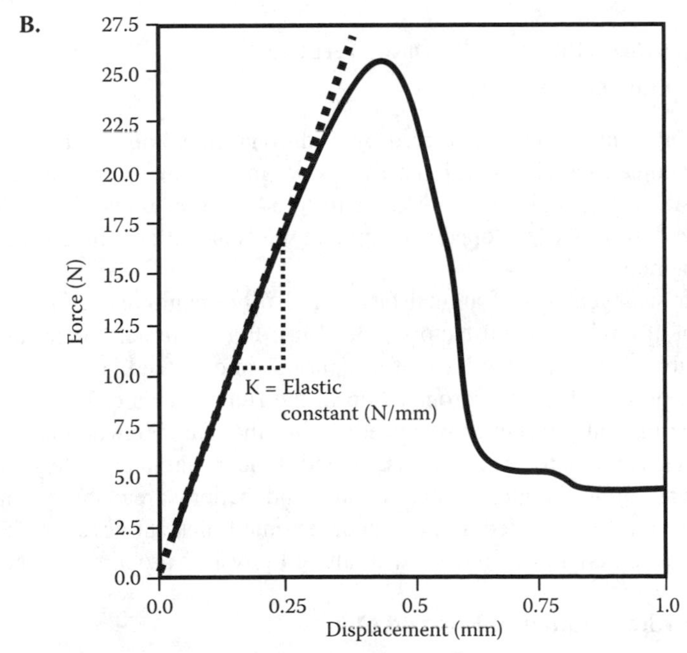

FIGURE 6.9 Typical load-displacement curve for a sample of fat compressed between two flat plates at a constant rate of deformation. Yield force, yield work, and yield deformation determinations are shown.

which a cone of particular mass and geometry sinks into a fat under the force of gravity and within a set period of time is determined. The depth of penetration is reported in units of 0.1 mm [124]. Penetration depth can also be converted to a "yield value" [125] or "hardness" [126]. Penetrometry can also be performed by driving a cone or a differently shaped probe into the fat at a constant speed. In this case, the maximum force at a specified depth is typically reported. Similarly, samples can be compressed under a constant speed between two parallel plates. Compression testing can be more sensitive and reveals more about the plastic nature of a fat than penetrometry [59].

6.3 MODIFYING THE PHYSICAL PROPERTIES OF LIPIDS

Fats and oils can be tailored to have specific functionalities and physical properties [127]. This can be accomplished by manipulating composition and/or crystallization and storage conditions. Changes in composition can be brought about by altering the original FA and TAG composition, by fractionation, hydrogenation, interesterification, or blending. Often a combination of these approaches is used to tailor specific properties.

6.3.1 Changes Through Breeding, Feeding, and Genetic Alterations

Vegetable oils and fats can be altered by traditional breeding practices or through genetic modifications [128–133]. These developments have implications for both oxidative stability and physical functionality [134]. For example, high-oleic, high-palmitic sunflower oils developed from mutant seeds had altered oxidative stabilities and melting ranges [135].

The physical features of animal fats can also be manipulated through feeding and control of environmental factors [136–138]. For example, butter spreadability is improved by increasing the level of unsaturates in cow feed [139–141], provided the oils are protected from hydrogenation in the rumen. This effect of unsaturated feed is seen naturally in butter produced during the summer when cows typically graze on fresh, and therefore more unsaturated, fodder. Summer butters have higher levels of unsaturates, higher iodine values, and better spreadability than winter butters. When cows were fed an algal-supplemented diet, the level of TFAs in the butter was increased and correlated with altered physical properties [106].

6.3.1.1 Hydrogenation of Fats and Oils

Hydrogenation involves the addition of hydrogen atoms to double bonds in unsaturated FAs. It serves to increase the oxidative stability of oils and increases the solid consistency by increasing the level of saturation. In most cases, vegetable oils are partially, rather than fully hydrogenated. This results in the formation of TFAs.

Both the saturates and TFA that form increase the melting temperature of a hydrogenated fat. Hydrogenation is a very complex reaction [142]. Initial oil composition, temperature, hydrogen pressure, type and amount of catalyst, agitation, and presence of poisons can affect the hydrogenation process [143,144]. The range of plasticity, polymorphic tendency, and TFA levels of hydrogenated oils are influenced by processing. By manipulating processing conditions, a wide range of product functionalities can be achieved [145].

Hydrogenation can be carried out under either selective or nonselective conditions. In selective hydrogenation, reaction conditions are tailored to preferentially saturate the most unsaturated FA present. This approach can be used to eliminate polyunsaturated FAs and improves an oil's oxidative stability. Nonselective hydrogenation results in a wider range of plasticity, although the partially hydrogenated products suffer from high levels of TFAs, which are of particular concern, at present, from a health perspective [146–148]. There is currently much interest in alternatives to partially hydrogenated fats for various food applications [149].

6.3.1.2 Interesterification of Fats and Oils

TAGs with desired functionalities can be structured using interesterification. This reaction involves the exchange of FAs within and between TAGs. The process can be used to improve the spreadability and baking properties of lard [150–153], to manufacture inexpensive confectionary fats [154–156], to replace hydrogenated fats in various applications [157–161], and to produce structured lipids with specific health benefits [162–164].

Interesterification can be carried out chemically using metal alcohols or enzymatically using random, FA specific, or positional specific (*sn*-1,3 - or 2-specific) lipases as catalysts. Chemical interesterification is random when the reaction is carried out at a temperature above that of the highest melting TAG present. The reaction is "directed" when carried out at a temperature below that of a certain TAG species. The temperature induces crystallization of the TAG in question and shifts the equilibrium toward producing more of that TAG.

Interesterification is not as established a technique as hydrogenation and is currently a more expensive option. However, the advantages of specificity and absence of TFA in interesterified oils make interesterification a promising tool for lipid modification [165].

6.3.1.3 Fractionation of Fats and Oils

Fractional crystallization accomplishes three objectives in fats [166]. It can remove a high melting species, which would otherwise crystallize in an oil during storage (i.e., winterization). It is also used to enrich an oil with more unsaturated species and to recover a TAG fraction with desirable melting properties for use in confectionary applications. Finaly, the functionality of fats and oils can be extended by fractionating to produce products with differing melting points, solid fat contents, and rheological properties [167].

The fractions of palm oil lend themselves to various applications [168]. Palm oleins demonstrate high oxidative stability for frying and dressing applications while the mid- and high-melting fractions have use as cocoa butter substitutes and hard stocks for margarine and shortening manufacture, respectively. Anhydrous milk fat can also be fractionated [169–174]. The fractions obtained can either be recombined in order to produce a butter with better cold spreadability [172] or used in other applications [170]. For instance, the higher melting milk fat fractions have use in pastry making and as bloom inhibitors in chocolate.

6.3.1.4 Blending of Fats and Oils and Combining Technologies

Blending of fat fractions or of fats and oils themselves is another approach to reformulating products with altered physical properties [175–178]. Blending a hard stock into a liquid oil increases the overall melting range and can extend the range of plasticity. For example, when milk fat is blended with vegetable oils, firmness is decreased [152,179]. In some cases, the eutectics, which form when some fats are blended, can help to achieve a desired range of plasticity [180,181]. Blending can also be used to manipulate polymorphism. To maintain the desirable beta prime polymorph in margarines and cake shortenings, a β'-tending fat like palm oil or hydrogenated palm oil is often incorporated [41].

Groups of TAGs classified according to their structure and melting temperatures can be related to the properties they impart on spreads [182]. The presence of non-TAG materials can also influence fat physical properties. Emulsifiers, for example, can modify fat crystallization and control polymorphism [3,44,62,183–185].

Often a combination of the above methods is used to tailor the desired functionalities of fats [176,186]. For example, plastic fats with no TFAs can be prepared by fractionating palm oil and recombining the fractions in different proportions [187]. Regional differences also exist in terms of fats and oils and modification technologies. Compared with North America, which relies heavily on soybean and canola oil basestocks, Europe uses a wider range of vegetable oils and fats. As a result, fractionation and interesterification technologies are used more extensively in the European market [188].

6.3.2 Changes Related to Processing of Fats and Oils

Some ways in which crystallization parameters affect fat crystal networks have already been discussed. Manufacturing considerations such as temperature treatments, mechanical working, and scale of operation are important [189]. Temperature has a large influence by determining the proportion of TAGs that will solidify. In margarine manufacture, temperature determines SFC, polymorphism, and consistency [190]. Cooling rate determines the rates of nucleation, crystal growth, and aggregation. It therefore determines crystal size, number, and polymorphism. Cooling rate also influences mixed crystal formation.

The importance of tempering in some fats emphasizes how critical temperature control and manipulations are to the physical properties of fat crystal networks. Tempering ensures that the desired crystalline form is present. Seeding with crystals of a desired polymorph is sometimes done in conjunction with temperature manipulations. Tempering is an invaluable tool in the confectionary and shortening industries. In shortenings, it influences firmness [191], polymorphic stability [90], and functionality [189,192]. During storage, temperature fluctuations can induce melting, recrystallization, and polymorphic transformations. Fluctuations are generally associated with an increase in crystal size due to Ostwald ripening, dissolution of smaller crystals, and growth of larger crystals.

Shear also affects a fat's physical properties. It influences heat and mass transfer in a crystallizing system. When introduced during crystallization, it breaks crystals, promoting secondary nucleation and a finer, more granular crystal network. This is the case when shortenings are manufactured in a votator and when butter is churned, although churning also accomplishes phase inversion. Mechanical working of shortenings and margarines is an important determinant of functionality [189]. In margarine production, agitating the blend after crystallization extends the product's plastic range by disrupting intermolecular forces. Stirring leads to a lower SFC and softer texture when fats are stored [110]. In various shortenings, the presence of shear alters polymorphism and hardness [193].

6.4 CONCLUSIONS

The functionalities of lipids in foods are determined by their physical properties. Complex relationships exist between the parameters that determine the structure of a fat. Because these parameters are interrelated, it can be difficult to isolate the effect of just one factor on texture. However, with continued research we are better able to understand, predict, and manipulate fat behavior to achieve desirable results.

REFERENCES

1. Aronhime, J., Sarig, S., and Garti, N., Emulsifiers as additives in fats: effect on polymorphic transformations and crystal properties of fatty acids and triglycerides, *Food Struct.,* 9, 337, 1990.
2. Smith, P.R., Cebula, D.J., and Povey, M.J.W., The effect of lauric-based molecules on trilaurin crystallization, *J. Am. Oil Chem. Soc.,* 71, 1367, 1994.
3. Wright, A.J. et al., The effect of minor components on milk fat crystallization, *J. Am. Oil Chem. Soc.,* 77, 463, 2000.
4. Jensen, R.G. and. Newburg, D.S., Milk lipids, in *Handbook of Milk Composition,* Jensen, R.G. Ed., Academic Press, New York, 1995, p. 545.
5. Small, D., Glycerides, in *The Physical Chemistry of Lipids: From Alkanes to Phospholipids,* 2nd ed., Small, D., Ed., Plenum, New York, 1986, p. 345.
6. deMan, J.M., Physical properties of milk fat, *J. Dairy Sci.,* 47, 1194, 1964.
7. Aronhime, J., Sarig, S., and Garti, N., Reconsideration of polymorphic transformations in cocoa butter using the DSC, *J. Am. Oil. Chem. Soc.,* 65, 1140,1988.
8. Sato, K., Yoshimoto, N., and Arishima, T., Crystallization phenomena in fats and lipids, *J. Disp. Sci. Technol.,* 10, 363, 1989.
9. Guy, A.G., *Introduction to Materials Science,* McGraw-Hill, New York, 1972, p. 310.
10. Timms, R.E., Crystallization of fats, in *Developments in Oils and Fats,* Hamilton, R.J., Ed., Blackie Academic, London, 1995, p. 204.
11. Garside, J., General principles of crystallization, in *Food Structure and Behavior; Food and Technology: A Series of Monographs,* Blanshard, J.M.W. and Lillford, P., Eds., Academic Press, New York, 1987, p. 35.

12. Lawler, P.J. and Dimick, P.S., Crystallization and polymorphism of fats, in *Food Lipids; Chemistry, Nutrition and Biotechnology,* Akoh, C.C. and Min, D.B., Eds., Marcel Dekker, New York, 1998, p. 229.
13. Larsson, K., *Lipids- Molecular Organization, Physical Functions and Technical Applications,* The Oily Press, Dundee, United Kingdom, 1994, p. 7.
14. Boistelle, R., Fundamentals of nucleation and crystal growth, in *Crystallization and Polymorphism of Fats and Fatty Acids,* Garti, N. and Sato, K., Eds., Marcel Dekker, New York, 1988, p. 189.
15. van den Tempel, M., *Surface-Active Lipids in Foods, S.C.I. Monograph No. 32,* Society of Chemical Industry, 1968, p. 22.
16. Walstra, P., van Vliet, T., and Kloek, W., Crystallization and rheological properties of milk fat, in *Advanced Dairy Chemistry, Vol. 2: Lipids,* Fox, P.F., Ed., Chapman Hall, London, 1994, p. 179.
17. Walstra, P., Fat crystallization, in *Food Structure and Behavior; Food and Technology: A Series of Monographs,* Blanshard, J.M.W. and Lillford, P., Eds., Academic Press, New York, 1987, p. 67.
18. Mulder, H. and Walstra, P., Crystallization behavior of milk fat, in *The Milk Fat Globule. Emulsion Science as Applied to Milk Products and Comparable Foods,* C.A.B. Pudoc, Wageningen, the Netherlands, 1974, p. 33.
19. Skoda, W. and van den Tempel, M., Growth kinetics of triglyceride crystals, *J. Crystal Growth,* 1, 207, 1967.
20. Knoester, M., deBruyne, P., and van den Tempel, M., Crystallization of triglycerides at low supercooling, *J. Crystal Growth,* 3,4, 776, 1968.
21. Walstra, P., Secondary nucleation in triglyceride crystallization, *Prog. Colloid Polym. Sci.,* 108, 4, 1998.
22. Grall, D.S. and Hartel, R.W., Kinetics of butterfat crystallization, *J. Am. Oil Chem. Soc.,* 69, 741, 1992.
23. Liang, B., Shi, Y., and Hartel, R.W., Phase equilibrium and crystallization behavior of mixed lipid systems, *J. Am. Oil Chem. Soc.,* 80, 301, 2003.
24. Campos, R., Narine, S.S., and Marangoni, A.G., Effect of cooling rate on the structure and mechanical properties of milk fat and lard, *Food Res. Int.,* 35, 971, 2002.
25. Rousset, P., Modeling crystallization kinetics of triacylglycerols, in, *Physical Properties of Lipids,* Marangoni, A.G. and Narine, S.S., Eds., Marcel Dekker, New York, 2002, p. 1.
26. Stapley, A.G.F., Tewkeswbury, H., and Fryer, P.J., The effects of shear and temperature history on the crystallization of chocolate, *J. Am. Oil Chem. Soc.,* 6, 677, 1999.
27. Mazzanti, G., Welch, S.E., Sirota, E.B., Marangoni, A.G., and Idziak, S.H.J., Crystallization of bulk fats under shear, in *Soft Materials: Structure and Dynamics,* Dutcher, J.R. and Marangoni, A.G., Eds., Marcel Dekker, New York, 2004.
28. Sperling, L.H., *Introduction to Physical Polymer Science,* 2nd eds., John Wiley Sons, New York, 1992, p. 279.
29. Larsson, K., Molecular arrangement in glycerides, *Fette. Seifen Anstrichm.,* 74, 136, 1972.
30. Cebula, D.J., McClements, D.J., Povey, M.J.W., and Smith, P.R., Neutron diffraction studies of liquid and crystalline trilaurin, *J. Am. Oil Chem. Soc.,* 69, 1992, 130.
31. Toro-Vazquez, J.F. and Gallegos-Infante, A., Viscosity and its relationship to crystallization in a binary system of saturated triacylglycerides in sesame seed oil, *J. Am. Oil Chem. Soc.,* 73, 1237, 1996.
32. Marangoni, A.G., Steady-state fluorescence polarization spectroscopy as a tool to determine microviscosity and structural order in lipid systems, in *Physical Properties of Lipids,* Marangoni, A.G. and Narine, S.S., Eds., Marcel Dekker, New York, 2002.

33. Rossell, J.B., Phase diagrams of triglyceride systems, in *Advances in Lipid Research, Vol 5,* Paolotti, R. and Kretchevsky, D., Eds., Academic Press, New York, 1967, p. 353.
34. Marangoni, A.G. and Lencki, R.W., Ternary phase behavior of milk fat fractions, *J. Agric. Food Chem.,* 46, 3879, 1998.
35. Mulder, H., Melting and solidification of milk fat, *Neth. Milk Dairy J.,* 7, 149, 1953.
36. Sherbon, J.W., Crystallization and fractionation of milk fat, *J. Am. Oil Chem. Soc.,* 51, 22, 1974.
37. Timms, R.E., The phase behavior and polymorphism of milk fat, milk fat fractions and fully hardened milk fat, *Aust. J. Dairy Technol.,* 35, 47, 1980.
38. Walstra, P. and van Beresteyn, E.C.H., Additional evidence for the presence of mixed crystals in milk fat, *Neth. Milk Dairy J.,* 29, 238, 1975.
39. Rye, G., Litwinenko, J., and Marangoni, A.G., Fat crystal networks — structure and rheology, in *Bailey's Industrial Oil and Fat Products,* Shaihidi, F., Ed., John Wiley & Sons, New York, 2004.
40. Nawar, W.W., Chemistry, in *Bailey's Industrial Oil and Fat Products,* Vol. 1, 1996, 397.
41. deMan, J.M. and deMan, L., Polymorphism and texture of fats, in *Crystallization and Solidification Properties of Lipids,* Widlak, N., Hartel, R.W., and Narine, S.S., Eds., AOCS Press, Champaign, IL, 2001, p. 225.
42. Chapman, D., The polymorphism of glycerides, *Chem. Rev.,* 62, 433, 1962.
43. Clarkson, C.E. and Malkin, T., Alternation in long chain compounds. Part II. An X ray and thermal investigation of the triglycerides, *J. Chem. Soc.,* 137, 666, 1934.
44. Sato, K., Ueno, S., and Yano, J., Molecular interactions and kinetic properties of fats, *J. Prog. Lipid Res.,* 28, 91, 1999.
45. Sato, K., Uncovering the structures of beta' fat crystals: what do the molecules tell us?, *Lipid Technol.,* 13, 36, 2001.
46. tenGrotenhuis, E.,van Aken, G.A., van Malssen, K.F., and Schenk, H., Polymorphism of milk fat studied by differential scanning calorimetry and real-time x-ray powder diffraction, *J. Am. Oil Chem. Soc.,* 76, 1031, 1999.
47. Hagemann, J.W., Thermal behavior and polymorphism of acylglycerides, in *Crystallization and Polymorphism of Fats and Fatty Acids,* Garti, N. and Sato, K., Eds., Marcel Dekker, New York, 1988, p. 9.
48. Sato, K., Polymorphism of pure triacylglycerols and natural fats, in *Advances in Applied Lipid Research,* Vol. 2, Padley, F.B., Ed., JAI Press, London, 1996, p. 213.
49. Abrahamsson, S., Dahlén, B., Löfgren, H., and Pascher, I., Lateral packing of hydrocarbon chains, *Prog. Chem. Fats Other Lipids,* 16, 125, 1978.
50. Gunstone, F.D., Harwood, J.L., and Padley, F.B., *The Lipid Handbook,* Chapman & Hall, London, 1986, chap. 1.
51. Amey, R.L. and Chapman, D., Infrared spectroscopic studies of model and natural biomembranes, in *Biomembrane Structure and Function, Topics in Molecular and Structural Biology,* Vol. 4, Chapman, D., Ed., 1984, p. 199.
52. Yano, J. et al., Structural analyses of triacylglycerol polymorphs with FT-IT techniques. S. b'1-form and 1,2-dipalmitoyl-3-myristoyl-sn-glycerol, *J. Phys. Chem. B,* 10, 8120, 1997.
53. Arishima, T. et al., ^{13}C cross-polarization and magic-angle spinning nuclear magnetic resonance of polymorphic forms of three triacylglycerols, *J. Am. Oil Chem. Soc.,* 73, 1231, 1996.
54. D'Souza, V., deMan, J.M., and deMan, L., Short spacings and polymorphic forms of natural and commercial solid fats: a review. *J. Am. Oil Chem. Soc.,* 67, 835, 1990.

55. Higami, M. et al., Simultaneous synchrotron radiation x-ray diffraction-dsc analysis of melting and crystallization behavior of trilauroylglycerol in nanoparticles in oil-in-water emulsion, *J. Am. Oil Chem. Soc.,* 80, 731, 2003.
56. Ollivon et al., Simultaneous examination of structural and thermal behaviors of fats by coupled x-ray diffraction to differential scanning calorimetry techniques: applications to cocoa butter polymorphism, in *Crystallization and Solidification Properties of Lipids,* Widlak, N., Hartel, R.W., and Narine, S.S., Eds., AOCS Press, Champaign, IL, 2001, chap. 3.
57. Hoerr, C.W., Morphology of fats, oils and shortenings, *J. Am. Oil Chem. Soc.,* 37, 539, 1960.
58. Chen, C.W. et al., Isothermal crystallization kinetics of refined palm oil, *J. Am. Oil Chem. Soc.,* 79, 403, 2002.
59. deMan, J.M. and deMan, L., Texture of fats, in *Physical Properties of Lipids,* Marangoni, A.G. and Narine, S.S., Eds., Marcel Dekker, New York, 2002, chap. 7.
60. Che-Man, Y.B. et al., A study on the crystal structure of palm oil-based whipping cream, *J. Am. Oil Chem. Soc.,* 80, 409, 2003.
61. Ziegler, G.R., Solidification processes in chocolate confectionary manufacture, in *Crystallization and Solidification Properties of Lipids,* Widlak, N., Hartel, R.W., and Narine, S.S., Eds., AOCS Press, Champaign, IL, 2001, chap. 19.
62. Schlichter-Aronhime, J. and Garti, N., Solidification and polymorphism in cocoa butter and the blooming of fats and fatty acids, in *Crystallization and Polymorphism of Fats and Fatty Acids,* Garti, N. and Sato, K., Eds., Marcel Dekker, New York, 1988, p. 363.
63. Wainwright, R.E., Oil and fats in confections, in *Bailey's Industrial Oil and Fat Products, Vol 3: Edible Oil and Fat Products: Product and Application Technology,* Hui, Y.H., Ed., John Wiley & Sons, New York, 1996, p. 353.
64. Loisel, C. et al., Phase transitions and polymorphism of cocoa butter, *J. Am. Oil Chem. Soc.,* 85, 425, 1998.
65. Van Malssen, K. et al., Phase behavior and extended phase scheme of static cocoa butter investigated with real-time x-ray powder diffraction, *J. Am. Oil Chem. Soc.,* 76, 669, 1999.
66. Wille, R.E. and Lutton, E.S., Polymorphism in cocoa butter, *J. Am. Oil Chem. Soc.,* 43, 491, 1966.
67. MacMillan, S.C. et al., In situ small angle x-ray scattering (SAXS) studies of polymorphism with the associated crystallization of cocoa butter fat using shearing conditions, *Cryst. Growth Des.,* 2, 221, 2002.
68. Mazzanti, G. et al., Orientation and phase transitions of fat crystals under shear, *Cryst. Growth Des.,* 3, 721, 2003.
69. Garti, N., Schlichter, J., and Sarig, S., Polymorphism of even monoacid triglycerides in the presence of sorbitan monostearate, studied by DSC, *Thermochim. Acta.,* 93, 29, 1985.
70. Sato, K., Molecular aspects in fat polymorphism, in *Crystallization and Solidification Properties of Lipids,* Widlak, N., Hartel, R., and Narine, S.S., Eds., AOCS Press, Champaign, IL, 2001, chap. 1.
71. Shen, C.F., deMan, L., and deMan, J.M., Effect of palm stearin and hydrogenated palm oil on the polymorphic stability of hydrogenated canola oils, *J. Oil Palm Res.,* 2, 143, 1990.
72. Schlichter, J., Sarig, S., and Garti, N., Mechanistic considerations of polymorphic transformations of tristearin in the presence of emulsifiers, *J. Am. Oil. Chem. Soc.,* 64, 529, 1987.

73. Takeuchi, M. et al., Binary phase behavior of 1,3-distearoyl-2-oleoyl-sn-glycerol (SOS) and 1,3-distearoyl-2-linoleoyl-sn-glycerol (SLS), *J. Am. Oil. Chem. Soc.,* 79, 627, 2002.
74. Humphrey, K.L., Moquin, P.H., and Narine, S.S., Phase behavior of a binary lipid shortening system: from molecules to rheology, *J. Am. Oil. Chem. Soc.,* 80, 1175, 2003.
75. Walstra, P. et al., Dairy *Technology: Principles of Milk Properties and Processes,* Marcel Dekker, New York, 1999, pp. 50–70, 501–510.
76. van den Tempel, M., Mechanical properties of plastic disperse systems at very small deformations, *J. Colloid Sci.,* 16, 284, 1961.
77. Chawla, P. and deMan, J.M., Crystal morphology of shortenings and margarines, *Food Struct.,* 9, 329, 1990.
78. Precht, D., Fat crystal structure in cream and butter, in *Crystallization and Polymorphism of Fats and Fatty Acids,* Garti, N. and Sato, K., Eds., Marcel Dekker, New York, 1988, p. 305.
79. Shama, F. and Sherman, P., An automated parallel-plate viscoelastometer for studying the rheological properties of solid food materials, in *Rheology and Texture of Foodstuffs. S.C.I. Monograph No. 27*, Society of Chemical Industry, London, 1968, p. 77.
80. Heertje, I., Microstructural studies in fat research, *Food Struct.,* 12, 77, 1993.
81. Johansson, D. and Bergenståhl, B., Sintering of fat crystal networks in oil during post-crystallization processes, *J. Am. Oil Chem. Soc.,* 72, 911, 1995.
82. deMan, J.M. and Wood, F.W., Hardness of butter. I. Influence of season and manufacturing method, *J. Dairy Sci.,* 41, 360, 1958.
83. Kulkarni, S. and Rama Murthy, M.K., Studies on changes in rheological characteristics of butter stored at different temperatures for different periods, *Ind. J. Dairy Sci.,* 40, 232, 1985.
84. Mortensen, B.K. and Danmark, H., Factors influencing the consistency of butter, *Milchwissenschaft,* 36, 393, 1982.
85. deMan, J.M. and Beers, A.M., Review: fat crystal networks: structure and rheological properties, *J. Text. Stud.,* 18, 303, 1987.
86. Heertje, I., van der Vlist, P., Blonk, J.C.G., Hendricks, H.A.C.M., and Brakenhof, G.J., Confocal laser scanning microscopy in food research: some observations, *Food Microstruct.,* 6, 115, 1987.
87. Shukla, A. and Rizvi, S.S.H., Relationship among chemical composition, microstructure and rheological properties of butter, *Milchwissenschaft,* 51, 144, 1996.
88. Narine, S.S. and Marangoni, A.G., Fractal nature of fat crystal networks, *Phys. Rev. E,* 59, 1908, 1999.
89. Marangoni, A.G. and Rousseau, D., Is plastic fat rheology governed by the fractal nature of the fat crystal network?, *J. Am. Oil Chem. Soc.,* 73, 991, 1996.
90. Moziar, C., deMan, L., and deMan, J.M., Effect of tempering on the physical properties of shortening, *Can. Inst. Food Sci. Technol.,* 22, 238, 1989.
91. van Aken, G.A. and Visser, K.A., Firmness and crystallization of milk fat in relation to processing conditions, *J. Dairy Sci.,* 83, 1919, 2000.
92. Herrera, M.L. and Hartel, R.W., Effect of processing conditions on physical properties of a milk fat model system: microstructure, *J. Am. Oil Chem. Soc.,* 77, 1197, 2000.
93. Martini, S., Herrera, M.L., and Hartel, R.W., Effect of processing conditions on microstructure of milk fat fraction/sunflower oil blends, *J. Am. Oil Chem. Soc,* 79, 1063, 2002.
94. Cornily, G. and leMeste, M., Flow behavior of lard and its fractions at 15°C, *J. Text. Stud.,* 16, 383, 1985.

95. Brooker, B.E., Low-temperature microscopy and x-ray analysis of food systems, *Trends Food Sci. Technol.,* 1, 100, 1990.
96. deMan, J.M., Microscopy in the study of fats and emulsions, *Food Microstruct.,* 1, 209, 1982.
97. Hicklin, J.D., Jewell, G.G., and Heathcock, J.F., Combining microscopy and physical techniques in the study of cocoa butter polymorphs and vegetable fat blends, *Food Microstruct.,* 4, 241, 1985.
98. Manning, D.M. and Dimick, P.S., Crystal morphology of cocoa butter, *Food Microstruct.,* 4, 249, 1985.
99. Marangoni, A.G. and Hartel, R.W., Visualization and structural analysis of fat crystal networks, *J. Food. Technol.,* 9, 46, 1998.
100. Xu et al., Multiphoton fluorescence excitation: new spectral windows for biological nonlinear microscopy, *Proc. Natl. Acad. Sci. USA,* 1996, 10763.
101. Kellens, M., Meeussen, W., and Reynaers, H., Study of the polymorphism and the crystallization kinetics of tripalmitin: a microscopic approach, *J. Am. Oil Chem. Soc.,* 69, 906, 1992.
102. Marangoni, A.G., The nature of fractality in fat crystal networks, *Trends Food Sci. Technol.,* 13, 37, 2002.
103. Narine, S.S. and Marangoni, A.G., Mechanical and structural model of fractal networks of fat crystals at low deformations, *Phys. Rev. E,* 60, 6991, 1999.
104. Narine, S.S. and Marangoni, A.G., Relating structure of fat crystal networks to mechanical properties: a review, *Food Res. Int.,* 32, 227, 1999.
105. Litwinenko, J.W., Rojas, A.M., Gerschenson, L.N., and Marangoni, A.G., Relationship between crystallization behavior, microstructure, and mechanical properties in a palm oil-based shortening, *J. Am. Oil Chem. Soc.,* 79, 647, 2002.
106. Pal Singh, A., Avramis, C.A., Kramer, J.K.G., Hill, A.R., and Marangoni, A.G., Algal meal supplementation of the cows' diet alters the physical properties of milk fat, *J. Dairy Res.,* 71, 66, 2004.
107. Rao, A.M., Measurement of flow and viscoelastic properties, in *Rheology of Fluid and Semisolid Foods: Principles and Applications,* Rao, A.M., Ed., Chapman & Hall, Gaithersburg, MD, 1999, p. 59.
108. Steffe, J.F., Viscoelasticity, in *Rheological Methods in Food Process Engineering,* 2nd ed., Freeman Press, East Lansing, MI, 1996, p. 294.
109. Herrera, M.L. and Hartel, R.W., Effect of processing conditions on physical properties of a milk fat model system: rheology, *J. Am. Oil Chem. Soc.,* 77, 1189, 2000.
110. Janssen, P.W.M., Modeling fat microstructure using finite element analysis, *J. Food Eng.,* 61, 387, 2004.
111. Shukla, A. and Rizvi, S.S.H., Viscoelastic properties of butter, *J. Food Sci.,* 60, 902, 1995.
112. Shellhammer, T.H., Rumsey, T.R., and Krochta, J.M., Viscoelastic properties of edible lipids, *J. Food Eng.,* 33, 305, 1997.
113. Borwanker, R.P. et al., Rheological characterization of melting of margarines and tablespreads, *J. Food Eng.,* 16, 55, 1992.
114. Shukla, A. et al., Physicochemical and rheological properties of butter made from supercritically fractionated milk fat, *J. Dairy Sci.,* 77, 45, 1994.
115. Toro-Vazquez, J.F. et al., Rheometry and polymorphism of cocoa butter during crystallization under static and stirring conditions, *J. Am. Oil Chem. Soc.,* 82, 2004, 2004.
116. Kapsalis, J.G. et al., Effect of chemical additives on the spreading quality of butter. I. The consistency of butter as determined by mechanical and consumer panel evaluation method, *J. Dairy Sci.,* 43, 1560, 1960.

117. Pompei, C. et al., Development of two imitative methods of spreadability evaluation and comparison with penetration tests, *J. Food Sci.,* 53, 597, 1988.
118. de Bruijne, W.D. and Bot, A., Fabricated fat-based foods, in *Food Texture Measurement and Perception,* Rosenthal, A.J., Ed., Aspen, Gaithersburg, MD, 1999, p. 185.
119. Rohm, H., Magnitude estimation of butter spreadability by untrained panelists, *J. Food Sci. Technol.,* 23, 550, 1990.
120. Mortensen, B.K. and Danmark, H., Consistency characteristics of butter, *Milchwissenschaft,* 37, 530, 1982.
121. Wright et al., Rheological properties of milkfat and butter, *J. Food Sci.,* 66, 1056, 2001.
122. Rohm, H., Rheological behavior of butter at large deformations, *J. Text. Stud.,* 24, 139, 1993.
123. Sone, T., The rheological behavior and thixotropy of a fatty plastic body, *J. Phys. Soc. Jpn.,* 16, 961, 1961.
124. American Oil Chemists' Society, *Official Methods and Recommended Practices of the American Oil Chemists' Society,* 4th edn., AOCS Press, Champaign, IL, 1993, Cd 16-60.
125. Haighton, A.J., The measurement of the hardness of margarine and fats with cone penetrometers, *J. Am. Oil Chem. Soc.,* 36, 345, 1959.
126. Vasic, I. and deMan, J.M., Effect of mechanical treatment on some rheological properties of butter, in *Rheology and Texture of Foodstuffs,* S.C.I. Monograph No. 27, Society Chem. Ind., London, 1968, p. 251.
127. Weyland, M., Confectionary oils and fats- profiling fat functionality, *Manufact. Confect.,* 79, 53, 1999.
128. Bennett, B., Oil's well. New research yields improved high-oleic oils. *Food Proc.,* 58, 39, 1997.
129. Cherrak, C.M. et al., Low-palmitic, low-linolenic soybean development, *J. Am. Oil Chem. Soc.,* 80, 539, 2003.
130. Gupta, M.K., NuSun-Healthy Oil at a Commodity Price, *Lipid Technol.* 12, 29, 2000.
131. Jalani, B.S. et al., Improvement of palm oil through breeding and biotechnology, *J. Am. Oil Chem. Soc.,* 74, 1451, 1997.
132. KeShun, L. and Brown, E.A., Enhancing vegetable oil quality through plant breeding and genetic engineering, *Food Technol.,* 50, 67, 1996.
133. Rouselin, P. et al., Modification of sunflower oil quality by seed-specific expression of a heterologous DELTA9-stearoyl-(acyl carrier protein) desaturase gene, *Plant Breed.,* 121, 2002.
134. Qing, L., Singh, S.P., and Green, A.G., High-stearic and high-oleic cottonseed oils produced by hairpin RNA-mediated post-transcriptional gene silencing, *Plant Phys.,* 129, 1732, 2002.
135. Guinda, A. et al., Chemical and physical properties of a sunflower oil with high levels of oleic and palmitic acids, *Eur. J. Lipid Sci. Technol.,* 105, 130, 2003.
136. Katsumata, M. et al., Influence of a high ambient temperature and dietary fat supplementation on fatty acid composition of depot fats in finishing pigs, *Ann. Sci. Technol,* 66, 225, 1995.
137. Palmquist, D.L., Beaulieu, A.D., and Barbano, D.M., Feed and animal factors influencing milk composition, *Dairy Sci.,* 76, 1753, 1993.
138. Skrivan, M. et al., Influence of dietary fat source and copper supplementation on broiler performance, fatty acid profile of meat and depot fat, and on cholesterol content in meat, *Br. Poultry Sci.,* 41, 608, 2000.
139. Ashes, J.R. et al., Manipulation of the fatty acid composition of milk by feeding protected canola seeds, *J. Dairy Sci.,* 75, 1090, 1992.

140. Lin, M.P. et al., Modification of fatty acids in milk by feeding calcium-protected high oleic sunflower oil, *J. Food Sci.,* 61, 24, 1996.
141. Mohamed, O.E. et al., Influence of dietary cottonseed and soybean on milk production and composition, *J. Dairy Sci.,* 71, 2677, 1988.
142. Allen, R.R., Principles and catalysts for hydrogenation of fats and oils. *J. Am. Oil Chem. Soc.,* 55, 792, 1978.
143. Karabulut, I., Kayahan, M., and Yaprak, S., Determination of changes in some physical and chemical properties of soybean oil during hydrogenation, *Food Chem.,* 81, 453, 2003.
144. Patterson, H.B.W., *Hydrogenation of Fats and Oils: Theory and Practice,* AOCS Press, Champaign, IL, 1994.
145. Bockisch, M., Modification of fats and oils, in *Fats and Oils Handbook,* AOCS Press, Champaign, IL, 1998, chap. 6.
146. Ascherio, A. et al., Trans fatty acids and coronary heart disease, *N. Engl. J. Med.,* 340, 1994, 1999.
147. Hu, F.B. et al., Dietary fat intake and the risk of coronary heart disease in women, *N. Engl. J. Med.,* 337, 1491, 1997.
148. Mensik, R.P. and Katan, M.B., Effect of dietary trans fatty acids on high-density and low-density lipoprotein cholesterol levels in healthy subjects, *N. Engl. J. Med.,* 323, 439, 1990.
149. Wilson, E., Trans fat update, *Manufact. Confect.,* 84, 73, 2004.
150. Hoerr, C.W. and Waugh, D.F., Some physical characteristics of rearranged lard, *J. Am. Oil Chem. Soc.,* 32, 37, 1955.
151. Kurashige, I., Matsuzaki, N., and Takahaski, H., Enzymatic modification of canola/palm oil mixtures: effects on the fluidity of the mixture, *J. Am. Oil Chem. Soc.,* 70, 849, 1993.
152. Rousseau, D. et al., Restructuring butterfat through blending and chemical interesterification. 3. Rheology, *J. Am. Oil Chem. Soc.,* 73, 963, 1997.
153. Rousseau, D. and Marangoni, A.G., The effects of interesterification on physical and sensory attributes of butterfat and butterfat-canola oil spreads, *Food Res. Int.,* 31, 381, 1999.
154. Abigor, R.D. et al., Production of cocoa butter-like fats by the lipase-catalyzed interesterification of palm oil and hydrogenated soybean oil, *J. Am. Oil Chem. Soc.,* 80, 1193, 2003.
155. Bloomer, S., Adlercreutz, P., and Mattiasson, B., Triglyceride interesterification by lipases: 1. Cocoa butter equivalents from a fraction of palm oil, *J. Am. Oil Chem. Soc.,* 67, 519, 1990.
156. Noor Lida, H.M.D. et al., TAG composition and solid fat content of palm oil, sunflower oil, and palm kernel olein blends before and after chemical interesterification, *J. Am. Oil Chem. Soc.,* 79, 1137, 2002.
157. Huang, K. and Akoh, C.C., Lipase-catalyzed incorporation of n-3 polyunsaturated fatty acids into vegetable oil, *J. Am. Oil Chem. Soc.,* 71, 1277, 1994.
158. List, G.R. et al., Effect of interesterification on the structure and physical properties of high-stearic acid soybean oils, *J. Am. Oil Chem. Soc.,* 74, 327, 1997.
159. Kok, L.L. et al., Trans-free margarine from highly saturated soybean oil. *J. Am. Oil Chem. Soc.,* 76, 1175, 1999.
160. Petrauskaite, V. et al., Physical and chemical properties of trans-free fats produced by chemical interesterification of vegetable oil blends, *J. Am. Oil Chem. Soc.,* 75, 489, 1998.
161. Seriburi, V. and Akoh, C.C., Enzymatic interesterification of lard and high-oleic sunflower oil with *Candida antarctica* lipase to produce plastic fats, *J. Am. Oil Chem. Soc.,* 75, 1339, 1998.

162. Akoh, C.C., Structured lipids, in *Food Lipids: Chemistry, Nutrition, and Biotechnology,* Akoh, C.C. and Min, D.B., Eds., Marcel Dekker, New York, 1998, p. 699.
163. Osborn, H.T. and Akoh, C.C., Structured lipids: novel fats with medical, nutraceutical, and food applications, *Crit. Rev. Food Sci. Food Saf.,* 1, 93, 2002.
164. Haumann, B.F., Structured lipids allow fat tailoring, *Inform,* 8, 1004, 1997.
165. Marangoni, A.G. and Rousseau, D., Engineering triacylglycerols: the role of interesterification, *Trends Food. Sci. Technol.,* 6, 329, 1995.
166. Hamm, W., Trends in edible oil fractionation, *Trends Food Sci. Technol.,* 6, 121, 1995.
167. Abeshima, T., Fractionation of edible oils and fats, *J. Jpn. Oil Chem. Soc.,* 47, 553, 1998.
168. Illingworth, D., Fractionation of fats, in *Physical Properties of Lipids,* Marangoni, A.G. and Narine, S.S., Eds., Marcel Dekker, New York, 2002, 411.
169. Deffense, E., Fractionated milk fat products in bakery products, in *New Uses for Milk,* Dairy Science Research Center, Laval University, Quebec, 1989, p. 79.
170. Kaylegian, K.E. and Lindsay, R.C., *Handbook of Milkfat Fractionation Technology and Applications,* AOCS Press, Champaign, IL, 1994.
171. McGillivary, W.A., Softer butter from fractionated fat or by modified processing, *N.Z. J. Dairy Sci. Technol.,* 7, 111, 1972.
172. Makhloauf, J. et al., Fractionnement de la matiere grasse laitiere par cristallisation simple et son utilization dans la fabrication de beurres mous, *Can. Inst. Food Technol. J.,* 20, 236, 1987.
173. Vanhoutte, B. et al., Monitoring milk fat fractionation: filtration properties and crystallization kinetics, *J. Am. Oil Chem. Soc.,* 80, 213, 2003.
174. Wright, A.J. et al., Solvent effects on the crystallization behavior of milk fat fractions, *J. Agric. Food Chem.,* 48, 1033, 2000.
175. Nor Aini, I. et al., Trans-free vanaspati containing ternary blends of palm oil-palm stearin-palm olein and palm oil-palm stearin-palm kernel olein, *J. Am. Oil Chem. Soc.,* 76, 643, 1999.
176. Pal, P.K., Bhattacharyya, D.K., and Ghosh, S., Modifications of butter stearin by blending and interesterification for better utilization in edible fat products, *J. Am. Oil Chem. Soc.,* 78, 31, 2001.
177. Roy, S. and Bhattacharyya, D.K., Comparative nutritional quality of palm stearin liquid oil blends and hydrogenated fat (vanaspati), *J. Am. Oil Chem. Soc.,* 73, 617, 1996.
178. Williams, S.D., Ransom-Painter, K.L., and Hartel, R.W., Mixture of palm kernel oil with cocoa butter and milk fat in compound coatings, *J. Am. Oil Chem. Soc.,* 74, 357, 1997.
179. Wilbey, R.A., Production of butter and dairy-based spreads, in *Modern Dairy Technology, Vol. 1: Advances in Milk Processing,* Chapman & Hall, New York, 1994, p. 107.
180. Danthine, S. and Deroanne, C., Blending of hydrogenation low-erucic acid rapeseed oil, low-erucic acid rapeseed oil, and hydrogenated palm oil or palm oil in the preparation of shortenings, *J. Am. Oil Chem. Soc.,* 80, 1069, 2003.
181. deMan, L. and deMan, J.M., Functionality of palm oil, palm oil products, and palm kernel oil in margarine and shortening, POIRM Occasional Paper No. 32, Palm Oil Research Institute of Malaysia, 1994.
182. Wiedermann, L.H., Margarine and margarine oil: formulation and control, *J. Am. Oil Chem. Soc.,* 55, 823, 1978.
183. Krog, N., Functions of emulsifiers in food systems, *J. Am. Oil Chem. Soc.,* 54, 124, 1977.
184. Martini, S. et al., Effect of sucrose esters and sunflower oil addition on crystalline microstructure of a high-melting milk fat fraction, *J. Food Sci.,* 67, 3412, 2002.

185. Nasir, M.I., Effects of sucrose polyesters and sucrose polyester-lecithins on crystallization rate of vegetable ghee, in *Crystallization and Solidification Properties of Lipids,* Widlak, N., Hartel, R.W., and Narine, S.S., Eds., AOCS Press, Champaign, IL, 2001, chap. 7.
186. Cho, F. and deMan, J.M., Physical properties and composition of low trans canola/palm blends partially modified by chemical interesterification, *J. Food Lip.,* 1, 53, 1993.
187. Jeyarani, T. and Yella Reddy, S., Preparation of plastic fats with zero trans FA from palm oil, *J. Am. Oil Chem. Soc.,* 80, 1107, 2003.
188. Hamm, W., Regional differences in edible oil processing practice. II. Refining, oil modification, and formulation, *Lipid Technol.,* 13, 105, 2001.
189. Ghotra, B.S., Dyal, S.D., and Narine, S.S., Lipid shortenings: a review, *Food Res. Int.,* 35, 1015, 2002.
190. Miskander, M.S. et al., Effect of emulsion temperature on physical properties of palm oil-based margarine, *J. Am. Oil Chem. Soc.,* 79, 1163, 2002.
191. Nor Aini, I., Effects of tempering on physical properties of shortenings based on binary blends of palm oil and anhydrous milk fat during storage, in *Crystallization and Solidification Properties of* Lipids, Widlak, N., Hartel, R.W. and Narine, S.S., Eds., AOCS Press, Champaign, IL, 2001, chap. 4.
192. Chrysam, M.M., Table spreads and shortenings, in *Bailey's Industrial Oil and Fat Products,* Vol. 3, Applewhite, T.H., Ed., John Wiley & Sons, New York, 1985, p. 41.
193. Narine, S.S. and Humphrey, K.L., A comparison of lipid shortening functionality as a function of molecular ensemble and shear: microstructure, polymorphism, solid fat content and texture, *Food Res. Int.,* 38, 28, 2004.

7 Beyond Hard Spheres: The Functional Role of Lipids in Food Emulsions

John N. Coupland

CONTENTS

7.1 THE FUNCTIONAL ROLE OF LIPIDS IN EMULSIONS

In most foods, the lipid and aqueous material coexist in some form of dispersion. Prevalent amongst of these are emulsions where the oil is present in fine droplets in an aqueous continuous phase (or vice versa, although in the context of this work only oil-in-water emulsions will be considered). The functionality of an emulsion depends on its microstructure, which in turn depends on the ingredient interactions as affected by process conditions and time. The development of quantitative relationships among emulsion composition, structure, and functionality is one of the major successes of food physical chemistry. Several books and reviews summarize the current state of knowledge focusing on the fundamental colloid science (e.g., [1]) or the application of these principles to foods (e.g., [2,3]).

A common approach to understanding structure–function relationships in emulsions is to consider the droplets as hard spheres. Fine droplets are certainly spherical because of the need to minimize surface area, and the shear forces under normal use conditions are rarely sufficient to deform them. The droplets are also inevitably coated with a layer of amphiphilic material, usually protein, which provides kinetic stability to the product. In purely physical terms, then, it is reasonable to be more concerned with the interactions between the coating layers and to regard the droplets themselves as inert "space-fillers" bound to surfactant. However, in foods, the lipid in the product plays an important functional role in the product beyond its nutritional value as fat. Defining functionality as the set of human responses to the overall food, we shall see that the flavor, texture, and stability of the food can be affected by the lipid composition in ways that cannot be explained by modeling the droplets as hard spheres. In this work I will examine the roles lipid chemistry and physics can play in controlling food emulsion functionality.

7.2 EFFECTS OF LIPID ON EMULSION STABILITY

At its simplest, the role of a lipid in an emulsion is to be water insoluble. The same phenomena that lead to the molecular insolubility of a lipid molecule in water (i.e., the hydrophobic effect of very strong water–water attractions dominating over relatively weak water–lipid attractions and the entropy of mixing) lead at a bulk scale to interfacial tension and, hence, phase separation. Strong oil–water interfacial tension (~27 mN m^{-1} for triglycerides and 50 mN m^{-1} for alkanes [4]) has three primary consequences for emulsion structure: (1) lipid droplets in oil tend to be spherical, (2) there is a strong thermodynamic pressure for lipid droplets to coalesce to minimize the interfacial area, and (3) amphiphilic molecules will adsorb at the surface.

A molecule will adsorb to an oil–water interface when the energetic gain achieved by solubilizing part of its structure in either phase outweighs the loss in translational entropy cost of restricting its mobility. The energetic "benefit" is largely due to moving the hydrophobic group of the amphiphilic molecule from water to an oil phase. There is limited evidence that the relatively small differences between triglyceride food oils cause significant difference in sorption behavior; however, Chanamai and others [5] noted that while decane droplets could be stabilized with gum arabic, decanol droplets could not. They argued that the relatively low decanol–water interfacial tension was inadequate to allow gum sorption and the droplets were consequently inadequately protected against coalescence.

7.2.1 Creaming

Despite the stabilizing effect of adsorbed material, emulsions are still thermodynamically unstable structures. Phase separation in an emulsion can proceed by a mixture of creaming, flocculation, coalescence, and Ostwald ripening. Creaming is the gravitational separation of oil from water due to the density difference between phases as retarded by the frictional drag on the moving droplets. At its simplest the terminal velocity of a creaming particle is given by [6]

$$v_{\text{Stokes}} = \frac{-2gr^2\Delta\rho}{9\eta} \tag{7.1}$$

where r is particle radius, $\Delta\rho$ is the density contrast between the phases, η is the continuous phase viscosity, and g is the acceleration due to gravity.

Food oils could potentially have different creaming rates due to their different densities; however Coupland and McClements [7] showed that the standard error in liquid oil density for a set of 14 published values of diverse composition was only 4%, and it seems unlikely this small difference would lead to any significant differences. Oil-soluble weighting agents (e.g., brominated vegetable oil and ester gums) can be used to increase the dispersed phase density and decrease the creaming rate. Chanamai and McClements [6] showed that the amount of these ingredients needed to match the aqueous-phase density (and hence stop creaming) was relatively large (25% for brominated vegetable oils and >45% for the oil-soluble gums selected) and their effects could readily be overcome by aqueous sucrose. The gums also reduced the homogenization efficiency by increasing dispersed phase viscosity.

7.2.2 Flocculation and Coalescence

Droplet flocculation and coalescence require two droplets to closely interact (either via Brownian or orthokinetic collision or by their close proximity in a concentrated emulsion or cream layer) and either a semipermanent covalent or noncovalent attraction to hold them in close proximity (i.e., flocculation) or the lamella separating the two droplets to rupture allowing their contents to merge (i.e., coalescence). (Partial coalescence, a phenomenon related to both coalescence and flocculation, is discussed in more detail in "Droplet Crystallization," section 7.3.) Smoluchowski's classic formulation of aggregation kinetics assumed the droplets were hard spheres with no forces acting between them and would completely, immediately, and irreversibly coalesce upon collision [8]. However, the rates predicted are much faster than observed in reality and many workers have used a collision efficiency term (<<1) to attempt to reconcile the differences. Collision efficiency is typically governed by the interfacial layer on the droplets so any differences on the type or extent of adsorption, particularly on relatively polar lipids [5], may make some differences in flocculation and coalescence rate.

7.2.3 Disproportionation

Ostwald ripening is the diffusion of lipid molecules from smaller to larger droplets and is driven by differences in interfacial curvature. The solubility of oil in water is so low [9] for most food oils that the rate of Ostwald ripening is usually negligible. However, solubility increases with decreasing carbon chain length [10] and presence of polar groups so for certain fine flavor emulsions it can significantly contribute to the rate of phase separation. For example, Chanamai and others [5] showed that while low-polarity, low-solubility (i.e., hexadecane) droplets were stable to Ostwald ripening, low-polarity, higher solubility (i.e., decane), higher polarity, and higher solubility (i.e., decanol) droplets showed significant changes in particle size over 100 h.

From the preceding discussion it can be seen that variations in oil structure can play a small but sometimes important role in altering the stability of food emulsions. In the next section, we will see how these effects become much more significant when the oil crystallizes.

7.3 LIPID CRYSTALLIZATION IN EMULSIONS AND ITS EFFECT ON STABILITY

When a liquid oil is cooled below its melting point, there is a thermodynamic pressure for crystallization. At moderate supercooling, the kinetic barrier to crystallization can be significant and the onset of crystallization particularly prolonged. The barrier to crystallization is the free energy cost of forming a stable nucleus in the melt and this is usually overcome by nucleating at some appropriate heterogeneous solid surface. In emulsion droplets, lipid crystallization is complicated because each droplet must nucleate independently and because the presence of the interfacial material (and indeed the curvature of the surface itself) may impact crystal formation [11]. For fine dispersions of relatively pure oils (e.g., simple alkanes), the number of catalytic impurities can be

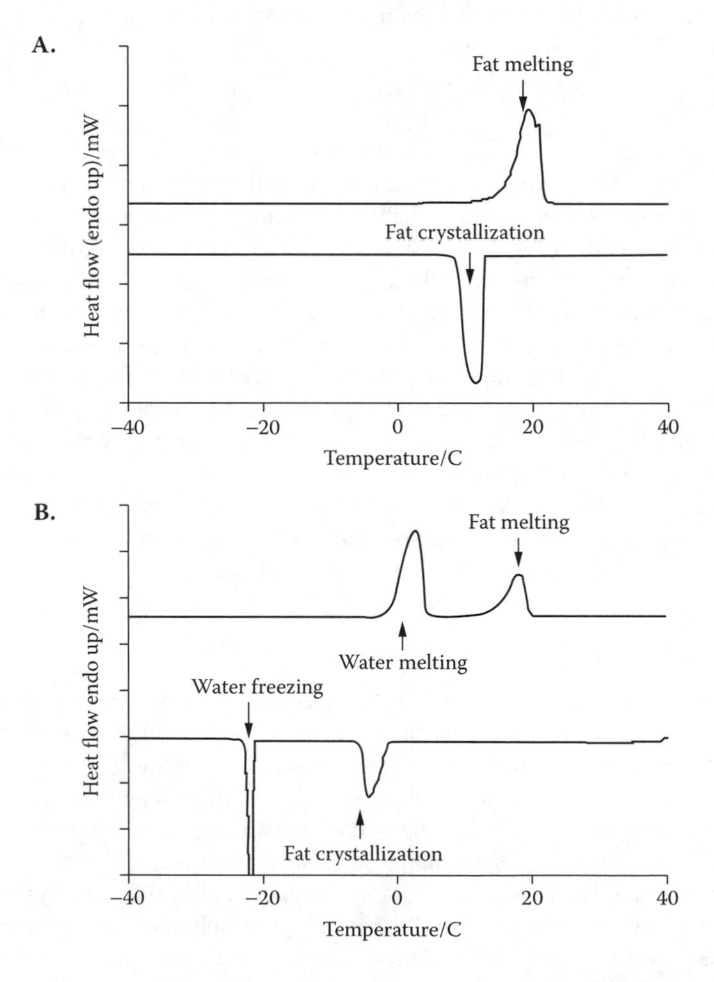

FIGURE 7.1 DSC thermograms for (a) bulk hexadecane, (b) emulsified hexadecane, (c) bulk salatrim, and (d) emulsified Salatrim™. Heating and cooling cycles were conducted at 5°C min^{-1} against an empty cell blank. In all cases the cooling thermogram is offset below the heating thermogram of the same sample and the *y*-axis is rescaled to allow easier differentiation of peaks.

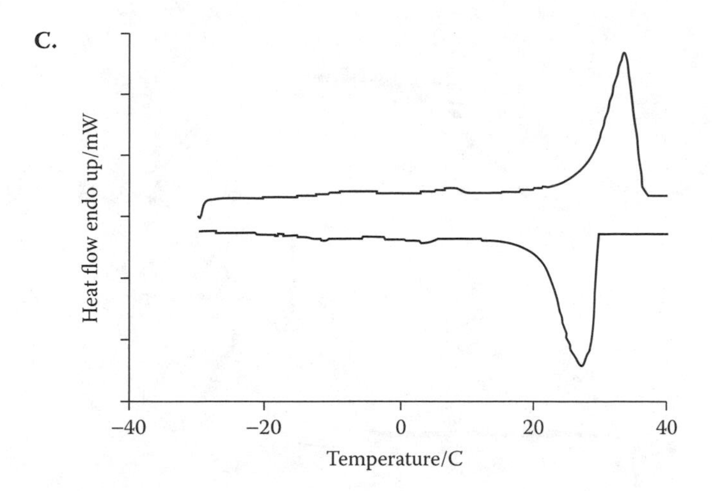

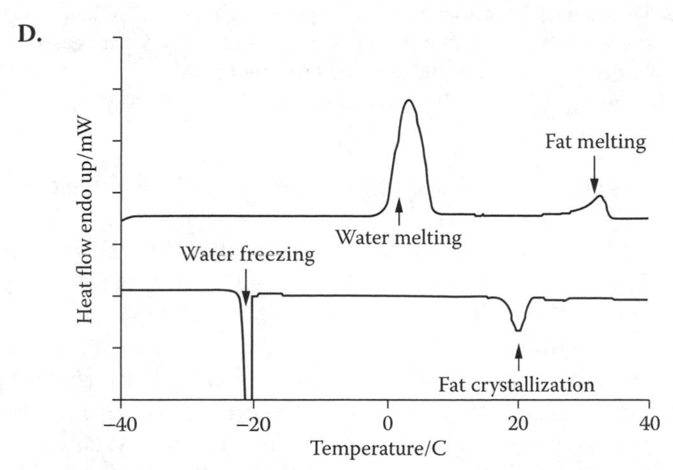

FIGURE 7.1 (*Continued*).

much less than the number of droplets, and the bulk of the oil must crystallize via homogeneous nucleation [12]. Walstra advanced an argument based on the temperature dependence of nucleation times that homogeneous nucleation is not reasonable in most food oils; and, in fact, except in the finest droplets of very pure oils, a heterogeneous mechanism dominates [13]. This is illustrated in Figure 7.1, which shows the heating and cooling thermogram of *n*-hexadecane and salatrim as bulk fat and fine emulsions. The amount of supercooling seen in the alkane in bulk is 2 to 3°C and in emulsion ~18°C, while for the Salatrim, the supercoolings are about 1 and 5°C, respectively. Deep supercooling is characteristic of homogeneous nucleation and consequently only the emulsified alkane is believed to crystallize via this mechanism. Note that the broad heating and melting curves in the Salatrim (due to multiple polymorphic transitions) make unambiguous definition of supercooling, from thermograms alone, difficult. (Some workers have argued that, in fact, many food oils crystallize homogeneously and extremely rapidly into the α-polymorph, and then slowly convert to more stable forms [14,15].)

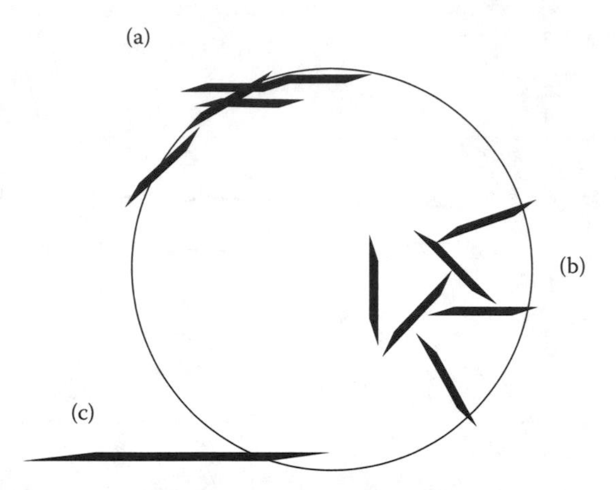

FIGURE 7.2 Diagrammatic representation of the potential alignment of fat crystals at a droplet interface: (a) crystals lie parallel to the surface; (b) a crystal network permeates the droplet; (c) a large crystal extends away from the surface. (Adapted from Walstra, P., in *Encyclopedia of Emulsion Technology*, Becher, P. Ed., Marcel Dekker, New York, 1996, p. 1.

Walstra also pointed out that in emulsion droplets there is typically more than one fat crystal present [13]. This is illustrated in Figure 7.2, which shows large (~10 μm) emulsion droplets before and after crystallization. This observation runs contrary to the view of nucleation-limited crystallization that would predict one nucleation event and hence one crystal per droplet. One interpretation is that for many food oils, the growth of the crystal is slow due to the difficulty of incorporating a triglyceride from a complex melt into the growing crystal face and that sometimes ordered lipid molecular clusters can detach from the existing crystals and act as secondary nucleation sites. However, measuring the fat crystal structure within emulsion droplets is difficult, as they are too small to allow conventional microscopy and often too dilute to allow conventional x-ray scattering. Making inferences from small to large droplets is problematic, as the effects of surface curvature and small volumes decrease with the square and cube of radius, respectively.

Using a purpose-built integrated calorimeter with synchrotron x-ray diffraction, Ollivon and others [16] have been able to reveal subtle changes in polymorphic form as the phase transition proceeds. Crystals in emulsions, particularly the more unstable polymorphs, are often more disordered than those in bulk and often remain in the unstable polymorphic form longer than the bulk fat [14,15]. In very fine droplets, crystallization leads to the formation of needle-shaped particles (solid lipid nanoparticles, Figure 7.3). These often have thicknesses of a few molecular dimensions with unusual melting characteristics [17,18].

A phase transition in the lipid causes little change to the functional properties of the emulsion, provided the hard-sphere model applies. For example, the diameter of *n*-octadecane (as measured by dynamic light scattering by the author) and *n*-eicosane (by electroacoustics [19]) droplets decrease about 5% upon crystallization (due to density differences between the solid and liquid phases). The resultant changes in

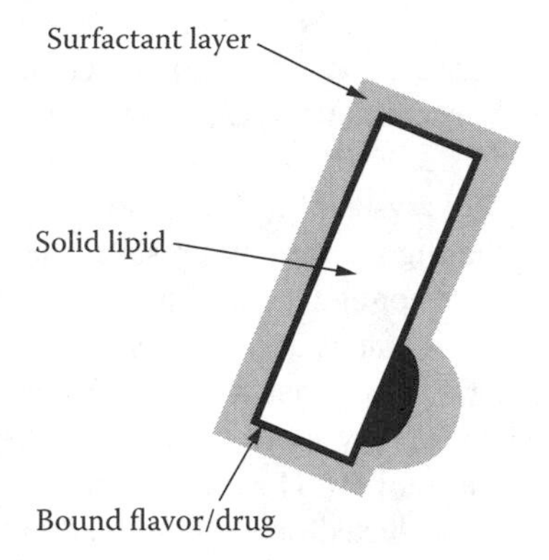

FIGURE 7.3 Diagrammatic representation of a solid lipid nanoparticle showing the bound drug solubilized between the solid fat and the surfactant layer. Typical end-to-end length would be ~100 nm. (After Bunjes, H., Koch, M.H.J., and Westesen, K., *Langmuir*, 16, 5234, 2000.)

volume fraction (and hence viscosity) would be expected to be minimal. However, in some cases the fat crystals can protrude through the interfacial layer and interact with other droplets, causing partial coalescence of the dispersed phase [20,21]. The crystal bridge between the droplets is reinforced by liquid oil flowing out to cover the link and the droplets are permanently linked (Figure 7.4b). Partial coalescence is similar to flocculation in that the effective (hydrodynamic) volume fraction increases as a result of the interaction, but is also similar to coalescence in that there is direct oil-to-oil contact between droplets. Partially coalesced droplets will completely coalesce if the solid fat supporting the double shape is melted.

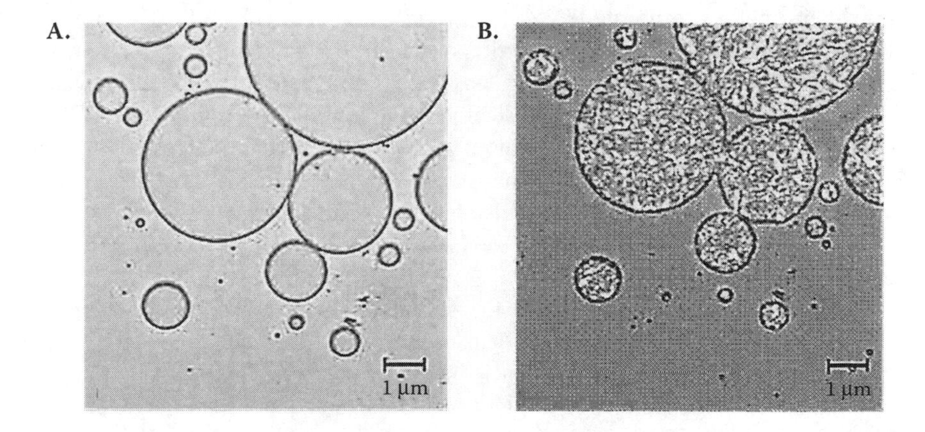

FIGURE 7.4 Optical micrograph of (a) liquid and (b) crystalline confectionary coating fat emulsion droplets. The polarizing filters were adjusted to allow crystalline and noncrystalline features to be seen simultaneously. At this scale several independent crystals can be seen in several droplets and there is limited protrusion of fat into the aqueous phase, but it is unclear whether a similar morphology will persist in submicron droplets.

Different fats have different tendencies to partial coalescence. First, the droplets need to be semicrystalline in order to react. Walstra [22] argued that a solid fat content of 10 to 50% was optimal, while Vanapalli et al. [23] showed that *n*-alkane droplets tended not to partially coalesce during steady-state cooling while coating fat droplets did, as the alkane suddenly and completely crystallized. A second factor affecting the tendency of a fat to partial coalescence is the alignment of the crystals relative to the interface. Walstra [21] observed three main types of configuration (Figure 7.2); which is preferred depends on the solid fat–liquid oil and solid fat–water interfacial energies. Crystals that protrude more into the aqueous phase are more prone to partial coalescence (Figure 7.2c). Confocal microscopy has shown distinct crystal protrusion from the surfaces associated with partial coalescence of the emulsion [24]. However, optical microscopy observations of fat crystals in droplets can be conducted only when the droplets and crystals are quite large. Electron microscopy of submicron droplets has revealed extensive partial coalescence in the absence of recognizable surface crystals [25].

7.4 EFFECTS OF LIPIDS ON SMALL-MOLECULE FUNCTIONALITY IN FOODS

The distribution of molecules in a multiphase system is described by their partition coefficients, the ratio of the activities of the components in either phase. In many cases, the concentration is low and so the activity coefficient is approximately 1 and the partition coefficient can be expressed as a ratio of concentrations. Partition coefficients depend on the molecular structure of the small molecule, with more polar molecules favoring the aqueous phase (i.e., low K_{ow}) and more nonpolar molecules accumulating in the droplets (i.e., high K_{ow}). This simple parameter reveals the role lipid droplets can have in the effective concentration and, hence, reactivity of small molecules in food emulsions. Consider as an example the partition of the antioxidants methyl carnosate ($K_{ow} = 96.2$) and gallic acid ($K_{ow} = 0.017$) in a 40% oil-in-water emulsion [26]. The proportion of the antioxidants in the oil phase would be 98.5 and 1.2%, respectively. In this case, it is anticipated that the effectiveness of the antioxidant depends on its physical partitioning into the lipid phase [27]. (In fact, partitioning in an emulsion is often more complex than suggested by these simple two-phase models. For example, Wedzicha and Ahmed [28] examined the partitioning of benzoic acid in an emulsion system and found a significant fraction of the bound acid [~6 mg m^{-2}] was in fact associated with the surface protein rather than the sunflower oil droplets.)

Partition is more complex when the solute of interest has ionizable groups. Charged molecules are effectively oil insoluble while the uncharged group may partition into the oil phase. This is particularly important for antimicrobial organic acids that have a negative charge at pH > pK. Only the aqueous acid is an effective antimicrobial, so its effectiveness can be expected to decrease in an emulsion at high pH. An effective partition coefficient can be calculated as follows:

$$K_{ow} = \frac{c_o}{c_w(1+10^{\mathrm{pH}-\mathrm{pK}})} \tag{7.2}$$

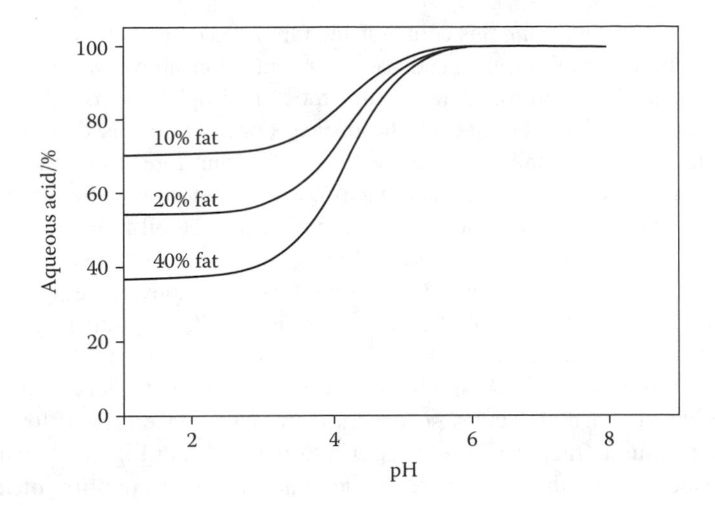

FIGURE 7.5 The effect of pH and oil volume fraction on the proportion of benzoic acid ($pK = 4.18$) in the aqueous phase of an emulsion. While the protonated acid has some tendency to partition into oil, the anionic form is exclusively water soluble.

where c is the concentration in the oil and water phases, respectively. The dramatic effect of pH and fat content is illustrated by considering benzoic acid as an example. Using Equation 7.2, the relative aqueous concentration of benzoic acid (pK = 4.18) was calculated as a function of pH for emulsions (Figure 7.5), assuming $K_{ow} = 5.27$ but neglecting effects due to dimerization of the acid [29]. Effectively all of the acid is aqueous above the pK value as the anionic form is exclusively water soluble. Below the pK there is some partitioning into the oil phase and so the higher the volume fraction of fat, the lower the amount in the aqueous phase.

Ingredient partitioning is particularly important in understanding the flavor of food emulsions. Our sense of flavor depends on a combination of three chemical senses: taste (e.g., sweet, sour, salt, and bitter), trigeminal (e.g., pungency) in the mouth, and aroma in the retronasal passages. (In fact, our sensory response to taste probably is affected by many other nonchemical factors. Interestingly, Bourne [30] in his book on food texture cites a study showing large numbers of people are unable to identify the flavor of a homogenized food product without the accompanying texture and visual cues.) Among these factors, the headspace concentration of flavor volatiles is the most directly affected by the presence of emulsion droplets. The headspace–food partition coefficient will be diminished if the volatiles can be incorporated into oil droplets [31]:

$$K_{ge} = \frac{K_{gw}}{1-(K_{dc}-1)\phi} \tag{7.3}$$

where K is the partition coefficient between the subscripted phases (g = headspace, e = overall emulsion, d = dispersed phase, and c = continuous phase) and ϕ is the volume fraction of fat. Using this equation and tabulated data [31] for some common food odorants, the equilibrium headspace concentration above various emulsions was calculated (Figure 7.6). Octanal (the most hydrophobic volatile) partitions strongly into any oil present and the headspace concentration is depressed, while butanol (the most hydrophilic) is excluded by the fat and forced into the headspace at high-volume fractions. Butanal and butan-2-one are more volatile, so the overall headspace concentration is higher, although increasing the oil fraction will bind up some of the former. Rabe and co-workers [32] showed that the type of liquid oil did not significantly affect the amount of headspace volatile above an emulsion.

While the effect of liquid oil droplets on volatile molecules can be explained in terms of partitioning (or exclusion), the interactions between solid fat droplets and volatiles is more complex. A simple approach is to treat the crystalline fats as impermeable to other molecules so crystallization has an effect of lowering the effective oil volume fraction (i.e., replace ϕ in Equation 7.3 or Figure 7.6 with ϕ_{liquid}). Crystallization would therefore increase the headspace concentration of lipophilic volatiles and have little effect on more hydrophilic volatiles. Roberts and others [33] studied the effect of solid fat content on the headspace concentration above a food emulsion. While this approach worked reasonably well, it seemed the effective volume fraction in Equation 7.3 was somewhat larger than the actual liquid oil content of the droplets, suggesting some volatile is bound by the solid fat.

The binding of small molecules by solid lipid droplets is the basis of the functionality of solid lipid nanoparticles as a drug-delivery system. Largely hydrophobic

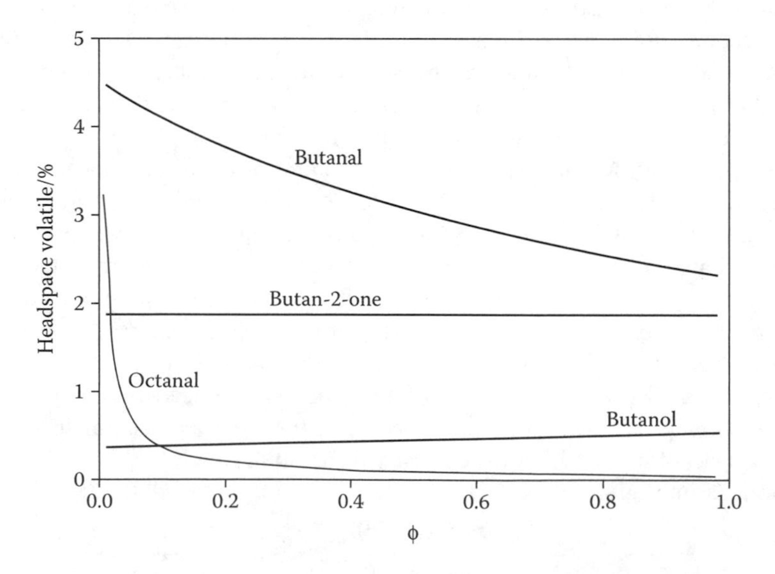

FIGURE 7.6 Calculated proportion of butanal, butan-2-one, butanol, and octanal in the headspace above emulsions with different volume fractions. The volume of the headspace was taken as ten times that of the emulsion and the partition coefficients used in the calculations were taken from the literature (McClements, D.J., *Food Emulsions. Principles, Practice, and Techniques*, 2nd ed., CRC Press, Boca Raton. FL, 2004, pp. 632.

drugs can be incorporated into liquid oil, which is then homogenized to form a fine lipid emulsion. The drug would diffuse rapidly out of the liquid droplets, but when they are crystallized, the release rate is slowed. The nanoparticle suspension can be applied, injected, or ingested and the pharmaceutical slowly released. Very fine solid droplets have a needle-like appearance and nuclear magnetic resonance studies have suggested the drug is adsorbed at their surface beneath the surfactant layer (Figure 7.3). Recent work on using solid lipid nanoparticles as flavor carriers in food revealed similar interactions. A series of *n*-aldehydes (hexanal, heptanal, octanal, and nonanal) were mixed with *n*-eicosane and emulsified (mean diameter 0.3 μm). Some of the emulsion was cooled directly to the measurement temperature (between the melting point and crystallization temperature), while some was cooled to induce crystallization before reheating to the measurement temperature. The headspace aldehyde concentration was measured as a function of time, and after several days the solid droplets were melted and the liquid droplets crystallized before continuing the kinetic measurements (Figure 7.7).

The concentration of hexanal was much greater above the solid than above the liquid droplets, consistent with the model of solid fat excluding volatiles. The behavior of the higher aldehydes was more complex. The headspace concentration was initially higher above the solid droplets, probably because the volatile was excluded by the solid fat (Figure 7.7a). However, the concentration progressively decreased over time, perhaps because of slow surface adsorption onto the solid droplets (Figure 7.3). When the solid droplets were melted, there was an instantaneous decrease in headspace concentration as the free and adsorbed volatile was absorbed. The headspace concentration above the liquid droplets was time independent but suddenly increased when the fat was crystallized as the absorbed volatiles were excluded (Figure 7.7b). After the initial peak, the headspace concentration of the higher aldehydes again decreased by the adsorption process. This study gives some indication of the complexity of the effects of fat crystallization on emulsion aroma. Work is underway to establish more clearly the subtleties of the interactions responsible.

These analyses depend on the volume fraction of oil present but not on the size of the droplets. Droplet size can play a role in slowing the release kinetics, but this is likely to be significant only for relatively nonpolar solutes in large emulsion droplets, and it is unclear how much of a practical effect these differences play in determining the small molecule functionality in real food emulsions. For example, Wedzicha and Couet [34] showed benzoic acid is released extremely rapidly from emulsion droplets (–ms). While the effect of pH on the absolute amount of aqueous acid (Figure 7.5) has a large effect on the availability of the acid, millisecond-scale diffusion is unlikely to be a crucial factor in most cases.

A final aspect of small molecule functionality affected by the lipids in emulsions is color. While food oils may be highly colored, emulsions are usually a uniform white. This is because the scattering of light by the droplets (as governed by relative refractive index, size, and number of particles) overcomes the effects of absorbance of light by the lipid pigments. In general, the color of the oil will affect only the color of the emulsion if the particles are relatively large (i.e., tens of micrometers, >wavelength of visible light). A full treatment of the effect of emulsion structure on color perception is given by McClements [35].

A.

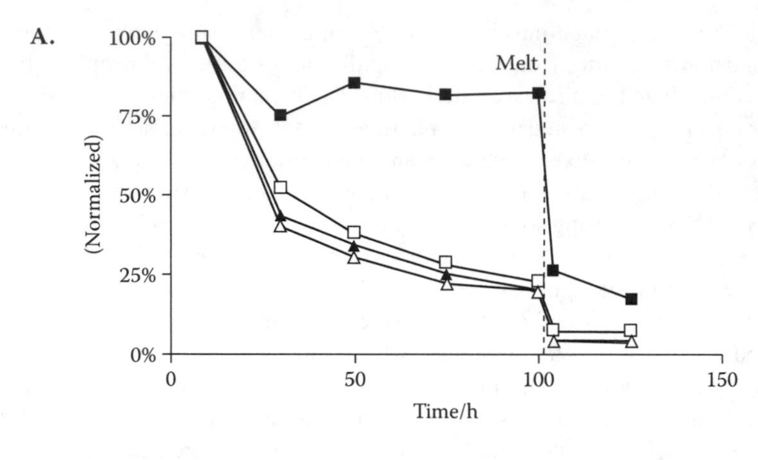

B.

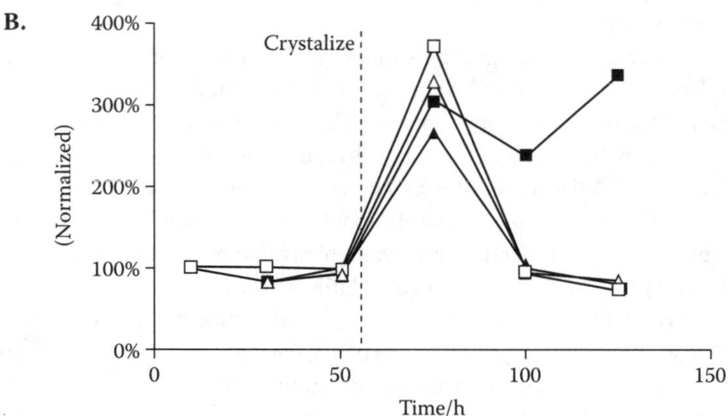

FIGURE 7.7 Headspace concentration (normalized to initial value) of hexanal (closed squares), heptanal (open aquares), octanal (closed triangles), and nonanal (open triangles) above suspensions of nanoparticles (a) initially liquid and crystallized after 50 h and (b) initially solid and melted after 100 h.

7.5 CONCLUSIONS

The physical functionality of most food emulsions can be reasonably described by considering their droplets as hard spheres. The differences in physical properties (i.e., density and interfacial tension) between most liquid triglycerides are too small to lead to differences in the functional properties of the droplets. This simplification becomes less reliable when the oil itself contains a significant amount of amphiphilic materials (e.g., free fatty acids, mono- and diglycerides, and oil-soluble surfactants), which can change the nature of the oil–water interface more significantly. Furthermore, highly functional food oils (e.g., flavor oils) may themselves have polar groups that can alter the interfacial properties of the droplets. The surface activity of the lipid can also change as a result of lipid oxidation, but the low flavor threshold of the products responsible are likely to spoil the taste of the products before visible changes in physical stability. Crystallization of the lipid phases also leads to more dramatic changes to the simple "hard sphere model." Semicrystalline droplets are vulnerable to partial coalescence. The extent of partial coalescence depends on the solid fat content but also the chemical nature of the fat and applied shear forces.

The nature of the oil present much more dramatically affects the chemical functionality of emulsions. First, the droplets are not "hard" with respect to small molecules, which can partition into the oil and change their overall functionality. Partitioning is mainly a function of the polarity (and hence the interfacial tension and permeability) of the oils, and, consequently, while most triglyceride oils are similar, more polar, functional lipids and semicrystalline droplets would be expected to have unusual properties.

In conclusion, troubleshooting physical problems with food emulsions should initially focus on the nature of the interface as affected by the properties of the aqueous phase. However, once the factors have been eliminated, or when more concerned with chemical functionality, the contents of the "hard spheres" should be considered.

ACKNOWLEDGMENTS

I am grateful to Supratim Ghosh for conducting the experiments reported in Figure 7.1 and 7.7. Julian McClements (University of Massachusetts) kindly provided me an early draft of the second edition of his book [31], and I am grateful to Jochen Weiss (University of Tennessee) for helpful discussion. The data reported in Figure 7.7 are the result of a collaboration with Devin Peterson, supported by a grant from the Center For Food Manufacturing (Penn State).

REFERENCES

1. Hunter, R.J., *Foundations of Colloid Science,* Oxford University Press, Oxford, 1986.
2. McClements, D.J., *Food Emulsions. Principles, Practice, and Techniques,* CRC Press, Boca Raton, FL, 1999.
3. Dickinson, E., *An Introduction to Food Colloids,* Oxford University Press, Oxford, 1992.
4. Walstra, P., Dispersed systems: basic considerations, in *Food Chemistry,* Fennema, O.R., Eds, Marcel Dekker, New York, 1996, pp. 95–156.
5. Chanamai, R., Horn, G., and McClements, D.J., Influence of oil polarity on droplet growth in oil-in-water emulsions stabilized by a weakly adsorbing biopolymer or a nonionic surfactant, *J. Colloid Interface Sci.,* 247, 167, 2002.
6. Chanamai, R. and McClements, D.J., Impact of weighting agents and sucrose on gravitational separation of beverage emulsions, *J. Agric. Food Chem.,* 48, 5561, 2000.
7. Coupland, J.N. and McClements, D.J., Physical properties of liquid edible oils, *J. Am. Oil Chem. Soc.,* 74, 1559, 1997.
8. Vanapalli, S.A. and Coupland, J.N., Orthokinetic stability of food emulsions, in *Food Emulsions,* Friberg, S.E., Larsson, K. and Sjoblom, J., Eds., 2004.
9. Coupland, J.N., et al., Solubilization kinetics of triacyl glycerol and hydrocarbon emulsion droplets in a micellar solution, *J. Food Sci.,* 61, 1114, 1996.
10. Weiss, J., et al., Influence of molecular structure of hydrocarbon emulsion droplets on their solubilization in nonionic surfactant micelles, *Colloids Surf. A.,* 121, 53, 1997.
11. Coupland, J.N., Crystallization in emulsions, *Curr. Opin. Colloid Interface Sci.,* 7, 445, 2002.
12. McClements, D.J., et al., Droplet size and emulsifier type affect crystallization and melting of hydrocarbon-in-water emulsions, *J. Food Sci.,* 58, 1148, 1993.

13. Walstra, P., Secondary nucleation in triglyceride crystallization, *Prog. Colloid Polym. Sci.,* 108, 1998.
14. Lopez, C., et al., Thermal and structural behavior of milk fat — 1. Unstable species of cream, *J. Colloid Interface Sci.,* 229, 62, 2000.
15. Lopez, C., et al., Thermal and structural behavior of milk fat — 2. Crystalline forms obtained by slow cooling of cream, *J. Colloid Interface Sci.,* 240, 150, 2001.
16. Keller, G., et al., DSC and x-ray diffraction coupling — specifications and applications, *J. Therm. Anal. Calor.,* 51, 783, 1998.
17. Unruh, T., et al., Investigations on the melting behavior of triglyceride nanoparticles, *Colloid Polym. Sci.,* 279, 398, 2001.
18. Bunjes, H., Koch, M.H.J., and Westesen, K., Effect of particle size on colloidal solid triglycerides, *Langmuir,* 16, 5234, 2000.
19. Djerdjev, A., Beattie, J.K., and Hunter, R.J., Electroacoustic study of the crystallization of n-eicosane oil-in-water emulsions, *Langmuir,* 19, 6605, 2003.
20. Boode, P. and Walstra, W., Partial coalescence in oil-in-water emulsions, *Colloids Surf. A.,* 81, 121, 1993.
21. Walstra, P., Emulsion stability, in *Encyclopedia of Emulsion Technology,* Becher, P. Ed., Marcel Dekker, New York, 1996, p. 1.
22. Walstra, P., *Physical Chemistry of Foods,* Marcel Dekker, New York, 2003.
23. Vanapalli, S.A., Palanuwech, J., and Coupland, J.N., Stability of emulsions to dispersed phase crystallization: effect of oil type, dispersed phase volume fraction, and cooling rate, *Colloids Surf. A.,* 204, 227, 2002.
24. Campbell, S.D., Goff, H.D., and Rousseau, D., Relating bulk-fat properties to emulsified systems: characterization of emulsion destabilization by crystallizing fats, in *Crystallization and Solidification Properties of Lipids,* Widlak, N., Hartel, R., and Narine, S. Eds., AOCS Press, Champaign, IL, 2001, pp. 176–189.
25. Palanuwech, J. and Coupland, J.N., Effect of surfactant type on the stability of oil-in-water emulsions to dispersed phase crystallization, *Colloids Surf. A.,* 223, 251, 2003.
26. Huang, S., et al., Partition of selected antioxidants in corn oil-water model systems, *J. Agric. Food Chem.,* 45, 1991, 1997.
27. McClements, D.J. and Decker, E.A., Lipid oxidation in oil-in-water emulsions: impact of molecular environment on chemical reactions in heterogeneous food systems, *J. Food Sci.,* 65, 1270, 2000.
28. Wedzicha, B.L. and Ahmed, S., Distribution of benzoic acid in an emulsion, *Food Chem.,* 50, 9, 1994.
29. Landy, P., Voilley, A., and Wedzicha, B., Interphase transport of benzoic acid in emulsions, *J. Colloid Interface Sci.,* 205, 505, 1998.
30. Bourne, M.C., *Food Texture and Viscosity. Concept and Measurement,* 2nd ed., Academic Press, London, 2002.
31. McClements, D.J., *Food Emulsions. Principles, Practice, and Techniques,* 2nd ed., CRC Press, Boca Raton, FL, 2004, pp. 632.
32. Rabe, S., et al., Lipid molarity affects liquid/liquid aroma partitioning and its dynamic release from oil/water emulsions, *Lipids,* 38, 1075, 2003.
33. Roberts, D.D., Pollien, P., and Watkze, B., Experimental and modelling studies showing the effect of lipid type on level on flavor release from milk-based liquid emulsions, *J. Agric. Food Chem.,* 51, 189, 2003.
34. Wedzicha, B. and Couet, C., Kinetics of transport of benzoic acid in emulsions, *Food Chem.,* 55, 1, 1996.
35. McClements, D.J., Theoretical prediction of emulsion color, *Adv. Colloid Int. Sci.,* 97, 63, 2002.

8 Medium-Chain Triglycerides

Jenifer Heydinger Galante and Richard R. Tenore

Medium-chain triglycerides, (MCTs), have a long history of use as a fat source in medical nutrition products. Since they first became commercially available in 1955, patients with fat malabsorption syndromes, premature infants, and the critically ill have benefited from this rapidly absorbed, concentrated source of energy. Their nutritional value is attributed to the unique metabolic pathway of MCTs compared with conventional fats and oils. MCTs are also valued for their unique physical attributes by food technologists and formulators.

The two most common sources for medium-chain fatty acids (MCFAs) are coconut oil and palm-kernel oil (Table 8.1). Both oils are highly prized for industrial uses, with coconut oil being the more important of the two. Their fatty acids are key raw materials across a broad range of chemical intermediates. Coconut or palm-kernel oils are hydrolyzed to liberate their fatty acids from glycerol, and then the fatty acids are separated by fractional distillation. The lower boiling or top fraction of the fatty acids contains primarily C8:0 (caprylic acid) and C10:0 (capric acid). Because these MCFAs are key contributors to irritation in detergent applications, they have traditionally been removed and were at one time considered valueless. As a consequence, the caprylic and capric acids were available for other uses.

MCTs were first synthesized by direct esterification of glycerol with MCFA in the late 1940s by Dr. Vigen Babayan of the Drew Chemical Company in an effort to find uses for caprylic and capric acids [1]. Today, MCTs continue to be produced by the esterification of glycerol with MCFAs from either coconut or palm-kernel oil, the only commercially important sources of MCFAs.

Prior to esterification, C8 and C10 acids are combined in various ratios to give different MCT products ranging from solid to liquid. The typical MCT will vary in C8:C10 ratio across the range from 95:5 to 5:95, most commonly 70:30, and contain less than 6% total of other fatty acids (C6 and C12). The esterification reaction is carried out at high temperatures (200°C and higher), usually without the use of a catalyst. Water in the reaction is removed continuously to drive the reaction to completion. When esterification is complete, excess fatty acids are removed from the reaction mixture by vacuum distillation and the crude MCTs are deodorized to remove volatile odor and flavor components as well as any residual fatty acids.

Initially, MCTs were mistakenly classified as "fractionated coconut oil" by the U.S Food and Drug Administration (FDA). In 1991, the FDA reversed this ruling. The FDA accepted for filing a generally recognized as safe (GRAS) affirmation

TABLE 8.1

Typical Weight % Fatty Distribution in Coconut and Palm-Kernel Oils

	6:0	8:0	10:0	12:0	14:0	16:0	18:0	18:1	18:2	18:3	20:0
Coconut oil	0.5	7.1	6.0	47.1	18.5	9.1	2.8	6.8	1.9	0.1	0.1
Palm-kernel oil	0.2	3.3	3.4	48.2	16.2	8.4	2.5	15.3	2.3		0.1

Weight % Fatty Acid Distribution in Typical 810 Cuts

	6:0	8:0	10:0	12:0
C8-10 acid	<7.0	55.0–60.0	35.0–40.0	<3.0
C8-10 acid	<1.0	55.0–62.0	35.0–45.0	<2.0
90% C8	<2.0	95.0–99.0	<1.0	
90% C10		<2.0	95.0–98.0	<1.0

petition for MCTs prepared by the Stepan Company, on June 17, 1994. Captrin was proposed as the common name for the randomized triglycerides of primarily C8 and/or C10 fatty acids. MCTs are now properly labeled as captrin, medium-chain triglycerides, glyceryl tri(caprylate/caprate), or capric/caprylic triglycerides. Use of the terminology "highly fractionated palm-kernel or coconut oil" is still common today, but this nomenclature is obsolete and does not reflect the nature of MCTs as they are actually manufactured.

MCT unique properties are a direct result of their shorter fatty acid chain length compared with those of other edible fats and oils, which typically contain C16 and C18 fatty acids. For example, the C8:0 fatty acid in MCT is almost 100 times more water soluble than C16:0 (68 mg/100 ml vs. 0.72 mg/100 ml at 20°C) [2]. As for the triglycerides themselves, MCTs are soluble in water to the extent that they form a stable emulsion at 0.01% weight, while long-chain triglycerides (LCTs) are insoluble in water. As a result of this difference in water solubility, MCTs are metabolized differently than LCTs [2–6].

Upon ingestion, LCTs are hydrolyzed to the *sn*-2 monoglyceride and long-chain fatty acids (LCFAs) by pancreatic lipases. After emulsification with bile salts, the LCFAs are absorbed into intestinal cells and reattached to the *sn*-2 monoglyceride template to reform LCT. LCT and cholesterol are then bound with phospholipids and proteins into units called chylomicrons. Chylomicrons enter the lymphatic system and later the circulatory system for distribution throughout the body. If not required for energy, the LCTs are stored in adipose tissues [7].

In contrast, MCTs are metabolized more like carbohydrates and can be absorbed and utilized as rapidly as glucose [5]. In the small intestine, pancreatic lipases hydrolyze MCTs five times faster than LCTs [3], rapidly releasing the MCFAs. MCTs are also hydrolyzed more completely than LCTs. The MCFAs are able to enter the circulatory system without emulsification by bile salts because they are sufficiently water soluble to do so. MCFAs bypass the lymph system and enter the portal vein directly. They travel rapidly to the liver where they are quickly oxidized for energy. They are cleared from the circulation twice as rapidly as LCTs [8]. Unlike LCTs, MCTs are not stored in body tissues and when consumed at normal levels, MCTs have little impact on serum cholesterol levels [9].

Studies have shown that MCTs provide a caloric reduction compared with conventional fats and oils due in part to higher heat energy losses upon metabolism and utilization (increased thermogenesis) [3,10,11]. Animal and human studies show MCTs have a 16% greater heat loss upon metabolism than LCTs; this results in MCTs contributing 6.8 kcal/g while LCTs contribute 9.0 kcal/g [12]. MCTs also have a greater satiating effect than LCTs. Recent clinical trials have suggested that MCTs may play a role in the prevention of obesity or as a stimulant for weight loss [13,14]. An MCT diet may decrease body weight and fat in overweight persons compared with an LCT diet because of their greater energy expenditure and oxidation [15]. These findings indicate the potential for use of MCTs in functional foods for weight management.

MCTs can enhance the absorption of certain minerals and fat-soluble vitamins. For example, it has been found that absorption of calcium [16,17] and vitamin E [18] can benefit by the presence of MCTs rather than LCTs.

Because of their rapid oxidation, it has been suggested that MCTs may play a role in the treatment or prevention of the dementia of Alzheimer's disease. MCTs rapidly produce ketone bodies upon oxidation; this may improve cognitive ability by increasing neuronal metabolism [19].

It has been shown that MCTs do not impair the function of the reticuloendothelial system, which controls intravenous clearance of bacteria from the body and plays a role in lipid clearance [20]. MCTs do not suppress the immune system, as a high intake of linoleic acid has been shown to do [21,22]. In fact, preliminary research suggests that MCTs may exert a positive effect on the immune system. Compared with LCTs, MCTs suppressed certain types of infection [23] and exhibited antitumor activity [24].

In addition to their nutritional benefits, MCTs have found utility in a number of applications because of their unique physical attributes. With their shorter fatty acid chain length and corresponding smaller molecular size, MCTs are more polar than LCTs, making them excellent solvents compared with other fats. MCTs are miscible with the same types of compounds as LCTs, such as hydrocarbons, esters, and natural oils. MCTs are also miscible with more polar compounds such as alcohols, acids, and ketones, while LCTs are not.

MCTs are highly regarded for their clean organoleptic quality. They are odorless and tasteless so they do not contribute any off-notes to products. MCTs are widely used in the flavor industry because of their superior organoleptic quality and solvent capabilities. Because they are virtually colorless, they are also used as a carrier for colors and essential oils and as a solvent for extracting flavorings. The pharmaceutical industry has also taken advantage of the solvency powers of MCTs in vitamin and drug delivery.

When MCTs are used to carry flavor to a beverage, they can also double as the beverage clouding agent. Often no emulsifiers are required because of MCT slight water solubility and specific gravity very close to unity. (For example, the specific gravity of Neobee® M-5 at 25°C is 0.95.)

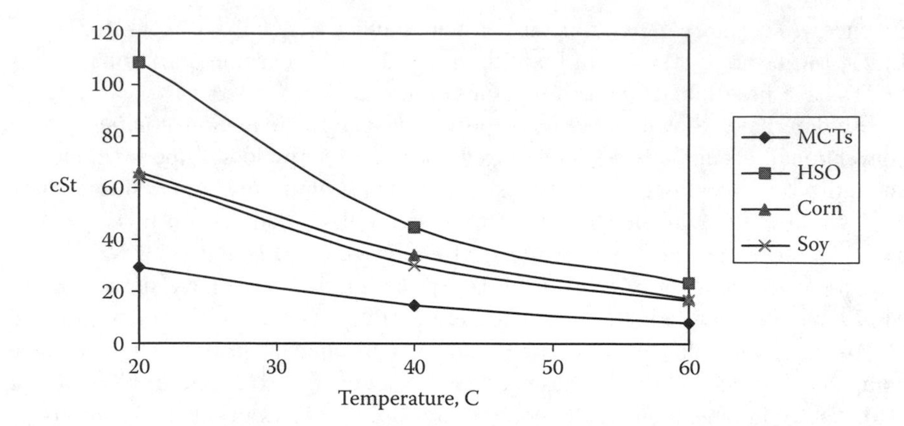

FIGURE 8.1 Viscosity of various oils.

MCTs typically have one half to one third the viscosity of conventional fats. For example, Neobee M-5 has a viscosity of about 15 cSt at 40°C, less than half the value of soybean oil at the same temperature (Figure 8.1). This characteristic makes them ideally suited for spray oil applications where use at low temperatures takes advantage of the lower viscosity and fluid nature of MCTs. They spread easily and adhere well to surfaces, providing uniform coverage. This makes them ideally suited for use in a number of applications. Examples include but are not limited to the following:

- As a moisture barrier, MCTs are applied at 0.25% by weight to coat dried raisins. In contrast, 0.5% by weight of a typical high-stability spray oil must be used.
- MCTs function as an antistick agent for high-sugar, high-protein systems such as confections or dried fruits.
- They may be used as a confectionery glaze for gummies and jellies to impart gloss.
- MCTs may be used as spray oil to coat cereal and crackers, reduce dusting, impart gloss, and retain freshness.
- They can be used for topical seasoning delivery to chips and as a carrier and stick agent.

MCTs adhere well to metals and can therefore be used as lubricants or release agents. As more and more regulatory agencies ban the use of mineral oil for food contact machinery, MCTs are the ideal replacement. MCTs show improved heat stability compared with conventional oils, as MCTs do not polymerize or blacken upon extended heating. For these reasons, MCTs are becoming ubiquitous among manufacturers of molded confections and are finding ever-increasing use in continuous baking processes. For prefried frozen and packaged foods, MCTs have recently been identified as the ideal spray oil for lubrication of the conveyer belts. Functioning as a lubricant and an antistick agent, MCTs are compatible with the food and the frying oil and do not readily degrade under the conditions of use.

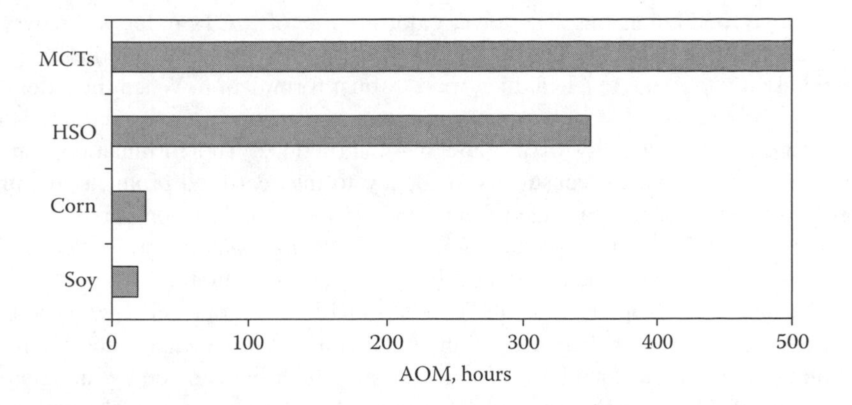

FIGURE 8.2 Oxidative stability of various oils.

The most outstanding attribute of MCTs is their exceptional stability to air oxidation, which can result in extended shelf life of finished products. MCTs have an active oxygen method (AOM) value of more than 500 h. In comparison, soybean oil has an AOM value of only 19 h. The high-stability oils also have a lower AOM value, about 300 to 350 h (Figure 8.2). While the oxidative stability of MCTs is important in any application, it is especially significant in applications involving high surface areas. For this reason, MCTs are ideal for use as antidust agents for powdered mixes and seasoning dry blends, and as spray oils to coat cereals, crackers, and chips.

The low viscosity of MCTs can also be exploited by blending them with conventional oils. Viscosity of the blend decreases as the amount of MCTs is increased (Figure 8.3). A formulator can gain several advantages by using MCTs as a fat extender. In addition to viscosity reduction, blending MCTs with any oil will usually result in a higher density, allowing for more easily formulated and

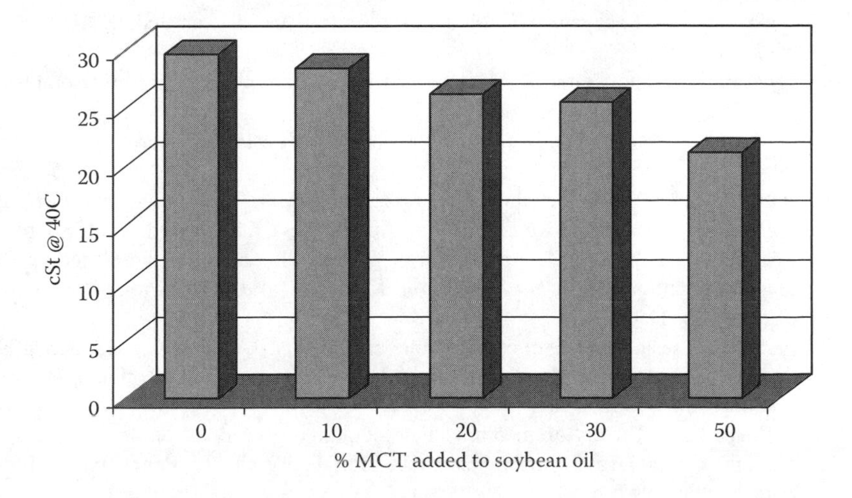

FIGURE 8.3 Viscosity of MCT/soybean oil blends.

more stable food emulsions. The lower caloric value of MCTs reduces the overall caloric intake per serving, while the improved stability and heat tolerance of the extended fat contribute to a healthier, more robust formulation. When the extended fat is incorporated into a consumer-prepared formulation, these characteristics contribute to the reliability of the cooking instructions. The formulation can be designed to overcome the consumers' tendency to mistreat food products by either improper storage of the packaged food or simply overcooking or otherwise mistreating the product during preparation. Thus, the organoleptic properties in the final food can be maintained as intended by the food formulator.

The functional properties of MCTs, both nutritional and physical, will increasingly provide a basis for their use in functional foods, nutraceuticals, and the manufacturing arena. It has long been known that structured lipids made by the interesterification of MCTs with conventional fats and oils improve the absorption of unsaturated fatty acids. To take advantage of this, "healthy oils" made from interesterified MCTs and canola, olive, soy, or fish oils are now entering the world's marketplace. One of the most promising areas for growth, however, has been brought about by growing consumer awareness of the detrimental effects of *trans* fatty acids. A recent introduction to the market is a *trans*-free product made by interesterifying MCTs with fully hardened soy: medium- and long-chain triglycerides (MLTs). MLTs have been shown to meet all of the required characteristics of commercially available shortening. As consumers become increasingly aware of the differences in the makeup of fat, this type of MCT-derived fat becomes a viable alternative to partially hydrogenated fats. The search for a healthier, more easily digested fat initiated by Dr. Babayan's decision to synthesize MCTs in the 1940s continues today.

REFERENCES

1. Babayan, V.K., MCTs – their composition, preparation, and application, *J. Am. Oil Chem. Soc.*, 45, 23, 1967.
2. Bach, A.C. and Babayan, V.K., Medium-chain triglycerides: an update, *Am. J. Clin. Nutr.,* 36, 950, 1982.
3. Johnson, R.C. and Cotter, R., Metabolism of medium chain triglyceride emulsion, *Nutr. Int.,* 2, 50, 1986.
4. Senior, J.R., *Medium Chain Triglycerides,* University of Pennsylvania Press, Philadelphia, 1968, p. 3.
5. Babayan, V.K., Medium chain triglycerides, in *Dietary Fat Requirements in Health and Development,* Beare-Rogers, J., Ed., AOCS Press, Champaign, IL, 1988, p. 73.
6. Bezard, J. and Bugaut, M., Absorption of glycerides containing short, medium and long chain fatty acids, in *Fat Absorption,* Kuksis, A., Ed., CRC Press, Boca Raton, FL, 1986, p. 119.
7. Flatt, J.P., Use and storage of carbohydrate and fat, *Am. J. Clin. Nutr.,* 61, 952S, 1995.
8. Sailer, D. and Muller, M., Medium chain triglycerides in parenteral nutrition, *J. Parenter. Enteral Nutr.,* 5, 115, 1981.
9. Furman, R.H., Effects of medium chain length triglycerides on serum lipids, in *Medium Chain Triglycerides,* Senior, J., Ed., University of Pennsylvania Press, Philadelphia, 1967, p. 51.

10. Baba, N., Brace, E.F., and Hashim, S.A., Enhanced thermogenesis and diminished deposition of fat in response to overfeeding with diet containing medium-chain triglyceride, *Am. J. Clin. Nutr.,* 35, 678, 1982.
11. Hill, J.O. et al., Thermogenesis in humans during overfeeding with medium-chain triglycerides, *Metabolism,* 38, 641, 1989.
12. Ingle, D.L. et al., Dietary energy value of medium-chain triglycerides, *J. Food Sci.,* 64, 960, 1999.
13. St.-Onge, M.P. and Jones, P.J.H., Greater rise in fat oxidation with medium chain triglyceride consumption is associated with lower initial body weight and greater loss of body weight and subcutaneous adipose tissue, *FASEB J.,* 17, Abstract No. 772.82003, 2003.
14. St.-Onge, M.P. and Jones, P.J.H., Physiological effects of medium-chain triglycerides: potential agents in the prevention of obesity, *J. Nutr.,* 132, 329, 2002.
15. Tsuji, H. et al., Dietary medium-chain triacylglycerols suppress accumulation of body fat in a double-blind, controlled trial in healthy men and women, *J. Nutr.,* 131, 2853, 2001.
16. Garcia-Lopez, S. and Miller, G.D., Bioavailability of calcium from four different sources, *Nutr. Res.,* 11, 1187, 1991.
17. Sulkers, E.J. et al., Comparison of two preterm formulas with or without addition of medium chain triglycerides, (MCTs): effects on mineral balance, *J. Pediatr. Gastroenterol. Nutr.,* 15, 42, 1992.
18. Gallo-Torres, H.E., Ludorf, J., and Brin, M., The effect of medium chain triglycerides on the bioavailability of vitamin E, *Int. J. Vitam. Nutr. Res.,* 48, 240, 1978.
19. Henderson, S.T., Use of medium chain triglycerides for the treatment and prevention of Alzheimer's disease and other diseases resulting from reduced neuronal metabolism, US Patent Application US 2002/0006959 A1, 2002.
20. Jensen, G.L. et al., Parenteral infusion of long- and medium-chain triglycerides and reticuloendothelial system function in man, *J. Parenter. Enteral Nutr.,* 14, 467, 1990.
21. Tappia, P.S. and Grimble, R.F., Complex modulation of cytokine induction by endotoxin and tumor necrosis factor from peritoneal macrophages of rats by diets containing fats of different saturated, monounsaturated, and polyunsaturated fatty acid composition, *Clin. Sci.,* 87, 173, 1994.
22. Gogos, C.A. et al., Medium- and long-chain triglycerides have different effects on the synthesis of tumor necrosis factor by human mononuclear cells in patients under total parenteral nutrition, *J. Am. Coll. Nutr.,* 13, 40, 1994.
23. Papavassilis, C., Use of medium-chain triacylglycerols in parenteral nutrition of children, *Nutrition,* 16, 460, 2000.
24. Kimoto, Y. et al., Antitumor effect of medium chain triglyceride and its influence on the self-defense system of the body, *Cancer Detect. Prev.,* 22, 219, 1998.

9 Frying Lipids

Sérgio D. Segall and William E. Artz

CONTENTS

9.1 INTRODUCTION

Deep-fat frying is a popular method of food preparation in many parts of the world. Despite the problems associated with the high caloric content of fat and the health concerns regarding the ingestion of *trans* fatty acids (TFAs), the flavor and textural attributes of fried food continue to be highly desired. The sound application of the appropriate technology for oil/fat extraction, purification, and process is important for the preparation of high-quality frying fats and oils [1]. Among the various sources of fats and oils, three that are used most extensively for deep-fat frying include tallow, hydrogenated vegetable oil, and palm oil. Triacylglycerols (TAGs) containing primarily saturated fatty acids would be well suited for deep-fat frying and more resistant to oxidation and thermal degradation than TAGs containing primarily unsaturated fatty acids, but solid fats can be rather difficult to handle. Melting large blocks of fat must be done relatively slowly and carefully to prevent problems from overheating. Starting in the mid-1900s, consumption of vegetable oils (particularly soybean, sunflower, and canola) has increased, due to the lower price and the perceived health benefits regarding unsaturated fats and oils [2]. However, the relatively high degree of unsaturation allows these oils to oxidize easily when used for deep-fat frying. To minimize this problem, hydrogenation was used to reduce the concentration of polyunsaturated fatty acids (linoleic and particularly linolenic acid) that oxidize rapidly. For example, linoleic acid oxidizes ~50 times faster than oleic acid [3]. The hydrogenation process allowed the development of a variety of shortenings, as well as margarine. Although hydrogenation provides numerous advantages, it induces the formation of significant amounts of TFAs, geometric isomers of their corresponding native *cis* fatty acids. TFAs occur naturally in certain animal fats, particularly butter, but they are also a by-product of some food oil processes, such as deodorization and, in particular, hydrogenation.

However, there are negative health impacts associated with an increase in TFA intake, such as an adverse effect on the ratio of low- and high-density lipoproteins in the blood. Indeed, some researchers assert that there is strong evidence that TFAs are more harmful than saturated fats [4]. This is reflected in the new labeling requirements regarding TFA content statements on labels in the United States and the new regulations implemented in some countries in Europe limiting the TFA content in food products. With the increased consumption of fried foods and the associated increase in the amount of fat in the diet, as well as the increased consumption of saturated and *trans* fatty acids, TFAs are considered undesirable due to the increased risks of coronary heart disease and myocardial infarction [5,6]. Due to the absorption of up to 40% fat by some products during frying, these fats/oils can have a substantial impact on the type of fat in our diet. As a result of the public's interest in reducing TFAs in the diet, the consumption of naturally high-oleic acid oils, such as olive oil, and some newly developed hybrid high-oleic vegetable oils, such as NuSun oil (high-oleic acid sunflower oil), has increased [7]. Although oils high in polyunsaturated fatty acids (PUFAs) are preferable in the diet, they degrade rapidly during frying, producing cyclic fatty acids, polymers, and a variety of polar oxidation products much more rapidly than oils high in saturated and monounsaturated fatty acids; hence, the preference for high-oleic acid oils. Furthermore, without special packaging, the shelf life of products fried with oils high in PUFAs may be substantially reduced as compared to products fried in hydrogenated vegetable oils or more saturated fats and oils, such as tallow or palm oil. No frying oil satisfies all of the objectives associated with fats and oils (low cost, good functionality, which would include good frying stability and long-term storage stability, and optimal nutritional characteristics), so compromises are necessary to find the best frying oil/product match.

The nutritional factors involved include the polyunsaturated fatty acid content and the TFA content, while the process characteristics include the type of product to be fried, frying temperature, oil storage, and cost. In the past, the cost had been the primary factor, although the widespread concern of the public regarding TFAs may change that emphasis.

The objective of this chapter is to present some of the latest information on frying fat and oil research, including techniques used to avoid the formation of undesirable compounds that affect oil quality.

9.2 THE FRYING PROCESS

Deep-fat frying is used in the home and in restaurants and by both small and large food processing companies. It is an ancient method of food processing, associated with the addition of substantial textural changes and very characteristic fried food flavors to the food product. The flavor changes are even more notable if flavored varietal oils are used.

Heat transfer, mass transfer, various organic reactions, competing reaction kinetics, inorganic catalysis, thermodegradation reactions, and many other physical–chemical events are involved in deep-fat frying. The high temperatures involved, the substantial amount of water released from the food, as well as the various polar materials that migrate from the food to the oil during deep-fat frying, necessitate the use of oil with relatively good stability. All of these factors warrant consideration when trying to comprehend how frying oils influence the quality of fried foods [8]. Even though a large body of research exists on this topic, the chemistry that occurs during frying is still not fully understood due to the complexity of the process.

During frying, the food is added to the oil at temperatures that can induce the rapid release of considerable amounts of water. Therefore, frying results in product dehydration (from slight to substantial) and rather intense heating and cooking at the product surface. The mass and heat transfer phenomenon (water release from the food and oil incursion into the food product) can be understood better by comparing fried chicken to potato chips. After frying, potato chips are essentially a dry product. Substantially less desiccation occurs with thicker food products, although extensive dehydration can occur if the food material remains in the frying oil for an excessive amount of time. Chicken is rather complex, since the product is typically breaded and the breading will absorb substantial amounts of oil during frying, yet chicken fat is simultaneously extracted and released into the frying oil. After frying substantial amounts of chicken, the frying oil composition is usually closer to that of chicken fat, rather than the original partially hydrogenated vegetable oil.

One can consider frying as a partial drying process, and when compared with other drying fluids (air or unsaturated steam), oil has a relatively large heat capacity (approximately 2 kJ kg^{-1} K^{-1}), particularly upon comparison to air. This enables the oil to store a relatively large amount of energy in a relatively small volume. For example, a 10°C decrease in the temperature of 1 L of oil will vaporize 7 L of water [9].

Vitrac et al. [9] described three stages for frying. The first stage is characterized by substantial water release from the food product. Immediately upon product immersion in the oil, steam bubbles exit the surface of the food product. As the product temperature increases, heat is conducted toward the center of the food product, inducing further moisture loss.

The next stage is characterized by mass transfer (oil absorption) onto and into the food. The internal temperature of the product approaches the boiling point of water, while the surface is at a temperature much closer to that of the hot oil. When the product (e.g., french fries) crust thickness reaches approximately one third of the thickness of the product, almost 50% of water has been removed from the product.

The third stage is characteristic of products that eventually reach very low water content due to extensive dehydration, such as the chip-type products. After frying, these products are relatively stable at room temperature, assuming the appropriate packaging is provided to prevent moisture absorption and oxidation. Careful control of the process is important since excessive residence time in the frying oil can result in very undesirable product characteristics. Likewise, if the frying time is insufficient, the moisture reduction may be incomplete and quality problems during storage can occur.

Frying generally produces substantial changes in the structure of the food product, particularly at the surface. While the textural changes that occur in fried food products are very important, they are beyond the scope of this chapter.

9.2.1 Degradation during Frying

The refined oils used for frying are designed to be relatively stable under frying conditions. Crude vegetable oil is refined to remove polar substances (phospholipids, free fatty acids (FFA), monoglycerides, trace metals, oxidation products, etc.) to produce a high-quality oil. In spite of careful refining, traces of undesirable components (e.g., ppb traces of transition metal ions) do remain in the oil. Another problem associated with oil stability is the migration of food components into the oil. These compounds leach from the food and can significantly alter the composition and quality of the oil. Dobarganes et al. [10] classified six main groups responsible for the most important aspects of frying oil quality. These include (1) phospholipids and oxidation products with emulsion functionality that can contribute to foaming; (2) Maillard browning products and other pigments (particularly from meat) that can cause darkening of the oil; (3) phenolic compounds from added spices or present naturally in the food that can alter, perhaps even improve, the oil stability; (4) off-flavor compounds derived from the food products (e.g., onions and fish) that can affect other food products; (5) cholesterol and cholesterol oxidation products (e.g., that migrate from meat into the oil) that can be absorbed by foods with low cholesterol contents during subsequent frying operations; and finally, (6) lipid-soluble vitamins and trace metals (e.g., iron) that can accelerate the oxidative process.

9.2.2 Used Oil Adsorbents

There is substantial published evidence that a variety of adsorbents are very effective in removing the majority of the polar oxidation products from used frying, thereby extending the life of the frying oil. However, the range in polar compound absorption efficacy among these various adsorbents can vary substantially. There is also evidence (unpublished data) that magnesium silicate-based adsorbents can reduce the transition metal content of used frying oils, as well.

During the high temperatures (170 to 190°C) used for frying and the hydrolysis reactions that occur due to the presence of moisture, there is a rapid increase in the FFA content of the oil, from <0.1 to 0.5% through 0.8 [11]. Near the end of the frying life of the oil, the FFA content often exceeds 2%. The FFAs, hydroperoxide decomposition products, and various volatile oxidation products formed during frying can induce off-flavor formation and other undesirable changes in food and oil.

Initially, the fresh oil is relatively bland in flavor. As frying is continued, the desirable fried food flavor intensity increases, primarily due to the accumulation of volatile compounds, such as decadienal isomers from the oxidation of linoleic acid. As the thermodegradation continues, the extent of oxidation can become excessive and undesirable flavor components can accumulate, adding a sharp and bitter note to the oil. To reduce these effects, Lin et al. [12] tested four different adsorbents and two adsorbent combinations using seasoned chicken as the test product. They found that the FFA content was reduced by 78.3% with a combination of 3% Hubersorb 600, 3% Magnesol, and 2% Frypowder, while a 68.1% reduction was achieved with a combination of 2% Hubersorb 600, 3% Magnesol, and 2% Britesorb. They also found that the total polar compound content was reduced up to 38% when the first blend (3% Hubersorb 600, 3% Magnesol, and 2% Frypowder) was used. Hubersorb 600 is a calcium silicate powder (6 μm particles). Frypowder is composed of porous rhyolite and citric acid. Britsorb is composed of silicon dioxide and aluminum hydroxide; and Magnesol is a form of magnesium silicate powder. It is important to note that atypically large amounts of adsorbent were used in this study and that the used frying oil was vacuum filtered and stored after the addition of 50 ppm of BHT and 50 ppm of PG.

Kalapathy and Proctor [13] investigated the efficacy of a silicate film to reduce the FFA content of used frying oils. Silicates have been used in powder form to reduce the FFA formed in frying oils. The addition of silicate powder, or powder of any type for that matter to the oil, adds a new step in the process since it must be removed subsequently by filtration. One may be able to avoid that processing step, by using a silica-based film produced from silica extracted from rice hull ash. Kalapathy and Proctor [13] placed the silicate film in a basket and immersed it in the oil. A 25% reduction in the FFA content of the frying oil was observed after a single treatment.

9.3 REDUCED PUFA WITH HYDROGENATION

Selective hydrogenation will reduce the content of linoleic and linolenic acids, while increasing the oleic acid content. Unfortunately, it also increases the TFA content. A limited number of hybrids (e.g., NuSun™, a sunflower hybrid) have been developed that are high in oleic acid and low in PuFAs that do not need hydrogenation. Another promising technique to reduce the TFA content is the production of genetically engineered oils, rich in oleic acid.

The pathway for triacylglycerol biosynthesis is complex. To find the specific enzyme that must be manipulated to produce a specific fatty acid composition is not an easy task. Most of the 16-acyl carbon chains and some 18-carbon chains are the result of a 16:0-ACP thioesterase (Fat B) activity. When the expression of the Fat B gene was repressed, it was possible to produce an oil with an increased oleate content and a saturated fatty acid content of less than 4% [14]. Polyunsaturated fatty acid biosynthesis generally occurs as a result of a seed-specific, omega-6 phospholipid desaturase (Fad 2-1). If expression of the Fad 2-1 gene was repressed, it is possible to produce a soybean oil with a total PUFA content of less than 5% and oleic acid content of 85%. The oil has a tenfold greater oxidative stability than refined, bleached and deodorized soybean oil. The oil contained no TFAs and its performance was similar to fully hydrogenated, heavy-duty frying shortening with respect to both its storage and cooking stability [15]. Warner and Mounts [16] compared the frying stability of genetically modified, low-linoleic, nonhydrogenated and hydrogenated soybean and canola oils to nonhydrogenated, traditional soybean and canola oils. The genetically modified oil had better room odor characteristics, less polar compound production, and less FFA formation, than traditional oils.

Tompkins and Perkins [17] examined the frying performance of low-linolenic acid soybean oil from genetically modified soybean using two types of food (shoestring potatoes and fish nuggets). Partially hydrogenated and unhydrogenated low-linolenic soybean oils were compared to two partially hydrogenated soybean foils. The hydrogenated low-linolenic soybean oil had greater amounts of FFA and lower *p*-anisidine values and polymer contents, than the other oils tested, but hydrogenated low-linolenic acid oil did not differ in total polar content, Lovibond red color, and maximum foam height compared to the others oils tested.

9.4 REACTIONS INDUCED DURING FRYING

The golden color of fried products is produced mainly by Maillard browning reaction products. Pokorny [18] found that the intensity of the brown color was correlated with the loss of lysine, histidine, and methionine. Nonenzymatic browning reactions will occur with both carbohydrates and carbonyl compounds derived from PuFAs, and free amino groups to produce imino Schiff bases that will polymerize as a result of aldol condensation reactions to produce polymeric compounds that are brown in color. Lipid oxidation products will react with proteins, amino acids, and amines during frying.

9.4.1 ACRYLAMIDE

Although there are many desirable flavor attributes, as well as an attractive golden-brown color formed during frying, one negative component formed during frying is acrylamide. The early research involving acrylamide was done partially due to the suspicion that this substance was involved with the development of peripheral neuropathy. Acrylamide can also be metabolized to epoxide glycidamide, which also has neurotoxic effects. In spite of the fact that acrylamide formation has not been fully elucidated, it is generally agreed that the Maillard reaction and temperatures greater than 120°C in starchy foods are correlated with acrylamide formation. Amino acids can react with acrylic acid, which originates from acrolein, to produce acrylamide. The precursor (acrolein) is formed when water is eliminated from glycerol as a result of the high temperatures associated with frying. Many foods have a relatively high water content that facilitates TAG hydrolysis, producing FFAs, mono- and diglycerides and glycerol [19]. Gertz and Klostermann [19] found that an increase in the acrylamide formation occurred with palm olein and oils containing silicone. Silicone alters the surface tension of the oil and functions as an antifoaming agent. It forms a monomolecular layer on the oil surface. They suggested that the monolayer formed by the silicone can hinder the evaporation of acrolein or acrylamide. The concentration of these compounds will therefore be increased in fried foods.

The mechanism [19] postulated for the elevated concentration of acrylamide in products fried with palm olein is related to relatively high concentrations of diacylglycerol in the oil (as much as 8%). This leads to an increase in the monoacylglycerol content (0.3% to 0.5%). Decomposition of monoacylglycerols to fatty acids, followed by dehydration of glycerol to acrolein as a result of pyrolysis can occur.

9.4.2 Thermal Acceleration of Lipid Oxidation

Heating fats and oils substantially increases the accumulation of lipid oxidation products in the oil. Consumption of thermally oxidized oils increased the lipid peroxidation product content in the tissue [20]. The presence of these products in the diet can enhance the development of atherosclerosis [21]. Since the Low-density lipoprotein (LDL) oxidization appears to be involved in the atherogenic process and the initiation of atherosclerotic lesion formation [22], Eder et al. [23] decided to investigate the effect of dietary oxidized fats on the atherogenicity of oxidized LDL. Fat heated at low temperatures for longer periods of time were less likely to increase oxidized LDL accumulation, while fats heated at higher temperatures, which produced more secondary lipid oxidation products, increased the accumulation of oxidized LDL. More of the secondary oxidation products were formed at the higher temperatures, and more of the primary oxidation products were formed at the lower temperatures. The physiological effect will differ, since primary products are very toxic when administered parenterally, but less toxic upon oral administered.

An *in vivo* study was conducted by Saka and co-workers [24] to evaluate the activity of the glutathione enzyme system in the rat liver. The evaluation included diets containing three different oil systems. The relationship with the lipid peroxide levels in the oils and the subsequent detoxifying reactions in the liver was interesting. It is well known that the glutathione system has significant detoxification activity. Saka et al. [24] showed that the enzyme system was altered when heated sunflower oil (heated continuously for 10 h at 220°C) was fed to the rats. Since there was an increased concentration of toxic lipid oxidation products in these oils, the glutathione system provides important contributions to the detoxifying process, transforming some of these compounds into less toxic primary alcohols. The activity of glutathione reductase is a defensive mechanism that is relatively active in certain organs (liver, kidneys, and intestine).

FAs and magnesium can react to form magnesium soaps in the lumen. The absorption of magnesium was studied to evaluate the influence of heated oil. Perez-Granados et al. [25] studied the effect of olive, sunflower, and palm olein frying oils (maximum 25% of TPC) on magnesium absorption in the rat. They found that the three oils did not show differences in absorption and retention efficiency of magnesium.

During the process of frying, oils undergo a series of reactions that produces many new compounds (polymers and polar compounds) that can have important physiological effects. Incorporation of oxidized lipids into membranes can increase the oxidation of adjacent membrane macromolecules. Some of these macromolecules have functions affecting the extent of mutagenesis [26,27]. Marquez-Ruiz et al. [28] studied the effect of thermo-oxidized oil on animals, although definitive results were not obtained. Soriguer et al. [29] evaluated the influence of the degradation of cooking oils and the relationship to hypertension. They found a polar compound (PC) concentration 20% in 10% of the cooking oil samples, which indicated excessive degradation of the oil. High PC concentrations (>20%) were found in 6.2% of the olive oil samples as compared to 11.9% of the sunflower oil samples.

An analysis of the plasma phospholipid fatty acid profiles indicated that the concentration of saturated fatty acids was greatest in those who only use sunflower oil [29]. They found that the consumption of sunflower based cooking oils was a risk factor for hypertension, and that monounsaturated fatty acid concentration in the serum was inversely related to hypertension.

9.5 ANTIOXIDANTS

Research involving lipid oxidation and its effect in human health has increased substantially in the past decade. It has been observed that the central nervous system, due to its high metabolic rate and minimal antioxidant protection, is vulnerable to oxidative stress [30]. Consumption of oxidized lipids can contribute to this oxidative stress.

One of the most important characteristics of frying oils is their high temperature oxidative stability. Other important characteristics that should be considered include a high smoke point, a bland flavor, the low foaming, and the nutritive value of the oil. Vegetable oils with a high percentage of polyunsaturated fatty acids oxidize very rapidly at elevated temperatures, with a concurrent rapid reduction in essential fatty acids, rapid rate of polymerization, and rapid darkening of the oil.

Oxidation is primarily a free-radical reaction. Although most free radicals are electrically neutral, they are still very reactive because of the unpaired electron. Free radicals can abstract a hydrogen atom from another unsaturated lipid, producing a new radical and continuing the reaction. Combination with oxygen, followed by hydrogen abstraction, produces a hydroperoxide. Homolytic decomposition of the hydroperoxide will accelerate the rate of the reaction, due to the formation of two free radicals, a hydroxy and an alkoxy radical.

If a new radical with an increased stability is formed (e.g., from an antioxidant), the new free radical will be less effective in abstracting a hydrogen from an unsaturated lipid, thereby reducing the rate of oxidation. The increased stability is provided, for example, by the resonance stabilization that occurs upon formation of the free radical in the phenolic-based antioxidants.

Lipid oxidation is recognized as one of the major deteriorative processes that affect the sensory and nutritional quality of food, so the use of synthetic antioxidants is widespread in the food industry. The addition of antioxidants such as butylated hydroxytoluene (BHT), butylated hydroxylanisol (BHA), propyl gallate (PG), and *tert*-butylhydroquinone (TBHQ) will certainly reduce the rate of oxidation at room temperature. They are relatively cheap, colorless, tasteless, and odorless. However, at the high temperatures (150 to 200°C) encountered during frying, they are either rapidly inactivated and/or rapidly evaporated, particularly BHA and BHT, primarily by a mechanism similar to steam distillation [31]. Therefore, the search continues for new antioxidants that can resist the high temperatures and conditions associated with frying that result in the rapid loss of any added antioxidants. In addition, the health effects concerning the use of synthetic antioxidants are of some concern. Although many investigators assert that antioxidants are harmless, some studies [32] have indicated that the synthetic antioxidants BHT and BHA can have potentially harmful effects, since they can promote the development of cancerous cells in rats.

Substantial research has been directed toward identifying plant extracts that have significant antioxidant activity. The development and availability of modern analytical techniques, such as HPLC/MS, have contributed significantly to this effort. The antioxidant activities of plants are generally closely correlated with the presence of phenolic compounds, which are mostly derivatives of flavanoids and related compounds, or the phenylpropanoid acids and related compounds. There is a second group of compounds with substantial antioxidant activity that has been long recognized for that activity, which includes the tocopherols and related compounds, such as the tocotrienols.

9.5.1 Tocopherols

Tocopherols are one of the most important natural antioxidants present in vegetable oils [33]. Many factors affect the efficacy of natural antioxidants during frying, such as the presence of synergists and pro-oxidant transition metals. The tocopheryl semiquinone radical formed after donation of a hydrogen atom from tocopherol to a peroxyl radical, is resonance stabilized. It is also possible to form a γ-tocopherol diphenylether dimer or a γ-tocopherol biphenyl dimer [34].

Since the fatty acid composition of triacylglycerols is the most important factor governing the rate of oxidation, the performance of antioxidants will be greatly influenced by the fatty acid composition of the oil. Jorge and co-workers [35] showed that tocopherol degraded faster in a saturated oil than in an unsaturated oil. Verleyen et al. [36] investigated the competitive oxidation between α-tocopherol and unsaturated TAG. They examined the stability of α-tocopherol at α-tocopherol concentrations of 0, 500, and 1000 ppm and at two temperatures (180 and 240°C). At 240°C α-tocopherol was more stable in the more unsaturated flaxseed oil than in palm oil, but this was not the case at 180°C. When tocopherol degradation was monitored during heating at 240°C and at a concentration of 1000 ppm, only 22.5% of the initial tocopherol was recovered in coconut oil, as compared to 85.7% in the flaxseed oil. At the higher temperature (240°C) the difference in the rate of oxidation of the two components is probably much less than at the lower temperature (180°C). The relative difference in concentration is probably the most important factor at 240°C; the tocopherol is partially protected by the unsaturated fatty acids, since the unsaturated fatty acids are present at a much greater concentration compared to the antioxidant. At lower temperatures, there may be sufficient difference in the rate of thermo-oxidation of the two components (tocopherol and polyunsaturated fatty acids), so that the more easily oxidized tocopherols are oxidized first. Others have also found that at very high temperatures, the rate and amount of tocopherol loss is greater, when the oil is less unsaturated [37,38].

To examine the effect of the various naturally occurring tocopherol isomers, the relative antioxidant efficacy of α-, β-, γ-, and δ-tocopherol [39] were compared. Triolein (OOO), trilinolein (LLL), and a 1:1 mixture of OOO and LLL were used to examine the antioxidant efficacy of tocopherol, when frying at 180°C for 10 h. They observed that the rate of loss for δ-tocopherol was less than for α-tocopherol. After 10 h of heating, only 5% of the α-tocopherol remained in the OOO, as compared to 37% of the δ-tocopherol. After heating, 37% of δ-tocopherol remained in the OOO oil samples, as compared to 14% in the mixture of OOO/LLL and 15% in the LLL samples. When polymer formation was investigated, the influence of δ-tocopherol was substantial with a reduction of more than 80% of the triacylglycerol polymer, as compared to the oil sample without added antioxidants. Although there was a reduction of polymer formation when α-tocopherol was added to OOO system, the reduction was less, 59%. In addition, the authors also observed a synergistic effect with α-, β-, γ-, and δ-tocopherol, when all four were added simultaneously to an oil sample.

9.5.2 Sterols

With the high demand for "natural" products by consumers, more companies are searching for new sources of natural antioxidants. One potential source of natural antioxidants is a class of substances called sterols. They are a common component in vegetable oils, and although some of them have no antioxidant activity or may even be pro-oxidants (e.g., sitosterol, stigmasterol, campesterol) [40], others such as Δ^5-avenasterol, citrostadienol, vernosterol, and Δ^7-avenasterol have antioxidant activity [41,42]. Gordon and Magos [40] suggested that the ethylidene group is responsible for the antioxidant effect of sterols. The sterols function by interrupting the oxidative chain reaction of fatty acids by formation of free-radicals at carbon 29, which then isomerizes to form a more stable free radical at a tertiary carbon. They outlined in detail the sterol structure and a suggested mechanism for free-radical formation. They include, as an example, the chemical structure of Δ^5-avenasterol and the ethylidene mechanism responsible for the antioxidant effect.

The heating conditions and various parameters involved in the investigation require careful consideration when comparing different experiments examining oil stability and plant sterols. Lampi et al. [43] compared the antioxidant effect of sitosterol, stigmasterol, α-tocopherol, and fucosterol, a geometric isomer of Δ^5-avenasterol, using high-oleic sunflower oil. Fucosterol had no antipolymerization activity. After 6 h of heating, the α-tocopherol was completely consumed, whereas 50% of the fucosterol remained after the same period of heating. There may be an explanation based on the differences in the heating methodology. The Δ^5-avenasterol did not stabilize the oils heated in an oven at 100 or 200°C, but there was an antioxidant effect when these oils were heated on a hot plate at the same temperatures. A comparison of heating using a hot plate and an oven suggests there may have been greater oil circulation, and consequently greater oxygen absorption, in the samples heated on the hot plate. The higher oxygen concentration may have increased the rate of oxidation, as compared to just thermal degradation, so that the antioxidant activity of the sterol became a more important factor.

9.5.3 Oryzanol, Sesamol, and Sesaminol

The other group of plant compounds that have substantial antioxidant activity is oryzanol and related compounds. Diack and Saska [44] found that the triterpene alcohol esters of ferulic acid comprise almost 70% of the oryzanol components. Together with the tocopherols, they can provide substantial oxidative stability to an oil system. The phenylpropanoid acid, ferulic acid, contains a phenolic group and the stabilization is achieved by resonance in the phenolic group that increases the stability of the free radical, when it is formed. The oryzanol helps reduce tocopherol degradation at high temperatures, which explains the synergistic effect of oryzanol and tocopherol.

Among the oils used for frying, sesame seed oil is one of the most heat stable, which is due to a group of compounds called sesaminol isomers [45,46]. Sesamol is formed by intermolecular transformation from sesamolin under anhydrous condition and by decomposition of sesamolin during the frying process in the presence of moisture, respectively. Sesamol and sesaminol both function as free-radical scavengers in vegetable oils.

9.5.4 Carotenoids

The carotenoids are an additional class of natural compounds that possesses antioxidant activity. They are the most ubiquitous group of pigments found in nature. Almost 600 carotenoids have been characterized, but only ~50 have any significant biological activity [47]. Approximately one third of all of the carotenoids are found in marine life [48].

Carotenoids are thought to react with peroxyl radicals to form resonance-stabilized radicals [49]. Alkyl peroxy radicals do not abstract a hydrogen atom from carotenoids, so the antioxidant mechanism is distinctly different from phenolic compounds. Carotenoids (e.g., β-carotene) act by trapping free radicals, such as a peroxy radical. The carbon-centered radical is resonance-stabilized by delocalization of an unpaired electron in the carotenoid polyene system leading to chain termination. Yanishlieva et al. [50] showed that the resulting β-carotene free radical (LOO-β-carotene$^\bullet$) can react in a pro-oxidative or antioxidative nature, depending on the oxygen pressure and interaction with other antioxidants. There are several reaction possibilities for β-carotene. It may react with the radical $X^\bullet$ and then with oxygen to yield a β-carotene peroxy radical (LOO-β-CarOO$^\bullet$), which can react with another radical to produce a stable product. The β-carotene peroxyl radical can also attack another β-carotene molecule to promote the auto-oxidation of β-carotene or attack a lipid substrate (LH) to produce a lipid radical ($L^\bullet$) and induce further oxidation. Also, α-tocopherol can scavenge a β-carotene peroxyl radical or the β-carotene radical can undergo β-scission to produce an epoxide and alkoxyl radical and continue the oxidation.

The β-carotene content of soybean oil stored at 25°C decreased from 20 to 15 ppm after 24 h of light exposure. After 2 h of frying, the concentration was reduced from 20 to 5 ppm and no β-carotene was detected after 10 h of heating [51].

At a low partial pressure for oxidation, the concentration of the reactive β-carotene peroxyl radical is dramatically reduced and, therefore, β-carotene can function as an effective antioxidant. However, at normal oxygen concentrations, β-carotene is generally a less effective antioxidant [52].

9.5.5 Plant Extracts and Spices with Antioxidant Properties

The effect of plant antioxidant blends have been studied by Irwandi et al. [71]. They investigated various mixtures of rosemary and sage oleoresin extracts in combination with citric acid using linoleic acid and palm olein model systems at room temperature. The extracts had a high protective index (PI) [53]. The PI index was calculated using an iron-supplemented linoleic acid model emulsion system. Some treatments had six to seven times more antioxidant activity than the control. Furthermore, these antioxidants showed good thermal resistance during frying.

One example of a product that has taken advantage of the presence of natural antioxidants is "*Good-Fry® Constituents*" (GFCs), which is a blend of sesame seed oil and rice bran oil designed to provide an oil of relatively high heat stability [45]. The sesamolin, oryzanol, tocotrienol, Δ^5-avenasterol, and γ-tocopherol content in GFCs are primarily responsible for the improvements in the frying oil stability. A comparison using french fries was done with high oleic acid soybean oil (HOSO), palm olein, and GFC. GFC showed satisfactory performance up to 65 h of frying (175°C), compared with 35 h for the HOSO and 40 h for the palm olein, based on the regulatory limit of 24 to 25% of total polar material in the oil (TPMs) used in many European Union countries. In this study, the range of TPM was 21.1 to 23.4%.

The antioxidant capacity of tea extracts is related to the phenolic concentration, particularly the flavanols. To investigate whether the antioxidant properties of tea extracts could be applied to frying oils, Zandi and Gordon [54] studied the effect of a methanol extract of tea leaves added to a frying oil. Their model system included low erucic acid rapeseed oil and potato slices fried at 180°C. After 12 frying operations, the *p*-anisidine values were determined. A rosemary extract (0.1%) and the tea leave extract (0.1%) were significantly different ($p < .01$) from the control, and the tea leave extract was at least as active as the rosemary extract.

Recently both herbs and spice have been receiving more attention concerning their antioxidant potential. Houhoula et al. [55] investigated the effect of ground oregano and an ethanol extract of oregano during frying (at 185°C) with cottonseed oil and sliced potatoes. The reduction in the *p*-anisidine value (*p*-AV) by the oregano was equivalent to that of the rosemary and the sage extract. After 12 h of frying, the *p*-AV increased from an initial value of 10 to 155 for the control, while the final values were 111 and 104 for the ground oregano and ethanol-derived extract, respectively. There was a twofold reduction in the polymer content with addition of ground oregano or ethanol extract. In addition, the potato slices had enhanced storage stability due to the addition of oregano to the frying oil.

Some researchers investigated the antioxidant activity of powdered vegetables. Lee et al. [56] added powdered spinach to flour to evaluate the extent of oxidation in fried products during frying and storage. The wheat flour was prepared by adding spinach powder at three levels (5, 15 and 25% on a dry-weight basis). The dough was immersed in soybean oil at 160°C for 1 min and this was repeated a second time. There was an increase in the carotene content of the oil from the spinach flour. The carotene content of the oil continued to increase upon repeated frying of the product. The total polar compound content of the oil was reduced as a result.

In spite of the claims associated with the stability and efficacy of natural antioxidants during frying, they are generally found only in small amounts, so that their regular use can be expensive. Synthetic antioxidants would be a good option, if they could be effective at frying temperatures. Zhang et al. [57] synthesized two antioxidants from TBHQ and *n*-lauryl alcohol in an effort to produce a more heat stable antioxidant and investigated their activity during frying. Lauryl *tert*-butylate hydroquinone (LTBHQ) and lauryl *tert*-butylate quinine (LTBQ) were synthesized from *n*-lauryl alcohol and TBHQ using phosphoric acid as a catalyst. Each antioxidant (0.02%) was added to a soybean oil sample, heated for 9 h at 190°C, and then cooled to room temperature for 15 h. This (heating/cooling) cycle was repeated seven times. Potato slices were used for the experiment. They observed that when the oxidative stability (via OSI) was used to evaluate the effect of antioxidant on the oil sample, the antioxidant powers of TBHQ, BHT, and BHA decreased substantially with an increase of temperature, whereas LTBHQ and LHBQ decreased only slightly. They also observed that LTBHQ and LTBQ were active at high temperature, but not at temperatures below 140°C.

9.6 CYCLIC FATTY ACID FORMATION

One of the many compounds formed during frying are the cyclic fatty acids, especially when oils rich in linoleic and linolenic acid are heated at very high temperatures (~200°C). It is believed that these compounds have toxic implications, due to their detrimental effects on the reproduction system of rats [58]. The concentration of monoenoic and dienoic cyclic fatty acids isolated from heated sunflower and linseed oil, respectively, was related to the concentration of linoleic and linolenic found in the oils. Two excellent reviews describing the structures of the molecular components involved in this process, including the proposed mechanisms of formation, have been published [59,60].

Lambelet et al. [61] examined low erucic acid rapeseed oil that had a high percentage of α-linoleic acid. During neutralization and bleaching, the oil was not altered significantly. However, during deodorization, when temperatures above 200°C were reached, significant changes were observed. Although the amount of cyclic fatty acids formed was much less than the amount of *trans* fatty acid formed during deodorization, cyclic fatty acids were present in significant amounts when more severe conditions were used (e.g., 6 h at 250°C). Under these conditions, as much as 650 mg of cyclic fatty acid monomers per kilogram of oil were found in deodorized low erucic acid rapeseed oil. Furthermore, under the same conditions, more than 50% of α-linolenic acid had been converted into *trans* isomers. Normally, conditions this severe are not used in industrial refining.

9.7 FLAVOR FORMATION IN FRIED FOODS

The quality of fried food is affected by parameters that include the quality and source of the oil. The main reactions during deep-fat frying include the thermolytic and oxidative reactions, which together produce the majority of the volatile and nonvolatile products formed. These reactions occur as a result of the combination of oxygen, heat, and moisture [62] and are directly dependent upon the fatty acid composition of the oil. Food fried in olive oil has a different flavor, than food fried in animal fat, for example. Nutty and buttery flavors are more pronounced in some oils than in others [63]. Pokorny [64] asserted that the following group of reactions (oxidative, hydrolytic, and pyrolytic) form many of the secondary and tertiary products responsible for the flavor of fried foods. These reactions affect the flavor quality—both positively and negatively—of fried foods. Ho et al. [65] found that (*E,E*)- and (*E,Z*)-2,4-decadienals are two of the primary oxidation products from linoleic acid. The 2-*trans*, 4-*trans* decadienal was identified as one of the primary compounds responsible for the rich flavor of fried potato chips.

Chyau et al. [66] studied the effects of soybean oil, corn oil, lard, and medium-chain triglyceride (MCT) based frying oil on flavor compound production in deep-fried shallots. Extracts from shallots fried in soybean oil, corn oil, or lard contained 28 flavoring compounds, whereas only 17 compounds were found in the extracts from shallots fried in MCT-based oil. The 2,5- and 2,6-dimethylpyrazines were found only in the MCT–shallot flavor extract and were absent in those prepared with other oils. They classified three flavoring compound categories: nitrogen-, oxygen- and sulfur-containing compounds. The MCT-based extract was low in oxygen-containing compounds, while the lard-based extract had predominately oxygen-containing compounds. When soybean oil and corn oil were compared, the soybean and cornextracts had similar volatile profiles, but there were differences in the concentrations of compounds produced.

To better understand the effect that various oxidation products have on food flavor, Warner et al. [67] investigated the effect of oleic and linoleic acid on flavor using pure TAG standards (containing a single fatty acid). They found eight key volatile compounds were formed upon heating the oil at 190°C, using MS–olfactometry analysis. (*E,E*)-2,4-Decadienal, (*E*)-2-heptenal, (*E*)-2-octenal, (*E,Z/Z,E*)-2,4-decadienal, (*E,E*)-2,4-nonadienal, (*E,Z*)-2,4-nonadienal, and (*E,E*)-2,4-octadienal were responsible for the deep-fried odor produced when LLL was heated. Upon frying in OOO, the fried food intensity was weak to moderate (<5), as compared to moderate to strong flavor intensities (4 to 10) that were detected after frying in LLL. Oils with a moderate content of linoleic acid, such as sunflower and cottonseed oil, produce a greater fried food flavor intensity than oils that are low in linoleic acid. Table 9.1 contains the volatile compounds formed, as well as their concentrations (ppm), during heating of OOO and LLL at 190°C for 1, 3, and 6 h of heating.

The highly unsaturated fatty acid content, particularly arachidonic acid, is associated with the development of oxidized off-flavors, such as warm-over flavor, in meat. More than 50% of the polyunsaturated fatty acids present in phospholipids consist of arachidonic acid. Artz et al. [68] investigated the thermal decomposition products formed from oxidized methylarachidonate using capillary gas chromatography-mass spectrometry. They found that of the identified volatile compounds, 43% were aldehydes, 24% were methyl esters, 13% were aliphatic hydrocarbons, 3.6% were ketones, and 2.5% were alcohols. Among the aldehydes, hexanal comprised 65% of the total with a ratio of hexanal to 2,4-decadienal of 4.5:1.

Warner et al. [69] examined the use of γ-tocopherol in potato chips using an OOO model oil system. Nonanal is one of the main volatile compounds formed upon the decomposition of oleic acid hydroperoxides and it is produced in relatively large quantities. Its formation was investigated in an effort to establish the effect of γ-tocopherol on potato chip quality. Potato chips were fried in OOO containing

TABLE 9.1
Fried Odor Volatile Concentrations (ppm) in Oil

	Heating Interval (h)					
	1		3		6	
Volatile	OOO[a]	LLL	OOO	LLL	OOO	LLL
(*E*)-2-heptenal	0.0	98.6	0.0	80.1	0.0	127.3
(*E*)-2-octenal	1.1	15.9	0.5	35.7	5.8	72.6
(*E,E*)-2,4-octadienal	0.0	0.2	0.0	0.4	0.0	0.6
(*E,Z*)-2,4-nonadienal	0.1	0.1	0.2	0.3	0.3	0.4
(*E,E*)-2,4-nonadienal	4.2	4.5	4.1	7.6	5.5	12.6
(*E,E*)-2,4-decadienal	12.6	221.1	10.7	423.7	14.1	632.9
(*E,Z/Z,E*)-2,4-decadienal	0.0	9.2	0.0	12.8	0.0	17.5
(*E,E*)-2,4-undecadienal	0.7	0.0	2.0	0.0	3.0	0.0

[a] OOO is the triacylglycerol triolein, while LLL is trilinolein.
[b] Heating temperature = 190°C.

Source: Modified from Warner, K. et al., *J. Agric. Food Chem.* 49, 899, 2001.

400 ppm of γ-tocopherol. The nonanal concentration of the potato chips stored for 4 d did not differ from that of the freshly fried potato chip, even though the γ-tocopherol concentration was only 12 ppm, suggesting that if sufficient γ-tocopherol is added to the oil, it would protect the potato chips during storage.

Fujisaki et al. [70] investigated deep-frying under various oxygen atmosphere concentrations. They monitored the generation of volatile aldehydes formed from high-oleic acid safflower oil (75% oleic acid and 16% linoleic acid) heated at 180°C. They found an interesting correlation between the oxygen concentration and the aldehyde formation. In the atmosphere containing 20% O_2, the total amount of aldehydes derived from oleic acid was comparable to the amount formed from linoleic acid. However, when the O_2 concentration was reduced to 4%, the sum of the aldehydes produced from linoleic acid (acetaldehyde, pentanal, hexanal, 2-heptanal, 2-octenal, 2-nonenal, and 2,4-decadienal) was greater than those derived from oleic acid (octanal, nonanal, decanal, 2-decenal, and 2-undecenal). This study emphasizes the relationship between the oil composition, the frying conditions, and the amount of flavor volatiles formed during frying; and suggests that under conditions of different oxygen availability, the relative rates of substrate oxidation will vary.

REFERENCES

1. Orthoefer, F.T., Oil used in the food service industry. *J. Am. Oil Chem. Soc.* 64, 795, 1987.
2. Sakurai, H. and Pokorny, J., The development and application of novel vegetable oils tailor-made for specific human dietary needs. *Eur. J. Lipid Sci. Technol.* 105, 769, 2003.

3. Brinkmann, B., Quality criteria of industrial frying oils and fats. *Eur. J. Lipid Sci. Technol.* 102, 539, 2000.
4. American Society of Clinical Nutrition, Position paper: task force on *trans* fatty acids. *Am. J. Clin. Nutr.* 63, 663, 1996.
5. Smith, L.M. et al., Lipid content and fatty acid profiles of various deep-fried foods. *J. Am. Oil Chem. Soc.* 62, 996, 1985.
6. Zock, P.L. and Katan, M.B., Hydrogenation alternatives: effects of *trans* fatty acids and stearic acid versus linoleic acid on serum lipids and lipoproteins in humans. *J. Lipid Res.* 33, 399, 1992.
7. Haumann, B.F., Modified oil may be key for sunflower future. *Inform* 5, 1198, 1994.
8. Blumenthal, M.M., A new look at frying science. *Cer. Foods World* 46, 352, 2001.
9. Vitrac, O., Trystram, G., and Raoult-Wack, A.L., Deep-fat frying food: heat and mass transfer, transformations and reactions inside the frying material. *Eur. J. Lipid Sci. Technol.* 102, 529, 2000.
10. Dobarganes, C., Márquez-Ruiz, G., and Velasco, J., Interactions between fat and food during deep-frying. *Eur. J. Lipid Sci. Technol.* 102, 521, 2000.
11. Orthoefer, F.T., Gurkin, S., and Liu, L., Dynamics of frying, in *Deep Frying: Chemistry, Nutrition and Practical Applications*. Perkins, E.G. and Erickson, M.D., Eds., AOCS Press, Champaign, IL, 1996, chap. 11.
12. Lin, S., Akoh, C.C., and Reynolds, A.E., Recovery of used frying oils with adsorbent combinations: refrying and frequent oil replenishment. *Food Res Int.* 34, 159, 2001.
13. Kalapathy, U. and Proctor, A., A new method for free fatty acid reduction in frying oil using silicate films produced from rice hull ash. *J. Am. Oil Chem. Soc.* 77, 593, 2000.
14. Kinney, A.J., Development of genetically engineered soybean oils for food applications. *J. Food Lipids* 3, 273, 1996.
15. Kinney, A.J. and Knowlton, S., Designer oils: the high oleic acid soybean, in *Genetic Modification in the Food Industry*. Roller, S. and Harlander, S., Eds., Blackie, London, 1998.
16. Warner, K. and Mounts, T.L., Frying stability of soybean and canola oils with modified fatty acid compositions. *J. Am. Oil Chem. Soc.* 70, 983, 1993.
17. Tompkins, C. and Perkins, E.G., Frying performance of low-linolenic acid soybean oil. *J. Am. Oil Chem. Soc.* 77, 223, 2000.
18. Pokorny, J., Browning from lipid-protein interaction. *Prog. Food Nutr. Sci.* 5, 421, 1981.
19. Gertz, C. and Klostermann S., Analysis of acrylamide and mechanisms of its formation in deep-fried products. *Eur. J. Lipid Sci. Technol.* 104, 762, 2002.
20. Liu, J.F. and Huang, C.J., Tissue α-tocopherol retention in male rats is compromised by feeding diets containing oxidized frying oil. *J. Nutr.* 125, 3071, 1995.
21. Cohn, J.S., Oxidized fat in the diet, postprandial lipaemia and cardiovascular disease. *Curr. Opin. Lipidol.* 13, 19, 2002.
22. Quinn, M.T. et al., Oxidatively modified low density lipoproteins: a potential role in recruitment and retention of monocyte/macrophages during atherogenesis. *Proc. Natl. Acad. Sci.* 84, 2995, 1987.
23. Eder, K. et al., Thermally oxidized dietary fats increase the susceptibility of rat LDL to lipid peroxidation but not their uptake by macrophages. *J. Nutr.* 133, 2830, 1993.
24. Saka, S., Aouacheri, W., and Abdennour, C., The capacity of glutathione reductase in cell protection from the toxic effect of heated oils. *Biochimie* 84, 661, 2002.
25. Pérez-Granados, A.M., Vaquero, M.P., and Navarro M.P., Effects of diets containing oils from repeated frying on magnesium absorption. *J. Sci. Food Agric.* 79, 699, 1999.

26. Hayam, I., Cogan, U., and Mokady, S., Enhanced peroxidation of proteins of erythrocyte membrane and of muscle tissue by dietary oxidized oil. *J. Biosci. Biotechnol. Biochem.* 61, 1011, 1997.
27. Hageman, G. et al., Assessment of mutagenic activity of repeatedly used deep-fat frying fats. *Mutat. Res.* 204, 593, 1988.
28. Marquez-Ruiz, G., Pérez-Camino M.C., and Dobarganes, M.C., Evaluación nutricional de grasas termooxidadas y de frituras. *Grasas y Aceites* 6, 432, 1990.
29. Soriguer, F. et al., Hypertension is related to the degradation of dietary frying oils. *Am. J. Clin. Nutr.* 78, 1092, 2003.
30. Aruoma, O.I., Editorial–Neuroprotection by dietary antioxidants: new age of research. *Nahrung/Food* 46, 381, 2002.
31. Hamama, A.A. and Nawar, W.W., Thermal decomposition of some phenolic antioxidants. *J. Agric. Food Chem.* 39, 1063, 1991.
32. Lindenschmidt, R.C. et al., The effect of dietary butylated hydroxyl toluene on liver and colon tumor development in mice. *Toxicology* 38, 151, 1986.
33. Kamal-Eldin, A. and Andersson, R., A mutative study of the correlation between tocopherol content and fatty acid composition in different vegetable oils. *J. Am. Oil Chem. Soc.* 74, 375, 1997.
34. Kochhar, S.P., Deterioration of edible oils, fats and foodstuffs, in *Atmospheric Oxidation and Antioxidants. Vol. II.* Scott, G., Ed., Elsevier, London, 1993.
35. Jorge, N. et al., Influence of diemethylpolysiloxane addition to edible oils: dependence of the main variables of the frying process. *Grasas Aceites* 12, 14, 1996.
36. Verleyen, T. et al., Oxidation at elevated temperatures: competition between α-tocopherol and unsaturated triacylglycerols. *Eur. J. Lipid Sci. Technol.* 104, 228, 2002.
37. Kajimoto, G. et al., Influence of fatty acid composition in oil on the thermal decomposition of tocopherols. *J. Am. Oil Chem. Soc.* 68, 196, 1991.
38. Yoshida, H., Hirooka, N., and Kajimoto, G. Microwave energy effects on quality of some seed oils. *J. Food Sci.* 55, 1412, 1990.
39. Barrela-Arellano, D. et al., Loss of tocopherols and formation of degradation compounds in triacylglycerol model systems heated at high temperature. *J. Sci. Food Agric.* 79, 1923, 1999.
40. Gordon, M.H. and Magos, P. The effect of sterols on the oxidation of edible oils. *Food Chem.* 10, 141, 1983.
41. White, P.J. and Armstrong, L.S., Effect of selected oat sterols on the deterioration of heated soybean oil. *J. Am. Oil Chem. Soc.* 63, 525, 1986.
42. Yan, P.S. and White, P.J., Lynalyl acetate and other compounds with related structures as antioxidants in heated soybean oil. *J. Agric. Food Chem.* 38, 1904, 1990.
43. Lampi, A., Dimberg, L.H., and Kamal-Eldin, A., A study on the influence of fucosterol on thermal polymerization of purified high oleic sunflower triacylglycerols. *J. Sci. Food Agric.* 79, 573, 1999.
44. Diack, M. and Saska, M., Separation of vitamin E and γ-oryzanols from rice bran by normal-phase chromatography. *J. Am. Oil Chem. Soc.* 71, 1211, 1994.
45. Kochhar, S.P., Stabilisation of frying oils with natural antioxidative components. *Eur. J. Lipid Sci. Technol.* 102, 552, 2000.
46. Suja, K.P., Jayalekshmy, A., and Arumughan, C., Free radical scavenging behavior of antioxidant compounds of sesame (*Sesamum indicum* L.) in DPPH center dot system. *J. Agric. Food Chem.* 52, 912, 2004.
47. Olson, J.A. and Krinsky, N.I., Introduction: the colorful fascinating world of the carotenoids: important physiologic modulators. *Fed. Am. Soc. Exp. Biol. J.* 9, 1547, 1995.

48. Matsuno, T. and Hirao, S., Marine carotenoids, in *Marine Biogenic Lipids, Fats, and Oils. Vol. I*. Ackman, R.G., Ed., CRC Press, Boca Raton, FL, 1989.
49. Burton, G.W. and Ingold, K.U., β-carotene: an unusual type of lipid antioxidant. *Science* 224, 569, 1984.
50. Yanishlieva, N.V., Aitzetmuller, K., and Raneva, V.G., β-carotene and lipid oxidation. *Fett/Lipid* 100, 444, 1998.
51. Warner, K. and Frankel, E.N., Effects of β-carotene on light stability of soybean oil. *J. Am. Oil Chem. Soc.* 64, 213, 1987.
52. Britton, G., Structure and properties of carotenoids in relation to function. *Fed. Am. Soc. Exp. Biol. J.* 9, 1551, 1995.
53. Lingnert, H., Vallentin, K., and Erikson, C.E., Measurement of antioxidative effect of model systems. *J. Food Process. Preserv.* 3, 87, 1979.
54. Zandi, P. and Gordon, M.H., Antioxidant activity of extracts from old tea leaves. *Food Chem.* 64, 285, 1999.
55. Houhoula, D.P., Oreopoulou, V., and Tzia, C., Antioxidant efficiency of oregano during frying and storage of potato chips. *J. Sci. Food Agric.* 83, 1499, 2003.
56. Lee, J. et al., Spinach (*Spinacia oleracea*) powder as a natural food-grade antioxidant in deep-fat-fried products. *J. Agric. Food Chem.* 50, 5664, 2002.
57. Zhang, C.X., Wu, H., and Weng, X.C., Two novel synthetic antioxidants for deep frying oils. *Food Chem.* 84, 219, 2004.
58. Martin, J.C. et al., Cyclic fatty acid monomers from heated oil modify the activities of lipid synthesizing and oxidyzing enzymes in rat liver. *J. Nutr.* 130, 1524, 2000.
59. Sébédio, J.L. and Grandgirard, A., Cyclic fatty acid: natural sources, formation during heat treatment, synthesis and biological properties. *Prog. Lipid Res.* 28, 303, 1989.
60. Christie, W.W. and Dobson, G., Formation of cyclic fatty acids during the frying process. *Eur. J. Lipid Sci. Technol.* 102, 515, 2000.
61. Lambelet, P. et al., Formation of modified fatty acids and oxyphytosterols during refining of low erucic acid rapeseed oil. *J. Agric. Food Chem.* 51, 4284, 2003.
62. Artz, W.E., Soheili, K.C., and Arjona, I.M. Esterified propoxylated glycerol soyate, a fat substitute model compound, and soy oil after heating. *J. Agric. Food Chem.* 47, 3816, 1999.
63. Stier, R.F., Chemistry of frying and optimization of deep-fat fried food flavour–an introductory review. *Eur. J. Lipid Sci. Technol.* 102, 507, 2000.
64. Pokorny, J., Flavor chemistry of deep fat frying in oils, in *Flavor Chemistry of Lipid Foods*. Smouse, T. and Perkins, E.G., Eds., AOCS Press, Champaign, IL, 1988, chap. 7.
65. Ho, C.T. et al., Flavor chemistry of Chinese foods. *Food Rev. Int.* 5, 253, 1989.
66. Chyau, C.C. and Mau, J.L., Effects of various oils on volatile compounds of deep-fried shallot flavouring. *Food Chem.* 74, 41, 2001.
67. Warner, K. et al., Effect of oleic and linoleic acids on the production of deep-fried odor in heated triolein and trilinolein. *J. Agric. Food Chem.* 49, 899, 2001.
68. Artz, W.E., Perkins, E.G., and Salvador-Henson, L., Characterization of the volatile decomposition products of oxidized methyl arachidonate. *J. Am. Oil Chem. Soc.* 70, 945, 1993.
69. Warner, K., Neff, W.E., and Eller, F.J., Enhancing quality and oxidative stability of aged fried food with γ-tocopherol. *J. Agric. Food Chem.* 51, 623, 2003.
70. Fujisaki, M., Endo, Y., and Fujimoto, K., Retardation of volatile aldehyde formation in the exhaust of frying oil by heating under low oxygen atmospheres. *J. Am. Oil Chem. Soc.* 79, 909, 2002.

10 *Trans* Fatty Acids and *Trans*-Free Lipids

Oi-Ming Lai and Seong-Koon Lo

CONTENTS

10.1 INTRODUCTION

In recent years, *trans* fatty acids, more commonly known as *trans* fats, have been gaining a lot of interest from the scientific and health professional communities primarily because of the potential role of *trans* fatty acids on cardiovascular disease risk. Publications on the adverse effects of *trans* fatty acids on coronary heart disease, low-density and high-density lipoprotein cholesterols, and blood lipid levels are extensive. In a 1994 report, it was estimated that approximately 30,000 annual deaths from premature coronary heart disease could be linked to the consumption of *trans* fatty acids [1]. Since then, more metabolic and epidemiologic studies have been reported. Due to the increasing evidence of the effects of *trans* fatty acids on public health, the U.S. Food and Drug Administration (FDA) has recently published a final rule on the labeling of *trans* fatty acids in food items.

TABLE 10.1
Comparison of Melting Points of *cis* and *trans* Isomers of Unsaturated Fatty Acids and Saturated Fatty Acids

Type of Fatty Acid		Melting Point (C)
cis Fatty acids	Linolenic acid (C18:3,*c*,*c*,*c*)	−11
	Linoleic acid (C18:2,*c*,*c*)	–5
	Oleic acid (C18:1,*c*)	14
trans Fatty acids	Linolenic acid (C18:3,*t*,*t*,*t*)	71
	Linolenic acid (C18:3,*c*,*t*,*t*)	49
	Linoleic acid (C18:2,*t*,*t*)	56
	Oleic acid (C18:1,*t*)	44
Saturated fatty acids	Behenic acid (C22:0)	80
	Arachidic acid (C20:0)	75
	Stearic acid (C18:0)	70
	Palmitic acid (C16:0)	63
	Myristic acid (C14:0)	54
	Lauric acid (C12:0)	44

10.2 CHEMICAL STRUCTURE

Fatty acid isomers are classified into two types, positional and geometric isomers. Positional isomers are formed when double bonds of the fatty acid molecule shift from their original position to other positions in the molecule. For example, fatty acids having double bonds at the Δ9 and Δ12 positions have been reported to shift to isomeric forms ranging from positions Δ4 to Δ16 [2], with the majority clustered in the vicinity of the original double bond [3–6]. Geometric isomers of unsaturated fatty acids are categorized into two forms: *cis* (*c*) and *trans* (*t*), as shown in Figure 10.1. The *cis* fatty

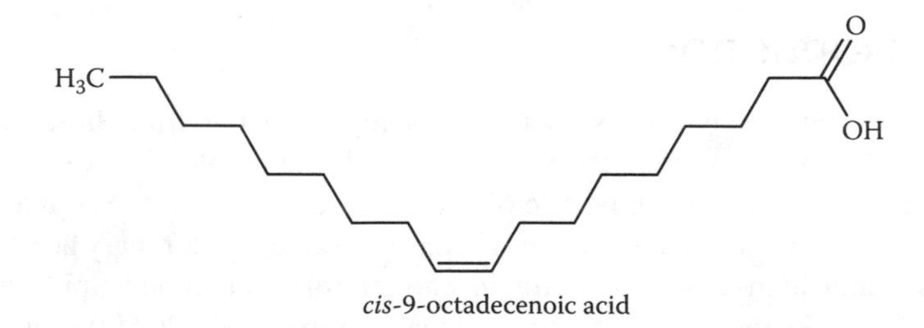

O
H_3C
OH
trans-9-octadecenoic acid

FIGURE 10.1 Geometric isomers of unsaturated fatty acids.

acids, which are commonly found in natural oils and fats, are relatively more reactive and require lower activation energy to be transformed to the *trans* isomer. On the other hand, *trans* fatty acids have a linear structural conformation that permits tighter stacking of molecules, hence, allowing them to have melting points similar to those of saturated fatty acids (Table 10.1). In addition, *trans* fatty acids can be formed at very high temperatures (e.g., deodorization process and frying) and through hydrogenation reactions and biohydrogenation in ruminant animals.

10.3 HYDROGENATION

Trans fatty acids are produced during hydrogenation of edible oils. The process of hydrogenation of edible oils is an important and widespread practice in the modification of oils, whereby all or part of the double bonds of the fatty acids are eliminated, thus producing a partially or completely hydrogenated fat, which possesses increased melting point, increased resistance to oxidation, and improved consistency [7–9]. By controlling the degree of hydrogenation, manufacturers can provide the consumer with fat products, such as margarines and shortenings, with the desired consistency and spreadability. The most prominent *trans* fatty acids formed from the partial hydrogenation of vegetable or fish oils are the $\Delta 9$ (elaidic) and $\Delta 10$ isomers [4,10].

10.4 BIOHYDROGENATION

In ruminant animals such as cows and sheep, unsaturated fatty acids are extensively hydrogenated in the rumen by bacteria [11–13]. Hay and Morrison [14] reported that complex enzyme systems of the rumen microflora are responsible for transforming the monounsaturated and polyunsaturated fatty acids in feedstuffs into saturated fatty acids and into geometric and positional isomers. The *trans* fatty acid produced by the rumen bacteria is predominantly vaccenic acid, which constitutes about 50% of all ruminant *trans* fatty acids [10,15]. As a result, dairy products and meats from ruminant animals contain small amounts of *trans* fatty acids [14,16–18].

10.5 DIETARY SOURCES OF *TRANS* FATTY ACIDS

The most important source of *trans* fatty acids in the food supply is from the commercial hydrogenation of edible oils and fats. These hydrogenated oils and fats are present in the form of margarines, shortenings, vegetable oils, and frying fats and in a variety of processed foods, snacks, fast foods, and bakery products that use these fats. Dietary sources of *trans* fatty acids can also be found in dairy products, meats, and animal fats, particularly from ruminant animals. This section summarizes available information on *trans* fatty acid content in foods.

10.5.1 MARGARINES

The *trans* fatty acid contents of margarines and spreads from different countries are summarized in Table 10.2. In the TRANSFAIR study conducted by Aro et al. [20], variable amounts of *trans* fatty acids were reported in margarines and low-fat spreads from 14 European countries. Typically, soft table margarines, hard household margarines,

TABLE 10.2
***trans* Fatty Acid Content of Margarines and Spreads from Different Countries**

Country	Food Item	Percentage *trans* Fatty Acids (%)			Total *trans* Fatty Acids/100 g Food Item			Reference
		Lowest	Highest	Average	Lowest	Highest	Average	
Denmark	Margarines, 1999, <10% C18:2 (n = 4)	0.0	2.9	1.0	—	—	—	Leth et al., 2003 [19]
	Margarines, 1999, 10–20% C18:2 (n = 16)	0.0	2.7	0.7	—	—	—	
	Margarines, 1999, 20–40% C18:2 (n = 10)	0.0	4.0	1.0	—	—	—	
Belgium	Soft table margarines	0.66	10.21	0.74	0.52	8.01	0.44	Aro et al., 1998 [20]
Denmark	Soft table margarines	—	—	1.51	—	—	1.23	
Finland	Soft table margarines	—	11.20	2.92	—	9.02	2.04	
France	Soft table margarines	0.11	1.88	0.12	0.07	1.52	0.09	
Germany	Soft table margarines	0.30	5.44	4.86	0.20	3.94	3.84	
Greece	Soft table margarines	0.26	3.20	3.20	0.20	2.56	2.56	
Iceland	Soft table margarines	—	—	16.51	—	—	10.65	
Italy	Soft table margarines	—	—	5.16	—	—	3.84	
Netherlands	Soft table margarines	0.42	8.16	0.42	0.29	6.53	0.29	
Norway	Soft table margarines	2.61	15.81	15.81	2.04	12.73	12.73	
Portugal	Soft table margarines	0.23	14.80	1.90	0.19	12.08	1.53	
Spain	Soft table margarines	0.33	16.64	1.03	0.27	13.40	0.83	
Sweden	Soft table margarines	0.11	0.47	0.13	0.09	0.37	0.10	
United Kingdom	Soft table margarines	0.38	15.87	13.02	0.26	11.98	0.09	
Belgium	Hard household margarines	—	—	—	—	—	—	Aro et al., 1998 [20]
Denmark	Hard household margarines	—	—	1.65	—	—	1.31	
Finland	Hard household margarines	—	—	4.49	—	—	3.59	
France	Hard household margarines	0.25	6.27	0.25	0.20	4.20	0.20	
Germany	Hard household margarines	—	—	5.96	—	—	4.77	

Greece	Hard household margarines	0.61	16.55	0.61	0.49	13.24	0.49	
Iceland	Hard household margarines	14.29	28.07	21.21	11.57	22.32	16.23	
Italy	Hard household margarines	—	—	0.42	—	—	0.34	
Netherlands	Hard household margarines	0.73	21.27	0.73	0.59	17.02	0.59	
Norway	Hard household margarines	6.82	19.85	18.58	5.39	15.28	14.86	
Portugal	Hard household margarines	1.91	13.13	1.91	1.56	10.77	1.56	
Spain	Hard household margarines	—	—	—	—	—	—	
Sweden	Hard household margarines	0.09	7.83	0.82	0.07	6.34	0.66	
United Kingdom	Hard household margarines	—	18.37	6.89	—	15.16	5.44	
Belgium	Low-fat spreads	0.34	10.26	0.91	0.11	4.16	0.34	Aro et al., 1998 [20]
Denmark	Low-fat spreads	—	—	1.87	—	—	0.79	
Finland	Low-fat spreads	1.24	—	3.44	0.75	—	1.48	
France	Low-fat spreads	0.13	3.66	0.52	0.05	1.50	0.21	
Germany	Low-fat spreads	0.70	1.10	0.82	0.28	0.23	0.28	
Greece	Low-fat spreads	0.34	1.88	0.34	0.14	1.02	0.14	
Iceland	Low-fat spreads	12.27	13.29	13.29	4.85	5.38	5.38	
Italy	Low-fat spreads	—	—	—	—	—	—	
Netherlands	Low-fat spreads	0.36	8.13	2.21	—	—	—	
Norway	Low-fat spreads	0.31	11.84	11.84	0.12	5.09	5.09	
Portugal	Low-fat spreads	—	—	0.16	—	—	0.07	
Spain	Low-fat spreads	—	—	—	—	—	—	
Sweden	Low-fat spreads	0.16	1.42	0.16	0.06	0.56	0.06	
United Kingdom	Low-fat spreads	8.23	15.98	12.37	3.42	4.07	4.64	
New Zealand	Margarines (n = 2)	14.33	14.66	14.50	—	—	—	Richardson et al., 1997 [21]
United States	Margarines, fat-free	—	—	1.52	—	—	0.04	Ali et al., 1996 [22]
Spain	Margarines (n = 32)	—	—	16.8	—	—	—	Fernandez San Juan, 1996 [23]

(continued)

TABLE 10.2 (Continued)
***trans* Fatty Acid Content of Margarines and Spreads from Different Countries**

Country	Food Item	Percentage *trans* Fatty Acids (%)			Total *trans* Fatty Acids/100 g Food Item			Reference
		Lowest	Highest	Average	Lowest	Highest	Average	
Austria	Stick margarines, 1991/1992 (n = 75)	—	—	21.30	—	—	—	Henninger and Ulberth 1996 [24]
	Tub margarines, 1991/1992 (n = 183)	—	—	15.72	—	—	—	
	Dietary margarines, 1991/1992 (n = 124)	—	—	0.42	—	—	—	
	Stick margarines, 1995 (n = 14)	—	—	8.65	—	—	—	
	Tub margarines, 1995 (n = 28)	—	—	9.95	—	—	—	
	Dietary margarines, 1995 (n = 4)	—	—	0.82	—	—	—	
New Zealand	Margarines (n = 7)	12.6	19.7	16.4	—	—	—	Lake et al., 1996 [25]
	Table spreads (n = 5)	14.3	16.9	15.7	—	—	—	
	Margarine/butter blend (n = 2)	6.1	13.1	9.6	—	—	—	
Germany	Margarines (n = 46)	0.17	25.90	9.32	—	—	—	Molkentin and Precht, 1996 [26]
	Dietary/reformatory margarines (n = 31)	0.03	2.94	0.65	—	—	—	
Denmark	Margarines, 1995, <20% C18:2	0.0	8.2	4.2	—	—	—	Ovesen et al., 1996 [27]
	Margarines, 1995, 20–40% C18:2	0.0	5.8	1.2	—	—	—	
	Margarines, 1995, >40% C18:2	0.0	5.6	1.3	—	—	—	
	Margarines, 1992, <20 C18:2	0.0	10.6	5.8	—	—	—	
	Margarines, 1992, 20–40% C18:2	0.0	22.3	9.8	—	—	—	
	Margarines, 1992, >40% C18:2	1.4	9.6	1.9	—	—	—	
France	Tub margarines, <5% *trans* (n = 5)	0.06	1.68	0.45	—	—	—	Bayard and Wolff, 1995 [28]
	Tub margarines, >5% *trans* (n = 7)	7.38	19.02	14.11	—	—	—	

United States	Stick margarines, corn oil	—	—	18.6	—	—	—	Michels and Sacks, 1995 [29]
United Kingdom	Tub margarines, corn oil	—	—	9.0	—	—	—	
Germany	Margarine blend, light	—	—	2.0	—	—	—	
	"No *trans*" margarine	—	—	0.5	—	—	—	
	Dietary margarines	—	—	1.2	—	—	—	
	Reformatory margarines	—	—	1.4	—	—	—	
Denmark	Hard margarines, >20% C18:2	17	23	21	14	19	17	Ovesen and Leth, 1995 [30]
	Soft margarines, >20% C18:2	0	15	8	0	12	7	
	Soft margarines, >55% C18:2	0	4	3	0	3	2	
Germany	Margarines, vegetable oils (n = 10)	0.7	6.4	4.1	—	—	—	Pfalzgraf and Steinhart, 1995 [31]
	Margarines, sunflower oil (n = 4)	8.6	21	16.1	—	—	—	
	Diet, reduced fat margarines (n = 4)	0.5	3.1	2.3	—	—	—	
	Dietary/reformatory margarines (n = 6)	0.4	2.4	1.3	—	—	—	
United States	Stick margarines (n = 10)	20.14	31.86	25.96	13.02	25.06	19.44	USDA, 1995 [32]
	Tub margarines (n = 8)	7.91	17.52	12.89	3.05	11.30	7.62	
	Spread (n = 5)	13.56	33.58	23.27	2.79	25.78	15.55	
	Spread, light (n = 2)	15.24	17.95	16.60	5.66	9.09	7.38	
New Zealand	Margarines (n = 8)	7.6	9.6	8.5	—	—	—	Ball et al., 1993 [33]
	Margarine blends (n = 3)	5.4	6.7	6.1	—	—	—	
	Reduced fat (n = 4)	7.9	11.8	10.5	—	—	—	
Spain	Margarines (n = 47)	—	—	10.8	—	—	—	Boatella et al., 1993 [34]
United States	Soft margarines	—	—	—	3.0	10.2	5.7	Litin and Sacks, 1993 [35]
	Stick margarines	—	—	—	7.9	19.8	13.2	
Australia	Margarines (n = 13)	8.01	14.54	12.16	—	—	—	Mansour and Sinclair, 1993 [36]
	Butter/dairy blends (n = 5)	3.44	4.75	4.26	—	—	—	
Germany	Margarines, undeclared composition (n = 1)	—	—	23.5	—	—	—	Pfalzgraf et al., 1993 [37]
	Margarines, hydrogenated vegetable oils (n = 6)	6.0	23.4	9.3	—	—	—	

(*continued*)

TABLE 10.2 (Continued)
***trans* Fatty Acid Content of Margarines and Spreads from Different Countries**

Country	Food Item	Percentage *trans* Fatty Acids (%)			Total *trans* Fatty Acids/100 g Food Item			Reference
		Lowest	Highest	Average	Lowest	Highest	Average	
	Margarines, hydrogenated vegetable oils	4.0	18.8	10.1	—	—	—	
	containing animal fat (*n* = 5)	0.6	3.6	1.6	—	—	—	
	Diet and reduced fat margarines (*n* = 4)							
Canada	Hard margarines, 6 types (*n* = 19)	—	—	34.2	—	—	—	Ratnayake and Pelletier 1992 [38]
	Soft margarines, 6 types (*n* = 31)	—	—	21.3	—	—	—	
Canada	Stick margarines, soybean oil (*n* = 3)	30.4	49.9	38.8	—	—	—	Ratnayake et al., 1991 [39]
	Stick margarines, vegetable oil (*n* = 5)	25.6	40.2	33.42	—	—	—	
	Stick margarines, may contain palm oil (*n* = 7)	32.8	40.6	35.2	—	—	—	
	Stick margarines, corn oil (*n* = 1)	—	—	28.2	—	—	—	
	Stick margarines, vegetable oil/animal fat (*n* = 1)	—	—	42.2	—	—	—	
	Stick margarines, unspecified (*n* = 2)	20.9	28.8	24.9	—	—	—	
	Tub margarines, soybean oil (*n* = 13)	12.4	27.3	18.7	—	—	—	
	Tub margarines, vegetable oil (*n* = 10)	22.1	35.2	26.4	—	—	—	
	Tub margarines, may contain palm oil (*n* = 2)	19.5	21.3	20.4	—	—	—	
	Tub margarines, corn oil (*n* = 3)	10.1	25.3	16.8	—	—	—	
	Tub margarines, sunflower oil (*n* = 2)	—	14.3	7.15	—	—	—	
	Tub margarines, olive oil (*n* = 1)	—	—	19.4	—	—	—	
United States	Margarines, weighted average	—	—	23.0	—	—	—	Enig et al., 1990 [40]
United States	Stick margarines, (*n* = 57)	16.1	34.8	23.7	—	—	—	Slover et al., 1985 [41]
	Stick margarines, lard (*n* = 1)	—	—	3.9	—	—	—	

	Tub margarines, (n = 26)	11.7	29.8	17.7	—	—	—	
	Block margarines, (n = 1)	—	—	21.7	—	—	—	
Israel	Hard margarines, (n = 3)	30.8	34.2	32.8	—	—	—	Enig et al., 1984 [42]
	Soft margarines, (n = 2)	13.4	14.3	13.9	—	—	—	
United States	Stick margarines, (n = 24)	17.4	36.0	24.2	—	—	—	Enig et al., 1983 [43]
	Tub margarines, (n = 13)	10.6	21.3	14.4	—	—	—	
	Diet margarines, (n = 3)	12.7	16.9	15.3	—	—	—	
Canada	Stick margarines, (n = 17)	21.5	39.3	30.0	—	—	—	Marchand, 1982 [2]
	Tub margarines, (n = 30)	9.4	27.4	15.6	—	—	—	
Australia	Margarines (n = 12)	4.0	14.0	11.0	—	—	—	Wills et al., 1982 [44]
Canada	Margarines (n = 8)	12.0	64.8	30.3	—	—	—	Beare-Rogers et al., 1979 [45]
Canada	Stick margarines, vegetable oil (n = 33)	27.2	32.9	30.7	—	—	—	Sahasrabudhe and Kurian, 1979 [46]
	Stick margarines, animal fat (n = 6)	4.2	25.3	18.4	—	—	—	
	Soft margarines, vegetable oil (n = 49)	8.7	28.5	15.0	—	—	—	
Germany	Regular margarines, 1976 (n = 58)	0.1	30.3	8.2	—	—	—	Heckers and Melcher, 1978 [47]
	Regular margarines, 1973/1974 (n = 61)	0.1	34.7	8.1	—	—	—	
	Low-calorie margarines, 1976 (n = 9)	0.2	11.2	3.9	—	—	—	
	Low-calorie margarines, 1973/1974 (n = 8)	0.2	5.3	1.9	—	—	—	
United States	Margarines (n = 5)	13.6	23.3	17.8	—	—	—	Smith et al., 1978 [17]
United States	Stick and brick margarines, (n = 77)	—	—	—	9.9	28.7	21.7	Weihrauch et al., 1977 [48]
	Tub margarines, (n = 13)	—	—	—	10.5	21.4	14.2	
Australia	PUFA margarines, (n = 40)	12.9	22.1	17.1	—	—	—	Parodi, 1976 [4]
	Regular margarines, (n = 7)	10.8	19.0	14.9	—	—	—	
United States	Stick margarines, (n = 6)	—	36.0	25.7	—	—	—	Carpenter and Slover, 1973 [5]
	Tub margarines, (n = 4)	14.0	28.0	21.8	—	—	—	

and low-fat spreads from Denmark, Finland, France, the Netherlands, Spain, and Sweden were reported to contain low amounts of *trans* fatty acids, while those from Iceland and Norway were relatively higher. However, in the United Kingdom, hard household margarines were shown to have relatively lower amounts than other varieties.

In earlier studies, generally, stick or hard margarines contain higher amounts of *trans* fatty acids compared with soft or tub margarines. Enig et al. [43] and Slover et al. [41] indicated that the average *trans* fatty acids in stick margarines from the United States is 24%, with minimum values of 16 to 17% and maximum values at 35 to 36%, while in soft or tub margarines, *trans* fatty acids averaged at 14 to 18% with maximum and minimum values of 10 and 30%, respectively. In addition, diet margarines contain about 15% *trans* fatty acids [43]. *Trans* fatty acid content in Canadian margarines was also reported on the high side [2,38,39,45,46].

The types of *trans* fatty acid isomers from margarines and spreads are tabulated in Table 10.3. Generally, *trans*-octadecenoic acid (C18:1*t*) is the predominant *trans* fatty acid isomer in these products.

10.5.2 Shortenings

Most shortenings have been reported to contain relatively high amounts of *trans* fatty acids (Table 10.4). Data from the TRANSFAIR study by Aro et al. [20] showed that frying, cooking, and baking fats obtained from these European countries contained *trans* fatty acids ranging from as low as 0.3 g to a staggering 50 g per 100 g food item. The low *trans* fatty acid content observed in the French and Italian frying fats are attributed to the relatively high amounts of saturated fatty acids (up to 98% of total fatty acids). In the United States, *trans* fatty acid content in shortenings was also relatively high, ranging from 10 to 42% [6,32,43,51–54].

Shortenings from Denmark, Germany, New Zealand, Portugal, Spain, and Sweden [19–21,26,49] were reported to contain somewhat lesser amounts of *trans* fatty acids compared with those from the United States and the other European countries.

10.5.3 Vegetable Oils

Vegetable oils are nonhydrogenated liquid oils and therefore do not contain significant amounts of *trans* fatty acids. Nevertheless, highly unsaturated vegetable oils (containing high amounts of C18:3 fatty acids) are often prone to rancidity caused by oxidative damage. It is known that these oils are often lightly hydrogenated in order to prolong their shelf lives. A comparison of *trans* fatty acid content of various vegetable oils from different countries is shown in Table 10.5. In general, vegetable oils sampled from these countries contain only minor quantities of *trans* fatty acids. Olive oil was reported to contain practically zero *trans* fatty acid content [20]. Aro et al. [20] also observed very low amounts of *trans* fatty acid content (below 0.9%) in refined soybean oil, sunflower oil, corn oil, rapeseed oil, and peanut oil sampled from 14 European countries, with the exception of Icelandic and Portuguese corn oil showing slightly higher amounts of *trans* fatty acids at 1.9 and 1.5%, respectively. The predominant type of *trans* fatty acid in these vegetable oils is the C18:2 *trans* isomers. Enig et al. [43] also reported a similar observation in a sample of partially hydrogenated soybean oil. Therefore, lightly hydrogenated oils appear

TABLE 10.3
***trans* Fatty Acid Composition of Margarines from Different Countries**

Country	Food Item	Percentage *trans* Fatty Acids (%)								Reference
		C16:1*t*	C18:1*t*	C18:2*t,c*	C18:2*c,t*	C18:2*t,t*	C18:2*t*	C20:1*t*	C22:1*t*	
Denmark	Margarines, 1999, <10% C18:2 (*n* = 4)	—	1.0 (0.0–4.0)	—	—	—	—	—	—	Leth et al. [19]
	Margarines, 1999, 10–20% C18:2 (*n* = 16)	—	0.7 (0.0–5.8)	—	—	—	—	—	—	
	Margarines, 1999, 20–40% C18:2 (*n* = 10)	—	1.0 (0.0–9.7)	—	—	—	—	—	—	
Iceland	Hard margarine, soybean oil	2.03	23.33	—	—	—	1.03	1.54	0.14	Aro et al. [20]
	Hard margarine, marine oil	2.15	4.40	—	—	—	0.39	3.51	3.70	
Norway	Margarines, vegetable oil	0.10	14.60	—	—	—	0.47	0.19	0.01	
	Margarines, vegetable/marine oils	4.83	11.60	—	—	—	0.44	4.37	2.83	
New Zealand	Margarines (*n* = 7)	—	14.6 (10.9–17.2)	1.2 (0.2–2.0)	—	0.2 (0.0–0.4)	—	—	—	Lake et al. [25]
	Table spreads (*n* = 5)	—	13.8 (12.5–14.7)	1.1 (0.4–2.5)	—	0.1 (0.0–0.5)	—	—	—	
	Margarine/butter blends (*n* = 2)	—	7.1 (3.5–10.6)	1.0 (1.0–1.1)	—	—	—	—	—	

(*continued*)

TABLE 10.3 (Continued)
***trans* Fatty Acid Composition of Margarines from Different Countries**

Country	Food Item	Percentage *trans* Fatty Acids (%)								Reference
		C16:1*t*	C18:1*t*	C18:2*t,c*	C18:2*c,t*	C18:2*t,t*	C18:2*t*	C20:1*t*	C22:1*t*	
France	Tub margarines, <5% trans (n = 5)	—	0.31 (0.0–1.53)	—	—	—	0.14 (0.06–0.27)	—	—	Bayard and Wolff [28]
	Tub margarines, >5% trans (n = 7)	—	13.53 (7.23–17.62)	—	—	—	0.59 (0.15–1.40)	—	—	
Germany	Margarines, vegetable oil (n = 10)	—	3.7 (0.46–6.1)	0.18 (0.1–0.4)	0.17 (0.1–0.4)	—	—	—	—	Pfalzgraf and Steinhart [31]
	Margarines, sunflower oil (n = 4)	—	15.2 (7.9–20.1)	0.43 (0.3–0.7)	0.5 (0.3–0.8)	—	—	—	—	
	Margarines, reduced fat (n = 4)	—	2.0 (0.1–2.9)	0.15 (0–0.3)	0.2 (0.1–0.3)	—	—	—	—	
	Margarines, diet/reformatory (n = 6)	—	0.58 (0–1.7)	0.33 (0.1–0.7)	0.35 (0.1–0.8)	—	—	—	—	
Canada	Stick margarines, soybean oil (n = 3)	—	31.13 (23.8–40.8)	6.27 (5.0–7.6)	—	1.23 (0.3–2.8)	—	—	—	Ratnayake et al. [39]
	Stick margarines, vegetable oil (n = 5)	—	30.16 (23.6–30.2)	2.4 (2.1–3.1)	—	0.5 (0–1.2)	—	—	—	
	Stick margarines, palm oil (n = 7)	—	32.9 (30.4–40.6)	2.33 (1.2–3.2)	—	0.44 (0.2–0.8)	—	—	—	
	Stick margarines, corn oil (n = 1)	—	26.4	1.7	—	0.2	—	—	—	
	Stick margarines, vegetable/animal fat (n = 1)	—	37.2	3.1	—	1.4	—	—	—	

	Stick margarines, unspecified (n = 2)	—	20.6 (17.8–23.4)	3.4	—	0.6	—	—	—	
	Tub margarines, soybean oil (n = 13)	—	16.36 (11.6–21.4)	1.25 (0.2–2.7)	—	0.2 (0–0.4)	—	—	—	
	Tub margarines, vegetable oil (n = 10)	—	23.53	1.45	—	—	—	—	—	
	Tub margarines, palm oil (n = 2)	—	17.95 (17.8–18.1)	1.25 (0.4–2.1)	—	0.2	—	—	—	
	Tub margarines, corn oil (n = 3)	—	15.93 (10.1–24.2)	0.97 (0–1.7)	—	—	—	—	—	
	Tub margarines, sunflower oil (n = 2)	—	6.65 (0–13.3)	0.45 (0–0.9)	—	—	—	—	—	
	Tub margarines, olive oil (n = 13)	—	19.2	0.2	—	—	—	—	—	
United States	Stick margarines, (n = 57)	—	22.23 (14.82–30.06)	0.61 (0–3.49)	0.69 (0.17–3.76)	0.16 (0–1.28)	1.46 (0.17–8.4)	—	—	Slover et al. [41]
	Tub margarines, (n = 26)	—	14.77 (10.74–18.44)	1.31 (0.10–4.78)	1.34 (0.19–5.31)	0.26 (0–1.47)	2.91 (0.35–11.56)	—	—	
	Overall margarines, (n = 83)	—	19.89 (10.74–30.06)	0.83 (0–4.78)	0.89 (0.17–5.31)	0.19 (0–1.47)	1.92 (0.35–11.56)	—	—	
United States	Stick margarines, (n = 24)	—	22.3 (15.9–31.0)	—	—	—	1.9 (0–5.2)	—	—	Enig et al. [43]
	Tub margarines, (n = 13)	—	12.7 (6.8–17.6)	—	—	—	1.7 (0–4.2)	—	—	
	Diet margarines, (n = 3)	—	12.0 (11.3–13.3)	—	—	—	3.3 (1.4–5.0)	—	—	

TABLE 10.4
***trans* Fatty Acid Content of Shortenings from Different Countries**

Country	Food Item	Percentage *trans* Fatty Acids (%)			Total *trans* Fatty Acids/100 g Food Item			Reference
		Lowest	Highest	Average	Lowest	Highest	Average	
Denmark	Shortenings, <10% C18:2 (*n* = 24)	2.4	11.2	6.8	—	—	—	Leth et al. [19]
	Shortenings, 10–20% C18:2 (*n* = 15)	1.0	10.0	5.5	—	—	—	
Belgium	Frying fats	—	—	35.34	—	—	35.34	Aro et al. [20]
Denmark	Frying fats	—	—	20.16	—	—	20.16	
Finland	Frying fats	—	—	15.92	—	—	15.84	
France	Frying fats	—	—	0.51	—	—	0.51	
Germany	Frying fats	—	—	7.75	—	—	7.71	
Greece	Frying fats	—	—	—	—	—	—	
Iceland	Frying fats	—	—	39.09	—	—	38.89	
Italy	Frying fats	—	—	0.30	—	—	0.30	
Netherlands	Frying fats	—	—	50.19	—	—	50.19	
Norway	Frying fats	—	—	28.65	—	—	28.51	
Portugal	Frying fats	—	—	5.58	—	—	5.55	
Spain	Frying fats	—	—	—	—	—	—	
Sweden	Frying fats	—	—	—	—	—	—	
United Kingdom	Frying fats	—	—	4.67	—	—	4.62	
Belgium	Cooking and baking fats	—	—	5.09	—	—	4.20	Aro et al. [20]
Denmark	Cooking and baking fats	—	—	—	—	—	—	
Finland	Cooking and baking fats	—	—	6.68	—	—	5.34	
France	Cooking and baking fats	—	—	0.55	—	—	0.42	
Germany	Cooking and baking fats	—	—	—	—	—	—	
Greece	Cooking and baking fats	—	—	5.75	—	—	5.75	

Iceland	Cooking and baking fats	—	—	15.97	—	—	12.46	
Italy	Cooking and baking fats	—	—	—	—	—	—	
Netherlands	Cooking and baking fats	—	—	0.40	—	—	0.39	
Norway	Cooking and baking fats	—	—	24.27	—	—	19.42	
Portugal	Cooking and baking fats	—	—	9.13	—	—	7.49	
Spain	Cooking and baking fats	—	—	1.12	—	—	1.11	
Sweden	Cooking and baking fats	—	—	0.36	—	—	0.28	
United Kingdom	Cooking and baking fats	—	—	10.90	—	—	10.85	
New Zealand	Pastry fat (n = 3)	5.42	7.02	6.33	—	—	—	Richardson et al. [21]
Austria	Shortenings, 1991/1992 (n = 86)	—	—	13.28	—	—	—	Henninger and Ulberth [24]
	Shortenings, 1995 (n = 15)	—	—	12.06	—	—	—	
Germany	Shortenings, cooking fats (n = 16)	0.04	32.51	9.79	—	—	—	Molkentin and Precht [26, 49]
Denmark	Shortenings	0	13.7	6.8	—	—	—	Ovesen et al. [27]
France	Shortenings (n = 3)	27.99	63.61	50.29	—	—	—	Bayard and Wolff [28]
Denmark	Shortenings, hardened, new	17	36	30	—	—	—	Ovesen and Leth [30]
	Shortenings, hardened, used	33	41	37	—	—	—	
	Shortenings, nonhardened, new	0	1	0	—	—	—	
	Shortenings, nonhardened, used	0	3	2	—	—	—	
United States	Shortenings (n = 12)	11.17	34.05	19.64	10.68	32.55	18.73	USDA [32]
United States	Vegetable shortenings	—	—	—	—	—	13.4	Litin and Sacks [35]
Germany	Shortenings, frying and baking fats	0.1	31.8	12.72	—	—	—	Pfalzgraf et al. [37]
Canada	Shortenings, unhydrogenated (n = 3)	2.0	3.2	2.5	—	—	—	Ratnayake et al. [50]
	Shortenings, hydrogenated (n = 3)	17.4	20.2	18.4	—	—	—	

(*continued*)

TABLE 10.4 (Continued)
***trans* Fatty Acid Content of Shortenings from Different Countries**

		Percentage *trans* Fatty Acids (%)			Total *trans* Fatty Acids/100 g Food Item			
Country	**Food Item**	**Lowest**	**Highest**	**Average**	**Lowest**	**Highest**	**Average**	**Reference**
United States	Shortenings, vegetable, 1960	—	—	26	—	—	—	Hunter and Applewhite [51]
	Shortenings, vegetable, 1970	—	—	19	—	—	—	
	Shortenings, vegetable, 1984	—	—	17	—	—	—	
	Shortenings, animal fats, 1960–1984	—	—	10	—	—	—	
United States	Shortenings, commercial, fresh (n = 9)	40.4	42.4	41.5	—	—	—	Smith et al. [52]
	Shortenings, commercial, used (n = 56)	12.8	41.4	30.7	—	—	—	
United States	Shortenings vegetable, (n = 7)	13.0	37.3	25.3	—	—	—	Enig et al. [43]
United States	Shortenings, vegetable oils (n = 6)	—	—	—	8.0	23.9	14.5	Lanza and Slover [53]
	Shortenings, meat fats + vegetable oils (n = 3)	—	—	—	2.8	6.6	4.3	
United States	Vegetable shortenings (n = 4)	—	—	10.7	—	—	—	Slover and Lanza [54]
United States	Vegetable shortenings (n = 1)	16.6	29.2	22.5	—	—	—	Scholfield et al. [6]

TABLE 10.5
***trans* Fatty A cid Content in Vegetable Oils from Different Countries**

Country	Food Item	Percentage *trans* Fatty Acids (%)			Reference
		Lowest	Highest	Average	
Unspecified	Sunflower oil, chemically refined	0.49	1.03	0.76	Tasan and Demirci [55]
	Sunflower oil, physically refined	2.31	2.81	2.56	
Hungary	Deodorized canola oil	—	—	6.10	Kemény et al. [56]
France	Olive oil, refined	—	—	0.00	Aro et al. [20]
	Sunflower oil, refined	—	—	0.22	
	Corn oil, refined	—	—	0.15	
	Rapeseed oil, refined	—	—	0.39	
	Peanut oil, refined	—	—	0.04	
	Vegetable oils, mixed	—	—	0.39	
Greece	Olive oil, refined	—	—	0.02	
Italy	Olive oil, extra virgin	—	—	0.00	
Spain	Olive oil, refined	—	—	0.11	
	Sunflower oil, refined	—	—	0.27	
	Reutilized vegetable oils, olive oil, beef	—	—	0.30	
	Reutilized vegetable oils, olive oil, fish	—	—	0.25	
	Reutilized vegetable oils, sunflower oil, beef	—	—	0.61	
	Reutilized vegetable oils, sunflower, fish	—	—	0.55	
	Reutilized vegetable oils, sunflower, discarded	—	—	0.89	
Iceland	Soybean oil, refined	—	—	0.40	
	Sunflower oil, refined	—	—	0.30	
	Corn oil, refined	—	—	1.91	
Netherlands	Soybean oil, refined	—	—	0.63	
	Sunflower oil, refined	—	—	0.45	
Portugal	Soybean oil, refined	—	—	0.86	
	Sunflower oil, refined	—	—	0.13	
	Corn oil, refined	—	—	1.55	
	Peanut oil, refined	—	—	0.63	
	Vegetable oil, mixed	—	—	0.25	
Germany	Sunflower oil, refined	—	—	0.89	
	Corn oil, refined	—	—	0.11	
	Coconut oil, refined	—	—	Trace	
	Vegetable oil, mixed	—	—	1.19	
Finland	Vegetable oil, mixed (4 brands)	—	—	0.42	

(*continued*)

TABLE 10.5 (Continued)
***trans* Fatty Acid Content in Vegetable Oils from Different Countries**

Country	Food Item	Percentage *trans* Fatty Acids (%)			Reference
		Lowest	Highest	Average	
Sweden	Vegetable oil, mixed (4 brands)	—	—	0.33	
Spain	Olive oil (*n* = 30)	—	—	0.1	Fernandez San Juan [23]
United States	Canola oil (*n* = 2)	0.17	0.23	0.20	USDA [32]
	Sunflower oil (*n* = 1)	—	—	0.5	
	Olive oil (*n* = 1)	—	—	0.09	
New Zealand	Rapeseed oil	—	—	0.9	Ball et al. [33]
	Safflower oil	—	—	0.4	
	Soybean oil	—	—	0.6	
	Corn oil	—	—	1.1	
Spain	Olive oil, refined (*n* = 12)	—	—	0.5	Boatella et al. [34]
	Seeds oils (*n* = 12)	—	—	2.3	
United States	Vegetable oil	0.00	1.06	0.42	Litin and Sacks [35]
Germany	Vegetable oils (*n* = 6)	0.0	1.5	0.28	Pfalzgraf et al. [37]
	Almond oil (*n* = 1)	—	—	0.1	
	Peanut oil (*n* = 1)	—	—	0.5	
	Walnut oil (*n* = 1)	—	—	0.2	

to contain more C18:2 *trans* isomers and fewer C18:1 *trans* isomers than heavily hydrogenated oils. Most brands of salad and cooking oils are composed of nonhydrogenated oils and, thus, contain very little or no *trans* fatty acids. Therefore, these products do not play a major role in increasing *trans* fatty acid intake in humans.

Data on *trans* fatty acid content in reutilized vegetable oils are also shown in Table 10.5. The oils had been reutilized for deep-frying of beef or fish. C18:2 *trans* isomers were observed to slightly increase after repeated usage of the oils, with the highest proportion (0.85%) found in reutilized sunflower oil.

10.5.4 Processed Foods

Processed foods such as soup concentrates, sauces, salad dressings, and mayonnaise are commonly found in supermarkets. *Trans* fatty acid contents in these products from different countries are tabulated in Table 10.6. Most soup concentrates and instant soups were found to contain moderate to high amounts, between 6 and 41% of *trans* fatty acids [57].

Generally, sauces appear to contain low amounts of *trans* fatty acids, except for sauce powders from Finland, tomato and curry sauces and gravy for roasts from

TABLE 10.6
***trans* Fatty Acid Content of Processed Foods from Different Countries**

Country	Food Item	Percentage *trans* Fatty Acids (%)	Total *trans* Fatty Acids/ 100 g Food Item	Reference
Finland	Soup, cubes (3 brands)	19.13	4.02	Aro et al. [57]
	Soup, dry (2 brands)	29.60	0.15	
Germany	Soup, dry with vegetables (n = 9)	24.71	3.78	
	Soup, dry with chicken (n = 3)	6.57	0.41	
	Soup, dry, instant (n = 4)	28.43	5.12	
Greece	Soup, cube	7.61	1.86	
Iceland	Soup, vegetable	41.25	4.70	
Italy	Soup, cubes (3 brands)	18.90	3.52	
	Soup, instant (7 brands)	30.13	7.98	
Spain	Soups, dry (n = 42)	15.4	—	Fernandez San Juan [23]
United States	Soup, cubes, beef bouillon (n = 3)	19.54	1.25	USDA [32]
	Soup, cubes, chicken bouillon (n = 3)	20.33	1.41	
Germany	Soup, clear	28.3	—	Pfalzgraf et al. [37]
	Soup, beef bouillon	9.9	—	
	Onion cream soup	34.9	—	
Finland	Sauce powders	16.02	2.42	Aro et al. [57]
France	Sauce, tomato with meat	3.85	0.13	
	Sauce, tomato with meat	2.24	0.10	
	Sauce, béarnaise	0.20	—	
	Sauce, tartare	0.48	0.30	
Germany	Sauce, tomato (n = 5)	15.27	1.69	
	Gravy for roasts	16.96	4.41	
Italy	Sauce, bechamelle (2 brands)	4.20	0.41	
	Sauce, ragout (3 brands)	0.43	0.03	
Sweden	Sauce, béarnaise	38.63	9.46	
United Kingdom	Sauce, mixes (20 brands)	18.69	4.02	
Germany	Sauce, curry	25.4	—	Pfalzgraf et al. [37]
	Sauce, instant, tomato	2.9	—	
	Mayonnaise	0.4	—	
United States	Salad dressing, blue cheese (n = 1)	5.47	1.21	Ali et al. [22]
United States	Salad dressing, French (n = 2)	0.64	0.24	USDA [32]

(*continued*)

TABLE 10.6 (Continued)
***trans* Fatty Acid Content of Processed Foods from Different Countries**

Country	Food Item	Percentage *trans* Fatty Acids (%)	Total *trans* Fatty Acids/ 100 g Food Item	Reference
	Salad dressing, Italian (n = 2)	0.94	0.40	
	Salad dressing, Italian, low calorie (n = 1)	0.91	0.19	
	Salad dressing, ranch (n = 1)	8.95	3.71	
	Salad dressing, ranch, low calorie (n = 2)	13.17	1.57	
	Mayonnaise (n = 2)	2.38	1.82	
United States	Mayonnaise, reduced calorie	—	0.21	Litin and Sacks [35]
United States	Salad dressings (n = 7)	n.d.	—	Enig et al. [43]
	Mayonnaise (n = 1)	4.5	—	
	Mayonnaise (n = 3)	n.d.	—	
	Burger sauce (n = 1)	4.6	—	
United States	Salad dressings (n = 2)	0.24	—	Slover et al. [58]
	Salad dressings (n = 1)	n.d.	—	
	Mayonnaise (n = 1)	0.34	—	
	Tartar sauce (n = 1)	0.37	—	
	Burger sauce (n = 2)	0.14	—	

Germany, béarnaise from Sweden, and sauce mixes from the United Kingdom with *trans* fatty acid contents exceeding 10%.

10.5.5 Fast Foods and Snack Items

Fast foods and snack items can contribute significant amounts of *trans* fatty acids in the diet. Table 10.7 summarizes the *trans* fatty acid contents of fast food items. french fried potatoes sampled from fast-food burger chains or snack bars from Finland, Germany, Iceland, the Netherlands, Norway, Spain, and Sweden contained relatively high amounts of *trans* fatty acids, ranging from 12 to 35%. In Belgium, Canada, Greece, New Zealand, Portugal, the United Kingdom, and the United States, *trans* fatty acids in french fries were observed to be lower [25,43,50,57,58].

Other fast-food items, such as hamburgers, fried fish, milkshakes, sandwiches, and muffins, are also sources of *trans* fatty acids. Ratnayake et al. [50] reported high amounts (26%) of *trans* fatty acids in Canadian hamburger buns made with hydrogenated fat. In the United States, fried chicken, pastries, pies, and turnovers were also shown to have high levels of *trans* fatty acids [40,43,58].

Snack items containing *trans* fatty acids are shown in Table 10.8. Frozen, prefried, and home-made french fries contained variable amounts of *trans* fatty acids,

TABLE 10.7
***trans* Fatty Acid Content of Fast Foods from Different Countries**

Country	Food Item	Percentage *trans* Fatty Acids (%)			Total *trans* Fatty Acids/100 g Food Item			Reference
		Lowest	Highest	Average	Lowest	Highest	Average	
Belgium	French fries (from 3 snack bars)	—	—	6.96	—	—	1.16	Aro et al. [57]
Finland	French fries (from 4 hamburger rest.)	—	—	34.84	—	—	4.56	
Germany	French fries (from 2 fast food rest.)	—	—	25.64	—	—	3.67	
Greece	French fries (from fast food stores)	—	—	0.45	—	—	0.05	
Iceland	French fries, deep fried	—	—	16.07	—	—	2.33	
Netherlands	French fries (from 4 snack bars)	—	—	30.00	—	—	4.50	
Norway	French fries (from hamburger rest.)	—	—	11.71	—	—	1.58	
Portugal	French fries (from fast food rest.)	—	—	2.94	—	—	0.47	
Spain	French fries (from fast food rest.)	—	—	33.57	—	—	6.04	
Sweden	French fries (from 2 fast food rest.)	—	—	28.95	—	—	4.43	
United Kingdom	French fries (from 12 burger rest.)	—	—	4.45	—	—	0.49	
Spain	French fries (n = 15)	—	—	20.9	—	—	20.4	Fernandez San Juan [23]
New Zealand	French fries (n = 2)	5.4	5.8	5.6	—	—	—	Lake et al. [25]
United States	French fries (n = 7)	7.3	34.2	20.6	1.0	5.2	3.0	USDA [32]
United States	French fries (n = 2)	—	—	—	2.12	3.02	2.60	Litin and Sacks [35]
Spain	French fries (n = 2)	22.5	32.8	27.7	—	—	—	Pfalzgraf et al. [37]
Canada	French fries (n = 1)	—	—	n.d.	—	—	—	Ratnayake et al. [50]
United States	French fries (n = 4)	3.2	25.8	—	—	—	—	Enig et al. [40]
United States	French fries (n = 6)	6.3	34.1	—	—	—	—	Smith et al. [59]
United States	French fries (n = 3)	4.6	5.1	4.8	—	—	—	Enig et al. [43]
	French fries (n = 4)	6.2	37.4	17.8	—	—	—	

(continued)

TABLE 10.7 (Continued)
***trans* Fatty Acid Content of Fast Foods from Different Countries**

Country	Food Item	Percentage *trans* Fatty Acids (%)			Total *trans* Fatty Acids/100 g Food Item			Reference
		Lowest	Highest	Average	Lowest	Highest	Average	
United States	French fries ($n = 3$)	—	—	—	0.52	0.72	0.60	Lanza and Slover [53]
United States	French fries ($n = 3$)	3.87	4.13	4.02	—	—	—	Slover et al. [58]
Spain	Hamburgers, beef ($n = 40$)	—	—	3.8	—	—	—	Fernandez San Juan [23]
	Hamburgers, burger ($n = 50$)	—	—	4.1	—	—	—	
United States	Hamburgers ($n = 10$)	—	—	—	0.38	0.79	0.55	Lanza and Slover [53]
United States	Hamburgers ($n = 10$)	3.00	5.16	3.73	—	—	—	Slover et al. [58]
United States	Cheeseburgers ($n = 9$)	—	—	—	0.45	0.69	0.59	Lanza and Slover [53]
United States	Cheeseburgers ($n = 9$)	2.66	4.38	3.44	—	—	—	Slover et al. [58]
Canada	Hamburger buns, unhydrogenated fat ($n = 2$)	0.4	2.6	1.5	—	—	—	Ratnayake et al. [50]
	Hamburger buns, hydrogenated fat ($n = 1$)	—	—	26.3	—	—	—	
United States	Sandwiches, ham/cheese ($n = 1$)	—	—	—	—	—	0.19	Lanza and Slover [53]
United States	Sandwiches, ham/cheese ($n = 1$)	—	—	1.62	—	—	—	Slover et al. [58]
United States	Sandwiches, fish ($n = 4$)	—	—	—	0.22	0.43	0.36	Lanza and Slover [53]
United States	Sandwiches, fish ($n = 4$)	1.39	3.58	2.75	—	—	—	Slover et al. [58]
United States	Fried fish ($n = 4$)	22.9	33.5	—	—	—	—	Enig et al. [40]
United States	Fried fish ($n = 4$)	5.8	29.9	—	—	—	—	Smith et al. [59]
United States	Fried fish ($n = 3$)	0.0	15.2	—	—	—	—	Enig et al. [43]
United States	Fish platter ($n = 1$)	—	—	—	—	—	0.76	Lanza and Slover [53]
United States	Fish platter ($n = 1$)	—	—	4.59	—	—	—	Slover et al. [58]
United States	Beef platter ($n = 1$)	—	—	—	—	—	0.69	Lanza and Slover [53]
United States	Beef platter ($n = 1$)	—	—	3.54	—	—	—	Slover et al. [58]
United States	Fried chicken ($n = 1$)	—	—	27.8	—	—	—	Enig et al. [40]

United States	Fried chicken($n = 5$)	7.7	16.4	—	—	—	—	Smith et al. [59]
United States	Onion rings ($n = 1$)	—	—	—	—	—	1.26	Lanza and Slover [53]
United States	Onion rings ($n = 1$)	—	—	3.78	—	—	—	Slover et al. [58}
United States	Milkshakes ($n = 4$)	3.0	4.9	3.7	0.01	0.07	0.04	USDA [32]
United States	Milkshakes ($n = 9$)	—	—	—	0.05	0.10	0.07	Lanza and Slover [53]
United States	Milkshakes ($n = 9$)	1.76	3.10	2.20	—	—	—	Slover et al. [58]
United States	Pies and turnovers ($n = 4$)	—	—	—	0.68	2.02	1.38	Lanza and Slover [53]
United States	Pies and turnovers ($n = 4$)	6.33	16.63	11.71	—	—	—	Slover et al. [58]
United States	Cookies ($n = 1$)	—	—	—	—	—	0.49	Lanza and Slover [53]
United States	Cookies ($n = 1$)	—	—	3.17	—	—	—	Slover et al. [58]
United States	Egg muffin ($n = 1$)	—	—	—	—	—	0.08	Lanza and Slover [53]
United States	Egg muffin ($n = 1$)	—	—	0.87	—	—	—	Slover et al. [58]
United States	English muffin with butter ($n = 1$)	—	—	—	—	—	0.19	Lanza and Slover [53]
United States	English muffin with butter ($n = 1$)	—	—	2.27	—	—	—	Slover et al. [58]
United States	Scrambled eggs ($n = 1$)	—	—	—	—	—	0.21	Lanza and Slover [53]
United States	Scrambled eggs ($n = 1$)	—	—	1.16	—	—	—	Slover et al. [58]
United States	Hot cakes with butter ($n = 1$)	—	—	—	—	—	0.52	Lanza and Slover [53]
United States	Hot cakes with butter ($n = 1$)	—	—	8.02	—	—	—	Slover et al. [58]
United States	Sausage biscuit ($n = 1$)	—	—	—	—	—	0.04	Lanza and Slover [53]
United States	Sausage biscuit ($n = 1$)	—	—	0.27	—	—	—	Slover et al. [58]
United States	Apple-cheese pastry ($n = 1$)	—	—	12.0	—	—	—	Enig et al. [43]
	Cheese Danish ($n = 1$)	—	—	34.6	—	—	—	

TABLE 10.8
***trans* Fatty Acid Content of Snack Items from Different Countries**

Country	Food Item	Percentage *trans* Fatty Acids (%)			Total *trans* Fatty Acids/100 g Food Item	Reference
		Lowest	Highest	Average		
Belgium	French fries, home-made, animal fat	—	—	5.93	0.83	Aro et al. [57]
	French fries, home-made, vegetable fat	—	—	34.06	5.65	
Denmark	French fries, prefried	—	—	21.15	1.50	
Finland	French fries, prefried (7 brands)	—	—	18.30	0.73	
France	French fries, frozen	—	—	2.84	0.12	
	French fries, frozen	—	—	12.23	0.18	
	French fries, home-made, peanut oil	—	—	0.01	0.00	
	French fries, home-made, sunflower oil	—	—	0.42	0.04	
	French fries, home-made, vegetaline	—	—	0.49	0.05	
	French fries, home-made, lard	—	—	0.81	0.11	
Germany	French fries, prefried (2 brands)	—	—	14.17	0.71	
Greece	French fries, frozen	—	—	16.69	0.60	
Iceland	French fries, frozen	—	—	0.57	0.03	
Italy	French fries, frozen (3 brands)	—	—	16.24	0.60	
Netherlands	French fries, prefried (3 brands)	—	—	12.49	0.50	
Norway	French fries, frozen (2 brands)	—	—	0.28	0.01	
Portugal	French fries, home-made	—	—	0.61	0.07	
	French fries, prefried, frozen	—	—	41.52	1.83	
Sweden	French fries, frozen (3 brands)	—	—	24.83	1.07	
United Kingdom	Frozen (10 brands)	—	—	13.97	1.31	
Spain	Potato chips (n = 40)	—	—	0.6	0.21	Fernandez San Juan [23]

New Zealand	Potato chips (n = 3)	0.3	0.8	0.5	0.16	Lake et al. [25]
Denmark	Potato chips (n = 1)	—	—	0.0	—	Ovesen and Leth [30]
United States	Potato chips (n = 11)	0.0	29.7	5.4	1.84	USDA [32]
Germany	Potato chips, unhydrogenated (n = 1)	—	—	0.7	—	Pfalzgraf et al. [37]
	Potato chips, hydrogenated (n = 2)	17.2	19.9	18.6	—	
Canada	Potato chips, unhydrogenated (n = 2)	0.4	2.0	1.2	0.43	Ratnayake et al. [50]
	Potato chips, hydrogenated (n = 3)	29.7	39.7	33.9	12.25	
United States	Potato chips (n = 9)	0.0	1.6	1.3	0.52	Smith et al. [59]
United States	Potato chips (n = 6)	0.0	27.4	4.6	—	Enig et al. [43]
Belgium	Kroepoek	—	—	0.47	0.13	Aro et al. [57]
France	Crisps	—	—	0.62	0.27	
Germany	Crisps (2 brands)	—	—	14.45	4.41	
	Sticks, pretzels (5 brands)	—	—	0.50	0.03	
	Chips, flips (4 brands)	—	—	4.38	1.03	
Iceland	Crisps, peanut oil	—	—	0.21	0.07	
	Potato crisps	—	—	0.67	0.12	
Italy	Crisps (4 brands)	—	—	0.48	0.16	
Netherlands	Crisps (5 brands)	—	—	11.89	3.92	
Portugal	Crisps (3 brands)	—	—	1.00	0.34	
Spain	Crisps	—	—	1.05	0.38	
Sweden	Crisps (2 brands)	—	—	0.48	0.17	
United Kingdom	Crisps (n = 21)	—	—	0.93	0.30	
Germany	Corn snack (n = 1)	—	—	19.4	—	Pfalzgraf et al. [37]
Canada	Corn snacks, unhydrogenated (n = 2)	0.9	1.6	1.3	0.33	Ratnayake et al. [50]
	Corn snacks, hydrogenated (n = 3)	29.9	33.9	32.3	9.56	

(continued)

TABLE 10.8 (Continued)
***trans* Fatty Acid Content of Snack Items from Different Countries**

Country	Food Item	Percentage *trans* Fatty Acids (%)			Total *trans* Fatty Acids/100 g Food Item	Reference
		Lowest	Highest	Average		
United States	Corn snacks ($n = 3$)	0.8	20.6	—	—	Enig et al. [40]
United States	Corn snacks ($n = 7$)	0.8	22.0	—	—	Smith et al. [59]
United States	Corn snacks ($n = 5$)	0.8	22.0	—	—	Enig et al. [43]
United States	Tortilla chips ($n = 1$)	—	—	17.5	4.48	USDA [32]
United States	Cheese snacks ($n = 2$)	10.0	25.8	17.9	5.68	USDA [32]
United States	Cheese snacks ($n = 4$)	23.5	53.9	—	—	Enig et al. [40]
United States	Cheese snacks ($n = 6$)	1.0	28.1	—	—	Smith et al. [59]
United States	Cheese snacks ($n = 1$)	—	—	33.4	—	Enig et al. [43]
Belgium	Popcorn	—	—	0.04	0.02	Aro et al. [57]
Denmark	Popcorn, oil and salt added	—	—	27.62	8.15	
Finland	Popcorn, microwave (2 brands)	—	—	33.15	8.95	
Iceland	Popcorn, microwave	—	—	34.82	9.05	
	Popcorn, ready-made	—	—	0.06	0.02	
Sweden	Popcorn, ready-made (2 brands)	—	—	0.56	0.10	
United States	Popcorn ($n = 5$)	26.9	35.2	31.3	7.54	USDA [32]
Germany	Popcorn ($n = 1$)	—	—	0.9	—	Pfalzgraf et al. [37]
New Zealand	Crackers ($n = 1$)	—	—	0.71	0.17	Richardson et al. [21]
New Zealand	Crackers ($n = 5$)	1.2	3.9	2.0	0.33	Lake et al. [25]
United States	Crackers ($n = 3$)	12.8	42.9	23.3	—	Ali et al. [22]
United States	Crackers ($n = 9$)	11.6	39.9	32.6	5.79	USDA [32]

Germany	Crackers (n = 2)	Trace	5.6	2.8	—	Pfalzgraf et al. [37]
Canada	Crackers, hydrogenated (n = 7)	13.8	35.4	25.8	5.44	Ratnayake et al. [50]
Germany	Crackers (6 brands)	—	—	2.15	0.51	Aro et al. [57]
Sweden	Peanut rings	—	—	0.16	0.04	Aro et al. [57]
	Noodles (2 brands)	—	—	13.85	5.61	
		—	—	0.73	0.23	
		—	—			
		—	—			
		—	—			
Finland	Snacks (12 brands)	—	—			
Spain	Snack	—	—	0.82	0.22	
	Snack	—	—	1.05	0.38	
United Kingdom	Snacks (n = 23)	—	—	0.72	0.18	

between 0.01 and 41.5%, indicating that different types of fats had been used for frying. On the other hand, in general, potato chips, crisps, popcorn, and crackers appeared to contain somewhat lower amounts of *trans* fatty acids.

10.5.6 Bakery and Confectionery Foods

Bakery and confectionery products generally use margarines, butter, bakery fats, and shortenings as the main sources of fat. Since *trans* fatty acids are known to be present in these fats, it is not surprising that bakery and confectionery products, too, contain *trans* fatty acids. A summary of *trans* fatty acid content in bakery and confectionery foods is shown in Table 10.9. In the study by van Erp-baart et al. [60], more than 200 different types of cookies and pastries from European countries were analyzed for *trans* fatty acids. Most of these products contain 10% and below of *trans* fatty acids. In general, Italian and Spanish cookies and biscuits contain the lowest amounts of *trans* fatty acids, while Icelandic, Norwegian, and American ones had the highest values [32,43,60].

Trans fatty acid contents in pastries from Germany contained *trans* fatty acids below 1%, while some pastries from Iceland, Norway, Sweden, the Netherlands, the United States, and Canada contained more than 20% of *trans* fatty acids [22,32,50,60]. French croissants had the lowest *trans* fatty acid content at about 3%, compared with Greek croissants at 15%. However, donuts from Greece, Spain, and the United Kingdom were reported to be low in *trans* fatty acids, while those from Iceland contained about 32% *trans* fatty acids [60]. In general, breads contained <1 to 8%, with the exception of Spanish and Norwegian breads having a relatively higher proportion of *trans* fatty acids at 17 and 14%, respectively.

In the United States, certain products such as breading mixes, cereals, candies, frostings, pizza crusts, and pretzels contained significant amounts of *trans* fatty acids ranging from 10 to 22% [22,32,43].

10.5.7 Dairy Products

Trans fatty acid contents of milk, butter, cheese, yogurt, and ice cream from various countries are compared in Tables 10.10 to 10.12. Generally, the amount of *trans* fatty acids in milk and butter from most European samples is below 5%. Australian and New Zealand butters have slightly higher proportions of *trans* fatty acids, at 6% and above [16,21,25]. *Trans* fatty acids in milk and butter are also affected by seasonal changes. Aro et al. [63] and Wolff et al. [65] reported that milk and butter obtained during summer contained more *trans* fatty acids than those obtained during winter.

A comparison of *trans* fatty acid content in cheeses revealed that most European and American cheeses contain 2 to 5% *trans* fatty acids. However, Swedish margarine cheese and Finnish modified-fat cheese were reported to contain about 12% *trans* fatty acids.

Trans fatty acid contents of ice cream samples made from dairy and vegetable fat are also compared in Table 10.12. Generally, ice creams made from dairy fats have a lower variation (2.6 to 6%) of *trans* fatty acids than those made from vegetable fats (0.2 to 31%). The high amounts of *trans* fatty acids as observed in some of these ice creams are predominantly from partially hydrogenated vegetable oils.

TABLE 10.9
***trans* Fatty Acid Content of Bakery and Confectionery Foods from Different Countries**

Country	Food Item	Percentage *trans* Fatty Acids (%)			Total *trans* Fatty Acids/100 g Food Item	Reference
		Lowest	Highest	Average		
Belgium	Cookie with chocolate	—	—	0.12	0.04	van Erp-baart et al. [60]
	Spiced cookies	—	—	6.93	1.31	
Denmark	Klejner (fried cookies)	—	—	2.66	0.51	
	Biscuit, sweet	—	—	9.75	1.82	
Finland	Filled biscuits	—	—	6.72	1.55	
	Plain sweet biscuits	—	—	12.27	2.38	
France	Biscuits, chocolate	—	—	0.42	0.12	
	Biscuits chocolate	—	—	7.36	1.73	
Germany	Butter cookies, chocolate, cream	—	—	1.62	0.28	
	Cookies (4 brands)	—	—	3.24	0.99	
Greece	White cookies, margarine	—	—	1.32	0.31	
	Cookies with eggs	—	—	8.28	1.95	
Iceland	Cookies, pooled	—	—	5.69	1.48	
	Biscuit, digestive	—	—	27.96	5.06	
Italy	Cookies, biscuits	—	—	0.49	0.08	
Netherlands	Cookies with butter	—	—	4.18	1.23	
	Dutch shortbread	—	—	15.02	4.51	
Norway	Maryland cookies	—	—	1.69	0.42	
	Biscuits, wheat marie	—	—	26.59	2.50	
	Biscuits, wholemeal	—	—	25.55	—	

(*continued*)

TABLE 10.9 (Continued)
***trans* Fatty Acid Content of Bakery and Confectionery Foods from Different Countries**

Country	Food Item	Percentage *trans* Fatty Acids (%)			Total *trans* Fatty Acids/100 g Food Item	Reference
		Lowest	Highest	Average		
Portugal	Biscuits, thin	—	—	0.72	0.13	
	Cookie, deer's tongue	—	—	6.78	1.80	
Spain	Biscuits, sandwich	—	—	0.13	0.03	
	Cookies	—	—	1.45	0.30	
Sweden	Cookies	—	—	12.63	3.22	
	Digestive	—	—	19.15	4.02	
United Kingdom	Biscuits, cream wafer	—	—	0.74	0.18	
	Biscuits, cream-filled	—	—	13.23	3.04	
Finland	Biscuits (2 brands)	—	—	11.25	1.97	Aro et al. [57]
Italy	Biscuits (6 brands)	—	—	7.32	1.54	
New Zealand	Cookies, chocolate-coated	—	—	1.7	—	Richardson et al. [21]
	Cookies, plain sweet	—	—	4.5	—	
United States	Cookies, peanut butter	—	—	1.7	—	Ali et al. [22]
	Biscotti	—	—	3.1	—	
Spain	Cookies (n = 42)	—	—	1.1	—	Fernandez San Juan [23]
New Zealand	Cookies and crackers (n = 5)	1.1	3.5	2.0	—	Lake et al. [25]
United States	Biscuits (n = 2)	22.9	36.5	29.7	—	USDA [32]
	Cookies (n = 7)	18.7	37.7	29.4	—	
Denmark	Cookies	3.0	17.0	8.0	—	Ovesen and Leth [30]
Canada	Cookies, unhydrogenated fat (n = 8)	2.6	6.4	4.3	—	Ratnayake et al. [50]
	Cookies, hydrogenated fat (n = 1)	7.6	38.7	22.1	—	
Korea	Cookies and cakes	—	—	1.3	—	Won and Ahn [61]

United States	Cookies (n = 25/21)	0/2.5	37.4	18.7/22.3	—	Enig et al. [43]
Greece	Cream crackers	—	—	9.10	—	van Erp-baart et al. [60]
Sweden	Cream crackers	—	—	29.08	—	
Norway	Cream crackers	—	—	24.66	—	
Finland	Cream cracker/biscuit	—	—	11.25	—	
Belgium	Cake (1 brand)	—	—	5.37	1.88	van Erp-baart et al. [60]
	Fruit pie	—	—	13.69	0.84	
Denmark	Cream pastry	—	—	5.38	1.10	
	Pastry, all types	—	—	9.67	2.32	
Finland	Fatty cake	—	—	3.16	0.63	
	Swiss roll	—	—	8.46	0.99	
France	Pastry puff, baked	—	—	16.77	4.36	
	Shortcrust pastry, baked	—	—	15.31	—	
Germany	Cake with almonds	—	—	0.00	0.00	
	Cake, mixture	—	—	15.11	3.70	
	Cake, marble	—	—	14.11	—	
Greece	Pastries	—	—	1.69	0.19	
	Cake, coffee	—	—	3.66	0.95	
Iceland	Danish pastry	—	—	5.40	1.27	
	Cruller, deep-fried	—	—	33.32	5.36	
	Sponge cake	—	—	12.03	—	
Italy	Plumcake	—	—	1.25	0.26	
	Krapfen	—	—	15.20	2.86	
Netherlands	Cake, almond	—		2.41	0.48	
	Cake, apple	—		22.54	1.69	
Norway	Danish pastry	—	—	16.63	3.49	
	Chrismas cake	—	—	23.25	1.79	
Portugal	Marble cake, chocolate	—	—	0.21	0.04	
	Rice cake	—	—	5.36	0.79	

(*continued*)

TABLE 10.9 (Continued)
***trans* Fatty Acid Content of Bakery and Confectionery Foods from Different Countries**

Country	Food Item	Percentage *trans* Fatty Acids (%)			Total *trans* Fatty Acids/100 g Food Item	Reference
		Lowest	Highest	Average		
Spain	Magdalenas	—	—	1.03	0.23	
	Cakes	—	—	15.35	2.86	
Sweden	Sponge cake	—	—	0.39	0.06	
	Pie, fruit	—	—	24.33	4.01	
United Kingdom	Teacakes	—	—	2.03	0.10	
	Scones	—	—	16.45	2.20	
	Shortcrust pastry, frozen	—	—	3.87	—	
New Zealand	Pastry	—	—	5.6	—	Richardson et al. [21]
	Pie, meat	—	—	3.9	—	
United States	Pie, chicken	—	—	25.5	—	Ali et al. [22]
Spain	Cakes (*n* = 50)	—	—	3.1	—	Fernandez San Juan [23]
New Zealand	Pastry (*n* = 5)	3.6	7.5	6.2	—	Lake et al. [25]
	Cakes (*n* = 5)	2.6	8.4	5.3	—	
United States	Cakes (*n* = 6)	3.9	28.3	18.5	—	USDA [32]
	Snack cakes (*n* = 1)	—	—	21.6	—	
	Danish pastry (*n* = 1)	—	—	8.5	—	
	Muffins (*n* = 1)	—	—	31.9	—	
	Rolls (*n* = 4)	2.2	25.6	10.0	—	
	Sweet rolls (*n* = 1)	—	—	14.3	—	
	Taco shells (*n* = 1)	—	—	31.5	—	
	Tortillas (*n* = 1)	—	—	16.6	—	
Denmark	Pastry	8.0	14.0	10.0	—	Ovesen and Leth [30]

Canada	Cakes, unhydrogenated fat ($n = 2$)	2.4	3.0	2.7	—	Ratnayake et al. [50]
	Cakes, hydrogenated fat ($n = 6$)	10.1	25.7	16.3	—	
	Muffins, unhydrogenated fat ($n = 2$)	0.5	1.3	0.9	—	
	Muffins, hydrogenated fat ($n = 2$)	16.5	24.2	20.4	—	
	Pie crusts, unhydrogenated fat ($n = 2$)	0.9	1.2	1.1	—	
United States	Cakes ($n = 4$)	0.1	24.0	13.9	—	Enig et al. [43]
	Pastry and pastry crusts ($n = 18/16$)	0/0.6	34.6	10.3/11.5	—	
Belgium	Croissants	—	—	7.18	2.26	van Erp-baart et al. [60]
Denmark	Croissants	—	—	8.82	2.69	
Finland	Croissants			7.59	1.94	
France	Croissants	—	—	11.60	2.55	
	Croissants	—	—	3.03	0.68	
Greece	Croissants	—	—	14.55	3.42	
Iceland	Croissants	—	—	8.32	1.91	
Italy	Croissants	—	—	13.17	3.56	
Netherlands	Croissants	—	—	5.54	1.33	
Portugal	Croissants	—	—	8.64	2.03	
Spain	Croissants	—	—	8.12	1.61	
United Kingdom	Croissants	—	—	6.61	1.72	
Denmark	Croissants	8.0	14.0	11.0	—	Ovesen and Leth [30]
Denmark	Donuts	—	—	4.76	0.98	van Erp-baart et al. [60]
Finland	Donuts	—	—	14.73	2.73	
Greece	Donuts	—	—	1.83	0.22	
	Donuts	—	—	2.94	0.59	
Iceland	Donuts	—	—	31.76	5.53	
Portugal	Donuts	—	—	6.00	1.20	
Spain	Donuts	—	—	1.12	0.26	
Sweden	Donuts	—	—	17.27	1.57	
United Kingdom	Donuts	—	—	1.18	0.25	

(continued)

TABLE 10.9 (Continued)
***trans* Fatty Acid Content of Bakery and Confectionery Foods from Different Countries**

Country	Food Item	Percentage *trans* Fatty Acids (%)			Total *trans* Fatty Acids/100 g Food Item	Reference
		Lowest	Highest	Average		
Belgium	Bread, brown	—	—	2.13	0.06	van Erp-baart et al. [60]
Denmark	Bread, white	—	—	3.03	0.11	
France	Bread, wheat	—	—	0.65	0.01	
	Bread, wholemeal	—	—	6.92	0.18	
Germany	Bread, whole grain with sunflower oil	—	—	0.05	0.00	
	Bread, whole grain (4 brands)	—	—	0.25	0.00	
Greece	Pita bread for souvlaki	—	—	0.66	0.06	
	Bread, white for sandwich	—	—	1.50	0.06	
Iceland	Bread, wholemeal	—	—	6.02	0.14	
Italy	Bread, wheat	—	—	0.23	0.00	
	Bread, wheat with oil	—	—	0.49	0.03	
Netherlands	Bread, wheat	—	—	2.21	0.04	
Norway	Bread, kneipp	—	—	14.41	0.43	
	Bread, fiberkneipp	—	—	7.96	0.22	
Spain	Bread, wheat	—	—	17.35	0.54	
Sweden	Bread, whole meal rye and wheat	—	—	1.89	0.03	
	Bread, white	—	—	2.74	0.09	
United Kingdom	Bread, white	—	—	0.63	0.01	
New Zealand	Bread, white	—	—	2.2	—	Richardson et al. [21]
United States	Bread (n = 6)	1.5	25.5	12.8	—	USDA [32]
Canada	Bread					Ratnayake et al. [50]
	Unhydrogenated fat (n = 6)	0.0	2.9	1.4	—	
	Hydrogenated fat (n = 2)	—	—	15.7	—	

United States	Breads and rolls (*n* = 10/9)	0/0.2	27.9	10.1/11.2	—	Enig et al. [43]
	Breading mixes, fried crusts (*n* = 8/5)	0/12.1	33.5	12.9/20.6	—	
Finland	Muesli (2 brands)	—	—	18.91	3.18	Aro et al. [57]
Sweden	Cereals with vegetable fat	—	—	23.67	4.73	
United States	Cereal with raisins	—	—	15.0	—	Ali et al. [22]
United States	Cereals, breakfast (*n* = 6)	4.2	40.3	20.7	—	USDA [32]
Canada	Cereals, unhydrogenated fat (*n* = 6)	0.0	1.6	0.5	—	Ratnayake et al. [50]
	Cereals, hydrogenated fat (*n* = 5)	9.2	33.7	16.5	—	
Spain	Assorted bakery products (*n* = 30)	—	—	9.4	—	Fernandez San Juan [23]
	Creams (*n* = 15)	—	—	1.8	—	
	Pizzas (*n* = 20)	—	—	3.1	—	
Denmark	Pizza	0.0	8.0	4.0	—	Ovesen and Leth [30]
Canada	Pizza crusts, unhydrogenated fat (*n* = 3)	0.0	1.7	0.6	—	Ratnayake et al. [50]
	Pizza crusts, hydrogenated fat (*n* = 3)	22.1	28.8	25.5	—	
United States	Pizza crusts and pretzels (*n* = 6/3)	0/14.4	31.4	10.4/20.2	—	Enig et al. [43]
United States	Candies (*n* = 7)	0.3	29.1	12.9	—	USDA [32]
	Frostings (*n* = 6)	19.7	24.7	21.7	—	
Germany	Bakery products (*n* = 13)	0.0	27.9	7.4	—	Pfalzgraf et al. [37]
	Sweets, nut/nougat creams (*n* = 5)	0.5	15.2	6.6	—	
	Sweets, other (*n* = 15)	0.2	15.7	3.2	—	
Canada	Candies/chocolates, unhydrogenated fat (*n* = 8)	—	—	0.0	—	Ratnayake et al. [50]
	Candies/chocolates, hydrogenated fat (*n* = 1)	—	—	11.1	—	
Germany	Nut-nougat crèmes (*n* = 12)	0.35	12.35	7.2	—	Laryea et al. [62]
United States	Candy and frostings (*n* = 9/7)	0/3.2	38.6	15.8/20.4	—	Enig et al. [43]
	Cream substitutes, cereals, puddings (*n* = 8/7)	0/0.4	36.1	10.7/12.2	—	
Spain	Assorted bakery products (*n* = 83)	—	—	1.6	—	Boatella et al. [34]

TABLE 10.10
***trans* Fatty Acid Content of Milk and Butter from Different Countries**

Country	Food Item	Percentage *trans* Fatty Acids (%) Average	Total *trans* Fatty Acids/ 100 g Food Item Average	Reference
Belgium	Milk	4.68	1.57	Aro et al. [63]
Denmark	Milk	4.07	0.16	
Finland	Milk	3.19	0.12	
France	Milk	5.09	0.08	
Germany	Milk	3.55	0.09	
Greece	Milk	3.90	0.11	
	Milk, goat	2.69	0.09	
	Milk, sheep	3.61	0.35	
Iceland	Milk, summer	5.24	0.21	
	Milk, winter	3.34	0.13	
Italy	Milk, summer	4.37	0.15	
	Milk, winter	3.94	0.16	
Netherlands	Milk, summer	4.41	0.15	
	Milk, winter	3.13	0.11	
Norway	Milk	3.72	0.15	
Portugal	Milk, summer	4.67	0.16	
	Milk, winter	4.07	0.12	
Spain	Milk	4.73	0.16	
Sweden	Milk	—	—	
United Kingdom	Milk	3.81	0.16	
Spain	Milk fat (n = 25)	3.4	—	Fernandez San Juan [23]
Germany	Milk fat (n = 100)	3.83	—	Precht and Molkentin [64]
Belgium	Milk fat	3.19	—	
Denmark	Milk fat (n = 4)	4.21	—	
Spain	Milk fat (n = 10)	4.04	—	
France	Milk fat (n = 10)	4.47	—	
Greece	Milk fat (n = 4)	4.01	—	
Italy	Milk fat (n = 12)	4.14	—	
Ireland	Milk fat (n = 22)	5.91	—	
Luxemburg	Milk fat	3.51	—	
Netherlands	Milk fat (n = 24)	4.09	—	
United Kingdom	Milk fat (n = 23)	4.78	—	
United States	Milk fat (U.S. composite sample, April)	2.94	—	USDA [32]
	Milk fat (U.S. composite sample, July)	3.39	—	

TABLE 10.10 (Continued)
***trans* Fatty Acid Content of Milk and Butter from Different Countries**

Country	Food Item	Percentage *trans* Fatty Acids (%) Average	Total *trans* Fatty Acids/ 100 g Food Item Average	Reference
	Milk fat (2 U.S. composite samples)	2.75	—	
Germany	Milk fat (n = 15)	3.64	—	Pfalzgraf et al. [37]
Spain	Milk fat (n = 34)	3.5	—	Boatella et al. [34]
United States	Milk fat (5 U.S. brands)	1.9	—	Smith et al. [17]
Belgium	Butter	5.43	4.51	Aro et al. [63]
Denmark	Butter	4.51	3.65	
Finland	Butter, summer	5.15	4.15	
	Butter, winter	4.01	3.25	
France	Butter	5.98	5.47	
Germany	Butter	4.04	3.23	
Greece	Butter	4.77	3.91	
Iceland	Butter	4.36	3.55	
Italy	Butter	4.18	3.49	
Netherlands	Butter	6.15	5.07	
Norway	Butter	4.84	3.90	
Portugal	Butter	—	—	
Spain	Butter	—	—	
Sweden	Butter	4.43	3.50	
United Kingdom	Butter	—	—	
New Zealand	Butter	6.72	—	Richardson et al. [21]
New Zealand	Butter (n = 5)	6.4	—	Lake et al. [25]
France	Butter, January (n = 12)	2.37	—	Wolff et al. [65]
	Butter, May/June (n = 12)	4.28	—	
United States	Butter	3.4	—	Michels and Sacks [29]
Austria	Butter	4.25	—	Henninger and Ulberth [24]
Germany	Butter (n = 5)	4.12	—	Pfalzgraf et al. [37]
Spain	Butter (n = 15)	5.3	—	Boatella et al. [34]
New Zealand	Butter, semisoft	3.2	—	Ball et al. [33]
	Butter, clarified	1.8	—	
United States	Butter (3 U.S. brands)	3.4	—	Enig et al. [43]
Australia	Butter (n = 116)	6.0	—	Parodi and Dunstan [16]

TABLE 10.11
trans Fatty Acid Content of Cheese and Yogurt from Different Countries

Country	Food Item	Percentage *trans* Fatty Acids (%)	Total *trans* Fatty Acids/ 100 g Food Item	Reference
Belgium	Cheese	5.00	1.08	Aro et al. [63]
	Cheese, sheep	5.73	1.66	
Denmark	Cheese	5.24	1.31	
Finland	Cheese	3.92	0.94	
	Cheese, unripened	4.35	1.20	
	Cheese, modified-fat	11.79	3.71	
	Cheese, modified-fat	1.15	0.32	
France	Cheese	3.85	0.85	
	Cheese, goat	3.57	0.86	
Germany	Cheese	3.91	0.94	
Greece	Cheese	4.65	1.23	
	Cheese, goat feta	5.24	1.28	
Iceland	Cheese	3.59	0.90	
Italy	Cheese	4.45	1.16	
	Cheese, sheep	7.11	2.24	
Netherlands	Cheese	4.37	1.40	
Norway	Cheese	3.83	1.03	
	Cheese, goat	4.57	1.37	
Portugal	Cheese	5.42	1.54	
	Cheese, sheep	5.25	1.47	
Spain	Cheese	5.68	1.45	
	Cheese, sheep	4.26	1.47	
Sweden	Cheese	—	—	
	Margarine cheese	11.60	3.54	
United Kingdom	Cheese	4.52	2.15	
Italy	Cheese, pecorino	4.99	—	Banni et al. [66]
	Cheese, ricotta	9.10	—	
	Cheese, Parmesan	3.22	—	
	Swiss cheese	3.70	—	
Greece	Cheese, feta	—	1.40	Boulous et al. [67]
France	Cheese, goat ($n = 8$)	2.68	—	Wolff et al. [65]
	Cheese, ewe ($n = 7$)	1.53	—	
United States	Cheese, cheddar	2.54	0.59	USDA [32]
	Cheese, processed ($n = 6$)	2.41–3.49	0.31–0.57	
New Zealand	Cottage cheese	0.90	—	Ball et al. [33]
	Cream cheese ($n = 2$)	1.2–2.4	—	
Germany	Cheese, regular (n-22)	3.01	—	Pfalzgraf et al. [37]
	Cheese, goat ($n = 2$)	3.15	—	
	Cheese, ewe ($n = 3$)	5.60	—	
Italy	Yogurt	4.34	—	Banni et al. [66]
United States	Yogurt, low-fat	2.39–3.18	0.01–0.02	USDA [32]
Germany	Yogurt ($n = 1$)	3.3	—	Pfalzgraf et al. [37]

TABLE 10.12
***trans* Fatty Acid Content of Ice Cream from Different Countries**

Country	Food Item	Percentage *trans* Fatty Acids (%)	Total *trans* Fatty Acids/ 100 g Food Item	Reference
Belgium	Ice cream, dairy fat	6.07	0.75	Aro et al. [63]
Finland	Ice cream, milk fat	4.58	0.55	
	Stick, milk fat	4.46	0.58	
	Cone, milk fat	3.14	0.37	
	Soft ice, dairy fat	3.93	0.32	
	Ice cream, vegetable fat	23.47	2.16	
	Stick, vegetable fat	30.58	2.72	
	Cone, vegetable fat	23.59	2.19	
Germany	Ice cream, dairy fat, premium	2.63	0.39	
Greece	Ice cream, dairy fat, vanilla	3.59	0.38	
	Ice cream, dairy fat, chocolate	2.84	0.27	
Iceland	Ice cream, vegetable fat, hard	21.01	2.04	
	Ice cream, vegetable fat, soft	0.75	0.05	
Italy	Ice cream, dairy fat, fruit	5.48	0.36	
	Ice cream, dairy fat, wafer	5.02	0.95	
	Ice cream, vegetable fat, biscuits	0.93	0.08	
Netherlands	Ice cream, vegetable fat	0.53	0.03	
Spain	Ice cream, vegetable fat	0.16	0.01	
	Ice cream, vegetable fat	0.17	0.01	
	Ice cream, vegetable fat	1.16	0.13	
Sweden	Ice cream, dairy fat	4.54	0.53	
	Ice cream, vegetable fat, cone	17.48	2.78	
	Ice cream, vegetable fat	19.89	2.05	
	Ice cream, vegetable fat, desert	13.77	1.85	
United Kingdom	Ice cream, dairy fat	3.06	0.54	
	Ice cream, vegetable fat	1.75	0.17	

10.5.8 Meats

Data on *trans* fatty acid content in meats and processed meats are summarized in Table 10.13. In general, *trans* fatty acids in pork (0.2 to 0.9%), chicken (0.2 to 1.7%), and turkey (0.3 to 3.6%) are relatively low. On the other hand, ruminant meats such as beef, lamb, and mutton have higher proportions of *trans* fatty acids, ranging from 2 to 9.5% [37,63].

Trans fatty acid contents of sausages from most European and American samples are generally low [23,32,37,63]. The *trans* fatty acid compositions of these sausages corresponded to that of pork, with the exception of the Icelandic lamb sausage and a beef-based Suçuk sausage from the Netherlands [63].

TABLE 10.13
***trans* Fatty Acid Content of Meats, Meat Products, and Animal Fats from Different Countries**

Country	Food Item	Percentage *trans* Fatty Acids (%)	Total *trans* Fatty Acids/ 100 g Food Item	Reference
Belgium	Beef	9.52	0.31	Aro et al. [63]
Denmark	Beef	3.02	0.59	
Finland	Beef	3.04	0.22	
France	Beef	3.20	0.46	
Germany	Beef	2.78	0.48	
Greece	Beef	4.58	0.98	
Iceland	Beef	4.11	0.34	
Italy	Beef	3.88	0.28	
Netherlands	Beef	4.32	0.76	
Norway	Beef	4.35	0.51	
Portugal	Beef	5.01	0.37	
Spain	Beef	4.32	0.19	
Sweden	Beef	3.40	0.25	
United Kingdom	Beef	4.68	0.16	
France	Beef (n = 10)	1.95	—	Wolff [68]
United States	Ground beef, raw (n = 2)	4.50	0.86	USDA [32]
	Ground beef, cooked (n = 2)	5.10	0.87	
Germany	Beef (n = 4)	2.73	—	Pfalzgraf et al. [37]
	Veal (n = 3)	1.37	—	
Spain	Beef (n = 45)	8.5	—	Boatella et al. [34]
United States	Beef	—	0.63	Litin and Sacks [35]
France	Beef tallow, refined (n = 10)	4.91	—	Bayard and Wolff [69]
France	Beef tallow (n = 2)	4.6	—	Wolff [68]
Germany	Beef tallow	1.9	—	Pfalzgraf et al. [37]
New Zealand	Beef fat	1.7	—	Ball et al. [33]
United States	Beef fat, raw (n = 12)	6.55	—	Slover et al. [41]
	Beef fat, braised (n = 6)	5.47	—	
United States	Beef fat (n = 1)	1.8	—	Enig et al. [43]
United States	Beef fat (n = 7)	4.2	—	Slover and Lanza [54]
Belgium	Lamb/mutton	9.19	1.33	Aro et al. [63]
Denmark	Lamb/mutton	8.52	0.86	
Finland	Lamb/mutton	4.32	0.36	
France	Lamb/mutton	8.49	1.32	
Germany	Lamb/mutton	8.30	1.21	
Greece	Lamb/mutton	5.16	0.98	
Iceland	Lamb/mutton	4.70	0.34	
Italy	Lamb/mutton	5.75	0.47	
Netherlands	Lamb/mutton	8.58	0.81	

TABLE 10.13 (Continued)
trans Fatty Acid Content of Meats, Meat Products, and Animal Fats from Different Countries

Country	Food Item	Percentage *trans* Fatty Acids (%)	Total *trans* Fatty Acids/ 100 g Food Item	Reference
Norway	Lamb/mutton	5.01	0.89	
Portugal	Lamb/mutton	5.52	0.67	
Spain	Lamb/mutton	4.39	0.30	
Sweden	Lamb/mutton	—	—	
United Kingdom	Lamb/mutton	5.23	0.23	
Germany	Lamb (n = 3)	7.53	—	Pfalzgraf et al. [37]
	Mutton (n = 3)	9.30	—	
United States	Lamb fat (n = 1)	6.6	—	Enig et al. [43]
Belgium	Pork	0.52	0.07	Aro et al. [63]
Denmark	Pork	0.83	0.03	
Finland	Pork	0.25	0.04	
France	Pork	0.52	0.08	
Germany	Pork	0.19	0.05	
Greece	Pork	0.26	0.05	
Iceland	Pork	0.56	0.10	
Italy	Pork	0.86	0.10	
Netherlands	Pork	0.69	0.22	
Norway	Pork	0.79	0.04	
Portugal	Pork	0.53	0.03	
Spain	Pork	2.23	0.14	
Sweden	Pork	0.29	0.04	
United Kingdom	Pork	0.76	0.02	
Spain	Pork (n = 35)	0.6	—	Boatella et al. [34]
United States	Pork	—	0.07	Litin and Sacks [35]
United States	Lard (n = 3)	1.00	1.07	USDA [32]
Germany	Lard	0.4	—	Pfalzgraf et al. [37]
Australia	Lard	0.73	—	Mansour and Sinclair [36]
Spain	Lard (n = 35)	0.7	—	Boatella et al. [34]
United States	Pork fat, separable (n = 6)	0.2	—	Slover et al. [70]
United States	Lard (n = 1)	0.3	—	Enig et al. [43]
Belgium	Chicken	1.44	0.25	Aro et al. [63]
Denmark	Chicken	0.40	0.08	
Finland	Chicken	0.24	0.02	
France	Chicken	0.45	0.06	
Germany	Chicken	0.58	0.10	
Greece	Chicken	0.38	0.05	
Iceland	Chicken	1.62	0.20	
Italy	Chicken	1.15	0.03	

(*continued*)

TABLE 10.13 (Continued)
***trans* Fatty Acid Content of Meats, Meat Products, and Animal Fats from Different Countries**

Country	Food Item	Percentage *trans* Fatty Acids (%)	Total *trans* Fatty Acids/ 100 g Food Item	Reference
Netherlands	Chicken	1.71	0.03	
Norway	Chicken	0.83	0.11	
Portugal	Chicken	0.44	0.07	
Spain	Chicken	0.66	0.08	
Sweden	Chicken	0.58	0.09	
United Kingdom	Chicken	0.73	0.06	
Germany	Rooster	0.5	—	Pfalzgraf et al. [37]
United States	Chicken	—	0.07	Litin and Sacks [35]
Belgium	Turkey, raw with skin	0.76	0.04	Aro et al. [63]
Denmark	Turkey, flesh and skin	0.57	0.03	
Finland	Turkey, raw	0.60	0.01	
France	Turkey, roasted	1.16	0.06	
Greece	Turkey, raw with skin	0.31	0.04	
Norway	Turkey, minced	1.11	0.07	
Portugal	Turkey, whole	0.73	0.02	
United Kingdom	Turkey, whole, raw	1.27	0.09	
United States	Turkey, raw (n = 2)	2.75	0.09	USDA [32]
	Turkey, ground and raw (n = 10)	3.58	0.27	
Germany	Turkey	1.4	—	Pfalzgraf et al. [37]
Denmark	Duck, flesh and skin, raw	0.67	0.34	Aro et al. [63]
France	Duck, roasted	0.33	0.04	
Germany	Duck	0.5	—	Pfalzgraf et al. [37]
	Pigeon, wild	0.2	—	
Belgium	Rabbit, unprepared	0.61	0.03	Aro et al. [63]
France	Rabbit, roasted	0.63	0.09	
Greece	Rabbit, raw	0.61	0.16	
Iceland	Horse, minced	0.45	0.06	Aro et al. [63]
Finland	Elk, raw	2.15	0.80	Aro et al. [63]
Norway	Moose, leg	1.70	0.80	
Finland	Reindeer, raw	2.19	0.08	Aro et al. [63]
Norway	Reindeer, leg	1.72	0.03	
Belgium	Sausages	0.83–1.81	0.14–0.40	
Denmark	Sausages	0.94–1.37	0.19–0.55	
Finland	Sausages	0.66–0.82	0.10–0.16	
France	Sausages	0.36	0.11	
Germany	Sausages	0.41–0.84	0.10–0.28	
Greece	Sausages	0.30–1.25	0.11–0.17	
Iceland	Sausages	0.89–5.30	0.36–1.11	

TABLE 10.13 (Continued)
***trans* Fatty Acid Content of Meats, Meat Products, and Animal Fats from Different Countries**

Country	Food Item	Percentage *trans* Fatty Acids (%)	Total *trans* Fatty Acids/ 100 g Food Item	Reference
Italy	Sausages	0.46–1.18	0.17–0.24	
Netherlands	Sausages	0.84–4.86	0.33–1.97	
Norway	Sausages	1.01–2.06	0.33–0.35	
Portugal	Sausages	0.27–0.79	0.04–0.13	
Spain	Sausages	0.73–1.40	0.13–0.55	
Sweden	Sausages	0.25–0.50	0.06–0.08	
United Kingdom	Sausages	0.63–1.57	0.25–0.38	
Spain	Sausages (*n* = 40)	0.7	—	Fernandez San Juan [23]
United States	Sausages, frankfurter	3.21	0.90	USDA [32]
	(*n* = 4)	4.55	1.27	
	Sausages, kielbasa, beef	0.35	0.09	
	Sausage links	0.41	0.11	
	Sausages, pork	0.93	0.36	
	Sausages, pepperoni	5.17	1.30	
	Bologna, beef (*n* = 2)	0.67	0.19	
	Bologna, pork (*n* = 2)	3.63	0.54	
	Turkey burger, cooked (*n* = 2)			
Germany	Sausages (*n* = 22)	0.68	—	Pfalzgraf et al. [37]
New Zealand	Meat patty	4.32	—	Richardson et al. [21]
	Luncheon meat	4.98	—	
Canada	Meat patty	3.5	—	Ratnayake et al. [50]
Germany	Pork, filet	0.2	—	Pfalzgraf et al. [37]
	Pork, bacon	0.4	—	
	Pork, cooked ham	0.2	—	
	Pork, smoked ham	0.5	—	
Spain	Meat products (*n* = 46)	0.5	—	Boatella et al. [34]

10.6 CLINICAL ASPECTS OF *TRANS* FATTY ACIDS

Coronary heart disease and stroke are the main causes of death in the United States [71]. These cardiovascular diseases are correlated to high serum cholesterol levels [72]. Apart from saturated fatty acids, *trans* fatty acids have also been demonstrated to raise cholesterol levels [73,74] and increase the risk of coronary heart disease [75,76].

Early studies [77–91] did not raise much concern about the health effects of dietary *trans* fatty acids until the work of Mensink and Katan [92]. The findings by Mensink and Katan [92] suggested that elaidic acid (C18:1*t*) resulted in higher blood total and low-density lipoprotein (LDL) cholesterol levels than oleic acid (C18:1*c*), and lower blood and LDL cholesterol levels than stearic acid (C18:0). They also found that high-density lipoprotein (HDL) cholesterol levels were comparable after subjects consumed high oleic acid and stearic acid diets, but were significantly lower after subjects consumed the high elaidic acid diet. Based on these findings, they concluded that the effects of *trans* fatty acids on serum lipoprotein profiles are similar to those of cholesterol-raising saturated fatty acids. Since the report by Mensink and Katan [92], numerous human trials have been conducted to confirm the effects of *trans* fatty acids on blood cholesterol levels [93–103].

In a study by Judd et al. [101], the effects of diets high in *cis* monounsaturated fatty acids and saturated fatty acids (C12:0–C16:0) relative to diets moderate (3.8%) or high (6.6%) in *trans* fatty acids were investigated. It was found that the high *trans* fatty acid diet resulted in the lowest HDL cholesterol level relative to other diets. Chisholm et al. [104] reported on the effect on blood lipid levels by substituting margarine containing unsaturated *trans* fatty acids for butter in moderately hypercholesterolemic subjects. They concluded that LDL cholesterol levels were lower after subjects consumed the margarine relative to the butter, while HDL cholesterol levels were similar regardless of the dietary fat consumed. In another study, Judd et al. [105] substituted 8% energy fat in a 39% energy fat diet with *trans* fatty acids, equal amounts of *trans* fatty acids and stearic acid, or a mixture of lauric, myristic, and palmitic acids. They noted that these diets resulted in higher LDL cholesterol levels than the diet having the same amounts of stearic acid, oleic acid, and carbohydrate. It was also observed that HDL levels were lowest after consumption of the diets containing *trans* fatty acids and/or stearic acid. Kim and Campos [106] conducted population studies on the effects of *trans* fatty acids on LDL size of 414 Costa Ricans. Results revealed that there is a correlation between *trans* fatty acid intake and large LDL particles, thus suggesting that effects of *trans* fatty acids on coronary heart disease may be mediated through their effects on LDL size.

In addition to modification of blood lipid and lipoprotein patterns, *trans* fatty acids were shown to increase lipoprotein (a) levels [97,107], while other reports indicate otherwise [99]. High levels of lipoprotein(a) have been associated with the increased risk of coronary heart disease [108]. Almendingen et al. [102] revealed that consumption of diets containing hydrogenated soybean or fish oils resulted in significantly higher lipoprotein(a) levels than diets containing butter. Similarly, Aro et al. [103] and Sundram et al. [109] observed that the consumption of diets high in *trans* fatty acids had increased lipoprotein(a) levels. However, studies by Chisholm et al. [104], Clevidence et al. [110], and Lichtenstein et al. [111] indicated that no significant changes in lipoprotein(a) levels were found when subjects consumed diets containing *trans* fatty acids. It has also been shown that lipoprotein(a) levels can be lowered by saturated fatty acids [110,111]. Therefore, it is important, when interpreting these data, to consider the magnitude of change and the potential of such changes to significantly affect the risk of cardiovascular diseases.

Isomeric fatty acids have been shown to affect cholesterol ester transfer protein (CETP) activity. In a study by Lagrost et al. [112], *in vitro* enrichment of HDL with *cis* fatty acids inhibited CETP activity, whereas enrichment with *trans* fatty acids increased CETP activity. Further investigations on the effects of *trans* fatty acids on

HDL cholesterol levels in human subjects by Abbey and Nestel [113] confirmed the inverse relationship between CETP activity and fall in HDL cholesterol levels. The authors postulated that the change in CETP activity could be related to fatty acid specificity. However, the study by Aro et al. [103] on the effects of diets high in *trans* fatty acids or stearic acid relative to dairy fat on CETP and phospholipid transfer protein (PLTP) activities reported that there was no significant effect of *trans* fatty acids or stearic acid on CETP activity. Nevertheless, a significantly higher reduction in PLTP activity was observed in the stearic acid diet compared with the *trans* fatty acid diet.

The effects of hemostatic factors of hydrogenated soybean oil and hydrogenated fish oil relative to butter were investigated by Almendingen et al. [114]. Based on the higher levels of plasminogen activator inhibitor type I antigen and increased activity of plasminogen activator inhibitor type I observed, the authors concluded that hydrogenated soybean oil had an undesirable antifibrinolytic effect relative to hydrogenated fish oil and butter. In another related investigation by Müller et al. [115], the effects of *trans* fatty acids on diurnal postprandial hemostatic variables were reported. Three different diets, one containing saturated fatty acids from palm oil, one based on partially hydrogenated soybean oil with 23% *trans* fatty acids, and one with a high proportion of polyunsaturated fatty acids, were randomly consumed by nine female participants. The study concluded that dietary *trans* fatty acids has an unfavorable effect on postprandial plasminogen activator activity and, thus, possibly on the fibrinolytic system. On the other hand, comprehensive reports by Mutanen and Aro [116] and Turpeinen et al. [117] revealed that diets high in *trans* fatty acids do not have a significant effect on blood clotting.

Trans fatty acids have also been shown to have an association with cancer. In an ecological study by Bakker et al. [118] across 11 European countries, a strong positive correlation between *trans* fatty acids and the incidence of colorectal cancer was observed. Nevertheless, this study did not account for several important risk factors, such as a positive family history, age, meat consumption, smoking, and alcohol consumption. In another related study, Slattery et al. [119] investigated on the association between colon cancer and the consumption of foods by women. A significant and positive correlation was found between *trans* fatty acid intake and the risk of colon cancer.

10.7 *TRANS*-FREE LIPIDS

It is well known that coronary heart disease is the number one killer of people. It is associated with a high level of LDL cholesterol, which, in turn, is linked to diet, particularly excess consumption of certain saturated and *trans* isomer fats. Consequently, a diverse array of "*trans*-free" or "low *trans*" fat products has been developed for the ever-increasing health-conscious population. Most of these *trans*-free lipids are margarines, spreads, and shortenings, which contain zero or low amounts of *trans* fatty acids and yet still possess desirable organoleptic and physical properties of hydrogenated fats. These patented *trans*-free fat products are summarized in Table 10.14.

There are several ways to produce *trans*-free lipids. One approach is to formulate these products by physically blending liquid oil with a high-melting solid fat rich in

TABLE 10.14
Patent Literature Survey of *trans*-Free and Low *trans* Fatty Acid Products

Applicant	Food Item	Method	*trans* Fatty Acid Claim	Reference
Land O'Lakes Inc. (United States)	Fat blend	Heating a mixture of liquid fat and solid fat, followed by rapid cooling the fat blend in ≤0.5 min to form a nucleated fat base blend	*trans*-Free	Landon [120]
ARCO Chemical Technology L.P. (United States)	Shortening products, spreads	Incorporate up to 100% of esterified propoxylated glycerin into partially hydrogenated vegetable oil	Less *trans* fatty acids	Mazurek and Ferenz [121]
Unilever N.V. (Netherlands)	Spreads	Aqueous dispersion of plant sterols and other high melting lipids	Minimal or no *trans* fatty acids	Patrick and Traska [122]
	Margarines, spreads	Interesterification of selectively fractionated nonhydrogenated high melting palm oil with dry fractionated nonhydrogenated palm-kernel oil	*trans*-Free	Sahasranamam [123]
Poul Moller Ledelses (Denmark)	Partially hydrogenated vegetable, animal or marine oils	Partial hydrogenation in supercritical or near-critical state	Low *trans* fatty acids	Härröd and Möller [124]
University of Guelph (Canada) and Caravelle Foods (Canada)	Margarines	Blend of 95% liquid oil and 5% hard fat	No *trans* fatty acids	Kakuda et al. [125]
E.I. Du Pont De Nemours and Company (U.S.)	Confectionary lipids	Fractionation of high stearic soybean oil	No *trans* fatty acids	Knowlton [126]
The Administrators of the Tulane Educational Fund (United States)	Partially hydrogenated oil	Partial electrochemical hydrogenation of edible oils	Low *trans* fatty acids	Pintauro [127]
Van den Bergh Foods Company (United States)	Stick margarine	Interesterified hard stock	< 10% *trans* Fatty acids	Reddy et al. [128]

Van den Bergh Foods Company (United States)	Spreads	Interesterification of 30–90% of high lauric rapeseed oil and 10–70% of an oil comprising at least 40% of saturated fatty acids of which at least 80% are C_{16}–C_{18} fatty acids and at most 60% are C_{16} fatty acid	*trans*-Free	Sassen and Wesdorp [129]
Van den Bergh Foods Company (United States)	Spreads	Blend of interesterified hard stock and liquid oil	<10% *trans* fatty acids	Schuurman et al. [130]
Fuji Oil Company Ltd. (Japan)	Hard butter	80% SUS type triglyceride, less than 2% SSS type TG, at least 1% polyglycerol fatty acid ester, at least 12% SUS and SSU type TG	*trans*-Free	Okada et al. [131]
Unilever Patent Holdings B.V. (Netherlands)	Spreads	5–80% fat phase containing 45–80% oleic acid, 5–20% saturated fatty acids	Low *trans* fatty acids	Livingston [132]
The Procter & Gamble Company (United States)	Shortening products	Mixture of 75–94% fat and 6–25% inert gas; fat phase contains 74 to 90% base oil, 10–20% hard stock having 20–80% beta phase tending component having IV ≤10	Low *trans* fatty acids	Roberts et al. [133]
Raisio Benecol Ltd. (Finland)	Margarines, mayonnaise, cooking oils, cheeses, butter and shortening	Substituting saturated and *trans* fatty acids with unsaturated fatty acid esters of sterols and/or stanols	*trans*-Free	Wester [134]
Kraft Foods Inc. (United States)	Fat	Triglycerides containing *cis*-asymmetrical monounsaturated fatty acids	Low *trans* fatty acids	Blaurock et al. [135]
Loders Croklaan B.V. (Netherlands)	Margarines	Less than 55% SUU type TG, less than 34% SUS type TG, SUS to SSU ratio below 2	Low *trans* fatty acids	Cain et al. [136]
Danisco A/S (Denmark)	Shortening products	Blend of vegetable oil with stearine fraction enriched with at least one diglycerides	No *trans* fatty acids	Doucet [137]
Cargill Inc. (United States)	Low *trans* fatty acid hydrogenated canola oil	Hydrogenation of high oleic and low polyunsaturated fatty acid canola oil	Low *trans* fatty acids	Kodali et al. [138]

(*continued*)

TABLE 10.14 (Continued)
Patent Literature Survey of *trans*-Free and Low *trans* Fatty Acid Products

Applicant	Food Item	Method	*trans* Fatty Acid Claim	Reference
CPC International Inc. (United States)	Dry soups, dry sauces	Interesterification of vegetable fat with animal fat	Low *trans* fatty acids	Kristott and Rossell [139]
Van den Bergh Foods Company (United States)	Plastic fat product	20–65% liquid oil, 35–80% structuring fat and having N_{35} of 0–5 and N_{20} of 12–40, and the difference between N_{20} and N_{35} of at least 12, preferably 18–30, structuring fat comprises 10–50% of a lauric fat	Low *trans* fatty acids	Broomhead and Huizinga [140]
Unilever N.V. (Netherlands)	Frying fat	Interesterification and fractionation to obtain an olein fraction comprising 0–6% SSS and at least 20% SSO, and N_{25}–N_{35} value of at least 25	*trans*-Free	Iburg and van den Oever [141]
Ogam Ltd. (Ireland)	Spreads	Comprises 35–80% of fat phase; fat phase contains 10% saturated fatty acids and 80% unsaturated fatty acids	No *trans* fatty acids	Blauel et al. [142]
Kraft Foods Inc. (United States)	Stick margarines, spreads	Comprises specific blends of co-interesterified liquid unsaturated vegetable oils hard stocks with incorporation of fully hydrogenated monoglyceride emulsifier	No *trans* fatty acids	Erickson and Boyington [143]
Van den Bergh Foods Company	Spreads	Interesterification of a behenic acid-rich fat with a palmitic and/or stearic acid-rich fat in a 60:40 ratio; interesterified mixture contains at least 5% behenic acid and sum of palmitic and stearic acids at least 50%	Low *trans* fatty acids	Lansbergen and Schijf [144]

Charleville Research Ltd. (Ireland)	Spread	Fat phase comprising olive oil and/or canola oil or high-oleic sunflower oil and an interesterified hard stock fat containing palm oil, palm oil fractions, and coconut oil	Low *trans* fatty acids	Maguire [145]
The Procter & Gamble Company (United States)	Shortenings	Comprises 75–94% fat phase and 6–25% inert gas; fat phase comprises 74–90% base oil having <16% C_4–C_{26} fatty acids, 10–20% β′ stable crystalline hard stock containing at least 65% PSP and PSS, and ratio of PSP:PSS of at least 0.8 to 1.0, 0–30% other triglycerides, and 0–5% monoglycerides or diglycerides	Reduced *trans* fatty acids	Scavone [146]
Nabisco Inc. (United States)	Bakery fat	Comprises a fat bearing C_2–C_4 fatty acids and fats bearing C_{16}–C_{22} saturated fatty acids	Low *trans* fatty acids	Sullivan [147]
Nabisco Inc. (United States)	Margarines and shortenings	Blend comprising 25–75% liquid oil, 25–75% fully hydrogenated oil bearing C_{16}–C_{24} fatty acids; at least 15% of C_{16} fatty acid is replaced by C_2–C_4 fatty acids	Reduced *trans* fatty acids	Wheeler et al. [148]
Kraft General Foods Inc. (United States)	Margarines	Enzymatic transesterification of stearic fat with a liquid vegetable oil using 1,3-positional specific lipase	Low *trans* fatty acids	Brown et al. [149]
Van den Bergh Foods Company	Filling fats	Comprises 35–80% SUS, <5% SSS, 7–60% (UUS + UUU), <40% SSU, and SUS:SSU ratio <6	No *trans* fatty acids	Cain et al. [150]
Kraft General Foods Inc. (United States)	Margarines	Enzymatically transesterified oil containing <6% C_8–C_{16} saturated fatty acids	Low *trans* fatty acids	Yayashi et al. [151]
Lever Brothers Company (United States)	Spreads	Interesterification of lauric fats, saturated fats, and a minor proportion of partially hydrogenated fats	Low *trans* fatty acids	Schmidt [152]

saturated fatty acids [120,125,132,133,137,140,142]. Margarines formulated from these fats are available in Canada and Europe, but are not extensively used in the United States. Another approach is to modify the oils by means of interesterification. Chemically or enzymatically interesterified blends of unsaturated and saturated oils will result in a product that possesses melting properties similar to those of physically blended oils, but comprises different triacylglycerol structures. Numerous patents [123,128–130,139,141,143–145,152] for such fat compositions have been published. Alternatively, there are patents whereby other compounds are used to formulate *trans*-free fats. In a patent application by Unilever N.V. [122], plant sterols were used to produce spreads with minimal or no *trans* fatty acids. Similarly, Raisio Benecol Ltd. obtained a U.S. patent on its method of producing *trans*-free margarines, mayonnaise, cooking oils, cheeses, butter, and shortening by substituting saturated and *trans* fatty acids with unsaturated fatty acid esters of sterols and/or stanols [134].

10.8 CONCLUSIONS

Trans fatty acids have been shown to be present in various food items that we consume every day. In order to make these data useful to the public, information on *trans* fatty acid composition, along with other fatty acids and the overall fat content, should be displayed on the food items. In addition, data on the number of calories consumed, total fat in the diet, percentage of calories provided by the fat, and relative proportions of saturated, monounsaturated, and polyunsaturated fat should be carefully considered. The role of the government is also imperative in that proper measures are taken to regulate *trans* fatty acids in food items and to help consumers better understand how to translate food choices into better quality of life.

REFERENCES

1. Willett, W.C. and Ascherio, A., Trans fatty acids: are the effects only marginal? *Am. J. Public Health,* 84, 722, 1994.
2. Marchand, C.M., Positional isomers of trans-octadecenoic acids in margarines, *Can. Inst. Food Sci. Technol. J.,* 15, 196, 1982.
3. Sampugna, J. et al., Rapid analysis of trans fatty acids on SP-2340 glass capillary columns, *J. Chromatogr.,* 249, 245, 1982.
4. Parodi, P.W., Composition and structure of some consumer-available edible fats, *J. Am. Oil Chem. Soc.,* 53, 530, 1976.
5. Carpenter, D.L. and Slover, H.T., Lipid composition of selected margarines, *J. Am. Oil Chem. Soc.,* 50, 372, 1973.
6. Scholfield, C.R., Davison, V.L., and Dutton, H.J., Analysis for geometrical and positional isomers of fatty acids in partially hydrogenated fats, *J. Am. Oil Chem. Soc.,* 44, 648, 1967.
7. Karabulut, I. and Kayahan, M., Investigations on changes in physical and chemical properties of palm kernel oil during hydrogenation, *Turkish J. Agric. Forestry,* 23, 891, 1999.
8. Sommerfeld, M., Trans unsaturated fatty acids in natural products and processed foods, *Prog. Lipid Res.,* 22, 221, 1983.

9. Coenen, J.W.E., Hydrogenation of edible oils, *J. Am. Oil Chem. Soc.,* 3, 382, 1976.
10. Aro, A. et al., Analysis of C18:1 cis and trans fatty acid isomers by the combination of gas-liquid chromatography of 4,4-dimethyloxazoline derivatives and methyl esters, *J. Am. Oil Chem. Soc.,* 75, 977, 1998.
11. Reiser, R., Hydrogenation of polyunsaturated fatty acids by the ruminant, *Fed. Proc.,* 10, 236, 1951.
12. Hartman, L., Shorland, F.B., and McDonald, I.R.C., Occurrence of trans-acids in animal fats, *Nature,* 174, 185, 1954.
13. Hartman, L., Shorland, F.B., and McDonald, I.R.C., The trans-unsaturated acid contents of fats of ruminants and nonruminants, *Biochem. J.,* 61, 603, 1955.
14. Hay, J.D. and Morrison, W.R., Isomeric monoenoic fatty acids in bovine milk fat, *Biochim. Biophys. Acta,* 202, 237, 1970.
15. Parodi, P.W., Distribution of isomeric octadecanoic fatty acids in milk fat, *J. Dairy Sci.,* 59, 1870, 1976.
16. Parodi, P.W. and Dunstan, R.J., The trans unsaturation content of Queensland milkfats, *Aust. J. Dairy Technol.,* 26, 60, 1971.
17. Smith, L.M. et al., Measurement of trans and other isomeric unsaturated fatty acids in butter and margarine, *J. Am. Oil Chem. Soc.,* 55, 257, 1978.
18. Wolff, R.L., Precht, D., and Molkentin, J., Occurrence and distribution profiles of trans-18:1 acids in edible fats of natural origin, in *Trans Fatty Acids in Human Nutrition,* Sebedio, L. and Christie, W.W., Eds., Oily Press, Dundee, United Kingdom 1998, p. 1.
19. Leth, T. et al., Trans FA content in Danish margarines and shortenings, *J. Am. Oil Chem. Soc.,* 80, 475, 2003.
20. Aro, A. et al., Trans fatty acids in dietary fats and oils from 14 European countries: the TRANSFAIR Study, *J. Food Comp. Anal.,* 11, 137, 1998.
21. Richardson, R.K., Fong, B.Y., and Rowan, A.M., The trans fatty acid content of fats in some manufactured foods commonly available in New Zealand, *Asia Pacific J. Clin. Nutr.,* 6, 2395, 1997.
22. Ali, L.H. et al., Determination of total trans fatty acids in foods: comparison of capillary-column gas chromatography and single-bounce horizontal attenuated total reflection infrared spectroscopy, *J. Am. Oil Chem. Soc.,* 73, 1699, 1996.
23. Fernandez San Juan, P.M., Study of isomeric trans-fatty acid content in the commercial Spanish foods, *Int. J. Food Sci. Nutr.,* 47, 399, 1996.
24. Henninger, M. and Ulberth, F., Trans fatty acids in margarines and shortenings marketed in Austria, *Z. Lebensm. Unters. Forsch.,* 203, 210, 1996.
25. Lake, R. et al., Trans fatty acid content of selected New Zealand foods, *J. Food Comp. Anal.,* 9, 365, 1996.
26. Molkentin, J. and Precht, D., Isomeric distribution and rapid determination of trans-octadecenoic acids in German brands of partially hydrogenated edible fats, *Nahrung,* 40, 297, 1996.
27. Ovesen, L., Leth, T., and Hansen, K., Fatty acid composition of Danish margarines and shortenings, with special emphasis on trans fatty acids, *Lipids,* 31, 971, 1996.
28. Bayard, C.C. and Wolff, R.L., Trans-18:1 acids in French tub margarines and shortenings: recent trends, *J. Am. Oil Chem. Soc.,* 72, 1485, 1995.
29. Michels, K. and Sacks, F., Trans fatty acids in European margarines, *N. Engl. J. Med.,* 332, 314, 1995.
30. Ovesen, L. and Leth, T., Trans fatty acids: time for legislative action?, *Nutr. Food Sci.,* 3, 16, 1995.
31. Pfalzgraf, A. and Steinhart, H., Trans-fettsaure-gehalte in margarinen, *Deutsche Lebensmittel-Rundschau.,* 91, 113, 1995.

32. USDA, U.S. Department of Agriculture, Agricultural Research Service, Nutrient Data Laboratory Food Composition Data, Special Purpose Table No. 1: Fat and fatty acid content of selected foods containing trans-fatty acids, http://www.nal.usda.gov/fnic/foodcomp, 1995.
33. Ball, M.J., Hackett, D., and Duncan, A., Trans fatty acid content of margarines, oils and blended spreads available in New Zealand, *Asia Pacific J. Clin. Nutr.,* 2, 165, 1993.
34. Boatella, J., Rafecas, M., and Codony, R., Isomeric trans fatty acids in the Spanish diet and their relationships with changes in fat intake patterns, *Eur. J. Clin. Nutr.,* 47, S62, 1993.
35. Litin, L. and Sacks, F., Trans-fatty acid content of common foods, *N. Engl. J. Med.,* 329, 1969, 1993.
36. Mansour, M.P. and Sinclair, A.J., The trans fatty acid and positional (sn-2) fatty acid composition of some Australian margarines, dairy blends, and animal fats, *Asia Pacific J. Clin. Nutr.,* 3, 155, 1993.
37. Pfalzgraf, A., Timm, M., and Steinhart, H., Gehalte von trans-fettsäuren in lebensmitteln, *Z. Ernährungswiss.,* 33, 24, 1993.
38. Ratnayake, W.M.N. and Pelletier, G., Positional and geometrical isomers of linoleic acid in partially hydrogenated oils, *J. Am. Oil Chem. Soc.,* 69, 95, 1992.
39. Ratnayake, W.M.N., Hollywood, R., and O'Grady, E., Fatty acids in Canadian margarines, *Can. Inst. Sci. Technol. J.,* 24, 81, 1991.
40. Enig, M.G. et al., Isomeric trans fatty acids in the U.S. diet, *J. Am. College Nutr.,* 9, 471, 1990.
41. Slover, H.T. et al., Lipids in margarines and margarine-like foods, *J. Am. Oil Chem. Soc.,* 62, 775, 1985.
42. Enig, M.G., Budowski, P., and Blondheim, S.H., Trans-unsaturated fatty acids in margarines and human subcutaneous fat in Israel, *Human Nutr.: Clin. Nutr.,* 38C, 223, 1984.
43. Enig, M.G. et al., Fatty acid composition of the fat in selected food items with emphasis on trans components, *J. Am. Oil Chem. Soc.,* 60, 1788, 1983.
44. Wills, R.B.H., Myers, P.R., and Greenfield, H., Composition of Australian foods 14. Margarines and cooking fats, *Food Technol. Aust.,* 34, 240, 1982.
45. Beare-Rogers, J.L., Gray, L.M., and Hollywood, R., The linoleic acid and trans fatty acids of margarines, *Am. J. Clin. Nutr.,* 32, 1805, 1979.
46. Sahasrabudhe, M.R. and Kurian, C.J., Fatty acid composition of margarines in Canada, *J. Inst. Can. Sci. Technol. Aliment.,* 12, 140, 1979.
47. Heckers, H. and Melcher, F.W., Trans-isomeric fatty acids present in West German margarines, shortenings, frying and cooking fats, *Am. J. Clin. Nutr.,* 31, 1041, 1978.
48. Weihrauch, J.L. et al., Fatty acid composition of margarines, processed fats, and oils: a new compilation of data for tables of food composition, *Food Technol.,* 31, 80, 1977.
49. Molkentin, J. and Precht, D., Determination of trans-octadecenoic acids in German margarines, shortenings, cooking and dietary fats by AgTLC/GC, *Z. Ernährungswiss.,* 34, 314, 1995.
50. Ratnayake, W.M.N. et al., Fatty acids in some common food items in Canada, *J. Am. Coll. Nutr.,* 12, 651, 1993.
51. Hunter, J.E. and Applewhite, T.H., Isomeric fatty acids in the U.S. diet: levels and health perspectives, *Am. J. Clin. Nutr.,* 44, 707, 1986.
52. Smith, L.M. et al., Changes in physical and chemical properties of shortenings used for commercial deep-fat drying, *J. Am. Oil Chem. Soc.,* 63, 1017, 1986.
53. Lanza, E. and Slover, H.T., The use of SP2340 glass capillary columns for the estimation of the trans fatty acid content of foods, *Lipids,* 16, 260, 1981.

54. Slover, H.T. and Lanza, E., Quantitative analysis of food fatty acids by capillary gas chromatography, *J. Am. Oil Chem. Soc.*, 56, 933, 1979.
55. Tasan, M. and Demirci, M., Trans FA in sunflower at different steps of refining, *J. Am. Oil Chem. Soc.*, 80, 825, 2003.
56. Keméńy, Z. et al., Deodorization of vegetable oils: prediction of trans polyunsaturated fatty acid content, *J. Am. Oil Chem. Soc.*, 78, 973, 2001.
57. Aro, A. et al., Trans fatty acids in French fries, soups, and snacks from 14 European countries: the TRANSFAIR Study, *J. Food Comp. Anal.*, 11, 170, 1998.
58. Slover, H.T., Lanza, E., and Thompson, R.H. Jr., Lipids in fast foods, *J. Food Sci.*, 45, 1583, 1980.
59. Smith, L.M. et al., Lipid content and fatty acid profiles of various deep-fat fried foods, *J. Am. Oil Chem. Soc.*, 62, 996, 1985.
60. van Erp-baart, M.-A. et al., Trans fatty acids in bakery products from 14 European countries: the TRANSFAIR Study, *J. Food Comp. Anal.*, 11, 161, 1998.
61. Won, J.-S. and Ahn, M.-S., A study on contents of trans fatty acids in foods served at university dormitory and their consumption, *Korean J. Nutr.*, 23, 19, 1990.
62. Laryea, M.D. et al., Trans fatty acid content of selected brands of West German nut-nougat cream, *Z. Ernährungswiss.*, 27, 266, 1988.
63. Aro, A. et al., Trans fatty acids in dairy and meat products from 14 European countries: the TRANSFAIR study, *J. Food Comp. Anal.*, 11, 150, 1998.
64. Precht, D. and Molkentin, J., Rapid analysis of the isomers of trans-octadecenoic acid in milk fat, *Int. Dairy J.*, 6, 791, 1996.
65. Wolff, R.L., Bayard, C.C., and Fabien, R.J., Evaluation of sequential methods for the determination of butterfat fatty acid composition with emphasis on trans-18:1 acids. Application to the study of seasonal variations in French butters, *J. Am. Oil Chem. Soc.*, 72, 1471, 1995.
66. Banni, S. et al., Characterization of conjugated diene fatty acids in milk, dairy products, and lamb tissues, *J. Nutr. Biochem.*, 7, 150, 1996.
67. Boulous, C. et al., Computed and chemically determined nutrient content of foods in Greece, *Int. J. Food Sci. Nutr.*, 47, 507, 1996.
68. Wolff, R.L., Content and distribution of trans-18:1 acids in ruminant milk and meat fats. Their importance in European diets and their effect on human milk, *J. Am. Oil Chem. Soc.*, 72, 259, 1995.
69. Bayard, C.C. and Wolff, R.L., Analysis of trans-18:1 isomer content and profile in edible refined beef tallow, *J. Am. Oil Chem. Soc.*, 73, 531, 1996.
70. Slover, H.T. et al., The lipid composition of raw and cooked fresh pork, *J. Food Comp. Anal.*, 1, 38, 1987.
71. National Center for Health Statistics, Advance report of final mortality statistics, 1992, *Monthly Vital Statistics Report,* U.S. Government Printing Office, Washington DC, 63, 1, 1995.
72. National Research Council, *Diet and Health: Implications for Reducing Chronic Disease Risk,* National Academy Press, Washington DC, 1989.
73. Huang, X. and Fang, C., Dietary trans fatty acids increase hepatic acyl-CoA: cholesterol acyltransferase activity in hamsters, *Nutr. Res.*, 20, 547, 2000.
74. Lichtenstein, A.H., Trans fatty acids and blood lipid levels, Lp(a), parameters of cholesterol metabolism, and hemostatic factors, *J. Nutr. Biochem.*, 9, 244, 1998.
75. Ascherio, A., Hennekens, C.H., and Buring, J.E., Trans-fatty acids intake and risk of myocardial infarction, *Circulation,* 89, 94, 1994.
76. Willett, W.C., Stampfer, M.J., and Manson, J.E., Intake of trans fatty acids and risk of coronary heart disease among women, *Lancet,* 341, 581, 1993.

77. Anderson, J.T., Grande, F., and Keys, A., Hydrogenated fats in the diet and lipids in the serum of man, *J. Nutr.,* 75, 388, 1961.
78. Grasso, S. et al., Effects of natural and hydrogenated fats approximately equal dienoic acid content upon plasma lipids, *Metabolism,* 11, 920, 1962.
79. Beveridge, J.M.R. and Connell, W.F., The effect of commercial margarines on plasma cholesterol levels in man, *Am. J. Clin. Nutr.,* 10, 391, 1962.
80. McOsker, D.E. et al., The influence of partially hydrogenated dietary fats on serum cholesterol levels, *J.A.M.A.,* 180, 380, 1962.
81. Erickson, B.A. et al., The effect of partial hydrogenation of dietary fatty acids, of the ratio of polyunsaturated to saturated fatty acids, and of dietary cholesterol upon plasma lipids in man, *J. Clin. Invest.,* 43, 2017, 1964.
82. de Longh, H. et al., The influence of some dietary fats on serum lipids in man, *Bibl. Nutr. Dieta.,* 7, 137, 1965.
83. Kummerow, F.A. et al., Swine as an animal model in studies on atherosclerosis, *Fed. Proc.,* 33, 235, 1974.
84. Kummerow, F.A. et al., The influence of three sources of dietary fats and cholesterol on lipid composition of swine serum lipids and aorta tissue, *Artery,* 4, 360, 1978.
85. Vergroesen, A.J. and Gottenbos, J.J., The role of fats in human nutrition: an introduction, in *The Role of Fats in Human Nutrition,* Vergroesen, A.J., Ed., Academic Press, New York, 1975, p. 8.
86. Mattson, F.H., Hollenbach, E.J., and Kligman, A.M., Effect of hydrogenated fat on the plasma cholesterol and triglyceride levels in man, *Am. J. Clin. Nutr.,* 28, 726, 1975.
87. Enig, M.G., Munn, R.J., and Keeney, M., Dietary fats and cancer trends: a critique, *Fed. Proc.,* 37, 2215, 1978.
88. Applewhite, T.H., Statistical "correlations" relating trans-fats to cancer: a commentary, *Fed. Proc.,* 38, 2435, 1979.
89. Bailar, J.C. III, Dietary fat and cancer trends: a critique, *Fed. Proc.,* 38, 2435, 1979.
90. Meyer, W.H., Further comments, *Fed. Proc.,* 38, 2436, 1979.
91. Laine, D.C. et al., Lightly hydrogenated soy oil versus other vegetable oils as a lipid-lowering dietary constituent, *Am. J. Clin. Nutr.,* 35, 683, 1982.
92. Mensink, R.P. and Katan, M.B., Effects of dietary trans fatty acids on high-density and low-density lipoprotein cholesterol levels in healthy subjects, *N. Engl. J. Med.,* 323, 439, 1990.
93. Flynn, M.A. et al., Effects of cholesterol and fat modification of self-selected diets on serum lipids and their specific fatty acids in normocholesterolic and hypercholesterolic humans, *J. Am. Coll. Nutr.,* 10, 93, 1991.
94. Fumeron, F. et al., Lowering of HDL_2-cholesterol and lipoprotein A-1 particle levels by increasing the ratio of polyunsaturated to saturated fatty acids, *Am. J. Clin. Nutr.,* 53, 655, 1991.
95. Zock, P.L. and Katan, M.B., Hydrogenation alternatives: effects of trans fatty acids and stearic acid versus linoleic acid on serum lipids and lipoproteins in humans, *J. Lipid Res.,* 33, 399, 1992.
96. Nestel, P.H. et al., Plasma cholesterol-lowering potential of edible oil blends suitable for commercial use, *Am. J. Clin. Nutr.,* 55, 46, 1992.
97. Nestel, P.H. et al., Plasma lipoprotein lipid and Lp[a] changes with substitution of elaidic acid for oleic acid in the diet, *J. Lipid Res.,* 33, 1029, 1992.
98. Wood, R. et al., Effect of butter, mono and polyunsaturated fatty acid enriched butter, trans fatty acid margarine, and zero trans fatty acid margarine on serum lipids and lipoproteins in healthy men, *J. Lipid Res.,* 34, 1, 1993.

99. Lichtenstein, A.H. et al., Hydrogenation impairs the hypolipidemic effect of corn oil in humans, *Arterioscler. Thromb.,* 13, 154, 1993.
100. Seppanen-Laakso, T. et al., Replacement of margarine on bread by rapeseed and olive oils: effects on plasma fatty acid composition and serum cholesterol, *Ann. Nutr. Metab.,* 37, 161, 1993.
101. Judd, J.T. et al., Dietary trans fatty acids: effects on plasma lipids and lipoproteins of healthy men and women, *Am. J. Clin. Nutr.,* 59, 861, 1994.
102. Almendingen, K. et al., Effects of partially hydrogenated fish oil, partially hydrogenated soybean oil, and butter on serum lipoproteins and Lp(a) in men, *J. Lipid Res.,* 36, 1370, 1995.
103. Aro, A. et al., Stearic acid, trans fatty acids, and dairy fat: effect on serum and lipoprotein lipids, apolipoproteins, lipoprotein(a), and lipid transfer proteins in healthy subjects, *Am. J. Clin. Nutr.,* 65, 1419, 1997.
104. Chisholm, A. et al., Effect of lipoprotein profile of replacing butter with margarine in a low fat diet: randomized crossover study with hypercholesterolaemic subjects, *Br. Med. J.,* 312, 931, 1996.
105. Judd, J.T. et al., Blood lipid and lipoprotein modifying effects of cis and trans monounsaturated fatty acids compared to carbohydrate, stearic acid and 12:0-16:0 saturated fatty acids in man fed controlled diets, *FASEB J.,* 12, A229, 1998.
106. Kim, M.K. and Campos, H., Intake of trans fatty acids and low-density lipoprotein size in a Costa Rican population, *Metabolism,* 52, 693, 2003.
107. Mensink, R.P. et al., Effect of dietary cis-trans-fatty acids on serum lipoprotein (a) levels in humans, *J. Lipid Res.,* 33, 1493, 1992.
108. Wild, S.H., Fortmann, S.P., and Marcovina, S.M., A prospective case-control study of lipoprotein (a) levels and apo(a) size and risk of coronary heart disease in Stanford Five-City Project participants, *Arterioscler. Thromb. Vasc. Biol.,* 17, 239, 1997.
109. Sundram, K. et al., Trans (elaidic) fatty acids adversely affect the lipoprotein profile relative to specific saturated fatty acids in humans, *J. Nutr.,* 127, 514S, 1997.
110. Clevidence, B.A. et al., Plasma lipoprotein (a) levels in men and women consuming saturated, cis or trans monounsaturated fatty acid enriched diets, *Arterioscler. Thromb. Vasc. Biol.,* 17, 1657, 1997.
111. Lichtenstein, A.H. et al., Hydrogenated vegetable oil result in higher lipoprotein levels than the naturally occurring oil and lower lipoprotein levels than butter, *Circulation,* 95, 1, 1997.
112. Lagrost, L. et al., Influence of apolipoprotein composition of high density lipoprotein particles on cholesteryl ester transfer protein activity, *J. Biol. Chem.,* 269, 3189, 1994.
113. Abbey, M. and Nestel, P.J., Plasma cholesteryl ester transfer protein activity increased when trans-elaidic acid is substituted for cis-oleic acid in the diet, *Atherosclerosis,* 106, 99, 1994.
114. Almendingen, K. et al., Effects of partially hydrogenated fish oil, partially hydrogenated soybean oil, and butter on hemostatic variables in men, *Arterioscler. Thromb. Vasc. Biol.,* 16, 375, 1996.
115. Müller, H. et al., Partially hydrogenated soybean oil reduces postprandial t-PA activity compared with palm oil, *Atherosclerosis,* 155, 467, 2001.
116. Mutanen, M. and Aro, A., Coagulation and fibrinolysis factors in healthy subjects consuming high stearic or trans fatty acids diets, *Thromb. Haemost.,* 77, 99, 1997.
117. Turpeinen, A.M. et al., Similar effects of diets rich in stearic acid or trans-fatty acids on platelet function and endothelial prostacyclin production in humans, *Arterioscler. Thromb. Vasc. Biol.,* 18, 316, 1998.

118. Bakker, N., van't Veer, P., and Zock, P.L., Adipose fatty acids and cancers of the breast, *Int. J. Cancer,* 72, 587, 1997.
119. Slattery, M.L., Benson, J., and Ma, K.N., Trans fatty acids and colon cancer, *Nutr. Cancer,* 39, 170, 2001.
120. Landon, T., Trans-isomer-free fat blend and a process for forming the trans-isomer-free fat blend, U.S. Patent No. US 6,544,579B1, 2003.
121. Mazurek, H. and Ferenz, M.R., Plastic and semisolid edible shortening products with reduced trans-fatty acid content, U.S. Patent No. US 6,495,188B2, 2002.
122. Patrick, M. and Traska, E.W., Aqueous dispersions or suspensions, European Patent Appl. No. EP, 1197,153A1, 2002.
123. Sahasranamam, U.R., Trans free hard structural fat for margarine blend and spreads, U.S. Patent Appl. No. US 2002/0,001,662A1, 2002.
124. Härröd, M. and Möller, P., Partially hydrogenated fatty substances with a low content of trans fatty acids, U.S. Patent No. US 6,256,596B1, 2001.
125. Kakuda, Y., Abraham, V., and Jahan-Aval, F., Process for preparing high liquid oil margarine, International Patent No. WO 01/80,659A2, 2001.
126. Knowlton, S., Fat products from high stearic soybean oil and a method for the production thereof, U.S. Patent No. US 6,229,033B1, 2001.
127. Pintauro, P.N., Synthesis of a low trans-content edible oil, nonedible oil, or fatty acid in a solid polymer electrolyte reactor, U.S. Patent No. US 6,218,556B1, 2001.
128. Reddy, P.R., Madsen, R.A., and Schuurman, J.H., Water in oil stick product, U.S. Patent No. US 6,322,842B1, 2001.
129. Sassen, C.L. and Wesdorp, L.H., Edible fat spread, U.S. Patent No. US 6,238,723B1, 2001.
130. Schuurman, J.H., Barmentlo, B., and Reckweg, F., Recirculation process for a fat continuous spread, U.S. Patent No. US 6,322,843B1, 2001.
131. Okada, T., Yamada, K., and Nago, A., Hard butter composition and its production, U.S. Patent No. US 6,258,398B1, 2001.
132. Livingston, R.M., Edible spread based on olive oil as the major fat component, U.S. Patent No. US 6,159,524, 2000.
133. Roberts, B.A., Scavone, T.A., and Riedell, S.P., Beta-stable low-saturate, low trans, all purpose shortening, U.S. Patent No. 6,033,703, 2000.
134. Wester, I., Fat composition for use in food, U.S. Patent No. US 6,162,483, 2000.
135. Blaurock, A.E., Krishnamurthy, R.G., and Huth, P.J., Nutritionally superior fat for food compositions, U.S. Patent No. US 5,959,131, 1999.
136. Cain, F.W., van der Struik, H.G.A.M., and Zwikstra, N., Non-trans, non-temper filling fats, U.S. Patent No. US 5,935,627, 1999.
137. Doucet, J., Shortening system, products therewith, and methods for making and using the same, U.S. Patent No. US 5,908,655, 1999.
138. Kodali, D.R., DeBonte, L.R., and Fan, Z., High stability canola oils, U.S. Patent No. US 455,885,643, 1999.
139. Kristott, J.U. and Rossell, J.B., Rapidly crystallizing fat having a low trans-fatty acid content, European Patent Appl. No. EP 0,823,473A1, 1998.
140. Broomhead, R.A. and Huizinga, H., Edible fat product, U.S. Patent No. US 5,667,837, 1997.
141. Iburg, J.E. and van den Oever, C.E., Frying fat, European Patent Appl. No. EP 0,806,146A1, 1997.
142. Blauel, F., Murphy, M.F., and Byrne, C.M., Spread, U.S. Patent No. 5,536,523, 1996.
143. Erickson, M.D. and Boyington, L.R., Stick-type margarines and spreads containing no trans fatty acid, U.K. Patent Appl. No. GB 2,292,949, 1996.

144. Lansbergen, A.J. and Schijf, R., Edible fats, U.S. Patent No. US 5,547,698, 1996.
145. Maguire, J., An edible fat blend, U.K. Patent Appl. No. GB 2,281,304, 1995.
146. Scavone, T.A., Beta-prime stable low-saturate, low trans, all purpose shortening, U.S. Patent No. 5,470,598, 1995.
147. Sullivan, J., Flaky pie shells that maintain strength after filling, U.S. Patent No. US 5,382,440, 1995.
148. Wheeler, E.L. et al., Low-palmitic, reduced-trans margarines and shortenings, U.S. Patent No. US 5,407,695, 1995.
149. Brown, P.H. et al., Enzymatic method for preparing transesterified oils, U.S. Patent No. US 5,288,619, 1994.
150. Cain, F.W. et al., Non-temper filling fats, U.S. Patent No. US 5,288,513, 1994.
151. Yayashi, D.K. et al., Margarine oils having both low trans-unsaturate and low intermediate chain saturate content, U.K. Patent Appl. No. GB 2,239,256, 1991.
152. Schmidt, W.J., Low-trans fats and oil- and water emulsion spreads containing such fats, U.S. Patent No. US 4,610,889, 1986.

11 Clinical Benefits of a Structured Lipid (Betapol^{TM}) in Infant Formula

Marianne O'Shea, Jeanet Gerritsen, and Inge Mohede

CONTENTS

11.1 INTRODUCTION

11.1.1 DIETARY FAT

Dietary fat is the main source of energy, particularly for newborn, preterm, and term infants. In addition, it provides essential nutrients, especially the fat soluble vitamins A, D, E, and K, and it is essential for their absorption. Dietary fat consists of triacylglycerols, which are composed of a glycerol backbone to which fatty acids are esterified. Three types of fatty acids can be distinguished based on the degree of saturation of the carbon chain: saturated, monounsaturated (containing one double bond), and polyunsaturated fatty acids (with two or more double bonds) (Figure 11.1). Fatty acids with at least 20 carbon atoms are referred to as long-chain fatty acids [1]. The fatty acid composition of triacylglycerols can vary, as well as the position on the glycerol backbone to which each fatty acid is esterified [2]. There are three different positions: two outer positions (*sn*-1 and *sn*-3) and one central position (*sn*-2) (Figure 11.2).

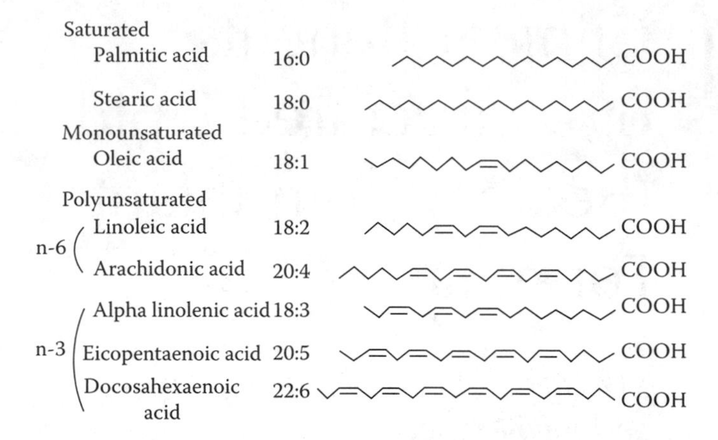

FIGURE 11.1 Different types of fatty acids.

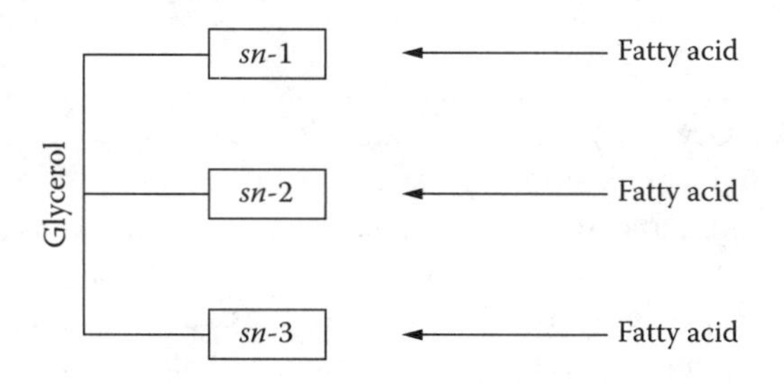

FIGURE 11.2 Triacylglycerol: a glycerol backbone with three fatty acids attached at different positions. The outer positions are *sn*-1 and *sn*-3; the central position is *sn*-2.

11.2 STRUCTURE AND COMPOSITION OF FATS: INFLUENCE ON FAT AND MINERAL ABSORPTION

In the first months of life, an infant relies completely on breast milk and/or infant formula. Triacylglycerols from human milk or infant formulas are the main source of energy for infants. About 50 to 60% of the energy consumed by infants is provided by this source of dietary fat. Although the composition of human milk is variable depending on the diet of the mother, it is generally considered the "gold standard," providing a well-balanced diet for optimamum growth and development of the infant [3]. The fatty acid composition of the infant diet influences the fatty acid composition of developing tissues [4]. Palmitic acid (16:0) accounts for a large proportion of the saturated fatty acids in human milk and for approximately 20 to 25% of the total human milk fatty acids, and therefore is responsible for approximately 10% of the energy intake in infants fed human milk [3]. In human milk, palmitic acid is esterified mainly (about 70%) to the central *sn*-2 position of the triacylglycerols. In contrast, in vegetable oils that are commonly used in infant formulas, palmitic acid is predominantly (>80%) esterified to the outer *sn*-1 and *sn*-3 positions [3]. The position of palmitic acid has consequences for the digestion, absorption, and subsequent metabolism of dietary triacylglycerols. After ingestion (milk) fat is emulsified with bile salts and triacylglycerols are hydrolyzed to release free fatty acids from the *sn*-1 and *sn*-3 positions [5].

Digestive enzymes (such as lingual and pancreatic lipase) and cofactors (such as colipase) act predominantly on the ester bonds at the *sn*-1 and *sn*-3 positions of triacylglycerol [6]. The products of this hydrolysis of dietary triacylglycerols are free fatty acids released from the *sn*-1 and *sn*-3 positions and monoacylglycerol from the *sn*-2 position. Monoglycerides together with bile salts are easily solubilized into mixed micelles and they are subsequently absorbed.

The position of the fatty acid at the glycerol backbone, as well as the degree of unsaturation and chain length, have a role in the rate of intestinal absorption of fats. Triacylglycerol blends in which the palmitic acid has been esterified to the *sn*-1 and *sn*-3 positions are less well absorbed compared with blends in which the palmitic acid is at the *sn*-2 position [3,6,7]. Similarly, longer chain, saturated fatty acids are generally less well absorbed than medium-chain (and unsaturated) fatty acids [6–12]. The reduction in absorbability with increasing chain length of saturated fatty acids has been proved to be the result of the formation of insoluble calcium soaps of these fatty acids. Insoluble calcium soaps cannot be absorbed from the intestine (Figure 11.3A). Instead, they are excreted with the feces, resulting in unnecessary loss of dietary energy and calcium, both of which are essential for the infant. On the one hand, calcium soaps result in harder stools and constipation. On the other hand, reduced calcium absorption affects bone formation. Calcium primarily affects the absorption of long-chain saturated fatty acids and has little effect on medium-chain or unsaturated fatty acids [10–13]. Fatty acids at the *sn*-2 position are preserved at this

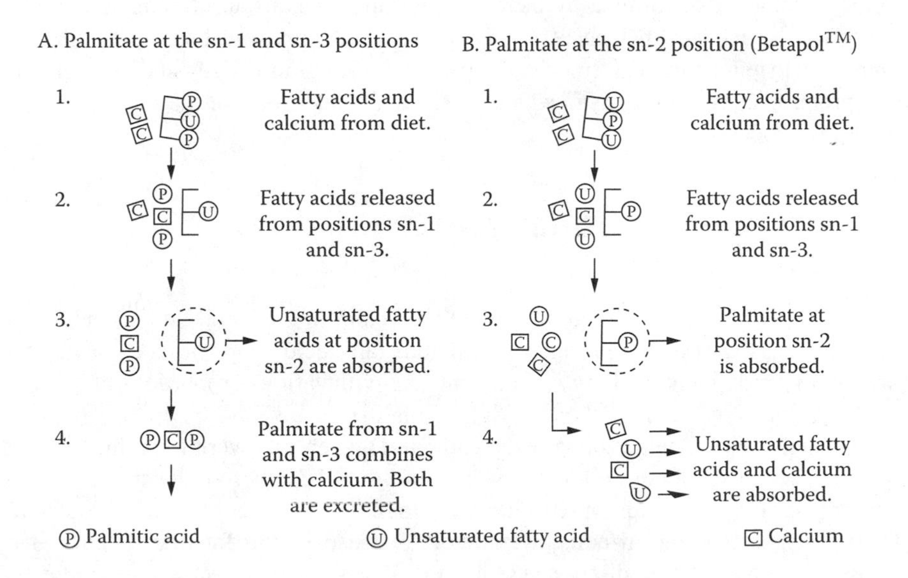

FIGURE 11.3 Lipid digestion and fatty acid absorption pathways.

position throughout the digestive processes and are absorbed as an *sn*-2 monoglyceride, preventing the possibility of the formation of fatty acid soaps (Figure 11.3B) [7].

Therefore, high levels of fat absorption from human milk are in part a result of the favorable positional distribution of palmitic acid, but have also been linked to the presence of bile-salt-stimulated lipase in human milk and the structure of the fat globule [7]. Malabsorption of saturated fatty acids can be crucial in newborn preterm and term infants as a result of low intraduodenal concentrations of pancreatic lipase and bile salts. Sometimes it is necessary to replace breast feeding with infant formula. Palmitic acid is absorbed better when it is located at the *sn*-2 position compared with the *sn*-1 and *sn*-3 positions. In human milk fat, most of the palmitic acid is actually located at the *sn*-2 position. In infant formulas that are produced from vegetable oils, palmitic acid is predominantly located at the *sn*-1 and *sn*-3 positions. An infant formula enriched with palmitic acid at the *sn*-2 position (therefore more similar to human milk) would provide more energy and result in a lower fecal loss of fat and calcium [8].

11.3 STRUCTURED LIPID FOR INFANT FORMULA

A structured vegetable fat has been developed for the infant formula Betapol (Loders Croklaan, Lipid Nutrition, Wormerveer, the Netherlands) (herein referred to as *structured lipid*), in which palmitic acid is predominantly esterified to the *sn*-2 position of the glycerol backbone, similar to human milk fat. The *sn*-1 and *sn*-3 positions are mainly occupied by unsaturated fatty acids, which are readily absorbed and do not form insoluble soaps (Figure 11.3B).

This structured lipid offers a vegetable fat that closely mimics the physical and chemical structure of human milk fat. As such, it may contribute to the health of infants taking infant formula by increasing fat and calcium absorption, resulting in softer stools, a lower incidence of constipation, and improved bone mineralization compared to regular infant formulas. The present review summarizes studies in animals and preterm and term infants in which these beneficial effects of the structured lipid were investigated.

11.4 STRUCTURED LIPID IN ANIMALS: *IN VIVO* STUDIES

In four experiments, Lien et al. examined the absorption of total fat and individual fatty acids, as well as the presence of calcium fatty acid soaps in feces of rats fed diets with nearly identical fatty acid contents but with various proportions of palmitic acid at the *sn*-2 position of the triacylglycerols [4].

For each experiment, young male Sprague-Dawley rats were fed a fat-free diet for 10 d, with 0.1 ml corn oil administered daily until 3 d before the start of the fat absorption assay. Groups of 10 rats were then fed diets containing the test fat (150 g/kg), replacing an equal weight of dextrose in the fat-free diet for 3 d, followed by the fat-free diet for another 3 d. Feces were collected throughout the 6-d assay. Consumption of the fat-containing food was measured for calculation of fatty acid intake. A control group was fed the fat-free diet for the 6-d period to allow a correction for the excretion of endogenous fatty acid.

- The first experiment compared the absorption of fat from the structured lipid (26.2% palmitic acid, of which 71.3% was at the *sn*-2 position) with that from a fat blend of a current proprietary infant formula (containing about half the amount of palmitic acid as the structured lipid, of which 8% was at the *sn*-2 position). The results showed that the fecal excretion of palmitic acid in the structured lipid group was significantly less than in the group fed the proprietary infant formula ($p < .05$), despite containing nearly twice as much palmitic acid. Fecal excretion of palmitic acid in the structured lipid group was only one fifth of that in the proprietary infant formula group.
- In the second experiment, Lien et al. showed that fecal excretion of palmitic acid after consumption of the structured lipid was approximately 1% of the total ingested amount of palmitic acid compared with approximately 18% in two similar blends (with 8 and 9% of the palmitic acid at the *sn*-2 position) [4]. A quantitative analysis showed that, in the structured lipid, 71.3% of the palmitic acid was at the *sn*-2 position (68% in human milk).
- The third experiment by Lien et al. was designed to determine the relationship between fat absorption and the proportion of palmitic acid at the *sn*-2 position of a fat blend [4]. For this experiment, a mixture of vegetable oils that had the same fatty acid composition as the structured lipid was blended [4]. This blend was mixed in various proportions with the structured lipid (100:0, 75:25, 50:50, 25:75, 0:100) to yield several fat blends with an *sn*-2 palmitic acid content of 5 to 79% of the total palmitic acid. The results of this third experiment showed that a lower proportion of palmitic acid at the *sn*-2 position resulted in an increased excretion of total fatty acids.
- In the fourth experiment, rats were fed diets containing the structured lipid (with 78.8% of the palmitic acid at the *sn*-2 position), the oil blend used in experiment 3 (with 4.8% of the palmitic acid at the *sn*-2 position and the same fatty acid profile as the structured lipid), or a 50:50 mixture of the two (with 41.8% of the palmitic acid at the *sn*-2 position) [4]. Analysis of the feces collected throughout this 6-d experiment showed that the total fat decreased from 350 to 160 and 70 mg as the percentage of palmitic acid at the *sn*-2 position increased from 4.8 to 41.8 and 78.8%.

These experiments demonstrated that the long-chain saturated fatty acids are best absorbed when present at the triacylglycerol *sn*-2 position [4]. Sanders et al. investigated the absorption, tissue distribution, and fate in the body of palmitic acid esterified to glycerol at the *sn*-1 (1-[1-^{14}C] palmitoyl, 2,3-dioleoyl glycerol [POO]) or *sn*-2 (1,3-dioleoyl, 2-[1-^{14}C]palmitoyl glycerol [OPO]) positions, particularly in neonatal and young animals [14]. These two palmitoyl triacylglycerols (both containing [1-^{14}C]-labeled palmitic acid ([^{14}C]P)) were synthesized. Whole-body autoradiography (WBA) and conventional radioanalytical techniques were used to monitor the absorption, distribution, and excretion in suckling and weanling rats.

From one litter, six suckling pups (four males and two females) were administered 0.1 ml [^{14}C]POO by oral gavage, and six (four males and two females) were administered 0.1 ml O[^{14}C]PO in the milk. From a second litter, six suckling pups (two males and four females) were administered 0.1 ml [^{14}C]POO by gavage, and six (two males and four females) were administrated 0.1 ml O[^{14}C]PO in the milk. At 2, 4, 8, 24, and 72 h after dosing, two rats of the same sex of each litter, one that received [^{14}C]POO and another that received O[^{14}C]PO, were sacrificed and a tissue assay was performed. At 8 h after dosing, one male pup from each diet group in each litter was sacrificed for WBA.

Of the weanling rats, five male rats were administered 1.5 ml [^{14}C]POO by oral gavage via a metal intubation tube and five were administered 1.5 ml O[^{14}C]PO in a dietary slurry. At 4, 8, 24, 48, and 96 h after dosing, one rat from each diet group was sacrificed for WBA. In the suckling rats, by using only two litters and taking only one rat from each diet group at each time point, it was not possible to determine whether there was any significant difference in the levels of [^{14}C] in the tissues. There was considerable interlitter variation in the [^{14}C] activity recorded in certain tissues taken from the similarly dosed rats at equivalent time points, for example, in the brown fat at 4 h after administration of [^{14}C]POO. However, levels of [^{14}C] in the brown fat of suckling and weanling rats were relatively high in both diet groups shortly after dosing and reached a peak after 4 h. The recoveries and rates of excretion of [^{14}C] in male weanling rats were similar for both of the ^{14}C-labeled palmitoyl triacylglycerols. Levels of [^{14}C] excreted in urine and feces were low, and excretion of [^{14}C]O_2 was high, providing evidence for extensive absorption. In suckling and weanling rats, the amount of [^{14}C] in the large intestine and fecal pellets appeared to be slightly (not significantly) higher in rats who received [^{14}C]POO compared with those who received O[^{14}C]PO. In rats, the *sn*-1 or *sn*-3 ester bond in the POO molecule is broken enzymatically by pancreatic lipases releasing free palmitic acid, which can then form insoluble calcium salts excreted in the feces. In contrast, the *sn*-2 ester bond of OPO is largely resistant to pancreatic lipases, and lipolysis by bile salt-stimulated lipase occurs but at a slower rate. As a result, 2-monopalmitin is produced, which is readily absorbed from the intestine. The slight (not statistically significant) differences found in fecal [^{14}C] activity between the groups of weanling rats in the study by Sanders et al. may reflect this difference in metabolism [14].

The effect of the positional distribution of stearic acid (18:0) and oleic acid (18:1 n-9) on the glycerol backbone, and the interaction with dietary calcium, on the apparent absorption of fat, energy, and nutrients were studied by Brink et al. [13]. It was hypothesized that saturated fatty acids esterified to the *sn*-1 and *sn*-3 positions may result in lower absorption of fat than saturated fatty acids esterified to the *sn*-2 position; that dietary calcium may further decrease absorption of fat by forming insoluble calcium soaps with the saturated fatty acids; and that this process may also result in decreased absorption of calcium. A total of 40 rats were fed purified diets containing either a fat blend high in 2-oleoyl-distearate (SOS; stearic acid esterified to the *sn*-1 and *sn*-3 positions) or a fat blend high in 1-oleoyl-distearate (SSO; stearic acid esterified to the *sn*-1 and *sn*-2 positions). Both diets were given at low (0.3 g/100 g) and high (1.0 g/100 g) concentrations of dietary calcium. This study showed that, at the low concentration of dietary calcium, the absorption of stearic acid in the rats fed the diet containing SOS

(84.7%) was significantly lower compared with those fed SSO (93.3%; $p < .001$). A high concentration of calcium in the diet decreased the absorption of stearic acid.

This effect was greater in the rats fed the diet containing SOS (45.8%) compared with those fed SSO (80.4%). The absorption of oleic acid was hardly influenced by its positional distribution. This is most likely a result of the presence of the double bond in this fatty acid. Total fat absorption was significantly lower ($p < .001$) in rats fed the diet containing SOS (low in calcium: 92.7%; high in calcium: 72.6%) compared with those fed SSO (low in calcium: 96.6%; high in calcium: 89.0%), with the effect being more pronounced with the high concentration of dietary calcium. At the low concentration of dietary calcium, magnesium absorption was decreased in rats fed the diet containing SOS (67.9%) compared with those fed SSO (71.9%), indicating that soaps of magnesium fatty acids were formed when luminal concentration of calcium was marginal [13]. The authors concluded that the decreased absorption of stearic acid from SOS compared with that of SSO may have been a result of increased formation of insoluble calcium and magnesium soaps in the intestine. Pufal et al. found an effect of positional distribution on the absorption of saturated fatty acids [2]. These investigators used the structured lipid to examine the effect of dietary triacylglycerol structure on lipoprotein metabolism. They compared the effect of two fats with similar fatty acid composition — but in which palmitic acid was primarily at the *sn*-1 and *sn*-3 (OOP) or *sn*-2 (OPO; the structured lipid) positions (107 disintegration per minute [dpm]/g of food for both OPO and OOP) — on chylomicron metabolism in the rat. The results indicated a relative enrichment of chylomicrons with palmitic acid when the rats were fed OPO compared with OOP. The investigators suggested that the greater amounts of palmitic acid found in chylomicrons from rats fed OPO may have been a result of more efficient absorption. In addition, a positive effect of the positional distribution of saturated fatty acids on their absorption has been shown by Innis et al. in another species [3]. They demonstrated that piglets fed synthesized triacylglycerol (with palmitic acid at the *sn*-2 position of the milk triacylglycerol) had significantly higher levels of palmitic acid cholesteryl ester (similar to that with piglets fed sow milk) compared with piglets fed the same fatty acids from conventional oils ($p \leq .08$). Furthermore, Innis et al. showed that the chain length of saturated fatty acids in infant formula also influenced the metabolism of dietary oleic acid, linoleic acid (LA), and α-linolenic acid in piglets [3]. Finally, when the formula contained amounts of palmitic acid similar to that in human milk, turnover of oleic acid and LA, and levels of arachidonic acid were improved.

These data show that the structured lipid (with its similarity to human milk fatty acid profile and in the positional distribution of palmitic acid) increased the absorption characteristics compared to fats with palmitic acid primarily at the *sn*-1 and *sn*-3 positions. Lien et al. reported that the excretion of palmitic acid and fecal fatty acid soaps was negatively correlated to the presence of palmitic acid at the *sn*-2 position [4]. As the proportion of palmitic acid at the *sn*-2 position was increased from 5 to 42 and 79%, the total fat in the feces decreased from 350 to 160 and 70 mg, respectively. Most of the decrease was observed in the amount of fatty acid soaps, which decreased more than tenfold from 267 to 23 mg. Neutral fat also decreased but to a lesser extent, so that the fatty acid soaps decreased from 76 to 55 and 33% of the total fat. The results of these animal studies on the structured lipid are summarized in Table 11.1.

TABLE 11.1
Summary of Animal Studies: Effect of the Positional Distribution of Fatty Acids on Absorption

Reference	Animal	Duration	Main Effects
[4]	Young male Sprague-Dawley rats, 90–110 g body weight	Fat-free diet for 10 d	*Experiment 1.* Fecal excretion of PA in Betapol' group was 1/5 of formula group ($p < .05$)
		Diets containing test fat for next 3 d	*Experiment 2.* Fecal excretion of PA from Betapol was ~1% compared with ~18% in the two fat blends *Experiment 3.* Decrease in the proportion of PA at the *sn*-2 position (78.8–4.8%), resulted in increased excretion of total fatty acids (0.15–1.65 meq) *Experiment 4.* No significant difference in fecal calcium content between the blend, the 50:50 mixture and Betapol; as the proportion of PA at the *sn*-2 position was increased from 5 to 42 and 79%, the total fat in the feces decreased from 350 to 160 and 70 mg; most of the decrease was observed in the amount of fatty acid soaps, which decreased more than tenfold from 267 to 23 mg
[14]	Suckling and weanling rats	Single administration	No significant difference between absorption of $[^{14}C]$POO and O$[^{14}C]$PO from the gut in suckling and weanling rats Radioactivity initially concentrated in brown fat with apparent migration to white fat of weanling rats by 96 h Levels of $[^{14}C]$ were low in blood, brain, and other tissues; excretion of $[^{14}C]$ was mainly by expiration of CO_2 (~72% in 96 h) Levels in urine and feces accounted for only ~6% of the excreted radioactivity In weanling and suckling rats the amount of $[^{14}C]$ in the large intestine and fecal pellets appeared to be slightly higher (n.s.) in rats administered $[^{14}C]$POO than those receiving O$[^{14}C]$PO
[13]	Male Wistar rats	4 weeks	Absorption of stearic acid in rats fed SOS (84.7%) was significantly lower compared with SSO (93.3%), in rats fed the low calcium diets ($p < .001$)

			Addition of calcium to the diet lowered the absorption of stearic acid; this effect was greater in the rats fed SOS (45.8%) compared with SSO (80.4%)
			Total fat absorption was significantly lower ($p < .001$) in rats fed SOS (92.7% in the low calcium diet, 72.6% in the high calcium diet) compared with rats fed SSO (96.6% in the low calcium diet, 89.0% in the high calcium diet), with the effect being more pronounced at the high level of dietary calcium
			At low levels of dietary calcium, magnesium absorption was decreased in rats fed SOS (67.9%) compared with those fed SSO (71.9%)
[2]	Male Wistar rats	Feeding for 1 week before chylomicron collection	Triacylglycerol in chylomicrons from animals fed OPO was relatively enriched in PA, at the expense of rats stearic acid, oleic acid, and LA
			No significant differences could be seen in the *in vitro* hydrolysis of chylomicron triacylglycerol from animals fed the two fats labeled with [^{14}C]PA
			Following hydrolysis, PA was released as free fatty acid from chylomicrons isolated from OOP-fed animals, but within 2-monoacylglycerol from those fed OPO
			The enrichment of chylomicrons with PA in animals fed O[^{14}C]PO resulted in increased delivery of [^{14}C]PA to the liver
[3]	Newborn male piglets	Bottle feeding and assignment to formula 1–4 until day 17th after birth	Proportions of oleic acid and LA in plasma were higher in piglets fed MCT (21.7 and 22.3%, male piglets, respectively) or coconut oil (17.7 and 21.0%, respectively) compared with Betapol (15.8 and 21.8%, respectively) or sow milk (13.3 and 13.2%, respectively; $p < .0125$)
		Blood sampling on day 18	Piglets fed Betapol™ had higher PA cholesterol ester (21.6%; identical to sow milk) compared with other formulas (9.3–12.2%; $p \leq .08$)

LA, linoleic acid; MCT, medium-chain triacylglycerols; n.s., not significant; O[^{14}C]PO, 1,3 dioleoyl,2-[1-^{14}C]palmitoyl glycerol; PA, palmitic acid; [^{14}C]POO, 1-[1-^{14}C]palmitoyl,2,3-dioleoyl glycerol; SOS, 2-oleoyl-distearate; SSO, 1-oleoyl-distearate; WBA, whole-body autoradiography.

11.5 STRUCTURED LIPID IN PRETERM INFANTS

Carnielli et al. investigated whether palmitic acid at the *sn*-2 position of dietary triacylglycerols in amounts similar to those found in human breast milk improved the absorption of fat, fatty acids, and calcium in preterm infants [15]. They used synthetic triacylglycerols that differed only in the positional distribution of palmitic acid. In a crossover study, 12 preterm infants born after a gestation of 28 to 32 weeks were randomly assigned to be fed an α formula (25.7% palmitic acid, of which 9.8% was at the *sn*-2 position) or the β formula (the structured lipid, 25.4% palmitic acid, of which 58.0% was at the *sn*-2 position) for 1 week at a postnatal age of 38 ± 7 d. Half the infants were first fed the α formula and then the β formula, and the other half were first fed the β formula and then the α formula. Fatty acids, fat, and mineral balances were measured at the end of each week. When fed the β formula, the infants had lower concentrations of myristic (14:0), palmitic, and stearic acids ($p < .001$) and higher concentrations of oleic acid ($p < .001$) and LA ($p = .01$) in feces compared with the α formula. The intestinal absorption of myristic, palmitic, and stearic acids was better with the β formula compared with the α formula ($p < .01$). With the β formula, total fat excretion was on average 0.2 g/kg/d lower and fat absorption was 5% higher compared with the α formula (results not statistically significant). Fecal excretion of calcium was lower (58.8 mg/kg/d) with the β formula compared with the α formula (82.0 mg/kg/d; $p < 0.05$), and urinary excretion of calcium was higher (4.0 vs. 2.3 mg/kg/d, respectively; $p < .05$). There was no significant difference in the intestinal absorption of phosphorus with either formula, but urinary excretion of phosphorus was lower with the β formula compared with the α formula (114 vs. 16.7 mg/kg/d, respectively; $p < .02$).

Carnielli et al. found correlations between fecal excretion of calcium and the excretion of fat and major fatty acids [15]. The fecal calcium content was significantly correlated with myristic ($r = .42$; $p = .04$), palmitic ($r = .46$; $p = .039$) and stearic ($r = .42$; $p = .025$) acids in the diet (i.e., with the major saturated fatty acids, but not the mono- or polyunsaturated fatty acids). The authors concluded that the positional distribution of fatty acids in dietary triacylglycerols had significant effects on the intestinal metabolism of the fatty acids. The β formula (which contained triacylglycerols with a structure similar to that in human milk triacylglycerols: 25.4% of the fatty acids was palmitic acid, predominantly esterified to the *sn*-2 position) was associated with an improvement in the absorption of myristic, palmitic, and stearic acids and in mineral balance.

The results of the study of Carnielli et al. [15] correspond with those found by Lucas et al. [16], which showed that a formula containing the structured lipid or a formula fat rich in palmitic acid at the *sn*-2 position improved the absorption of palmitic and stearic acids. In this 5-d, randomized study, 24 preterm infants were fed one of the following formulae: a formula containing the synthetic fat — the structured lipid with 73.9% of palmitic acid at the *sn*-2 position, a diet containing 8.4% of palmitic acid at the *sn*-2 position (diet A), or a diet containing 27.8% of palmitic acid at the *sn*-2 position (diet B). The results of this study showed that the proportion of palmitic acid absorbed from the formula was greatest in those fed the diet containing the structured lipid (91%) and was significantly less in infants receiving diets A or B (both 79%; $p < .03$, $p < .01$, respectively). In addition, this study also showed a significant difference between the treatment groups in the proportion of consumed fat that was excreted in the stools as calcium soaps ($p < .02$, by analysis of variance): in infants fed diet B (7.2%) it was more than twofold greater compared with those fed the structured lipid diet (3.3%, $p < .01$). The estimated mean fraction of dietary calcium absorbed from the diet was lower when fed diets A (43.9%) and B (40.0%) compared with the structured lipid diet (57.0%). Considered individually, these differences were not significant. However, when the data from infants fed diets A and B were combined, the calcium absorption (42.1%) was significantly lower compared with the structured lipid diet ($p < .03$). The authors concluded that the stereoisomeric structure of triacylglycerols in formulas for preterm infants may have substantial effects on the absorption of fatty acids and calcium.

Insufficient absorption and retention of calcium clearly leads to problems in preterm infants. They often have periods of illness when adequate calcium nutrition is difficult to achieve, and they are frequently discharged from hospital with bones that are undermineralized and remain so for months. Therefore, the results of Lucas et al. are of considerable clinical relevance [16].

The data of Carnielli et al. [17] showed that preterm infants (born after a gestation of 28 to 32 weeks) fed formulae in which palmitic acid was predominantly located at the *sn*-2 position of triacylglycerol — rather than at the *sn*-1 and *sn*-3 positions — had changes in the plasma levels of fatty acids that were consistent with enhanced absorption of palmitic acid from the *sn*-2 position compared with the *sn*-1 and *sn*-3 positions [7]. In their crossover study, seven preterm infants were fed with α and β formulas (1 week per formula) [17]. The β formula contained triacylglycerols resembling the stereoisomeric structure of human milk fat (25.4% palmitic acid by weight, 58% of which was at the *sn*-2 position). In the α formula, 25.7% was palmitic acid by weight, of which 87.3% was at the *sn*-1 and *sn*-3 positions.

No significant effects of these formulae on the essential fatty acid content of plasma phospholipids and sterol esters were observed. The sterol esters, free fatty acids and triacylglycerols in the plasma of the infants when fed the β formula contained significantly higher proportions of palmitic acid (18.90, 31.90, and 29.27%, respectively) compared with the α formula (14.96, 28.14, and 24.86%, respectively; $p \leq .05$). Lower proportions of LA were found in triacylglycerols of the infants fed with the β formula compared with those fed the α formula (β: 12.99%, α: 15.13%). The authors explained this increase of plasma triacylglycerols and free fatty acids when fed the structured lipid formula as a result of better fat absorption of the major dietary saturated fatty acids because of the higher plasma concentrations of triacylglycerol and free fatty acids of palmitic acid during this period. The lower proportion of LA may have been, according to the authors, a result of a displacement effect or percentage effect by the greater contribution of palmitic acid and other saturated fatty acids to plasma triacylglycerols and free fatty acids. Despite lower proportions of LA in plasma triacylglycerols, Carnielli et al. considered it unlikely that the

TABLE 11.2
Overview of Human Studies: Effect of Betapol on Absorption in Preterm Infants

Reference	Duration	Main Effects of Betapol
[15]	1 week per formula	The β formula resulted in lower concentrations of myristic, palmitic, and stearic acids ($p < .001$) and higher concentrations of oleic acid ($p < .001$) and LA ($p = .01$) in feces compared with the α formula
		Intestinal absorption of myristic, palmitic, and stearic acid was better after the β formula compared with the α formula ($p < .01$)
		Fecal excretion of calcium was lower after the β formula (58.8 mg/kg/d) compared with the α formula (82.0 mg/kg/d; $p < .05$)
		Total fat excretion was lower (n.s.) and fat absorption higher (n.s.) with the β formula compared with the α formula
		Fecal excretion of calcium was found to correlate with fecal excretion of fat and major saturated fatty acids: myristic ($r = .42$; $p = .04$), palmitic ($r = .46$; $p = .039$), and stearic ($r = .42$; $p = .025$)
[16]	5 d	The proportion of PA absorption from formula was greatest in those fed Betapol (91%) and significantly less compared with diet A or B (both 79%; $p < .03$; $p < .01$, respectively)
		The proportion of consumed fat excreted in stools as calcium soaps was higher with diet B (7.2%), and was more than twofold greater than with the Betapol diet (3.3%, $p < .01$)
		The estimated mean fraction of dietary calcium absorbed was lower in diet A (43.9%) and diet B (40.0%) compared with the Betapol diet (57.0%) (n.s.). When the results for diets A and B were combined, calcium absorption (42.1%) was significantly lower compared with the Betapol diet (57.0%; $p < .03$)
[17]	1 week per formula	No significant effects were found of study formulas on the essential fatty acid content of plasma phospholipids and sterol esters
		The plasma levels of sterol esters, free fatty acids, and triacylglycerols in the infants when fed the β formula contained significantly higher proportions of PA (18.90, 31.90, and 29.27%, respectively) compared with the α formula (14.96, 28.14, and 24.86%, respectively; $p \le .05$)
		LA was found in lower proportions in triacylglycerols (β: 12.99%, α: 15.13%) in the infants when fed β formula compared with the α formula

LA, linoleic acid; n.s., not significant; PA, palmitic acid.

α formula: 25.7% PA, of which 9.8% at *sn*-2; β formula (containing Betapol): 25.4% PA, of which 58.0% at *sn*-2; Diet A, 8.4% of the PA at *sn*-2; Diet B, 27.8% of the PA at *sn*-2.

structured lipid formula would result in a significant reduction in the supply of LA to developing tissues and brain, because LA was present in ample amounts in both formulas. Fecal excretion of calcium was significantly reduced in preterm infants fed formulas containing Betapol or formulas with palmitic acid at the *sn*-2 position compared with standard formula [7,15].

In line with these data, Lucas et al. [16] demonstrated that formulas that contained Betapol or were rich in palmitic acid at the *sn*-2 position improved calcium absorption in preterm infants [7] and reduced the formation of insoluble calcium soaps in the stool.

Calcium fatty acid soaps have been implicated in the development of milk bolus obstruction in preterm infants and may also be related to the observation that breast-fed infants have softer stools than formula-fed infants. Quinlan tested this hypothesis [7]. Stools were collected from 30 infants and were visually classified according to hardness and then chemically analyzed.

Formula-fed infants had significantly harder stools compared with breast-fed infants. The absolute levels of fatty acid soaps in stools from breast-fed infants were nine times lower than in formula-fed infants. In the latter, an average of 90% of the total fatty acids excreted was present as soaps, representing nearly 30% of the dry weight of the stool. By contrast, breast-fed infants excreted an average of only 3% of the dry weight of the stool as soaps.

Quinlan [7] suggested that these data provide convincing evidence that stool hardness is related to the extent of excretion of calcium soaps. Thus, manipulation of the structure of dietary triacylglycerol to reduce the excretion of soaps may therefore result in softer stools and a reduction in the incidence of constipation. In this study of term infants, softer stools were indeed reported in infants fed the structured lipid compared with those fed the control formula [7].

By minimizing fecal excretion of calcium, triacylglycerol in which palmitic acid is predominantly at the *sn*-2 position not only influences the softness of stools, but can also influence bone development as a result of biologically significant improvements in calcium absorption [7].

The results of these studies on the effect of Betapol on absorption and excretion of fatty acids in preterm infants are summarized in Table 11.2.

11.6 STRUCTURED LIPID IN TERM INFANTS

It has been speculated that the advantage of the *sn*-2 position over the *sn*-1 and *sn*-3 positions may be even greater in term infants [15]. Lipid digestion and absorption are reported to be better in term infants than in preterm infants because of a greater contribution of pancreatic lipase and more abundant bile salts. The latter are essential for monoglyceride absorption. Carnielli et al. studied the effect of three formulas (with identical fatty acid composition that differed only in triacylglycerol structure) on the intestinal absorption of fat and calcium in term infants [18]. Healthy male term infants were randomly assigned to receive one of the three study formulas from birth until at least an age of 5 weeks (nine infants per group). The structured lipid formula was designed to resemble human milk fat most closely: 23.9% of the formula was palmitic acid, predominantly (66%) esterified to the *sn*-2 position. The regular formula was identical to a currently marketed infant formula (19% palmitic acid, mainly esterified to the *sn*-1 and *sn*-3 positions). The third formula was designed to be an intermediate between the structured lipid and regular formulas; it contained a lower amount of palmitic acid in the *sn*-2 position compared with the structured lipid formula. The intermediate formula consisted of 24% palmitic acid of which 39% was at the *sn*-2 position. Excretion of fat in the infants fed the structured lipid formula (0.15 g/kg/d) was significantly lower compared with the other two formulae (intermediate: 0.44 g/kg/d; regular 0.68 g/kg/d; $p < .05$). The intestinal absorption of the major saturated fatty acids, that is, lauric acid (95.5%), myristic acid (83.7%), palmitic acid (96.5%), and stearic acid (91.2%), was significantly greater in infants fed the structured lipid formula compared with the other two formulae (intermediate: 81.3, 30.6, 86.3, and 82.3%, respectively; regular: 76.1, 11.0, 78.1, and 75.5%, respectively; $p < .05$). The fecal excretion of calcium in infants fed the structured lipid formula (43.3 mg/kg/d) was significantly lower compared with the other two formulae (intermediate: 59.9 mg/kg/d; regular: 68.4 mg/kg/d; $p < .05$), resulting in significant improvement in calcium absorption (the structured lipid: 53.1%; intermediate: 35.4%; regular: 32.5%; $p < .05$) and better mean retention values (the structured lipid: 42.8 mg/kg/d; intermediate: 26.9 mg/kg/d; regular: 27.4 mg/kg/d; not significant). Significant correlations between fecal excretion of calcium and excretion of fat and major fatty acids were found: palmitic acid showed the highest correlation coefficient ($r = .84$), followed by oleic acid ($r = .60$) and LA ($r = .51$).

In their article, Carnielli et al. [18] compared their results with those of Filer et al. [19] who had studied the influence of triacylglycerol structure on the absorption of fatty acids in term infants. In this study, a formula based on natural lard (with palmitic acid primarily at the *sn*-2 position) was compared with a formula based on a random lard (fatty acids randomized and thus equally distributed between the *sn*-1, *sn*-3, and *sn*-2 positions). The infants fed the formula based on natural lard showed an improved absorption of all fatty acids (most markedly palmitic and stearic acids) but calcium absorption did not improve.

The fact that no effect on calcium absorption was observed by Filer et al. was explained as follows: Carnielli et al. used synthetic triacylglycerol (the structured lipid; with palmitic acid predominantly at the *sn*-2 position) and compared this with fat in which palmitic acid was predominantly at the *sn*-1 and *sn*-3 positions [18]. In the reference fat of the study by Filer et al., the fatty acids were randomized and more or less equally distributed among the *sn*-1, *sn*-3, and *sn*-2 positions. In addition, the differences in calcium concentration between the formulas used by Filer et al. (69 and 72 mg/100 ml) [19] and those used by Carnielli et al. (52.5 and 54.0 mg/100 ml) [18] may have had a role. In the study by Carnielli et al., the fecal excretion of calcium exceeded the excretion of palmitic acid of 0.5 mmol/kg/d. If the formulae had a lower calcium content, the fecal calcium concentration would have been less, and the ratio of palmitic acid to calcium would have been higher (nearer 1:1). In such a case, the effect of fecal palmitic acid (and of palmitic acid at the *sn*-2 position) on the excretion of calcium might be more pronounced because of the larger proportion of calcium available for the formation of calcium palmitate and thus for fecal losses. Conversely, when the calcium concentration in formulae is more than double the amount in human breast milk, the effects of fecal palmitic acid and palmitic acid at the *sn*-2 position on calcium absorption would no longer be measurable.

Carnielli et al. also found that the stool samples from infants fed the structured lipid formula were softer compared with the intermediate and the regular formula [18]. In addition, Carnielli et al. reported that the variation in stool hardness was not due to a significant difference in water content of the infants' feces [18]. The authors suggested that the lipid content (including calcium soaps) of the stools has a much more pronounced effect on stool hardness compared with its water content.

Data from a study by Kennedy et al. [20] showed that infants receiving the structured lipid had softer stools at 6 to 12 weeks and a lower proportion of fatty acid soaps in their stool compared with infants fed the control formula [7]. The results from breast-fed infants were similar to infants fed the structured lipid. In the same study, the effects of the positional distribution of fatty acids on bone mineralization in healthy term neonates was also evaluated. These infants were randomly assigned to receive standard formula (12% of palmitic acid at the *sn*-2 position; $n = 103$) or formula high in palmitic acid at the *sn*-2 position (50% of palmitic acid at the *sn*-2 position; $n = 100$) for 12 weeks. A total of 120 breast-fed infants was also studied. Infants receiving the formula high in palmitic acid at the *sn*-2 position had significantly higher whole-body bone mineral content compared with the control formula (128.1 ± 9.7 vs. 122.7 ± 10.1 g, respectively, adjusted for size and sex; $p < .05$). The adjusted whole-body bone mineral content values of breast-fed infants (128.3 ± 9.1 g) were similar to those of infants fed formula high in palmitic acid at the *sn*-2 position and significantly higher than those of infants fed the control formula ($p < .02$). These studies on the effect of the structured lipid on absorption and excretion of fatty acids in term infants are summarized in Table 11.3.

11.7 CONCLUSIONS

Dietary lipids have an essential role as a source of energy. Both the fatty acid composition of the diet and the structure of triacylglycerols are important. Palmitic acid is the most abundant saturated fatty acid in human milk, in which it is esterified to the *sn*-2 position of triacylglycerols. In current blends of vegetable oil, palmitic acid is positioned at the *sn*-1 and *sn*-3 positions and thus forms insoluble calcium soaps in the infant's intestine, leading both to a loss of energy because of the reduced absorption of fat and the formation of hard stools due to reduced absorption of calcium, as well as affecting bone mineralization.

In the structured lipid Betapol, palmitic acid is predominantly esterified to the *sn*-2 position (similar to human milk). Studies in animals, and preterm and term infants have shown that infant formulae containing structured triacylglycerol with palmitic acid at the *sn*-2 position (Betapol) improve the absorption of dietary fat and calcium, and soften the stools as a result of the reduced formation of calcium soaps. These differences have biological and clinical significance in the infant (i.e., improvement of bone mineralization and reduction in the incidence of constipation).

TABLE 11.3
Overview of Human Studies: Effect of Betapol on Absorption in Term Infants

Reference	Duration	Main Effects of Betapol
[18]	From birth until at least an age of 5 weeks	Excretion of fat after β formula (0.15 g/kg/d) was lower compared with the other two formulae (intermediate 0.44 g/kg/d; regular 0.68 g/kg/d; $p < .05$)
		Absorption of fat in the intestines was significantly higher in infants receiving Betapol (97.6%) compared with those receiving the intermediate (93%) or regular formulae (90%; $p < .05$)
		Intestinal absorption of lauric acid (95.5%), myristic acid (83.7%), PA (96.5%), and stearic acid (91.2%) was significantly greater with the β formula compared with the other two formulae (intermediate 81.3, 30.6, 86.3, and 82.3%, respectively; regular 76.1, 11.0, 78.1, and 75.5%, respectively; $p < .05$)
		Fecal excretion of calcium in infants fed the β formula (43.3 mg/kg/d) was significantly lower compared with those fed the other two formulae (intermediate 59.9 mg/kg/d; regular 68.4 mg/kg/d; $p < .05$)
		Correlation between the fecal excretion of calcium and the excretion of PA ($r = .84$) and oleic acid ($r = .60$) and LA ($r = .51$; $p < .05$)
[20]	12 weeks	Infants receiving formula high in PA at the *sn*-2 position had significantly higher whole-body bone mineral content (128.1 ± 9.7 vs. 122.7 ± 10.1 g, adjusted for size and sex) compared with control formula ($p = .05$)
		Breast-fed infants had adjusted whole-body bone mineral content values (128.3 ± 9.1 g) similar to those of infants fed formula high in PA at the *sn*-2 position and significantly higher than those of infants fed the control formula ($p < .02$)
		Infants receiving formula high in PA at the *sn*-2 position had softer stools at 6 and 12 weeks and lower proportions of fatty acid soaps in their stool compared with the control formula; the softer stools were similar to those of breast-fed infants

LA, linoleic acid; PA, palmitic acid.

β formula (containing Betapol): 23.9% PA, of which 66% at *sn*-2; Intermediate formula: 24% PA, of which 39% at *sn*-2; Regular formula: 19% PA, mainly esterified to *sn*-1 and *sn*-3, only 13% at *sn*-2; Standard formula: 12% of PA at *sn*-2 position; High in PA at *sn*-2 formula: 50% of PA at *sn*-2 position.

REFERENCES

1. Fats and oils in human nutrition. Report of a joint expert consultation. FAO Food and Nutrition Paper 57, 1998.
2. Pufal D.A., Quinlan, P.T., and Salter, A.M. Effect of dietary triacylglycerol structure on lipoprotein metabolism: a comparison of the effects of dioleoylpalmitoylglycerol in which palmitate is esterified to the 2- or 1(3)-position of the glycerol. *Biochim Biophys Acta,* 1258, 41, 1995.
3. Innis S.M., Quinlan, P., and Diersen-Schade, D. Saturated fatty acid chain length and positional distribution in infant formula: effects on growth and plasma lipids and ketones in piglets. *Am J Clin Nutr,* 57, 382, 1993.
4. Lien, E.L., Boyle, F.G., Yuhas, R., Tomarelli, R.M., and Quinlan, P. The effect of triglyceride positional distribution on fatty acid absorption in rats. *J Pediatr Gastroenterol Nutr,* 25, 167, 1997.
5. Meijer, G.W. and Korver, O. Interesterification: the nutritional consequences, in *Fat in the Diet. Proceedings of the 21st World Congress of the International Society for Fat Research (ISF),* Oct 1995; The Hague, Netherlands. PJ Barnes & Associates, Bridgwater, UK, 1996, p. 17.
6. Bracco, U. Effect of triglyceride structure on fat absorption. *Am J Clin Nutr,* 60 Suppl, 1002S, 1994.
7. Quinlan P. Structuring fats for incorporation into infant formulas, in *Fat in the Diet. Proceedings of the 21st World Congress of the International Society for Fat Research (ISF),* Oct 1995; The Hague, Netherlands. PJ Barnes & Associates, Bridgwater, UK, 1996, p. 21.
8. Mu, H. and Høy, C.E. Intestinal absorption of specific structured triacylglycerols. *J Lipid Res,* 42, 792, 2001.
9. Heird, W.C., Grundy, S.M., and Hubbard V.S. Structured lipids and their use in clinical nutrition. *Am J Clin Nutr,* 43, 320, 1986.
10. Jensen, M.M., Christensen, M.S., and Høy, C.E. Intestinal absorption of octanoic, decanoic, and linoleic acids: effect of triglyceride structure. *Ann Nutr Metab,* 38, 104, 1994.
11. Porsgaard, T. and Høy, C.E. Lymphatic transport in rats of several dietary fats differing in fatty acid profile and triacylglycerol structure. *J Nutr,* 130, 1619, 2000.
12. Jandacek, R.J. The solubilization of calcium soaps by fatty acids. *Lipids,* 26, 250, 1991.
13. Brink, E.J., Haddeman, E., De Fouw, N.J., and Weststrate, J.A. Positional distribution of stearic acid and oleic acid in a triacylglycerol and dietary calcium concentration determines the apparent absorption of these fatty acids in rats. *J Nutr,* 125, 2379, 1995.
14. Sanders, D.J., Howes, D., and Earl, L.K. The absorption, distribution and excretion of 1- and 2-[^{14}C]palmitoyl triacylglycerols in the rat. *Food Chem Toxicol,* 39, 709, 2001.
15. Carnielli, V.P. et al. Feeding premature newborn infants palmitic acid in amounts and stereoisomeric position similar to that of human milk: effects on fat and mineral balance. *Am J Clin Nutr,* 61, 1037, 1995.
16. Lucas, A. et al. Randomised controlled trial of a synthetic triglyceride milk formula for preterm infants. *Arch Dis Child Fetal Neonatal Ed,* 77, F178, 1997.
17. Carnielli, V.P. et al. Effect of dietary triacylglycerol fatty acid positional distribution on plasma lipid classes and their fatty acid composition in preterm infants. *Am J Clin Nutr,* 62, 776, 1995.

18. Carnielli, V.P. et al. Structural position and amount of palmitic acid in infant formulas: effects on fat, fatty acid, and mineral balance. *J Pediatr Gastroenterol Nutr,* 23, 553, 1996.
19. Filer, L.J. Jr, Mattson, F.H., and Fomon, S.J. Triglyceride configuration and fat absorption by the human infant. *J Nutr,* 99, 293, 1969.
20. Kennedy, K. et al. Double-blind, randomized trial of a synthetic triacylglycerol in formula-fed term infants: effects on stool biochemistry, stool characteristics, and bone mineralization. *Am J Clin Nutr,* 70, 920, 1999.

12 Cocoa Butter, Cocoa Butter Equivalents, and Cocoa Butter Substitutes

Vijai K.S. Shukla

CONTENTS

12.1 INTRODUCTION

Chocolate is associated with an imported commodity, the cocoa bean. Approximately 3.5 million tons of cocoa beans are produced worldwide each year. Cocoa beans come from the cacao (ka-ka-o) tree, *Theobroma cacao. Theobroma* means "food of the gods." The tree is cultivated in West Africa, South America, Central America, and the Far East. Ivory Coast is the leading producer, contributing 25% of the world's production.

12.2 COCOA BUTTER

The total fat content of the whole bean on a dry basis is around 48 to 49% and triglyceride is the major storage component. A mature cocoa bean can store up to 700 mg of cocoa butter. Since a tree may produce as many as 2000 seeds a year, a single tree could yield up to 15 kg of cocoa butter annually.

Cocoa butter is the most expensive constituent of the chocolate formulations and an extremely important component. It is composed predominantly of (>75%) symmetrical triglycerides with oleic acid in the 2-position. Approximately 20% of triglycerides are liquid at room temperature and cocoa butter has a melting range of 32 to 35°C and softens at around 30 to 32°C. This is essential to the functionality of cocoa butter in its applications. It contains only trace amounts of the unsymmetrical triglycerides (PPO, PSO, and SSO). P = palmitic acid, O = oleic acid, and S = stearic acid; the order of letters indicates the position of the acids in the triglyceride molecule.

The unique triglyceride composition together with the extremely low level of diglycerides gives cocoa butter its desirable physical properties and its ability to recrystallize during processing in a stable crystal modification. The complexity of the crystallization of cocoa butter is because triglycerides can crystallize in a number of different crystal modifications, dependent on triglyceride composition and on crystallizing and tempering conditions during manufacturing and storage.

The fatty acid composition, various analytical constants, and triglyceride composition of different cocoa butters are presented in Tables 12.1 to 12.3, respectively [1,2]. These results show that Malaysian cocoa butter contains maximum amounts of monounsaturated triglycerides. Brazilian cocoa butter contains a minimum amount of monounsaturated triglycerides and maximum amounts of other unsaturated triglycerides. Cocoa butters from India and Sri Lanka are quite close to Malaysian in terms of the hardness and triglyceride composition.

There is a good correlation between the triglyceride composition and solid fat content of these cocoa butters. Malaysian, Sri Lankan, and Indian cocoa butters are the hardest and Brazilian is the softest. The quality of the Brazilian cocoa butter can be improved by mixing it with Malaysian cocoa butter, which will result in higher solid fat content

TABLE 12.1
Fatty Acid Composition of Various Cocoa Butters by GLC (wt%)

Sample Cocoa Butter	C14	C16	C16:1	C17	C18	C18:1	C18:2	C18:3	C20	C20:1	C22	C24
Ghana	0.1	24.8	0.3	0.3	37.1	33.2	2.6	0.2	1.1	Trace	0.2	0.1
India	0.1	25.3	0.3	0.2	36.2	33.5	2.8	0.2	1.1	0.1	0.2	Trace
Brazil	0.1	23.7	0.3	0.2	32.9	37.4	4.0	0.2	1.0	0.1	0.2	Trace
Nigeria	0.1	25.5	0.3	0.3	35.8	33.2	3.1	0.2	1.1	0.1	0.2	0.1
Ivory Coast	0.1	25.4	0.3	0.2	35.0	34.1	3.3	0.2	1.0	0.1	0.2	0.1
Malaysia	0.1	24.8	0.3	0.3	37.1	33.2	2.6	0.2	1.1	Trace	0.2	0.1

TABLE 12.2
Analytical Constants of Various Cocoa Butters

Sample Cocoa Butter	IV	C3[a]	% DAG	% FFA	Pulsed NMR Extended[b]				Pulsed NMR BS 684 Method[c]			
					20 C	25 C	30 C	35 C	20 C	25 C	30 C	35 C
Ghana	35.8	32.2	1.9	1.53	84.0	78.0	36.0	0.1	76.0	69.6	45.0	1.1
India	34.9	32.4	1.5	1.06	88.1	83.3	44.7	1.8	81.5	76.8	54.9	2.3
Brazil	40.7	32.0	2.0	1.24	67.7	56.6	18.5	0.6	62.6	53.3	23.3	1.0
Nigeria	35.3	33.1	2.8	1.95	83.7	77.3	35.4	0.1	76.1	69.1	43.3	0
Ivory Coast	36.3	32.0	2.1	2.28	82.3	74.8	32.7	0.9	75.1	66.7	42.8	0
Malaysia	34.2	34.3	1.8	1.21	89.3	83.7	49.6	1.8	82.6	77.1	57.7	2.6
Sri Lanka	35.2	33.2	1.1	1.58	—	—	—	—	79.7	74.2	50.4	1.1

IV, iodine value; DAG, diacylglycerols; FFA, free fatty acids.

[a] Melting point stabilization 64 h at 25°C.
[b] Tempering 64 h at 20°C.
[c] BS 684: British Standard method 684; tempering 40 h at 26°C.

at various temperatures. At the International Food Science Centre we have measured diglyceride levels varying between 1.5 and 2.8%. Higher diglyceride levels affect the crystallization of cocoa butters remarkably, and thus all efforts should be made to reduce these levels in good quality cocoa butters. The main drawbacks of Malaysian cocoa beans are their excessive acidic flavor, weak chocolate flavor, and certain other off-flavors. Several attempts have been made to improve these characteristics.

The deodorization of cocoa butter is necessary to reduce free fatty acid content and to give a product that satisfies the present-day requirement of neutral bland flavor. Deodorization is at least partially a suitable method for eliminating chlorinated insecticides from cocoa butter. The normal deodorization temperatures are in the range of 160 to 180°C. The oxidative stability of various cocoa butters listed in Table 12.4 shows extremely high values and these are unaffected during the deodorization process. Stability against oxidation depends on natural antioxidants present in cocoa butters. The tocopherol composition in Table 12.5 shows a predominance of γ-tocopherol and total tocopherol levels ranging between 100 and 300 mg/kg. Dimick and colleagues have found phospholipid levels ranging from 3.62 to 4.72 μg per 500 μg in cocoa butters of differing origin (Table 12.6) [3]. They are extending the scope of their research for determining the influence of phospholipids during the crystallization of hard and soft cocoa butters.

The results of thermorheographic (TRG) experiments presented in Table 12.7 show very high correlation with the pulsed nuclear magnetic resonance (NMR) data in Table 12.2. Thus, Malaysian cocoa butter crystallizes quickly and Brazilian is the slowest crystallizing, which correlates with the hardness of these cocoa butters as measured by pulsed NMR technology.

TABLE 12.3
Triglyceride Composition (mol%) of Various Cocoa Butters by High-Speed Liquid Chromatography

	Tri-glyc	Ghana	India	Brazil	Ivory Coast	Malaysia	Sri Lanka	Nigeria
Trisaturated	PPS	0.3	0.6	Trace	0.3	0.8		0.3
	PSS	0.4	0.5	Trace	0.3	0.5	1.9	0.5
Total		0.7	1.1	Trace	0.6	1.3	1.9	0.8
Monounsat.	pS	40.1	39.4	33.7	39.0	40.4	40.2	40.5
	SOS	27.5	29.3	23.8	27.1	31.0	31.2	28.8
	POP	15.3	15.2	13.6	15.2	15.1	14.8	15.5
	SOA	1.1	1.3	0.8	1.3	1.0	1.0	1.0
Total		84.0	85.2	71.9	82.6	87.5	87.2	85.8
Diunsat.	PLiP	2.5	2.0	2.8	2.7	1.8	2.5	2.2
	POO	2.1	1.9	6.2	2.7	1.5	2.3	1.7
	PLiS	3.6	3.1	3.8	3.6	3.0	1.4	3.5
	SOO	3.8	3.3	9.5	4.1	2.7	3.9	3.0
	SLiS	2.0	1.7	1.8	1.9	1.4		1.8
	AOO		0.8		0.5	0.5		0.5
Total		14.0	12.8	24.1	15.5	10.9	10.1	12.7
Polyunsat.	PliO	0.6	0.5	1.5	0.8	0.3	0.8	0.4
	000	0.4	Trace	1.0	Trace			Trace
	SLiO	0.3		1.2	0.5	Trace		0.3
	Alia		0.4					
	liOO			0.3				
Total		1.3	0.9	4.0	1.3	0.3	0.8	0.7

The addition of milk fat to cocoa butter results in marked lowering of the melting point adversely affecting the crystallization behavior and the hardness as shown in Table 12.8. An obvious decrease is clearly evident in the solid fat content and deterioration in solidification properties as shown in the values of the Jensen curve.

TABLE 12.4
Oxidative Stability of Cocoa Butters (CBs)

Samples	Induction Time at 120°C (h)
CB Trinidad	42.3
CB Brazil	35.3
CB Colombia	38.4
CB Venezuela	41.3
CB Ecuador	19.1
CB Ivory Coast	42.9
CB Ghana	42.2

TABLE 12.5
Tocopherol Composition of Various Cocoa Butters by High-Speed Liquid Chromatography

	Tocopherols (mg/kg)							
Sample	Total	α-T	β-T	γ-T	δ-T	α-TT	-TT	δ-TT
Brazil	176	0.7	1.2	164	6.9	—	2.0	0.7
Ghana	198	2.7	1.5	183	6.8	0.6	2.3	0.7
India	265	6.5	2.2	245	9.1	—	2.3	—
Ivory Coast	126	0.4	0.4	117	6.2	—	2.3	—
Malaysia	149	0.5	—	140	7.4	—	0.6	—

T, tocopherol; TT, tocotrienol.

TABLE 12.6
Phospholipid Quantities of Origin Cocoa Butter Samples (% by Weight)

Cocoa Butter	Phospholipid (μg)	% by Weight[a]
Malaysia	3.62	0.72
Ivory Coast	4.35	0.87
Ghana	4.72	0.94
Ecuador	3.80	0.76
Dominican Republic	4.72	0.94
Brazil	4.54	0.91

[a] Ratio of phospholipid mass divided by sample mass (500 μg) times 100.

TABLE 12.7
Thermorheographic (TRG) Values of Different Types of Cocoa Butters

	Measuring Temperature: 22 C		
	Times of Different Melting Points (min)		
	30 mp	50 mp	80 mg
Malaysia	13	16	20
Sri Lanka	12	14	15
Ghana	31	36	42
Brazil	178	187	199

TABLE 12.8
Analytical Constants of the Blends of Cocoa Butters (CB) and Milk Fats (MF)

Sample	IV (Wijs)	Pulsed NMR BS684 Method 2				Solidification Curve (Jensen)		
		20 C	25 C	30 C	35 C	Max. Temp. (C)	Time min/ max (min)	Temperature Rise (C)
CB Malaysia	35.6	82.1	78.3	57.9	2.1	31.0	40	7.3
CB Malaysia (90%) MF (10%)	35.4	69.1	63.9	43.0	1.2	29.5	39	6.0
CB Malaysia (85%) MF (15%)	35.3	61.4	56.7	37.1	1.0	29.0	37	6.1
CB Malaysia (80%) MF (20%)	35.2	53.9	49.3	31.0	1.0	28.5	35	5.7
CB Malaysia (75%) MF (25%)	35.1	46.0	42.1	25.5	1.3	27.5	32.5	5.0
CB Brazil	39.7	62.8	53.5	29.9	0.4	29.0	30.5	4.6
CB Brazil (90%) MF (10%)	39.1	53.5	42.9	19.8	0	27.5	37	5.9
CB Brazil (85%) MF (15%)	38.8	46.1	36.1	13.0	0.5	26.5	34.5	4.9
CB Brazil (80%) MF (20%)	38.4	36.3	29.9	11.5	0	26.0	37.5	4.7
CB Brazil (75%) MF (25%)	38.1	27.2	17.9	6.3	0	25.5	45	4.6
CB Ghana	35.6	77.8	72.9	49.2	0.6	30.5	36.5	5.6
CB Ghana (90%) MF (10%)	35.4	64.2	58.6	35.2	0	29.0	35	5.0
CB Ghana (85%) MF (15%)	35.3	56.8	51.7	29.7	0.3	28.5	37	5.2
CB Ghana (80%) Brazil (20%)	35.2	48.3	43.3	24.5	0	27.5	42	5.2
CB Ghana (75%) MF (25%)	35.1	39.6	34.9	18.4	0.1	27.0	47.5	4.4

These results are further confirmed by comparing the curves for milk fat and cocoa butter in various proportions. There are two reasons for this strong decrease in hardness:

- Liquid oil components of the milk fat soften the cocoa butter due to their fluidity.
- The solid fat components form eutectics with the triglycerides of cocoa butter.

The analytical results of the fractionation of Malaysian cocoa butter are given in Table 12.9. The stearin thus produced is primarily a mixture of POP, POS, and SOS and has virtually no components that are liquid at or near room temperature. With the removal of the moderating influence of the more liquid components of cocoa butter, the cocoa butter fraction (CBF) becomes more crystalline and complex, the melting range becomes shorter, and the heat of fusion increases. This stearin is extremely hard and can be used effectively to improve the quality of soft cocoa. Attempts have been made to improve the quality of Brazilian cocoa butter by fractionation.

12.3 CONFECTIONERY FATS

The historical uncertainty of cocoa butter supply and the volatility of cocoa butter prices depending on fluctuating cocoa bean prices have forced confectioners to seek other alternatives, which may have had a stabilizing influence on the prices of cocoa butter. Ever-increasing demand [4] of chocolate and chocolate-type products has increased the demand for cocoa beans from year to year. However, it is difficult to predict the supply of cocoa beans. This ensures continuing need for economical vegetable fats to replace cocoa butter in chocolate and confectionery products. Attempts by confectioners to use fats other than cocoa butters in their formulations were made as early as 1930. These experiments did not succeed because of the incompatibility of the fat blends used, which resulted in discoloration and fat bloom. However, these experiments demonstrated the need for the cocoa butter-type fats in the chocolate and confectionery industry.

Continued research in the field of confection science has resulted in the development of fats resembling the characteristics of cocoa butter. These fats have become known as hard butters [5]. These fats were developed using palm-kernel, coconut, palm, and other exotic oils, such as sal, shea, and illipe, as raw materials. The processes involved in producing such fats include hydrogenation, interesterification, solvent or dry fractionation, and blending. The most elementary hard butters are manufactured by combining the processes of hydrogenation and fractionation.

12.3.1 Hard Butters

The hard butters can be divided into the following three main groups based upon their characteristics and the raw materials used to produce them:

1. Lauric cocoa butter substitutes (lauric CBSs). These are fats that are incompatible with cocoa butter, but that have physical properties resembling those of cocoa butter.
2. Nonlauric cocoa butter substitutes (nonlauric CBSs). These are fats that are partly compatible with cocoa butter.

TABLE 12.9
Fractionation of Malaysian Cocoa Butter

Sample	% Yield	IV (Wijs)	Pulsed NMR BS684 Method 2				Triglyceride Composition (mol%) (HPLC)										
							Monounsaturated				Diunsaturated						
			20 C	25 C	30 C	35 C	POP	POS	SOS	Total	PliO	PliP	POO	PLiS	SOO	SLiS	Total
Malaysian cocoa butter			82.1	78.7	58.3	2.4	12.5	45.3	37.2	95.0	Trace	2.2	0.7	0.6	1.6		5.1
Cocoa butter stearine	79.4	29.7	96.6	95.7	89.1	13.7	11.4	51.3	37.3	100			Trace	Trace			
Cocoa butter oleine	20.6	52.2	1.2	0	0	0	11.3	19.1	17.5	47.9	3.6	16.3	6.6	9.0	15.6	1.0	52.1

3. Cocoa butter equivalents (CBEs) or extenders. These are fats that are fully compatible with cocoa butter (chemical and physical properties similar to those of cocoa butter).

Other terms used to describe hard butters include cocoa butter partial replacers, total replacers, modifiers, and extenders. All these categories can be further subdivided into a range of speciality fats, tailored to suit particular purposes.

12.3.1.1 Lauric CBS

This category offers a range of confectionery fats with different levels of physical properties, but all having triglyceride compositions, which make them incompatible with cocoa butter; that is, they are all used in formulations with cocoa powder, mainly for compound coating.

CBSs are produced from lauric fats, which are obtained from various species of palm-tree, the main varieties being palm, which produces palm-kernel oil and coconut. These fats differ from nonlaurics in that they contain 47 to 48% lauric acid, together with smaller amounts of other medium- and short-chain fatty acids. This gives the fats a solid consistency at cool ambient temperatures, but they nevertheless melt below 30°C. From a practical point of view, cost economy probably has been the main incentive behind the search for suitable and reliable substitutes for cocoa butter. The introduction of hydrogenation technique added another dimension to the alteration of lauric fats, but it must be emphasized that the palm-kernel stearins exhibit characteristics considerably better than those of hydrogenated palm-kernel oil, all dependent on the sharpness of the fractionation.

Palm-kernel stearins today have functional properties similar to those of cocoa butter (i.e., a steep NMR curve, a very brittle texture, and a narrow melting range that ensures a quick melt down and a pleasant feeling in the mouth). The interval between setting and melting points is short. This offers a technological advantage over cocoa butter. The tempering is simplified or may be omitted for normal coating purposes. Vegetable fats can crystallize in several polymorphic forms, the most common being alpha, β prime, and beta, which in the same order display an increasing stability, melting point, heat of fusion, and density. In general, lauric fats are stable in β prime form. The rate of crystallization of the α form is higher than that of the β form, which in turn crystallizes faster than the β form. The other sources for the manufacture of lauric CBS includes coconut, South American palm-kernel oils, tucum, cohune, babassu, and ouricury. Most of these minor varieties are seldom encountered in Europe, except in the country of origin, and they do possess specific properties.

The principal advantages and disadvantages of the lauric CBS are as follows:

1. Good oxidative stability; long shelf life [6]
2. Excellent eating quality and flavor release; no waxy aftertaste
3. Texture very similar to that of cocoa butter (i.e., excellent hardness and snap and not greasy to the touch)
4. Solidify quickly tempered or untempered
5. Excellent gloss and gloss retention
6. Available at a cost far less than cocoa butter

The principal disadvantages of the lauric CBS include

1. Mixing with cocoa butter results in a eutectic state [7]. If the manufacturer is to change from chocolate to confectionery coatings, an absolute cleanout of all tanks and enrobing systems is required. Separate production lines are preferred. These fats do not tolerate more than 6% cocoa butter.
2. When lauric CBS is exposed to moisture and fat splitting enzymes (lipase), there is a danger of fat hydrolysis and the liberated lauric acid has a distinct soapy flavor, which can be detected even at low concentrations [8]. These liberated fatty acids also have a lower flavor threshold as compared to the longer chains:
   a. Butyric acid (C4) 0.6 ppm
   b. Caproic acid (C6) 2.5 ppm
   c. Caprylic acid (C8) 350 ppm
   d. Capric acid (C10) 200 ppm
   e. Lauric acid (C12) 700 ppm
   f. Stearic acid (C18) 15,000 ppm
3. Relatively low milk fat tolerance

Generally, these fats are produced by fractionating palm-kernel oil and then hydrogenating stearine to iodine value less than 1 (Table 12.10). The resultant product as shown in Table 12.11 clearly shows dissimilarity with cocoa butter. Figure 12.1 illustrates the incompatibility of palm-kernel oil with cocoa butter. This is because the triglycerides of palm-kernel oil and cocoa butter are both physically and chemically

TABLE 12.10
Fractionation of Palm-Kernel Oil

	Palm Kernel	Fraction 1	Fraction 2
Wiley melting point (°F)	83.6	87.7	72.4
Iodine value	15.3	5.2	25.6
Yield, wt/%	50.0	50.0	
Fatty Acid Composition (wt/%)			
Caproic	0.3	0.1	0.6
Caprylic	4.3	2.1	6.9
Capric	3.9	3.1	4.6
Lauric	49.5	57.5	42.5
Myristic	16.1	21.5	11.0
Palmitic	7.8	8.3	7.0
Stearic	2.3	1.9	1.9
Oleic	13.7	4.8	21.4
Linoleic	1.9	0.6	4.0
Linolenic	0.1	Trace	0.1
Arachidic	0.1	0.1	Trace

TABLE 12.11
CBS Fatty Acid Composition: Typical Fatty Acid Composition, Weight%

	Carbon Number								
	8:0	**10:0**	**12:0**	**14:0**	**16:0**	**18:0**	**18:1**	**18:2**	**Others**
CBS	2%	3%	54%	21%	9%	9%			2%
Cocoa butter					25.5%	23.5%	34.5%	3.5%	2%

different, which leads to the development of eutectic. Thus, factories have to be properly cleaned if one is dealing with either of these products in production.

12.3.1.2 Nonlauric CBS

Nonlauric CBSs consist of fractions of hydrogenated oils: soybean, cotton, corn, peanut, safflower, and sunflower oils. These oils are hydrogenated under selective conditions to promote the formation of *trans* fatty acids, thereby increasing the solid contents considerably. The melting point of oleic acid — the *cis* configuration — is 14°C, whereas the isomer elaidic acid melts at 51.5°C.

Due to the similarity in the chain length and the molecular weight, products of this type can tolerate up to 25% cocoa butter on a fat basis when used in a confectionery coating (Table 12.12). Nonlauric CBSs possess good flavor and odor and color properties and do not need tempering. However, there is a tendency toward

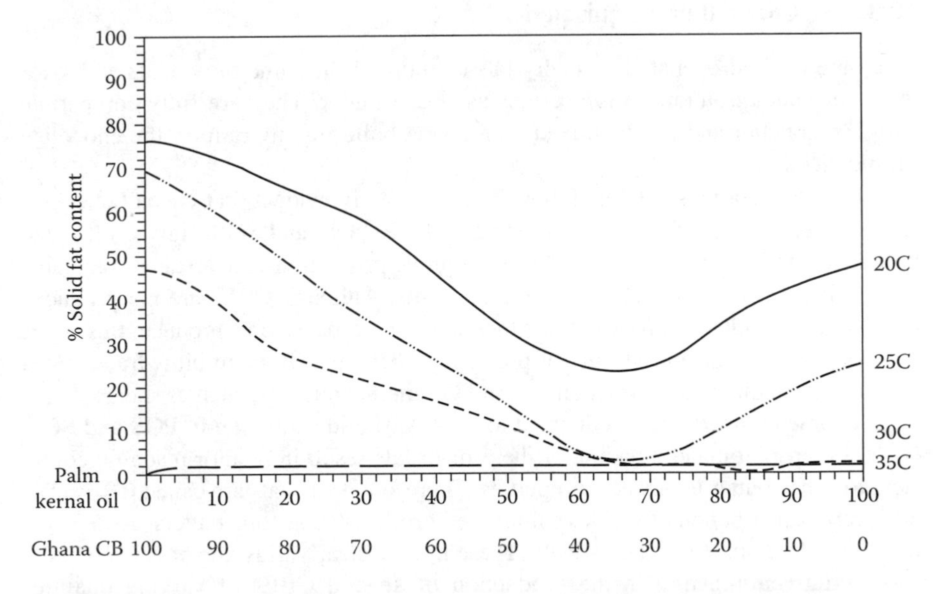

FIGURE 12.1 Percentage solid fat content of various blends of palm-kernel oil and cocoa butter Ghana at various temperatures (pulse NMR, BS 684 method 2).

TABLE 12.12
CBR Fatty Acid Composition: Typical Fatty Acid Composition, Weight%

	Carbon Number				
	16:0	**18:0**	**18:1**	**18:2**	**Others**
CBR	12%	13%	67%	6%	2%
Cocoa butter	25.5%	34.5%	34.5%	3.5%	

Conclusions: CBRs do not contain any lauric acid (C12:0) at all and cannot give any soapy flavor. CBRs and cocoa butter show resemblance.

bloom formation upon long-term storage, especially in products that are poorly formulated. Some of the advantages are as follows:

- Very low price
- Stable in price
- Nontempering
- Can be used with different cocoa powders
  - High fat content (i.e., 20%)
  - Low fat content (i.e., 12%)
- Can be used with some amounts of cocoa liquor
- Can be used with some amounts of milk fat
- No risk of soapy flavor

12.3.1.3 Cocoa Butter Equivalents

CBEs are nonhydrogenated specialty fats containing the same fatty acids and symmetrical monounsaturated triglycerides as cocoa butter. They are fully compatible with cocoa butter and can be mixed with cocoa butter in any ratio in the chocolate formulations.

From the data presented in Table 12.3 it is clearly evident that cocoa butter is a simple three-component system consisting of POP, POS, and SOS triglycerides; and if these three triglycerides are mixed in appropriate proportions, the resultant vegetable fat will behave as 100% cocoa butter equivalents. Although CBEs are not produced by mixing individual triglycerides, as they are very expensive to produce, this is the logic behind the whole procedure of producing CBEs [9,10]. Palm oil is fractionated to produce middle-melting fraction rich in POP; and exotic fats, such as shea, sal, and illipe (Borneo tallow), are fractionated to get triglyceride cuts rich in POS and SOS. Careful preparation and blending of these materials result in a tailor-made fat equivalent to cocoa butter in physical properties. Therefore, these fats are called CBEs. The triglyceride distribution of various palm-based products and exotic butters are depicted in Figures 12.2 and 12.3, respectively. These figures clearly illustrate how to combine various fats culminating in the production of several CBEs of varying qualities matching the financial economics of various consumer companies. The formulation of a suitable CBE is the greatest art in fat technology [11].

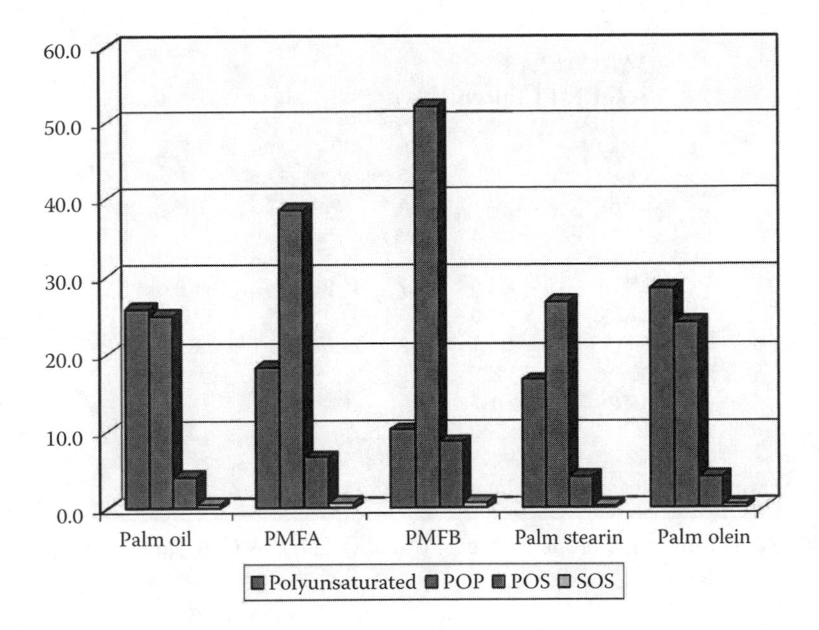

FIGURE 12.2 Triglyceride composition (in wt%) of palm products.

The main drawbacks of cocoa butter are

1. Low milk fat tolerance
2. Lack of stability at elevated temperatures
3. Tendency to bloom

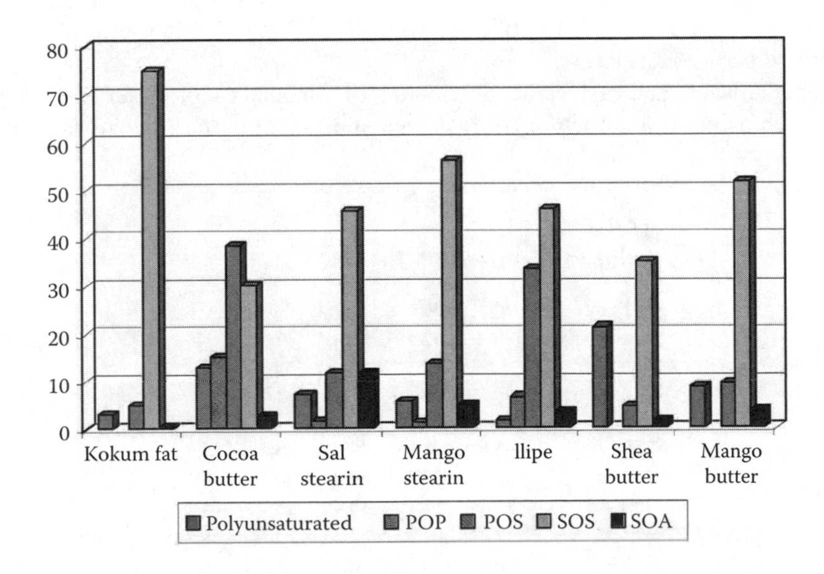

FIGURE 12.3 Triglyceride distribution (in wt%) of exotic fats.

TABLE 12.13
Solid Fat Content (%)

Milk fat	30	30	30	30	30
Cocoa butter	70	52.5	35	17.5	—
CBE	—	17.5	35	52.5	70
20°C	54.0	54.2	54.5	54.8	54.6
25°C	45.6	45.7	45.9	46.2	47.1
27.5°C	33.0	33.5	33.9	35.1	37.0
30°C	15.1	16.8	18.6	21.1	23.5
32.5°C	2.2	3.0	4.5	7.0	9.8
35°C	0.6	0.6	0.7	0.9	1.1

The principal advantages of incorporating CBEs are

1. Reduction in the production cost of chocolates, as CBE, are cheaper than cocoa butter
2. Stabilizing influence on fluctuating prices of cocoa butter
3. Improvement of the milk fat tolerance
4. An increasing resistance to storage at high temperatures
5. Bloom control

The greatest technological advantage lies in the compatibility of these fats with cocoa butter. These technological advantages are illustrated in Tables 12.13 and 12.14. The compatibility of these fats with cocoa butter is examined in the following section. In order to study the compatibility of Ghanian cocoa butter with several different raw materials used in the manufacture of CBEs and with CBEs themselves, several blends were prepared and the solid fat content of these blends was measured as described earlier [2,12].

The fatty acid and triglyceride distribution of Ghanian cocoa butter and selected fats, such as palm midfraction (PMF), shea starine, illipe fat, mango fat, sal oil

TABLE 12.14
Solid Fat Content (%)

Milk fat	30	30	30	30	30
Cocoa butter	70	52.5	35	17.5	—
CBE B	—	17.5	35	52.5	70
20°C	54.0	54.0	53.6	52.7	50.6
25°C	45.6	45.0	44.2	43.5	42.3
27.5°C	33.0	31.8	30.8	30.5	30.2
30°C	15.1	15.1	15.1	15.7	16.2
32.5°C	2.2	2.3	2.5	3.1	3.4
35°C	0.6	0.7	0.8	1.2	1.3

TABLE 12.15
Cocoa Butter Compared with Selected Oils and Fats Related to the Confectionery Industry: Typical Fatty Acid Composition (wt%)

Sample	C6	C8	C10	C12	C14	C16	C18	C18:1	C18:2	C18:3	C20
Cocoa butter				Trace	0.1	24.4	33.6	37.0	3.4	0.1	0.1
Palm mid-fraction				0.1	1.0	44.9	4.0	40.2	9.3	0.2	0.3
Shea stearine						4.0	56.0	34.4	3.7		1.9
Illipe fat					0.1	15.4	47.3	34.3	1.2		1.7
Mango fat						9.3	49.0	38.2	0.6		2.9
Sal oil (low quality)				0.1		6.2	50.0	42.5	2.1	0.3	7.9
Sal oil (high quality)						5.6	44.2	39.9	2.6	0.4	7.2
CBE A					0.4	30.4	30.1	34.5	3.6		1.0
CBE B				0.1	0.7	40.4	20.1	33.9	4.1		0.7
Coberine				0.1	0.5	31.6	30.8	32.6	3.3		1.1
Palm-kernel oil	0.2	3.2	3.1	46.2	16.1	8.5	2.3	17.0	3.1		0.2

(low quality), sal oil (high quality), CBE A, CBE B, Coberine, and palm-kernel oil, are shown in Tables 12.15 and 12.16, respectively. Coberine was the first CBE generated through the cooperation of Unilever and Cadbury's.

The ISO-NMR curve presented in Figure 12.4 shows the restriction in using PMF as an extender because of its limited compatibility with cocoa butter and especially with cocoa butter and milk fat blends. These results reveal that this fat has to be mixed with an exotic fat, thus improving its composition with POS- and SOS-type triglycerides. The incorporation of SOS in PMF advantageously improves the crystallization and tempering characteristics. The eutectic shown in Figure 12.4 deepens in the presence of undesirable triglycerides, such as PPP, PPO, POO, and POP; thus the fractionations of palm oil should be achieved with the minimum amounts of these undesirable triglycerides. A good-quality PMF can be produced only in a solvent fractionation plant. PMF produced in dry and detergent fractionation plants has to be refractionated to upgrade its quality for the formulation of CBE.

The ISO-NMR curve in Figure 12.5 shows higher solid fat content for the blends of cocoa butter and shea stearine. This shows good compatibility. However, once again, one has to concentrate on the removal of undesirable triglycerides, such as SSS, SSO, SOO, and SOS. Although the presence of large amounts of PPP and SSS are not desirable in view of the melting characteristics in the mouth, the presence of 1 to 5 mol%, particularly 1 to 3 mol%, is preferred, since it is apt to accelerate the solidification velocity of chocolate by cooling and to impart fat bloom resistance [13,14].

The content of SOS is higher in illipe fat and its triglyceride composition approaches that of cocoa butter; therefore, it shows very high compatibility, as shown in Figure 12.6. This is a preferred fat for producing chocolates for tropical regions.

TABLE 12.16
Cocoa Butter Compared with Selected Oils and Fats Related to the Confectionery Industry: Typical Triglyceride Composition (wt%)

Sample	C24	C26	C28	C30	C32	C34	C36	C38	C40	C42	C44	C46	C48	C50	C52	C54	C56	C58	C60
Cocoa butter													0.1	16.5	45.8	36.1	1.5		
Palm mid-fraction												0.5	3.4	49.1	37.4	9.3	0.3		
Shea stearine													0.	0.9	10.9	82.7	5.2	0.2	
Illipe fat														6.7	35.6	54.2	3.5		
Mango fat												0.2	0.1	2.1	16.5	68.7	11.2	1.0	
Sal oil (low quality)														1.1	14.1	64.7	17.8	2.3	
Sal oil (high quality)												0.2	0.3	1.0	13.2	66.4	16.5	2.1	0.1
CBE A												0.4	2.5	36.6	17.4	40.8	2.1	0.2	
CBE B												0.6	4.5	49.1	17.0	27.3	1.5		
Coberine												0.3	2.5	32.0	17.3	5.0	2.9		
Palm-kernel oil	0.1	0.8	0.8	1.3	6.3	8.2	20.9	16.0	9.4	9.1	6.7	5.5	6.3	2.6	2.8	3.2			

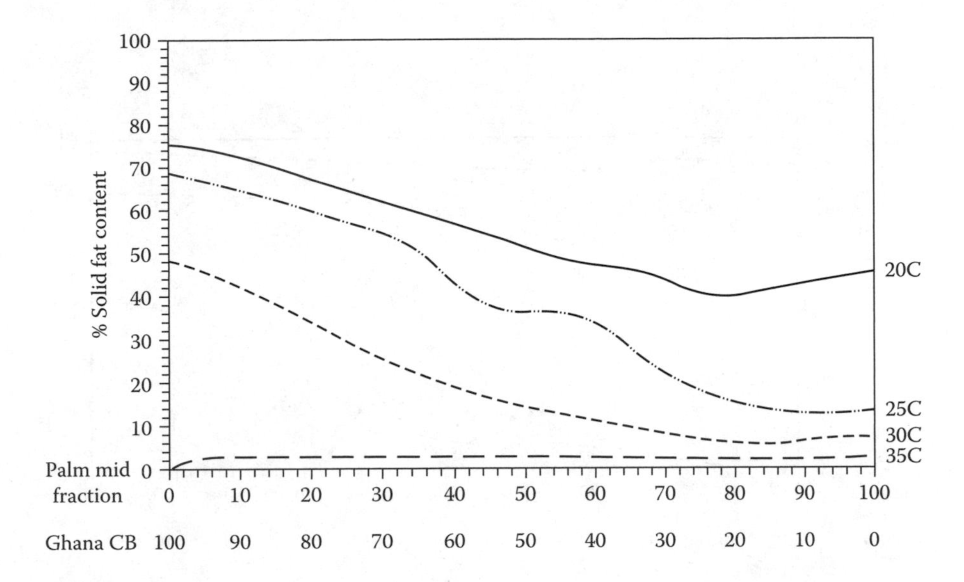

FIGURE 12.4 ISO-NMR (pulse NMR, BS 684 method 2) using palm midfraction as an extender.

Mango fat contains a large amount of SOO triglyceride, and therefore it has limited compatibility with cocoa butter, as illustrated in Figure 12.7. It has to be fractionated to improve its quality as an exotic fat. The sources of many of the SOS-containing fats are not regularly cultivated but are jungle crops with restricted availability; therefore, they are called exotic fats.

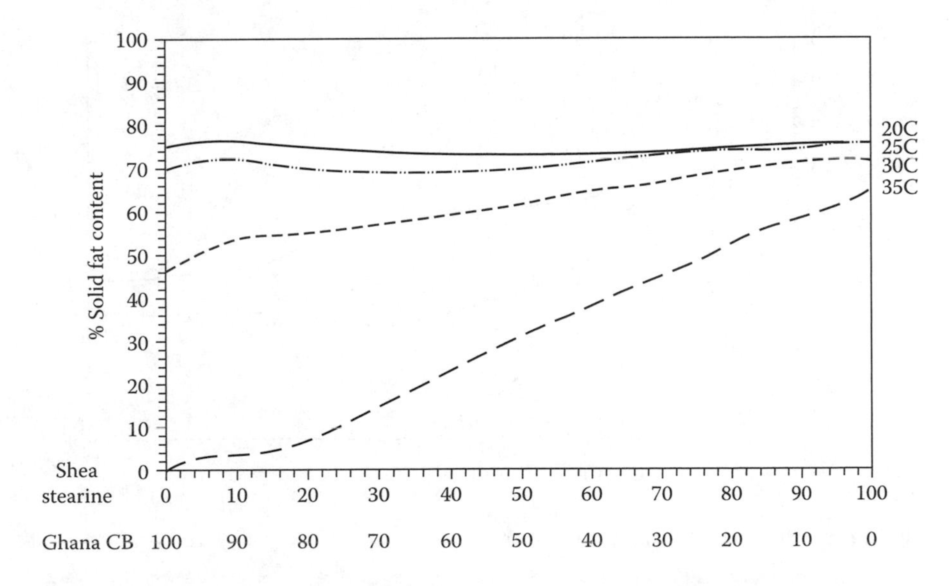

FIGURE 12.5 ISO-NMR (pulse NMR, BS 684 method 2) for blends of cocoa butter and shea stearine.

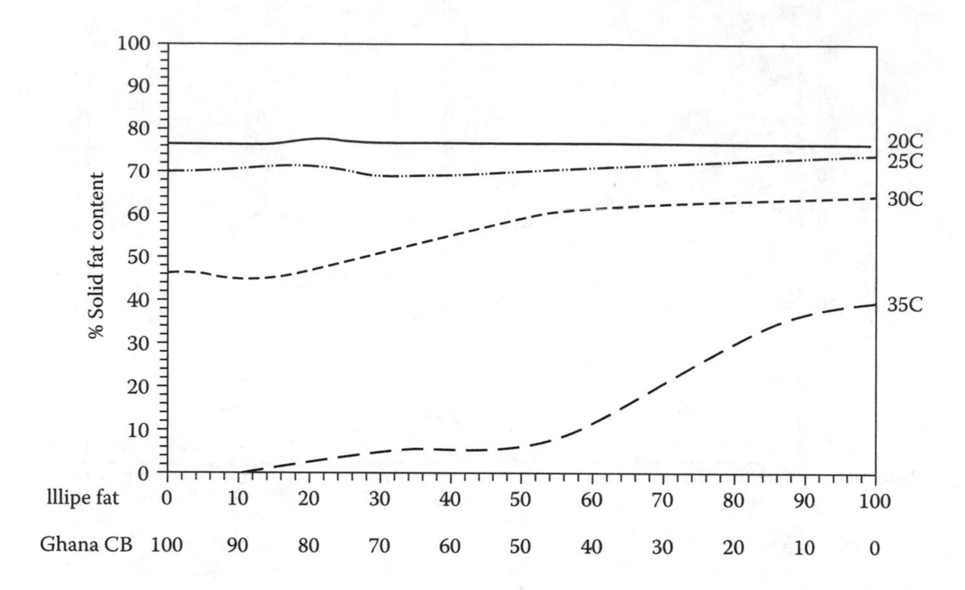

FIGURE 12.6 Percentage solid fat content vs. temperature of Illipe fat (pulse NMR, BS 684 method 2).

Figures 12.8 and 12.9 show the compatibility curves of a low- and a high-quality sal oil with cocoa butter. The quality of the sal fat depends upon the presence of triglycerides containing 9,10-dyhydrosetaric acid [15]. Thus, low-quality sal oil, which contains an appreciable amount of this triglyceride, has a limited compatibility with cocoa butter compared with high-quality sal oil.

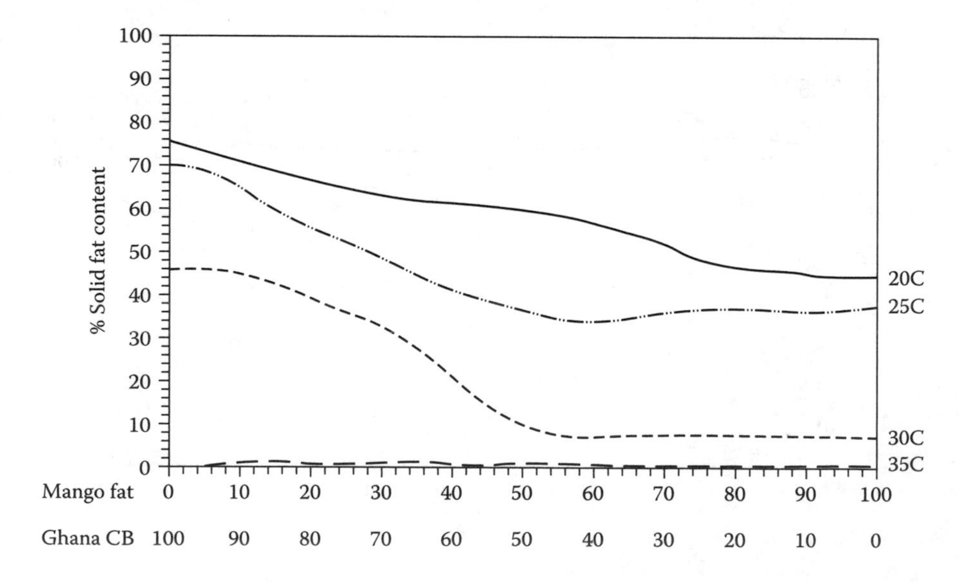

FIGURE 12.7 Percentage solid fat content vs. temperature of mango fat (pulse NMR, BS 684 method 2).

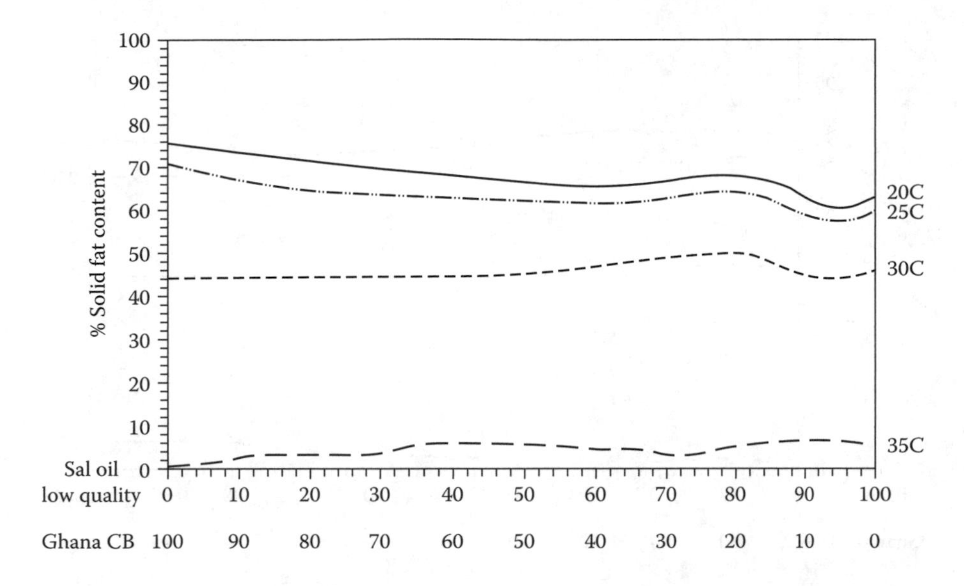

FIGURE 12.8 Compatibility curve of low-quality sal oil (pulse NMR, BS 684 method 2).

The compatibility curves of three different CBE, CBE A, CBE B, and Coberine, with cocoa butter are shown in Figures 12.10 to 12.12, respectively. These curves clearly show that these fats are fully compatible with cocoa butter. In addition, they also add desirable properties to it, such as the ability to raise the melting point and increase the tendency to "seed" at high temperatures.

The results illustrated in Figures 12.13 and 12.14 once again prove the suitability of various CBE in replacing 15% of the cocoa butter, which equates to approximately

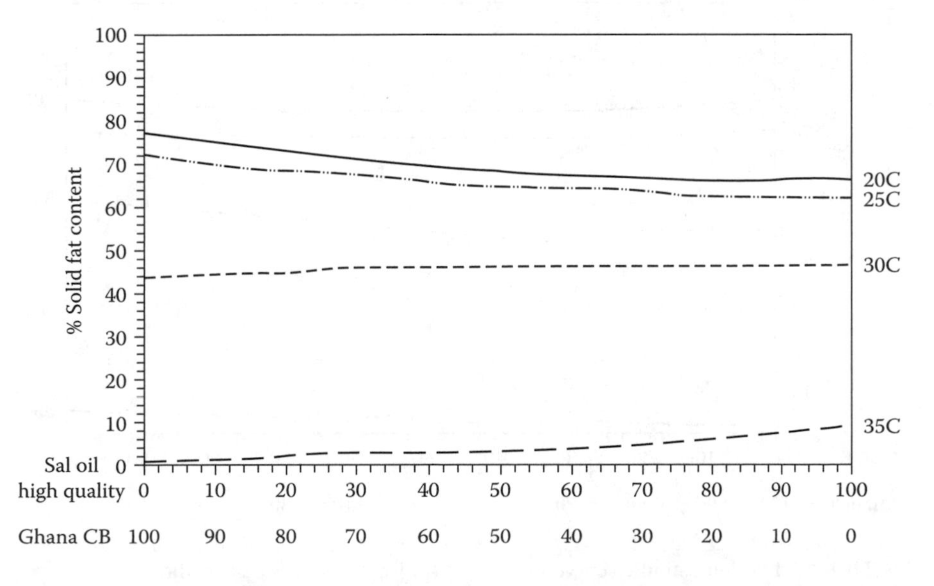

FIGURE 12.9 Compatibility curve of high-quality sal oil (pulse NMR, BS 684 method 2).

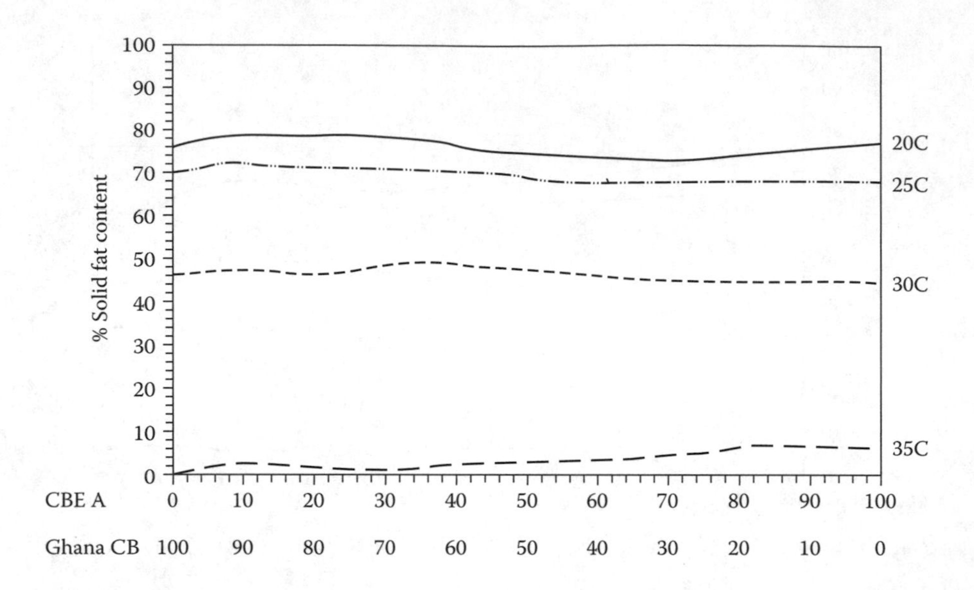

FIGURE 12.10 Compatibility curve of CBE A (pulse NMR, BS 684 method 2).

5% in chocolate. The results presented in Table 12.17 and Figure 12.15 show the variation in the quality of three different CBEs and their ability to add desirable properties to cocoa butter.

A number of fat blends were prepared in a three-component fat model system, including cocoa butter, milk fat, and selected oils and fats. The analytical constants

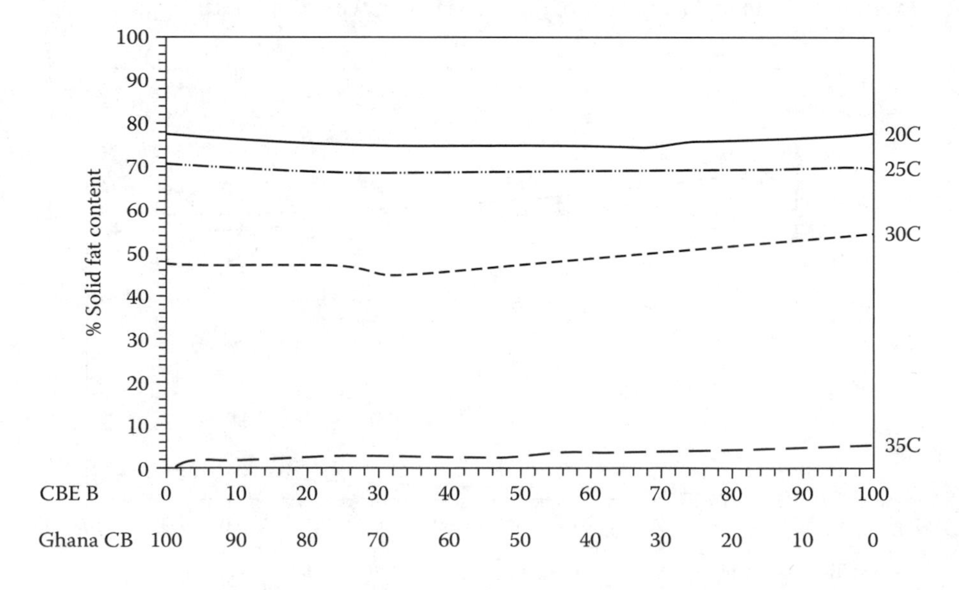

FIGURE 12.11 Compatibility curve of CBE B (pulse NMR, BS 684 method 2).

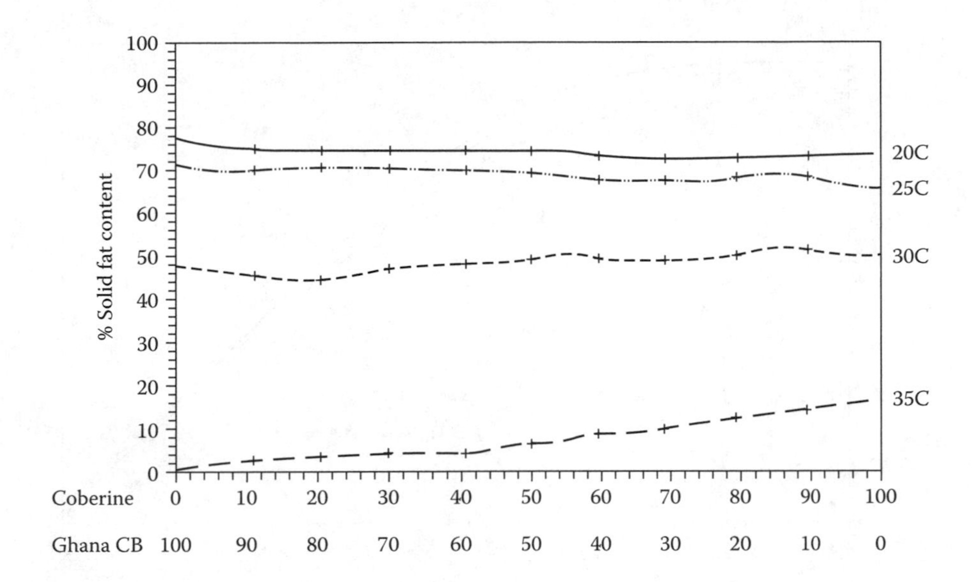

FIGURE 12.12 Compatibility curve of Coberine (pulse NMR, BS 684 method 2).

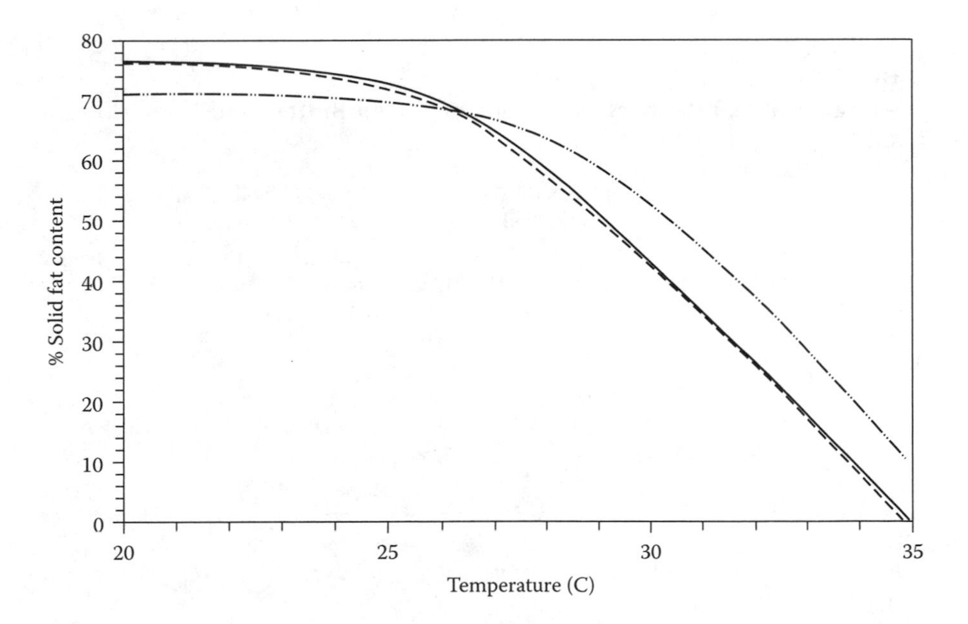

FIGURE 12.13 Solid fat content vs., temperature for cocoa butter Ghana (solid line), CBE A (dotted dashed line), and an 85:15 mixture of the two (dashed line) (pulse NMR, BS 684 method 2).

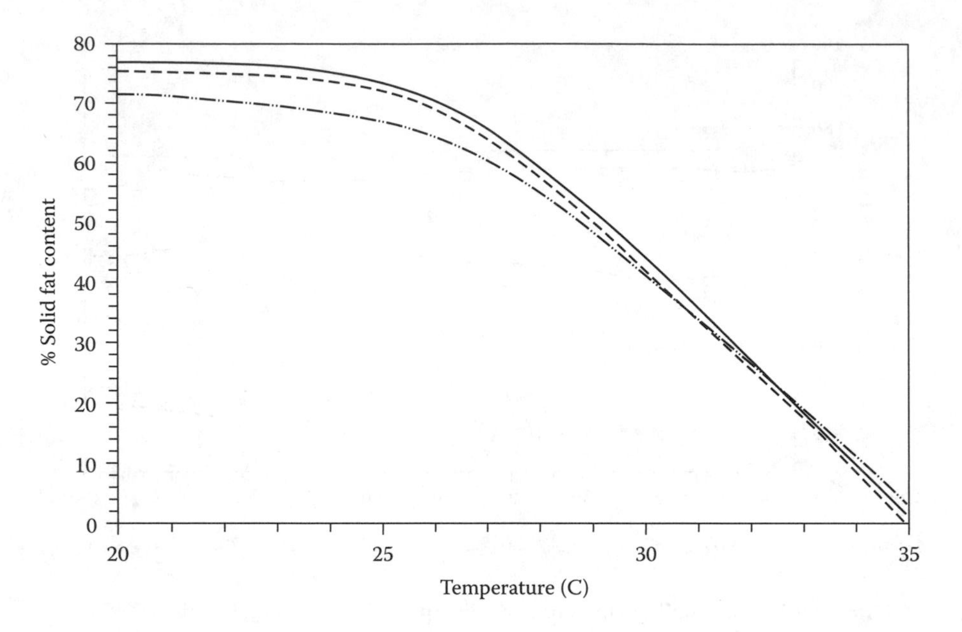

FIGURE 12.14 Solid fat content vs. temperature for cocoa butter Ghana (solid line), CBE B (dotted dashed line), and an 85:15 mixture of the two (dashed line) (pulse NMR, BS 684 method 2).

TABLE 12.17
The Analytical Constants of the Blends of Cocoa Butters and Three Illexao Types

		Pulse NMR BS 684 Method 2				Solidification Curve (Jensen)		
Sample	**IV (Wijs)**	**20 C**	**25 C**	**30 C**	**35 C**	**Max. Temp. (C)**	**Time min/ max (min)**	**Temp. Rise (C)**
Ghana cocoa butter	36.4	76.9	69.8	47.1	2.0	30.0	38	5.8
CBE A	35.7	74.1	68.6	53.7	9.0	29.5	44.5	5.0
CBE B	35.4	73.1	62.4	45.4	6.5	28.0	58	4.9
CBE C	36.5	77.4	74.4	65.6	33.8	32.5	35	7.5
Ghana cocoa butter (50%) CBE A	36.1	73.5	68.0	49.6	4.5	29.5	45	5.3
Ghana cocoa butter (50%) CBE B	35.9	68.8	63.8	45.9	3.6	28.5	47.5	4.1
Ghana cocoa butter (50%) CBE C	36.5	75.0	71.4	57.4	16.6	31.5	40	6.3

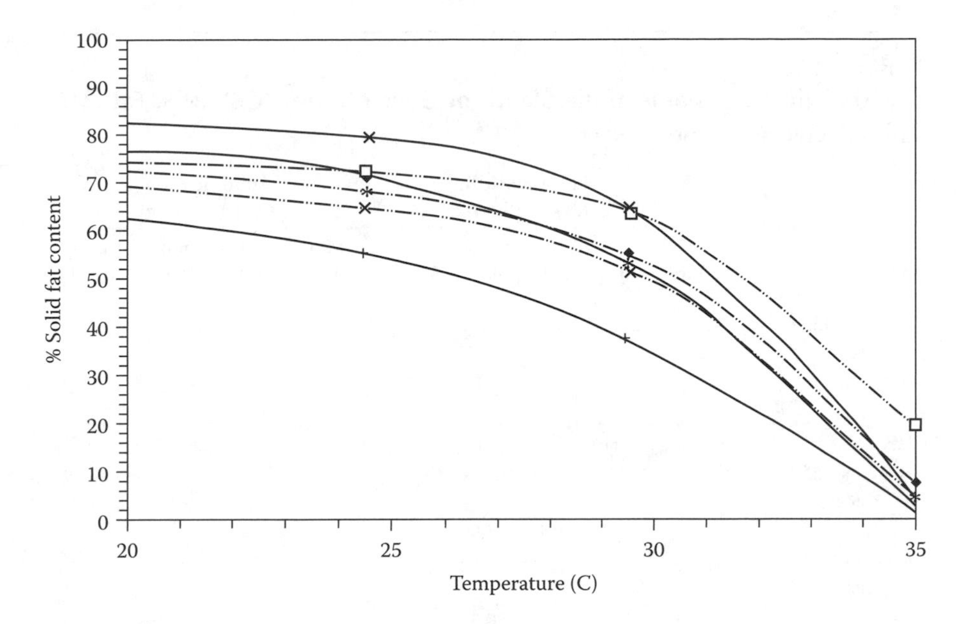

FIGURE 12.15 Solid fat content vs. temperature for a number of fat blends: cocoa butter Ghana (diamond), cocoa butter Brazil (plus sign), cocoa butter Malaysia (multiplication sign), cocoa butter Ghana/CBE A 30–67 (50:50) (solid square), cocoa butter Ghana/CBE B 30–67 (50:50) (asterisk), cocoa butter Ghana/CBE C 30–96 (50:50) (open square) (pulse NMR, BS 684 method 2).

of these blends are presented in Table 12.18 [16,17]. A critical evaluation of these results leads to the following conclusions:

1. Once again the high milk fat tolerance of three CBE types is confirmed.
2. A high compatibility of shea stearine and illipe fat together with no loss of hardness exists, even in the presence of milk fat.
3. Severe eutectic in the case of PMF and mango fat are due to the presence of milk fat, as explained earlier.
4. Remarkable softening occurs in the mixtures of palm-kernel oil and hydrogenated soybean oil, which renders them unacceptable.

Triglycerides are able to exist in several crystalline structures; for cocoa butter and most CBEs they exist in three main groups α, β', and β. Only one of these is fully stable. This phenomenon is called polymorphism. Through proper tempering, the chocolate manufacturer is aiming at a quick crystallization in the most stable modification. However, our x-ray investigations have shown that, so far, no tempering technique, however sophisticated, has been able to secure solidification of liquid chocolate in the highest melting and stable β' form in cocoa butter. CBEs of specific triglyceride composition also decrease the transformation rate, thereby intensifying the positive effect of milk fat.

TABLE 12.18
The Analytical Constants of the Blends of Cocoa Butter (CB), Milk Fat (MF) and Selected Oils and Fats

		Pulse NMR BS 684 Method 2				Solidification Curve (Jensen)		
Sample	IV (Wijs)	20 C	25 C	30 C	35 C	Max. Temp. (C)	Time min./ max. (min)	Temp. Rise (C)
CB Ghana (50%) MF (50%)	34.9	32.2	17.6	9.3	1.1	25.5	32	4.0
CB Ghana (55%) MF (30%)	35.4	30.9	17.0	8.7	0.2	25.0	28	3.6
CBE A CB Ghana (55%) MF (30%)	35.4	27.4	13.6	7.1	1.6	25.0	28	3.5
CBE B CB Ghana (55%) MF (30%)	35.5	33.0	19.1	11.9	1.4	25.5	28	3.6
CBE C CB Ghana (55%) MF (30%) Shea sterine (15%)	35.7	38.1	31.9	20.3	1.6	26.5	44.5	4.0
CB Ghana (55%) MF (30%) Coberine (15%)	35.4	30.5	17.5	9.6	0.6	25.0	26	3.4
CB Ghana (55%) MF (30%) Sal oil (15%)	36.2	32.8	17.7	11.8	0.7	25.5	37	3.5
CB Ghana (55%) MF (30%) Palm midfraction (15%)	37.6	18.8	9.5	3.7	1.0	24.0	28	3.3
CB Ghana (55%) MF (30%) Mango fat (15%)	35.6	24.2	4.9	3.9	0.4	24.5	37	3.7
CB Ghana (55%) MF (30%) Illipe fat (15%)	35.2	36.0	26.9	15.9	3.7	26.0	36.5	4.1
CB Ghana (55%) MF (30%) Hydrogenated soya-bean oil (15%)	41.5	24.2	10.2	5.1	1.0	24.0	27	2.2
CB Ghana (55%) MF (30%) Palm-kernel oil (15%)	32.8	18.0	7.3	4.8	0	24.5	34	3.3

TABLE 12.19
Effect of Triglycerides on the Crystallization Behavior of CBEs

	Temperature			
	20 C	25 C	30 C	35 C
CBEs	78.1	75.3	71.6	66.4
CBEs mixed with component A triglycerides	67.3	61.7	56.8	50.2
CBEs mixed with component B triglycerides	61.7	53.0	45.9	39.0

12.3.1.3.1 Effect of Diglycerides on the Crystallization Behavior of CBEs and Replacers

Diglycerides play a major role in inhibiting crystallization, as is shown clearly in Tables 12.19 and 12.20, respectively. 1,3-Diglycerides are less dangerous than 1,2-diglycerides due to their very symmetrical structure as of triglycerides. Therefore, it is of the utmost importance to catch the development of diglycerides sight from crude, which is extremely technical and requires careful production planning and control. These fit reasonably well with the crystal structures of triglycerides, producing slight hindrance in crystal formation and resultant crystal lattice. The dynamic crystallization velocity experiments performed for cocoa butter replacers (CBRs) with two different levels of diglycerides (4.2% and 9.5%, respectively) show a drastic deteriorating effect, as beautifully exemplified in Figure 12.16.

12.4 CHOCOLATE AS A FRIEND

Nutrition is an essential part of health. The top three leading causes of death in Western nations — heart disease, cancer, and stroke — all have significant nutrition-associated risk factors. Despite several new exciting nutritional findings about chocolate, researchers reluctantly agree that eating chocolate will never help you lose weight, and therefore chocolate treats should be balanced with other food choices throughout the day.

Chocolate is produced in many different recipes and contains other ingredients besides cocoa products [18,19]. Therefore, nutritional value will vary with the products.

TABLE 12.20
Effect of Triglycerides on the Crystallization Behavior of Non-*trans* CBRs

	Temperature			
	20 C	25 C	30 C	35 C
Non *trans* CBRs	63.0	41.2	26.3	5.2
Non *trans* CBRs mixed with 4.2% triglycerides	59.4	39.1	24.2	5.2
Non *trans* CBRs mixed with 9.5% triglycerides	50.0	34.5	19.8	4.8

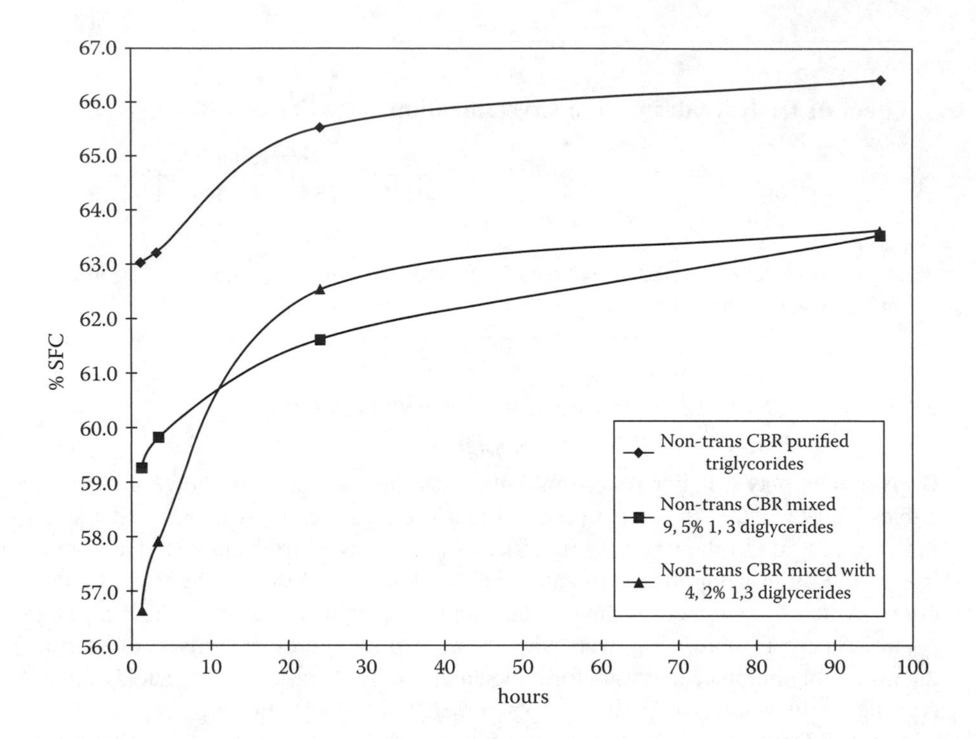

FIGURE 12.16 Dynamic crystallization velocity measurements at 20°C.

For example, dark chocolate has a high proportion of cocoa solids and will therefore retain more of the nutritional value of cocoa than milk chocolate, which has a high proportion of milk solids. The milk in milk chocolate provides a rich source of proteins and so its protein value is higher than that of dark chocolate.

The fat content in chocolate is high 30%, of which 62% is saturated; 34%, monounsaturated; and 3%, polyunsaturated. Chocolate also contains cholesterol from milk solids but not from cocoa.

Cocoa butter is the main lipid in dark chocolate and consists of oleic acid (monounsaturated) and stearic and palmitic acids (saturated). Stearic acid, unlike other saturated fatty acids, appears to have a neutral effect on total low-density lipoprotein (LDL) and high-density lipoprotein (HDL) cholesterol. Cocoa butter appears to decrease platelet activity, which can be a potential contributor to heart diseases.

Chocolate provides major source of copper, magnesium, zinc, potassium, calcium (specifically milk chocolate), and key vitamins such as riboflavin, thiamine, and niacin. Milk chocolate contains the methylxanthine stimulants, caffeine and theobromine, and the combined effects of theobromine and caffeine may be synergistic. Chocolate also contains biogenic amines, "feel-good" body chemicals, which when converted to dopamine act like adrenaline in the regulation of mood, food intake, and compulsive behaviors.

Chocolate is a rich source of phytochemicals, flavonoid antioxidants that are found in soya, black tea, red wine, broccoli, and cranberry juice. These inhibit oxidation reactions, which can cause damage leading to heart diseases, cancer, and aging. Research is still in its infancy, but chocolate as an antioxidant food eaten in moderation, as for all foods, may contribute to a healthy, balanced diet.

12.4.1 Nuts

As a complex plant food, nuts are a rich source of many nutrients and other bioactive compounds. There is consistent evidence, especially from epidemiologic data, that nuts have a strong protective effect against coronary heart disease. Nuts, besides being a good source of protein, are high in unsaturated fats and contain folate, vitamin E, dietary fiber, plant sterols, antioxidants, and arginine, which as a package may provide heart and cancer protection.

12.4.2 Chocolate as a Foe: Countering Areas of Concern — Obesity, Cholesterol, Tooth Decay, and *Trans* and Saturated Fatty Acids

Chocolate has been accused of making people fat. Obesity results from an imbalance between energy intake and energy expenditure — excess body fat is stored because intake exceeds usage. There is no direct link between sugar and obesity, and confectionery is not a significant contributor to dietary fat intake.

Cholesterol is another area that often concerns consumers. Modern research reveals that cocoa butter does not raise cholesterol because of the neutral effects of stearic acid on blood cholesterol.

Confectionery has also been accused of causing tooth decay or dental caries. Any carbohydrate food is potentially cariogenic, as bacteria in the mouth metabolize fermentable carbohydrates, which leads to acid formulation and decrease in pH; this in turn can result in demineralization of the tooth enamel, ultimately leading to caries. Sugars tend to be less cariogenic than starch because they dissolve in water and saliva and are removed from the mouth rapidly. Thus, the cariogenicity of food is not necessarily related to its sugar content.

The word "chocolate" is often used as a misnomer — anything that is sugary and brown is termed as chocolate. Many of these products are not pure chocolate and are produced with the cheapest available vegetable fats with excessive amounts of *trans* fatty acids. *Trans* fatty acids do not make a positive contribution to human nutrition. Pure chocolate produced with cocoa butter can enhance human nutrition if it is consumed in moderation with other foods.

12.5 CHOCOLATE AS A FUNCTIONAL FOOD

Generally, chocolate is regarded as junk food, but recently published research indicates it can be upgraded to the health-promoting category. Functional foods, also known as nutraceuticals, designer foods, therapeutic foods, super foods, foodiceuticals, and medifoods, are loosely defined as "food that contain some health-promoting components beyond traditional nutrients." It is becoming clear that a single nutrient may be less effective in promoting health and that a balance of several nutrients may be required to provide optimal nutrition and further improve health.

We can define these interactions as nutrient clusters. Chocolate can play a dominant role in providing these nutrient clusters and should be designed to include the following:

1. Nutritionally balanced triglycerides with specific positions occupied by essential fatty acids
2. Natural antioxidants to provide enhanced *in vitro* and *in vivo* protection of lipids
3. Essential nutrients such as selenium and zinc to provide corrective nutrition

12.6 RECIPE ENGINEERING AND OIL PROCESSING

Recipe engineering plays a vital role in the manufacturing of speciality fats. The major analytical techniques, such as gas–liquid chromatography, high-performance liquid chromatography, low-resolution pulsed NMR spectroscopy, differential scanning calorimetry, various static and dynamic crystallization measurement techniques, and rheology, are the tools employed to define the final product and its application in the confectionery industry.

Finally, fats not only provide functionality and structure to products but also can be dominant in their contribution to taste and flavor, including off-flavors. Thus, it is of the utmost importance that in processing specialty products, minute details must be taken into account in order to produce a product that not only tastes fresh but also lasts longer.

12.7 CONCLUSIONS

Confection science is extremely complicated due to the great diversity of triglyceride molecules and the presence of several surface-active components that play a major role in defining the quality of the final products. Sophisticated analytical methodologies will continue to have a major part in unfolding the mysteries of specialty fats in chocolate.

REFERENCES

1. Shukla, V.K.S., Schiøtz-Nielsen W., and Batsberg W., A simple and direct procedure for the evaluation of triacylglycerol composition of cocoa butters by high performance liquid chromatography — A comparison with the existing TLC-GC method, *Fette Seifen Anstrichm.* 85: 274–278 (1983).
2. Shukla, V.K.S., Studies on the crystallization behavior of the cocoa butter equivalents by pulsed Nuclear Magnetic resonance — Part I, *Fette Seifen Anstrichm.* 85: 467–471 (1983).
3. Dimick, P.S., Department of Food Science, Penn State University, College of Agricultural Sciences, 111 Borland Lab, University Park, PA 16802-2501, U.S. (personal communication).
4. U.K. Sweets Market, *Manufacturing Confectioner* 84:19–21 (2004).

5. Paulicka, F.R., Specialty fats, *J. Am. Oil Chem. Soc.* 53: 421–424 (1976).
6. Young, F.V.K., Palm kernel and coconut oils: analytical characteristics, process technology and uses, *J. Am. Oil Chem. Soc.* 60: 374–379 (1983).
7. Rossell, J.B., in *Advances in Lipid Research,* Paoletti, R. and Kritchevsky, D., Eds., Academic Press, New York, 1967, pp. 353–408.
8. Rossell, J.B., In *Rancidity in Foods,* 3rd ed., Allen, J. and Hamilton, R.J., Eds., Blackie Academic & Professional, Glasgow, 1994, pp. 22–53.
9. Shukla, V.K.S., and Nielsen, I.C., *Diversity of Minor Tropical Tree Crops and Their Importance for the Industrialized World. Tropical Forests,* Academic Press, New York, 1989.
10. Shukla, V.K.S., Studies on the crystallization behavior of the confectionery fats by pulse nuclear magnetic resonance employing various tempering modes, *J. Am. Oil Chem. Soc.* 64: 658 (1987).
11. Haumann, B. and Fitch, Confectionery Fats — For Special Uses, *J. Am. Oil Chem. Soc.* 61: 468–472 (1984).
12. Kheiri, M.S.A., *Formulation, Evaluation and Marketing of Cocoa Butter Replacer Fats,* PORIM, 1982.
13. Okawachi, T. and Sagi, N., Confectionery fats from palm oil, *J. Am. Oil Chem. Soc.* 62: 421–425 (1985).
14. Shukla, V.K.S., Confectionery fats, in *Proc. World Conference on Oilseed Processing and Utilization,* Wilson, R.F., Ed., AOCS Press, Champaign, IL, 2001, p. 47.
15. Yella Reddy, S. and Prabhakar, J.V., Effect of triglycerides containing 9,10-dihydroxystearic acid on the solidification properties of sal (*Shorea robusta*) fat, *J. Am. Oil Chem. Soc.* 62: 1126–1130 (1985).
16. Shukla, V.K.S., Milkfat in sugar and chocolate confectionery, in *Fats in Food Products,* Moran, D.P.J. and Rajah, K.K., Eds., Blackie Academic & Professional, Glasgow, 1994, pp. 255–276.
17. Shukla, V.K.S., Milkfat and its applications. The world of ingredients, *J. Practicing Food Technologist,* Jan–Feb: 30–33 (1995).
18. Shukla, V.K.S., Chocolate — The chemistry of pleasure, *INFORM* 8: 152–162 (1997).
19. Shukla, V.K.S., Chocolate — A food from the past with a future, *INFORM* 13: 764–766 (2002).

Part III

Lipids with Health and Nutritional Functionality

13 Marine Lipids and Omega-3 Fatty Acids

R.G. Ackman

CONTENTS

13.1 INTRODUCTION

The word "anecdote" is defined as "a short account of an interesting incident or event, often biographical." has Anecdotes have been a driving force in the growth of the functional foods and nutraceutical markets, which were worth $20.2 billion in the United States in 2002 [1]. The marketing pressures have been gradually increasing as large corporations have begun to apply commercial advertising on a large scale instead of the older small box advertising in newspapers or pamphlets from health and wellness stores.

13.2 MARINE OILS

Strangely, the oldest nutraceutical of all, cod liver oil, has emerged in a new disguise. At the same time, one of the oldest foods of *Homo sapiens*, fish and shellfish, has become a functional food type, the benefit of both fish oils and fish as food reflecting their content of long-chain omega-3 fatty acids. The original medicinal use of fish oils began with cod liver oil in the 1780s in England for arthritis and rheumatism. By the 1800s, it came to be used to prevent rickets [2]. The latter prevention depended on the vitamin D content, but quite recently the vitamin A content, valuable in its own right, became of concern because of the wide availability of this vitamin in many enriched foods. Vitamin D functions in the body in the deposition of calcium and phosphorus in bones at all ages. The important point is that cod liver oil was even recently only thought of as a carrier for these vitamins. In fact, it is a typical fish oil in terms of fatty acid composition, as shown in Table 13.1; the farmed salmon was fed on a diet of fish meal and fish oil, and so the muscle fats resemble the three wild marine fish oils.

Of these fatty acids, the saturated group (14:0, 16:0, 18:0) is dietary but can also be biosynthesized by all marine animals if and when required. The monoethylenic fatty acids can be from food or the fish can biosynthesize the 16:1n-7 and 18:1n-9, but the 18:1n-9 may be accompanied by 10 to 40% of 18:1n-7, made simply by adding an acetate unit to the 16:1n-7. The 20:1 and 22:1 each include up to three or four isomers each, but most are prepared by small marine crustacea called copepods, which convert their phytoplankton food partly into wax esters. When these zooplankton are eaten by predatory fish, such as capelin or herring, the long-chain fatty alcohols of the wax esters are then oxidized to the corresponding fatty acids,

TABLE 13.1
Important Fatty Acids in a Few Commercial Marine Fish Oils as w/w% of Total Fatty Acids

Fatty Acid	Atlantic Cod Liver[a]	Atlantic Menhaden[b]	South American Anchovy[b]	Farmed Atlantic Salmon[c]
14:0	3.3	7.3	7.5	7.9
16:0	13.4	19.0	17.5	16.0
18:0	2.7	4.2	4.0	3.0
16:1	9.6	9.1	9.0	10.8
18:1	23.4	13.2	11.6	15.2
20:1	7.8	2.0	1.6	4.5
22:1	5.3	0.6	1.2	3.9
18:2n-6	1.4	1.3	1.2	2.1
18:3n-3	0.6	1.3	0.8	0.9
18:4n-3	1.0	2.8	3.1	1.6
20:4n-6	1.4	0.2	0.1	0.7
20:5n-3	11.5	11.0	17.0	10.0
22:5n-3	1.6	1.9	1.6	3.4
22:6n-3	12.5	9.1	8.8	7.3

[a] Average; from Ackman, R.G., in *Nutritional Evaluation of Long- Chain Fatty Acids in Fish Oil*, Barlow, S.M. and Stansby, M.E., Eds., Academic Press, London, 1982, p. 25.

[b] Unilever Research, personal communication, 1980.

[c] Total muscle lipid. Ackman, R.G., unpublished data, 1998.

so the oils from herring, capelin, and other small carnivorous fish have fatty acid compositions with substantial proportions (10 to 30%) of 20:1 and 22:1 acids [3,4].

Like the 20:1n-9 and 22:1n-11 of rapeseed oil, the human body simply accepts these fatty acids in all isomers if from fish oil or fish muscle lipids and may use them for energy or convert them to 18:1 for various purposes [5]. The polyunsaturated fatty acids of Table 13.1 can all be biosynthesized by the small plant cells called phytoplankton (Figure 13.1). Many of the thousands of species of these include the 18:2n-6 and 18:3n-3 found in terrestrial plants, but Tables 13.1 and 13.2 reveal that these are of little use to fish. Since many other species of phytoplankton produce the highly unsaturated 20:5n-3 (EPA) and 22:6n-3 (DHA), it can be predicted that the C18 polyunsaturated fatty acids (except for 18:4n-3) are relatively useless and even simple small animals need the readily available EPA and DHA. Accordingly these are passed up the animal food chain, conserved as desirable fatty acids at each step. Growth requires new cells and new membranes that are built up of bilayers of phospholipids for which EPA and DHA are necessary (Table 13.2), the latter also reflecting development of vision and muscle control.

Fish muscle lipids are essentially based on cellular phospholipids distributed in nuclei, mitochondria, and membranes surrounding the whole cell. These are provided with a mixture of fatty acids peculiar to their membrane packing, structures, and functions. Among North Atlantic fish, the cod provides a ubiquitous reference mixture

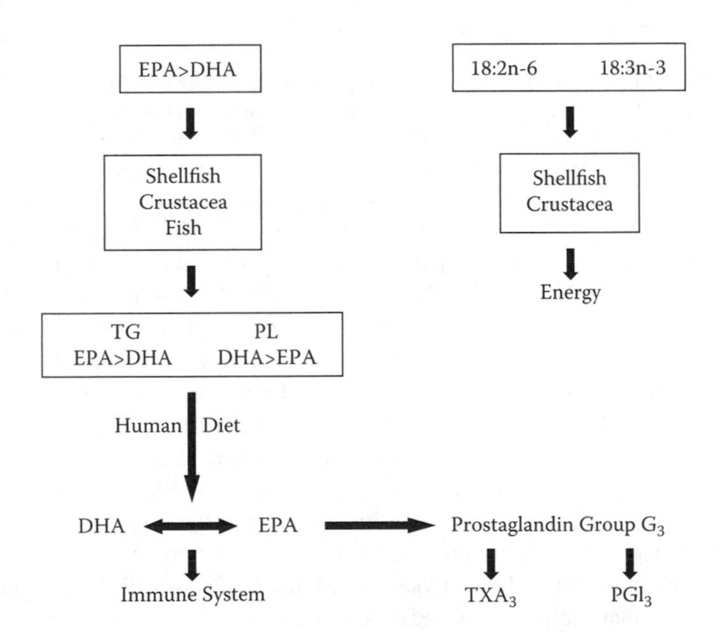

FIGURE 13.1 Outline of the utilization of different chain lengths of polyunsaturated fatty acids produced by marine phytoplankton. Sunlight creates everything necessary for life in the uppermost ocean layers. There are greater proportions of the small plant cells producing EPA than of DHA. Mammals heavily depend on prostaglandins produced from both the omega-6 and omega-3 families of polyunsaturated fatty acids.

TABLE 13.2
Important Fatty Acids in Edible Muscle Phospholipids of Cod (*Gadus morhua*) and Smolting Atlantic Salmon (*Salmo salar*), Compared with those of Eviscerated Bodies and Heads of Capelin (*Mallotus villosus*) and Scallop (*Placopecten magellanicus*) Adductor Muscle, Given in w/w% of Total Fatty Acids

Fatty Acid	Cod[a]	Smolting[b] Atlantic Salmon,Farmed[b]	Smolting Atlantic Salmon, Wild[b]	Capelin (body and head) [c]	Atlantic Scallop (edible muscle) [d]
14:0	1.4	1.6	1.1	2.1	2.1
16:0	19.6	17.0	18.7	20.9	19.8
18:0	3.8	2.7	5.2	2.6	4.2
16:1	3.5	2.3	2.7	6.5	2.6
18:1	13.8	8.3	9.0	11.4	4.7
20:1	3.0	2.5	0.4	3.1	2.0
22:1	1.0	0.7	0.1	2.7	—
18:2n-6	0.7	3.7	1.7	1.0	0.5
18:3n-3	0.1	0.3	1.5	0.7	0.3
18:4n-3	0.4	—	—	0.6	2.3
20:4n-6	2.5	1.6	10.4	0.5	1.2
20:5n-3	17.0	7.7	5.5	13.5	18.5
22:5n-3	1.3	1.9	3.4	2.1	1.1
22:6n-3	29.8	35.1	21.8	26.5	22.9

[a] From Addison, R.F. et al., *J. Fish. Res. Bd. Canada*, 25, 2083, 1968. With permission.
[b] From Ackman, R.G. and Takeuchi, T., *Lipids*, 21, 117, 1986. With permission.
[c] From Ackman, R.G. et al., *J. Fish. Res. Bd. Canada*, 26, 2037, 1969. With permission.
[d] Linder, Timmins and Ackman, unpublished results.

for studying fish muscle since almost all of the depot fat is in the liver. A thorough study [6] showed the flesh lipid to be 0.59 to 0.62% by weight, and the classes of that lipid were 2.5% sterol esters, 3.8% triglycerides, 6.4% cholesterol, 16.9% ethanolamine phosphatide, and 60.6% choline phosphatide, with a little serine phosphatide. The fatty acids of total cod phospholipid given in Table 13.2 are the same as those listed in Table 13.1, and this stresses the fact that about 80 to 85% of marine oils and lipids are made up of fatty acids of biochemical importance or interest, while a variety of minor odd-chain, branched chain, or unsaturated fatty acids, perhaps another 10 to 15%, are of little interest. Partly there is a historical reason, since many older analyses were conducted with packed column gas-liquid chromatography. Modern open-tubular analyses reveal much more detail [9]. The important point to note is that the phospholipids are always 0.6 to 0.8% of fish muscle and contain 30 to 40% of C20 and C22 polyunsaturated long-chain omega-3 fatty acids, with DHA (22:6n-3) > EPA (20:5n-3) in a ratio of about 6:1. In the triacylglycerol, fish oils the total for these two "omega-3" fatty acids of current interest may be thought of as 12 to 14% in Atlantic herring oil, or even lower, or as high as 30% of fatty acids in other fish oils with very little 20:1 and 22:1 [3].

Starting about 1980, encapsulated fish oils other than cod liver oil were sold as sources of omega-3 fatty acids. Originally U.S. menhaden oils, but now especially South American anchovy oils, provided the basis of the nutraceutical label mystique of "180/120" for fish oil capsules, referring to EPA and DHA in mg/g of oils. These oils were fully refined, including deodorization to improve flavor, which fortuitously stripped out much of the organochlorine materials present in the crude oils. A recent publication from the United Kingdom exposed cod liver oil contents of brominated hydrocarbons used as fire retardants in paints. This was a localized problem, but typical of the sensitivity of modern analytical methods and of the glee with which such problems are picked up by environmentalists and the news media.

Many plants have been proposed as alternative sources of omega-3 fatty acids [10]. Regrettably, these are all producing 18:3n-3 (alpha-linolenic acid) and not 20:5n-3 (eicosapentaenoic) and 22:6n-3 (docosahexaenoic) acids. One or two plant oils contain 18:4n-3, which is present in most fish lipids and is readily converted to 20:5n-3 in our bodies. It is indisputable that vegetarians do not consume much of the long-chain marine omega-3 fatty acids, better known to the public as EPA (20:5n-3) and DHA (22:6n-3), and appear to function as normal humans. Thus, in one extreme view, there appears to be no need for functional foods or nutraceutical supplements providing EPA and DHA if 18:3n-3 intake is adequate. This position is at odds with the current human drive for wellness and smarter children. The bookstore racks impressed a recent speaker who said that if the human race was in as much peril as the writers of the many books suggested, it would have been extinct long ago.

13.3 OMEGA-3 FATTY ACIDS

It is now established, even in the mind of the lay public, that there are two kinds of unsaturated fatty acids, mono- and polyunsaturated. The latter further divide into the structurally different omega-6 and omega-3. In Table 13.1, there are included only two fatty acids of the omega-6 family, but 18:2n-6 is the most common plant seed oil fatty acid, essential because our bodies cannot produce it, but we can convert it into 20:4n-6 (arachidonic acid), long known to be a vital source of eicosanoids. These are chemicals that are powerful agents affecting many human bodily functions, notably the clotting of blood through platelet aggregation. The slow discovery of such essential fatty acids has been reviewed by Holman [11] and the importance of fatty acid nutrients generally, and EPA and DHA in particular, as they affect the development of atherosclerosis have been much discussed in many conferences. Although there are thousands of articles to choose from, a definitive review on omega-3 fatty acids and on atherosclerosis appeared as early as 1988 [12] and this topic was again thoroughly reviewed from a nutrition point of view in 1994 [13]. Subsequently, omega-3 fatty acids and sudden cardiac death through arrhythmia have also been reviewed [14], and there is also a short, but critical, contemporary review of the field of postmyocardial infarction clinical trials with fatty acids [15]. Women should be especially alert to the mortality risks of myocardial conditions, which are the leading cause of death in women.

This review will not do more than point out that the other "essential" fatty acid has long been the omega-3 alpha-linolenic acid (18:3n-3). It was originally classed as that on the basis that we could not biosynthesize it ourselves [11]. Now it is publicized by the flax industry and some other plant oil industrial associations as if it were the only "omega-3" fatty acid. Not unexpectedly, the definition of essential fatty acid has had to be readjusted, but not without some argument and discussion [16–18]. A moderate view is that of Cunnane who would restrict the term "essential" for use with AA, EPA, and DHA, and coined "conditionally essential" for linoleic and alpha-linolenic [19]. In view of the major problems affecting world fisheries [20] and the great wastage of alpha-linolenic acid when consumed (see also below), these matters are not inconsequential, but redressing the misunderstandings is subject to terrible inertia. Unfortunately, all kinds of press releases and anecdotal publications, especially on web pages, have made the longer-chain, highly unsaturated omega-3 fatty acids appear to be a panacea for every ailment known to man. In fact, Lands was one of the first to bring the attention of the world that from about 1950 Western diets were deliberately overloaded with 18:2n-6 in an attempt to reduce serum cholesterol. Often this was achieved by reducing food levels of 18:3n-3 to avoid autoxidation of salad and frying oils. Corn oil, with only ≤1% of 18:3n-3, was considered ideal! His recent short review [21] is highly recommended reading.

13.4 FATTY FISH

In fish muscle, the basic phospholipids are often supplemented by depot fat triacylglycerols, leading to marine food fish with anywhere from 1 to 16% of edible muscle total lipids. There may be a fatty layer under the skin, and, as shown in Figure 13.2

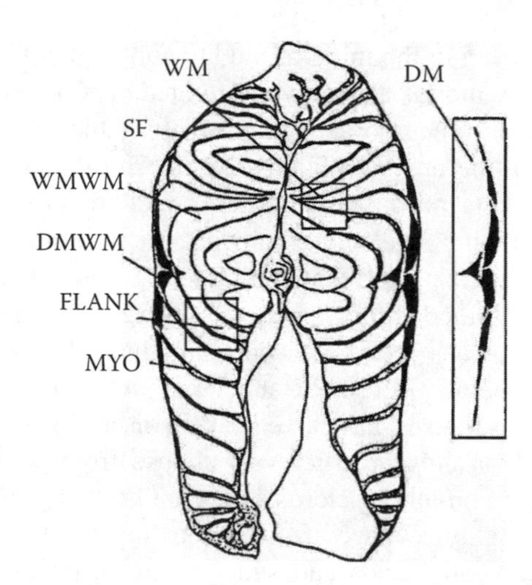

FIGURE 13.2 Cross section of salmon body muscle. WM, dorsal white muscle; DM, dark muscle; MYO, myosepta; SF, subdermal fat tissue. In salmon, the triacylglycerols are distributed in tiny droplets in the dark muscle and in the white muscle in larger fat cells (adipocytes) occurring along the myosepta, which is mainly made up of connective tissue. (From Zhou, S. et al., Storage of lipids in the myosepta of Atlantic salmon (*Salmo salar*), *Fish Physiol. Biochem.*, 14, 171, 1995. With permission.)

for a salmon steak, there will be a basic white (dorsal) muscle and an accompanying dark muscle layer on the flanks. In the lower part of Figure 13.2, one can see that there is a different arrangement of the myosepta (bands of connective tissue) and accompanying adipocytes (wholly fat-containing cells), leading to an excessively fatty deposition in the area called the belly flaps. In the salmon industry and some others, these are trimmed off and discarded, with the viscera, for animal feed or fish meal and oil manufacture.

Wild fish are prone to seasonal fluctuations in fat content, usually associated with spawning. The capelin examined for the Table 13.2 entry had body/head lipid ranging from 1.9 to 9.3% fat. This species is usually harvested only for fish meal and oil production. The common Atlantic mackerel *Scomber scombrus* may be a better example for food use. These fish migrate northwest past Nova Scotia in the spring, with fillet fat as low as 1.89%, and after feeding heavily all summer return with 20% fillet fat [22]. The EPA content of this lipid was then 7.1% and the DHA content was 10.6%, and like other fatty fish the fillet could be smoked to give a very popular product [23]. The brining and smoking processes actually stabilized the highly unsaturated lipids.

13.5 OFFICIAL RECOMMENDATIONS FOR LONG-CHAIN OMEGA-3 FATTY ACIDS

The first official medical recommendation that people should eat fish regularly was made in the United Kingdom in 1994 [24]. Twice a week was specified, with one meal being fatty fish. The American Heart Association (AHA) followed belatedly with committee statements in 2002 [25] and again in 2003 [26] with the following recommendations:

Population	Recommendation
Patients without documented chronic heart disease (CHD)	Eat a variety of (preferably oily) fish at least once a week, include oils and foods rich in α-linolenic acid (flaxseed, canola, and soybean oils; flaxseed and walnuts)
Patients with documented CHD	Consume ≈1 g of EPA + DHA per day, preferably from oily fish, EPA + DHA supplements could be considered in consultation with the physician
Patients needing triglyceride lowering	Consume 2 to 4 g of EPA + DHA per day provided as capsules under a physician's care

The AHA has published [26] a table of EPA + DHA contents of selected U.S. common fishery products, in grams per 3 oz (100 g) serving, and of the amount required of fishery product (or fish oil) to provide 1 g of EPA + DHA per day. This may be more easily understood than the usual w/w% percentages of fatty acid in muscle (edible) fat tables found for fish in scientific publications. Reference books may be in either format.

13.6 ORIGIN OF THE LONG-CHAIN OMEGA-3 FATTY ACIDS

This is a question that applies to both humans and fish. In humans, there is a rather inefficient conversion of vegetable 18:3n-3 to EPA and DHA [27–32]. Since Lands began to alert us to the problem of an excess of omega-6 over omega-3 fatty acids in Western diets, it has gradually been realized that the 18:2n-6 and 18:3n-3 can compete for enzymes in their respective elongations to 20:4n-6 and 20:5n-3 [28]. Figure 13.1 shows that this is not a problem for fish. There are thousands of species of phytoplankton, some producing no long-chain omega-3 fatty acids, and others small or large proportions of EPA and DHA [33–35]. Small marine animals also need these fatty acids for particular purposes. To take only one example from a vast literature, the popular scallop as typical of shellfish, it can be seen from Table 13.2 that the plentiful algal supply of EPA makes it a particularly useful omega-3 fatty acid for such animals [36], although DHA can have a role as well, especially in functional lipids such as phospholipids [37].

13.7 FISH AND SHELLFISH *ARE* FUNCTIONAL FOODS

It is not surprising that critics of omega-3 fatty acid sources abound [38], but aside from shellfish toxins, mercury in the foods of pregnant women [39–41], and sanitation concerns [42], there are few specific dangerous materials present at high levels in domestic seafood usually judged for quality on other standards [42]. Occasionally some alarmist groups complain about contaminants (usually heavy metals, pesticides, and organochlorine compounds) in fish or fish oils. This can be a general food problem and one very recent scientific paper compares the safety of dietary fish oils and vegetable oil supplements. In addition to reporting the polybrominated diphenyl ethers newly recognized in fish and fish oils, the paper discusses the "nutritional relevance for human essential n-3 fatty acid requirements" [43]. This makes risks a separate case from the functionality of fish in regard to long-chain omega-3 fatty acids. Similarly, a "certified organic" grading and claim may soon apply to farmed fish on rather feeble grounds unless cheap aquaculture feeds low in organic contaminants can be produced.

In parallel with Table 13.2 for phospholipids in the smolting Atlantic salmon, the triacylglycerols of the farmed fish and the wild fish had similar totals of the omega-6 polyunsaturated fatty acid AA and of EPA and DHA:

	20:4n-6	20:5n-3	22:6n-3
Farmed	0.3	1.7	6.2
Wild	2.8	2.1	4.1

The totals for omega-3 long-chain polyunsaturates are roughly the same, but in the wild fish living in freshwater, a greater part of the total is the 20:4n-6, also obvious in Table 13.2 in the phospholipids of the wild salmon taken in freshwater, and probably based on consumption of terrestrial insects, which have fatty acid compositions biased toward the omega-6 fatty acids. Freshwater fish do have EPA and DHA in reasonable amounts but there may be localized risks from contaminants, although probably not enough omega-6 fatty acids to do any harm if consumed occasionally. In conclusion, the marine fish probably have no need to biosynthesize EPA and DHA. In fact, it does not matter whether we eat fish as a functional food or take fish oil capsules as a nutraceutical: the origin of the EPA or DHA is probably the same, despite attempts to promote nonfish sources on a "vegetarian" basis.

The leading firm in enriching infant supplements with DHA, Martek Bioscience Corp., is said to have spent $99 million developing and testing their product over several years, originally using the marine microalgae *Crypthecodinium cohnii* in a fermentation facility [10]. It is therefore interesting to see such products as EPA and DHA promoted in advertising as vegetarian in origin. In fact, the EPA and DHA of fish oils probably are also primarily of vegetarian origin. One clinical trial of the algal type of DHA (no EPA) affected thrombogenic factors in vegetarian subjects favorably, and in the same way as fish oil DHA [45]. There may be some physiological differences in blood lipid responses between volunteers taking ethyl esters of fish oil fatty acids in concentrates and those eating fish. This has been reported recently [46], but is to be expected given that in fish the EPA and DHA are in the 2-position of triacylglycerols and as 2-monoacylglycerols are absorbed rapidly. They therefore tend to be returned to the liver as triacylglycerols via the lymph and distributed in the body in new triacylglycerols, with the EPA and DHA in the same positions. The single molecules of ethyl esters of EPA and DHA are absorbed by the intestinal wall much more slowly and hydrolyzed, and the fatty acids returned to the blood also in new triacylglycerols, but presumably in less specific positions on the glycerol.

13.8 NUTRACEUTICALS VS. FUNCTIONAL FOODS

An "official complete encyclopedia" on the subject of nutraceuticals, published in 2001 [47], devotes about 3 pages to essential fatty acids and 2 pages to fish oils, out of a total of 669 pages. Each of these sections lists nearly a dozen body conditions considered adverse states of human health. The two lists duplicate these conditions almost completely and are non-life-threatening except for diabetes and atherosclerosis or heart disease. Curiously, the latter two terms are listed under the texts for essential fatty acids and not under fish oils. A concise but mostly fair comparison of the common supply of dietary omega-6 vs. omega-3 fatty acids is included. This meager summary has to be compared with the probably 10,000 scientific papers reporting trials of animals or humans taking fish oil supplements and the much fewer trials with consumption of "oily" fish, or indeed any fish, shellfish, or even seal oil. The latter fat, it may be recalled, started the whole omega-3 cascade just over 25 years ago when Bang and Dyerberg conducted heart disease field research on the natives in Greenland and published a short paper in *The Lancet* [48]. Despite British Cod Liver Oils soon promoting MAXEPA fish oil capsules with pictures of stalwart fur-clad Eskimos, seal fats were soon forgotten, as fish oils were the only practical way to access EPA and DHA in quantity. Nutraceuticals appear to have initially "missed that boat," although fish oil capsules were available all over the world within a few years of the benefits of omega-3 fatty acids becoming apparent. One reason was, of course, the ready availability almost all over the world of truly functional foods in the form of fish and shellfish. Unfortunately, the population of that world divided into those who did not like eating fish or could not access it and those who did, and international opinions on desirable long-chain omega-3 fatty acid intakes vary widely [49]. With the appearance of the first official recommendations that fish should be consumed twice weekly, one serving being a portion of oily fish [24,25], there was confusion as to what constituted an "oily" fish. Through spring to fall, oily fish in northern latitudes, for example, mackerel, often go from lean (1 to 2% body fat) to very fat and suitable for smoking (20% body fat in fillets). This does no harm to the nutritive value of the omega-3 fatty acids in the fish [23,50]. One attempt to quantify the benefits of eating fish is to list their omega-3 value in "salmon units," since salmon is the most popular fish marketed, with some competition from tuna. Regrettably, salmon is a large fish with a remarkably nonhomogeneous distribution of fat in the edible muscle [51], so this concept is unlikely to succeed.

To make modest levels of long-chain omega-3 fatty acids widely available, microencapsulated fish oils or esters are now considered or marketed by various food product companies. These are typically ultrafine powders with a firm but digestible shell that can stand up to mixing and baking when added at a modest level into popular foods, raw or cooked. At present, the carrying capacity (oil content) is about 50%, but 75% is the target of several companies. According to Ocean Nutrition Canada, whose product received generally recognized as safe (GRAS) status from the U.S. Food and Drug Administration (FDA) in April 2004, the average daily consumption of long-chain omega-3 fatty acids is only 71 mg/d, or a tenth of the daily recommended intake [49]. Now that the fish smell problem, mostly from oil on the outside, has been mastered by Roche and BASF, the firms introducing microencapsulated supplements also includes Clover in Australia, DSM in Europe and the United States, and Biodar in Israel.

As may be gathered from this international review of national positions as of December 2003, opinions are rather divided on the minimum dietary intake useful or necessary, but clinical trials have frequently exceeded the FDA limit of 3.1 g/d of fish oil omega-3 EPA and DHA with no obvious adverse health effects.

The introduction of new products is carried out only after mature reflection by the companies concerned. Nutraceuticals are no exception. To make the magnitude of this process a little clearer, Table 13.3 presents a selection of new product counts by category for the United States tabulated for the whole of the year 2003. What will cause the consumer to try one of these products? There often has to be a health claim or, as one student put it, the text that is on the label is really a "health assertion." The impacts of these terminologies have recently been evaluated [52]. All foods are functional. Even if a basic product is modified by additives or processing to enhance flavor or appearance it may be promoted or advertised as "functional," provided the consumer can be convinced that it conveniently provides or augments a perceived health need.

Children are a special case. They are normally at the mercy of their parents. Particularly, obesity has become rampant and is a subject of concern outside this review, especially in regard to physical inactivity. To date, the omega-3 fatty acids have been of major interest in infant nutrition, where controversial literature abounds [53]. The food industry should be taking over after this phase, as discussed in a recent food technology magazine [54], but it is an uphill job in the face of clever merchandising technology.

TABLE 13.3
Some Food-Oriented New Product Counts by Category with Stock-Keeping Units (SKUs), the Number of Different Sizes, Flavors, Varieties and Packages

	2003 Full Year		2002 Full Year	
Foods	**Reports**	**SKUs**	**Reports**	**SKUs**
Baby food	17	90	16	74
Bread products	119	331	103	264
Cake mixes, frosting, and decorations	27	53	16	42
Cereals	115	189	91	182
Cheese	157	570	138	590
Chips	118	312	101	222
Cookies	260	846	210	695
Crackers	73	150	64	129
Dairy case foods	23	48	37	75
Desserts	43	109	58	150
Entree mixes	19	53	21	62
Fish	86	211	61	197
Fruits and fruit side dishes	66	216	61	165
Ice cream, novelties, and frozen yogurt	268	728	197	562
Margarine, butter, and spreads	24	35	14	25
Mayonnaise and imitation mayonnaise	14	33	16	30
Meal replacements and special diet foods	34	112	29	124
Meals and entrees, pizza, hot snacks, and sandwiches	272	884	249	807
Mixes, other baking and nonbaking	79	200	68	135
Oil, shortening, and cooking sprays	70	171	59	200
Pasta and pasta side dishes	123	515	108	349
Pastry and baked products	120	486	111	357
Poultry	99	257	83	275
Rice and rice side dishes	35	103	52	105
Salad dressings	76	218	65	183
Sauces and gravies	166	380	186	453
Snack bars	143	355	123	323
Snacks, other	197	670	210	525
Soup	85	293	84	293
Yogurt and yogurt imitations	51	200	37	163
Actual total in original article	4727	14,812	4335	13,452

Source: Modified from *Stagnito's New Products Magazine*, January 2004. With permission.

In February 2004, *Nutrition Reviews* carried two very up-to-date articles pertinent to this review. One [55] clarifies the position of the FDA in the United States, closely followed by a second in an unusual format: an executive summary from a committee set up by the Institute of Medicine deals with the principles behind the decisions [56]. Other countries and the European Union may be ahead of, or behind, some aspects of these discussions. In other words, nothing is final.

13.9 POSTSCRIPT

The latest (April 2004) report from the U.S. Institute of Medicine and National Research Council of the Nutritional Academies states that within the regulatory parameters of the Dietary Supplement Health and Education Act (DSHEA), supplement makers and others need to increase their reporting of health problems related to supplement use.

REFERENCES

1. Anon., Functional foods market growing, *Food Technol.,* 57,12, 2003.
2. Stansby, M.E., Nutritional properties of fish oil for human consumption-early developments, in *Fish Oils in Nutrition,* Stansby, M.E., Ed., Van Nostrand Reinhold, New York, 1990, p. 268.
3. Ackman, R.G., Fatty acid composition of fish oils, in *Nutritional Evaluation of Long-Chain Fatty Acids in Fish Oil,* Barlow, S.M. and Stansby, M.E., Eds., Academic Press, London, 1982, p. 25.
4. Ratnayake, W.N. and Ackman, R.G., Fatty alcohols in capelin, herring and mackerel oils and muscle lipids. 1. Fatty alcohol details linking dietary copepod fat with certain fish depot fats, *Lipids,* 14, 795, 1979.
5. Kramer, J.K.G., Sauer, F.D., and Pigden, W.J., Eds., *High and Low Erucic Acid Rapeseed Oils,* Academic Press, Toronto, 1983.
6. Addison, R.F., Ackman, R.G., and Hingley, J., Distribution of fatty acids in cod flesh lipids, *J. Fish. Res. Bd. Can.,* 25, 2083, 1968.
7. Ackman, R.G. and Takeuchi, T., Comparison of fatty acids and lipids of smolting hatchery fed and wild Atlantic salmon *Salmo salar, Lipids,* 21, 117, 1986.
8. Ackman, R.G. et al., Newfoundland capelin lipids fatty acid composition and alterations during frozen storage, *J. Fish. Res. Bd. Can.,* 26, 2037, 1969.
9. Ackman, R.G., The gas chromatograph in practical analyses of common and unknown fatty acids for the 21st century, *Anal. Chim. Acta,* 465, 175, 2002.
10. Haumann, B.F., Alternative sources for n-3 fatty acids, *Inform,* 9, 1108, 1998.
11. Holman, R.T., The slow discovery of the importance of 3 essential fatty acids in human health, *J. Nutr.,* 128, 427S, 1998.
12. Leaf, A. and Weber, P.C., Cardiovascular effects of n-3 fatty acids, *N. Engl. J. Med.,* 318, 549, 1988.
13. Hennig, B., Toborek, M., and Cader, A.A., Nutrition, endothelial cell metabolism, and atherosclerosis, *Crit. Rev. Food Sci. Nutr.,* 34, 253, 1994.
14. Leaf, A. et al., Clinical prevention of sudden cardiac death by n-3 polyunsaturated fatty acids and mechanism of prevention of arrhythmias by n-3 fish oils, *Circulation,* 107, 2646, 2003.
15. Grundy, S., N-3 fatty acids: Priority for post-myocardial infarction clinical trials, *Circulation,* 107, 1834, 2003.
16. Lauritzen, L. and Hansen, H.S., Which of the n-3 FA should be called essential?, *Lipids,* 38, 889, 2003.
17. Sinclair, A.J. and Attar-Bashi, N.M., Which of the n-3 PUFA should be called essential?, *Lipids,* 38, 1113, 2003.
18. Muskiet, F.A.J. et al., Is docosahexaenoic acid (DHA) essential? Lessons from DHA status regulation, our ancient diet, epidemiology and randomized controlled trials, *J. Nutr.,* 134, 183, 2004.

19. Cunnane, S., Recent studies on the synthesis, β-oxidation, and deficiency of linoleate and α-linolenate: Are essential fatty acids more aptly named indispensable or conditionally dispensable fatty acids?, *Can. J. Physiol. Pharmacol.,* 74, 629, 1996.
20. Arts, M.T., Ackman, R.G., and Holub, B.J., Essential fatty acids in aquatic ecosystems: a crucial link between diet and human health and evolution, *Can. J. Fish. Aquat. Sci.,* 58, 126, 2001.
21. Lands, W.E.M., Primary prevention in cardiovascular disease: Moving out of the shadow of truth about death, *Nutr. Metab. Cardiovasc. Dis.,* 13, 154, 2003.
22. Aminullah Bhuiyan, A.K.M., Ratnayake, W.M.N., and Ackman, R.G., Effect of smoking on the proximate composition of Atlantic mackerel (*Scomber scombrus*), *J. Food Sci.,* 51, 327, 1986.
23. Aminullah Bhuiyan, A.K.M., Ratnayake, W.M.N., and Ackman, R.G., Stability of lipids and polyunsaturated fatty acids during smoking of Atlantic mackerel (*Scomber scombrus L.*), *J. Am. Oil Chem. Soc.,* 63, 324, 1986.
24. Anon., U.K. Department of Health, Report on Health and Social Subjects no.46, Nutritional Aspects of Cardiovascular Disease, Report of the Cardiovascular Review Group Committee in Medical Aspects of Food Policy, HMSO, London, 1994, 136 pp.
25. Kris-Etherton, P.M. et al., Fish consumption, fish oil, omega-3 fatty acids, and cardiovascular disease, Committee Scientific Statement, *Circulation,* 106, 2747, 2002.
26. Kris-Etherton, P.M. et al., Fish consumption, fish oil, omega-3 fatty acids, and cardiovascular disease, *Arterioscler. Thromb. Vasc. Biol.,* 23, e20, 2003.
27. Emken, E.A., Adlof, R.O., and Gulley, R.M., Dietary linoleic acid influences desaturation and acylation of deuterium-labelled linoleic and linolenic acids in young adult males, *Biochim. Biophys. Acta,* 1213, 277, 1994.
28. Gerster, H., Can adults adequately convert α-linoleic acid (18:3n-3) to eicosapentaenoic acid (20:5n-3) and docosahexaenoic acid (22:6n-3)?, *Intern. J. Vitam. Nutr. Res.,* 68, 159, 1998.
29. Burdge, G. and Wouton, S.A., Conversion of α-linolenic acid to eicosapentaenoic and docosahexaenoic acids in young women, *Br. J. Nutr.,* 88, 411, 2002.
30. James, M.J., Ursin, V.M., and Cleland, L.G., Metabolism of stearidonic acid in human subjects: Comparison with the metabolism of other n-3 fatty acids, *Am. J. Clin. Nutr.,* 77,1140,2003.
31. Vos, E., Linoleic acid, >vitamin F6'- is the Western World getting too much? Probably, *Lipid. Technol.,* July, 81, 2003.
32. Francois, C.A. et al., Supplementing lactating women with flaxseed oil does not increase docsahexaenoic acid in their milk, *Am. J. Clin. Nutr.,* 77, 226, 2003.
33. Ackman, R.G., Tocber, C.S., and McLachlan, J., Marine phytoplankter fatty acids, *J. Fish. Res. Bd. Can.,* 25, 1603, 1968.
34. Volkman, J.R. et al., Fatty acid and lipid composition of 10 species of microalgae used in mariculture, *J. Exp. Mar. Biol. Ecol.,* 128, 219, 1989.
35. Dunstan, G.A. et al., Biochemical composition of microalgae from the green algal classes Chlorophyceae and Prasinophyceae. 2. Lipid classes and fatty acids, *J. Exp. Mar. Biol. Ecol.,* 161, 115, 1992.
36. Hall, J.M., Parrish, C.C., and Thompson, R.J., Eicosapentaenoic acid regulates scallop (*Placopecten magellanicus*) membrane fluidity in response to cold, *Biol. Bull.*, 202, 201, 2002.
37. Whyte, J.N.C., Bourne, N., and Hudson, C.A., Influence of algal diets on biochemical composition and energy reserves in *Patinopecten yessoensis* (Jay) larvae, *Aquaculture,* 78, 333, 1989.

38. Anon., Hooked on fish? There might be some catches, *Harvard Health Online,* http://www.health.harvard.edu/medline/Health/L0103b.htn, 4 pages, accessed 27/02/2003.
39. Wooltorton, F., Facts on mercury and fish consumption, *CMAJ,* 167, 897, 2002.
40. Anon., America's fish. Fair or foul?, *Consumer Rep.,* February, 25, 2001.
41. Swarc, S., Mothers, babies and mercury AND Risk-free at what cost?, http://www.techcentralstation.com/042604.html.
42. Nielsen, J., Hyldig, G., and Larsen, E., "Eating quality" of fish—a review, *J. Agric. Food Prod. Technol.* 11, 125, 2002.
43. Jacobs, M.N. et al., Time trend investigation of PCBs, PBDEs, and organochlorine pesticides in selected n-3 polyunsaturated fatty acid rich dietary fish oil and vegetable oil supplements; nutritional relevance for human essential n-3 fatty acid requirements, *J. Agric. Food Chem.,* 52, 1780, 2004.
44. Bourn, O. and Prescott, J., A comparison of the nutritional value, sensory qualities, and food safety of organically and conventionally produced foods, *Crit. Rev. Food Sci. Nutr.,* 42, 1, 2002.
45. Conquer, J.A. and Holub, B.J., Supplementation with an algae source of docosahexaenoic acid increases (n-3) fatty acid status and alters selected risk factors for heart disease in vegetarian subjects, *J. Nutr.,* 126, 3032, 1996.
46. Visioli, F. et al., Dietary intake of fish vs. formulations leads to higher plasma concentrations of n-3 fatty acids, *Lipids,* 38, 415, 2003.
47. Roberts, A.J., O'Brien, M.E., and Subak-Sharpe, G., *Nutraceuticals. The Complete Encyclopaedia of Supplements, Herbs, Vitamins, and Healing Foods,* A Perigee Book, Berkeley Publishing Group, Penguin Putnam, New York, 2001, pp. 295, 298.
48. Dyerberg, J. et al., Eicosapentaenoic acid and prevention of thrombosis and atherosclerosis? *Lancet* ii, July 15, 117, 1978.
49. Anon., Collected recommendations for long-chain polyunsaturated fatty acid intake, *Inform,* 14, 762, 2003.
50. Cardinal, N. et al., Sensory characteristics of cold-smoked Atlantic salmon (*Salmo salar*) from European market and relationships with chemical, physical and microbiological measurements, *Food Res. Int.,* 37, 181, 2004.
51. Katikou, P., Hughes, S.I., and Robb, D.H.F., Lipid distribution within Atlantic salmon (*Salmo salar*) fillets, *Aquaculture,* 202, 89, 2001.
52. Urala, N., Arvola, A., and Lahteenmahi, L., Strength of health-related claims and their perceived advantage, *Int. J. Food Sci.,* 38, 815, 2003.
53. Tolley, E.A. and Carlson, S.E., Considerations of statistical power in infant studies of visual acuity development and docosahexaenoic acid status, *Am. J. Clin. Nutr.,* 71, 1, 2000.
54. Ohr, L.M., Meeting children's nutritional needs, *Nutraceut. Funct. Foods,* 58, 4, 65, 2004.
55. Taylor, C.L., Regulatory frameworks for functional foods and dietary supplements, *Nutr. Rev.,* 62, 55, 2004.
56. Rosenberg, I.H. et al., Dietary reference intakes—Guiding principles for nutrition labelling and fortification, *Nutr. Rev.,* 62, 73, 2004.

14 Nutraceuticals, Aging, and Food Oxidation

Jae Hwan Lee and David B. Min

CONTENTS

14.1 INTRODUCTION

Aging is the accumulation process of diverse detrimental changes in the cells and tissues with advancing age resulting in the increase of the risks of disease and death [1]. Aging is influenced by many factors, including lifestyle, environmental conditions, and genetic predisposition [2]. With increasing age, the oxidation products from lipids, nucleic acids, proteins, sugars, and sterols are found to increase [3]. The main causes of the aging process seem to be related to reactive oxygen species and free radicals, such as superoxide anion, hydrogen peroxide, hydroxyl radicals, and singlet oxygen. Mitochondria, which consume more than 90% of the oxygen in aerobic living organisms, are the main reactive oxygen species and free-radical sources. Oxygen in mitochondria is reduced to water by four sequential steps [4]. Perhydroxyl radical ($HO_2^{\bullet}$) or its ionized form, superoxide anion ($^{\bullet}O_2^-$), is the first reduced intermediate of oxygen. Hydrogen peroxide (H_2O_2) and hydroxyl radical ($^{\bullet}OH$) are inevitable intermediates from oxygen to water reduction steps in the body. Approximately 1 to 5% of the oxygen consumed by mitochondria is reduced and converted to these reactive oxygen species [4].

Harman [1] suggested that initially generated superoxide anion and hydrogen peroxide are the main reactive oxygen species causing the oxidation of cells and tissues. Superoxide anion itself is not a strong oxidant but it reacts with proton in water solution to form hydrogen peroxide, which can serve as a substrate for the generation of hydroxyl radicals and singlet oxygen [5]. Hydroxyl radicals are strong oxidants and can abstract a hydrogen atom from any carbon–hydrogen bond and oxidize the compounds. For example, linoleic acids are mainly located in glycerolipids and phospholipids of cell membranes; therefore, cell membranes are easily oxidized and lose their functionality during the aging process.

Pro-oxidative enzymes such as lipoxygenase can generate free radicals [2]. Lipoxygenase can react with free forms of fatty acids, which can be released from glycerides by membrane-bound phospholipase A_2. Environmental sources, such as ultraviolet irradiation, ionizing irradiation, and pollutants, also produce reactive oxygen species [6]. Injured cells and tissues can stimulate the generation of free radicals [2]. Reactive oxygen species can be formed in foods through lipid oxidation and photosensitizers exposed to light [7]. Nonenzymatic lipid oxidation requires the presence of free forms of bivalent metal ions such as copper and iron, which are not common for healthy adults [8]. It has been assumed that free forms of irons are generated by the decompositions of iron-containing natural sources, such as hemoglobin and ferritin. [6]

Enzymatic and nonenzymatic antioxidant systems in the body, including superoxide dismutase, catalase, glutathione peroxidase, lipid-soluble vitamin E and carotenes, and water-soluble vitamin C, regulate the balance of reactive oxygen species with antioxidants [9,10]. As aging proceeds, the efficiency of antioxidant defense systems lowers and the ability to remove deleterious reactive oxygen species and free radicals decreases. The prevalent free-radical states, or so-called oxidative stress, initiate the oxidation of polyunsaturated fatty acids (PUFAs), proteins, deoxyribonucleic acid (DNA), and sterols. The age-associated increases in oxidized proteins, oxidized DNA, sterol oxidation products, and lipid oxidation products support the theory that reactive oxygen species and free radicals are involved in the aging process [6,11]. Consumption of fruits and vegetables containing high amounts of antioxidative nutraceuticals has been associated with the balance of the free radicals/antioxidants status, which helps to minimize the oxidative stress in the body and to reduce the risks of cancers and cardiovascular diseases [12].

14.2 REACTIVE OXYGEN SPECIES

Reactive oxygen species can be classified into oxygen-centered radicals and oxygen-centered nonradicals. Oxygen-centered radicals are superoxide anion ($^{\bullet}O_2^-$), hydroxyl radical ($^{\bullet}OH$), alkoxyl radical ($RO^{\bullet}$), and peroxyl radical ($ROO^{\bullet}$). Oxygen-centered nonradicals are hydrogen peroxide (H_2O_2) and singlet oxygen ($^1O_2$). Another reactive species are nitrogen species such as nitric oxide ($NO^{\bullet}$), nitric dioxide ($NO_2^{\bullet}$), and peroxynitrite ($OONO^-$) [13]. Reactive oxygen species in biological systems are related to free radicals, even though there are nonradical compounds in reactive oxygen species such as singlet oxygen and hydrogen peroxide. A free radical exists with one or more unpaired electron in atomic or molecular orbital. Free radicals are generally unstable, highly reactive, and energized molecules. Reactive oxygen species or free radicals in biological systems can be formed by pro-oxidative enzyme systems, lipid oxidation, irradiation, inflammation, smoking, air pollutants, and glycoxidation [5,6]. Clinical studies reported that reactive oxygen species are associated with many age-related degenerative diseases, including atherosclerosis, vasospasms, cancers, trauma, stroke, asthma, hyperoxia, arthritis, heart attack, age pigments, dermatitis, cataractogenesis, retinal damage, hepatitis, liver injury, and periodontitis (Figure 14.1) [14]. Reactive oxygen species also have been known to induce apoptosis of cells [14].

Benign functions of free radicals have been reported, including activation of nuclear transcription factors, gene expression, and a defense mechanism to target tumor cells and microbial infections [14]. Superoxide anion may serve as a cell growth regulator [6]. Singlet oxygen can attack various pathogens and induce physiological inflammatory response [5]. Nitric oxide is one of the most widespread signaling molecules and participates in every cellular and organ function in the body. Nitric oxide acts as a neurotransmitter and an important mediator of the immune response [15].

14.2.1 SUPEROXIDE ANION ($^{\bullet}O_2^-$)

Superoxide anion is a reduced form of molecular oxygen that develops when the oxygen receives one electron (Figure 14.2). Superoxide anion is an initial free radical formed from mitochondrial electron transport systems. Mitochondria generate energy using four electron chain reactions, reducing oxygen to water. Some of the

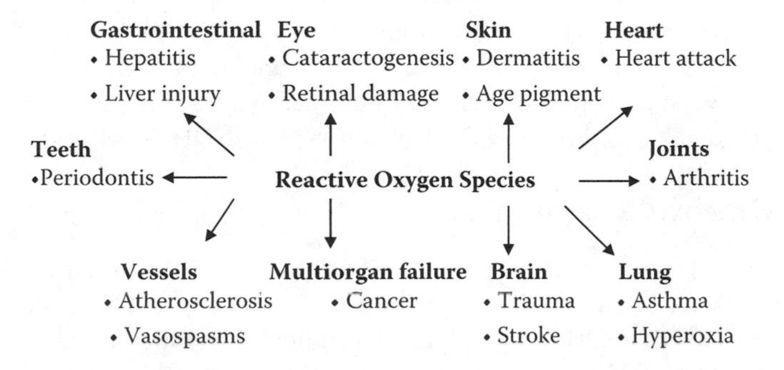

FIGURE 14.1 Clinical conditions involving reactive oxygen species.

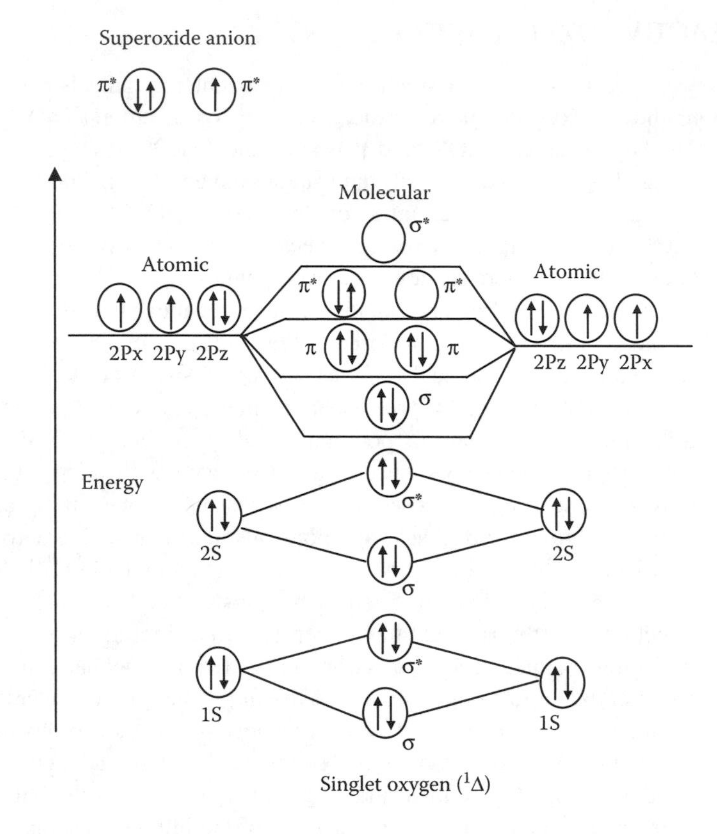

FIGURE 14.2 Molecular orbitals of singlet oxygen and superoxide anion.

electrons escaping from the chain reaction of mitochondria directly react with oxygen and form superoxide anions [1].

$$O_2 \xrightarrow[H^+]{e^-} \begin{matrix} HO_2^{\bullet} \\ {}^{\bullet}O_2^- \end{matrix} \xrightarrow[H^+]{e^-} H_2O_2 \xrightarrow[H^+]{e^-} {}^{\bullet}OH \xrightarrow[H^+]{e^-} H_2O$$

Superoxide anion plays an important role in the formation of other reactive oxygen species such as hydrogen peroxide, hydroxyl radical, or singlet oxygen ($2{}^{\bullet}O_2^- + 2H^+ \rightarrow H_2O_2 + O_2$) in living systems [5]. Superoxide anion can react with nitric oxide ($NO^{\bullet}$) and form peroxynitrite ($ONOO^-$), which can generate toxic compounds such as hydroxyl radical and nitric dioxide ($ONOO^- + H^+ \rightarrow {}^{\bullet}OH + {}^{\bullet}NO_2$) [6].

14.2.2 Hydroxyl Radical (${}^{\bullet}OH$)

Hydroxyl radical is the most reactive free radical and can be formed from superoxide anion and hydrogen peroxide in the presence of metal ions such as copper or iron (${}^{\bullet}O_2^- + H_2O_2 \rightarrow {}^{\bullet}OH + OH^- + O_2$). Hydroxyl radicals have the highest one-electron reduction potential (2310 mV) and can react with everything in living organisms at the second-order rate constants of 10^9 to 10^{10} M^{-1} s^{-1} [16]. In general, aromatic compounds or compounds with carbon–carbon multiple bonds undergo addition reactions with hydroxyl radicals, resulting in the hydroxylated free radicals. In saturated compounds, a hydroxyl radical abstracts a hydrogen atom from the weakest C–H bond to yield a free radical [16]. The resulting radicals can react with oxygen and generate other free radicals.

Hydroxyl radicals react with lipid, polypeptides, proteins, and DNA, especially thiamine and guanosine [3]. Hydroxyl radicals also add readily to double bonds. The barrier to the addition of hydroxyl radicals to double bonds is less than that of hydrogen abstraction, so that in competition addition is often favored. When a hydroxyl radical reacts with aromatic compounds, it can add on across a double bond, resulting in hydroxycyclohexadienyl radical [17]. The resulting radical can undergo further reactions, such as reaction with oxygen, to give peroxyl radical or decompose to phenoxyl-type radicals by water elimination.

14.2.3 Hydrogen Peroxide (H_2O_2)

Hydrogen peroxide can be generated through dismutation reaction from superoxide anion by superoxide dismutase. Enzymes such as amino acid oxidase and xanthine oxidase also produce hydrogen peroxide from superoxide anion. Hydrogen peroxide is highly diffusible and crosses the plasma membrane easily.

Hydrogen peroxide is the least reactive molecule among reactive oxygen species and is stable under physiological pH and temperature in the absence of metal ions. Hydrogen peroxide is a weak oxidizing and reducing agent and is thus regarded poorly reactive. Hydrogen peroxide can generate hydroxyl radical in the presence of metal ions and superoxide anion ($^{\bullet}O_2^- + H_2O_2 \rightarrow {^{\bullet}OH} + OH^- + O_2$) [6]. Hydrogen peroxide can produce singlet oxygen through reaction with superoxide anion or with HOCl or chloroamines in living systems [5,18]. Hydrogen peroxide can degrade certain heme proteins, such as hemoglobin, to release iron ions.

14.2.4 Singlet Oxygen ($^{\bullet}O_2$)

Singlet oxygen is nonradical and in excited status. Molecular orbital of singlet oxygen is shown in Figure 14.2. The electrons in the π antibonding orbitals of singlet oxygen are paired. Takayama et al. [19] reported that metastable phosphatidylcholine hydroperoxides present in the living organism produced singlet oxygen during their breakdown in the presence of Cu^{2+} in the dark. Singlet oxygen can be formed from hydrogen peroxide, which reacts with superoxide anion or with HOCl or chloroamines in cells and tissues [5].

Compared with other reactive oxygen species, singlet oxygen is rather mild and nontoxic for mammalian tissue [5]. However, singlet oxygen has been known to be involved in cholesterol oxidation [20]. Oxidation of cholesterol by singlet oxygen results in formation of 5α-OOH (3β-hydroxy-5α-cholest-6-ene-5-hydroperoxide), 5β-OOH (3β-hydroxy-5β-cholest-6-ene-5-hydroperoxide), 6α-OOH, and 6β-OOH [21,22]. Oxidation and degradation of cholesterol by singlet oxygen were observed to be accelerated by the copresence of fatty acid methyl ester. In humans, singlet oxygen is both a signal and a weapon with therapeutic potency against various pathogens, such as microbes, viruses, and cancer cells [5].

14.2.5 Peroxyl ($ROO^\bullet$) and Alkoxyl Radicals ($RO^\bullet$)

Peroxyl radicals ($ROO^\bullet$) are formed by a direct reaction of oxygen with alkyl radicals ($R^\bullet$), for example, reaction between lipid radicals and oxygen. Decomposition of alkyl peroxides (ROOH) also results in peroxyl ($ROO^\bullet$) and alkoxyl ($RO^\bullet$) radicals. Irradiation of UV light or presence of a transition metal ion can cause homolysis of peroxides to produce peroxyl and alkoxyl radicals ($ROOH \rightarrow ROO^\bullet + H^\bullet$, $ROOH + Fe^{3+} \rightarrow ROO^\bullet + Fe^{2+} + H^+$).

Peroxyl and alkoxyl radicals are good oxidizing agents, having more than 1000 mV of standard reduction potential [23]. They can abstract hydrogen from other molecules with lower standard reduction potential. This reaction is frequently observed in the propagation stage of lipid peroxidation. Very often, the alkyl radical formed from this reaction can react with oxygen to form another peroxyl radical resulting in chain reaction. Some peroxyl radicals break down to liberate superoxide anion or can react with each other to generate singlet oxygen [23]. Aromatic alkoxyl and peroxyl radicals are less reactive than respective open chain radicals due to the delocalization of electrons in the ring.

14.2.6 Nitric Oxide ($NO^\bullet$) and Nitric Dioxide ($NO_2^\bullet$)

Nitric oxide ($NO^\bullet$) is a free radical with a single unpaired electron. Nitric oxide is formed from L-arginine by NO synthase [15]. Nitric oxide itself is not a very reactive free radical, but the overproduction of nitric oxide is involved in ischemia-reperfusion and neurodegenerative and chronic inflammatory disease, such as rheumatoid arthritis and inflammatory bowel disease. Nitric oxide, exposed in human blood plasma, can deplete the concentration of ascorbic acid and uric acid and initiate lipid peroxidation [24].

Nitric dioxide ($NO_2^\bullet$) is formed from the reaction of peroxyl radical and NO, polluted air and smoking [25]. Nitric dioxide adds to double bonds and abstract labile hydrogen atoms, initiating lipid peroxidation and production of free radicals. It also oxidizes ascorbic acid [26].

14.2.7 Peroxynitrite ($OONO^-$)

Reaction of nitric oxide and superoxide anion can generate peroxynitrite ($^\bullet O_2^- + NO\cdot \rightarrow OONO^-$). Peroxynitrite is a cytotoxic species and causes tissue injury and oxidizes low-density lipoprotein (LDL) [6]. Peroxynitrite appears to be an important tissue-damaging species generated at the sites of inflammation [26]. Peroxynitrite ($OONO^-$) can cause direct protein oxidation and DNA base oxidation and modification acting as a "hydroxyl radical-like" oxidant [27].

14.2.8 Enzymatic Formation

Pro-oxidative enzymes, including reduced nicotinamide adenine dinucleotide phosphate (NADPH) oxidase [28], NO synthase [29], and the cytochrome P450 chain [18], can generate reactive oxygen species. Lipoxygenase generates free radicals. Lipoxygenase needs free PUFAs, which are not present in healthy tissue. Membrane-bound phospholipase produces PUFAs and lysolecithins. Lysolecithins change the cell membrane structures, and free PUFAs are oxidized to form lipid hydroperoxides.

Lipoxygenase with Fe^{2+} ion is in inactivated status. Once Fe^{2+} oxidized to Fe^{3+}, lipoxygenase can convert PUFAs into hydroperoxides [2].

14.3 LIPID OXIDATION AND AGING

Theories explaining aging processes are diverse, including disease, environment, immune dysfunction, inborn processes, and free radicals. Generally accepted aging theories are based on either developmentally programmed aging or damage-accumulation aging, both involving generation of free radicals [6]. Free-radical theories are based on the chemical characteristics of free radicals, which are ubiquitous in body systems. Free radicals or reactive oxygen species in the body can cause lipid oxidation, protein oxidation, DNA strand break and base modification, and modulation of gene expression.

Oxidative stress is a serious imbalance between reactive oxygen species and antioxidants. Oxidative stress is due to antioxidant-deficient diets or increased production of reactive oxygen species by environmental toxins like smoking or by inappropriate activation of phagocytes such as with chronic inflammatory disease [6]. Clinical studies reported that reactive oxygen species are associated with many degenerative diseases, which are associated with aging (Figure 14.1) [16].

14.3.1 Triplet Oxygen Oxidation with Fatty Acids

Lipid oxidation is a free-radical chain reaction and reactive oxygen species can accelerate lipid oxidation [7]. Triplet oxygen is a diradical compound, which can react with radical compounds. Food components are not radicals. In order to react with triplet oxygen, one hydrogen atom should be removed from food components to become radicals. The removal of hydrogen from saturated fatty acid requires approximately 100 kcal/mol but that of hydrogen between double bonds such as hydrogen at position 11 of linoleic acid is only about 50 kcal/mol [7]. Therefore, hydrogen between double bonds is most easily removed due to low energy requirement.

Triplet oxygen oxidation or auto-oxidation has three steps: initiation, propagation, and termination [30]. Initiation is a step for the formation of free alkyl radicals. The pentadienyl radical is formed from linoleic acid by hydrogen abstraction at the C11 position and rearranged to the C9 or 13 positions of conjugated radicals at equal concentration. During rearrangement, *cis* double bond changes to *trans* double bond. Triplet oxygen can react with conjugated double bond radicals of linoleic acid and produce peroxyl radical at C9 and 13 positions. The peroxyl radicals abstract hydrogen from other fatty acids and become hydroperoxide and generate free alkyl radicals, which is called the propagation step. The chain reactions of free alkyl radicals and peroxyl radicals accelerate the auto-oxidation. Radicals react with each other to form nonradical products, which is a termination step.

Triplet oxygen oxidation produces only the conjugated diene hydroperoxides in linoleic and linolenic acids. The relative reaction ratios of triplet oxygen with oleic, linoleic, and linolenic acid are 1:12:25 [30]. Linolenic acid reacts twice faster than linoleic acid due to the two pentadienyl groups. The number of double bonds does not change in triplet oxygen oxidation, even though *cis* conformation of double bond changes to *trans* conformation.

14.3.2 Singlet Oxygen Oxidation with Fatty Acids

Singlet oxygen is a nonradical compound and can directly react with nonradical compounds without radical intermediates. One of the highest degenerate π antibonding molecular orbitals in singlet oxygen is vacant and singlet oxygen tries to fill the empty molecular orbital with electrons (Figure 14.2). Singlet oxygen can directly react with electron-rich compounds such as unsaturated fatty acids with double bonds [31]. Singlet oxygen can react with any compounds with double bonds through the mechanisms of cycloaddition and "ene" reaction.

The reaction rates of singlet oxygen with oleic, linoleic, linolenic, and arachidonic acids are 0.7, 1.3, 1.9, and $2.4 \times 10^5\ M^{-1}\ s^{-1}$, respectively [32]. Singlet oxygen oxidation depends on the number of double bonds instead of types of double bonds, such as conjugated or nonconjugated double bonds.

14.3.3 Lipid Oxidation and Aging

Cell membranes in human beings are phospholipid bilayers with extrinsic proteins and are the direct target of lipid oxidation [20]. As lipid oxidation of cell membranes increases, the polarity of lipid phase, surface charge, and formation of protein oligomers increase, and molecular mobility of lipids, number of SH groups, and resistance to thermodenaturation decrease.

Malonaldehyde, one of the lipid oxidation products, can react with the free amino group of proteins, phospholipid, and nucleic acids leading to structural modification, which induces dysfunction of immune systems. A high level of lipid oxidation products can be detected in cell degradation after cell injury or disease. The increases in lipid oxidation products are found in diabetes, atherosclerosis, liver disease, apoplexy, and inflammation. LDL are complicated structures and oxidative modification of LDL has been reported to be involved with the development of atherosclerosis and cardiovascular disease [33]. Oxidized cholesterol or fatty acid moieties in the plasmatic LDL can develop atherosclerosis [11,20,34].

14.4 ANTIOXIDATIVE NUTRACEUTICALS

Nutraceuticals or functional foods are any food or food ingredients that may provide beneficial health effects beyond the traditional nutrients they contain [35]. Nutraceuticals are also known as medical food, nutritional supplements, and dietary supplements. Nutraceuticals range from isolated nutrients, dietary supplements, genetically engineered "designer" foods, herbal products, and processed products like cereals and soups. The functional food market has increased due to the fast growth of the older generation in the United States and their concerns with health beneficial foods [36]. Nutraceuticals can be grouped in different ways depending on the food sources, mode of action, and chemical structures [37].

As oxidative stress increases, the level of the pro-oxidants against antioxidants increases and the aging process accelerates. If reactive oxygen species and free radicals are the major causes of aging processes, antioxidative nutraceuticals can reduce the level of reactive oxygen species and free radicals, slow down the aging process, and increase life span. It has been reported that levels and activities of antioxidant enzymes, including superoxide dismutase, catalase, and glutathione peroxidase, are much higher in long-living species than in short-living ones. The concentration of vitamin E in elderly people over 65 years old is lower than that in younger adults. Consumption of optimum amounts of vitamins A and E increased the average life expectancy of animals [38,39].

Consumption of foods containing oxidized components has been implicated with acceleration of the aging process and increased risk of disease in humans. Animal and human studies showed that dietary lipid oxidation products cause the atherosclerotic process and accelerate the accumulation of oxidized lipids in macrophages and monocytes [40].

Antioxidative nutraceuticals can inhibit or slow down the formation of free alkyl radicals in the initiation step and interrupt the free-radical chain reactions in the propagation step during lipid oxidation. Antioxidative nutraceuticals can be antioxidative enzymes, hydrogen-donating compounds, metal chelators, and singlet oxygen quenchers.

14.4.1 Antioxidative Enzymes

Antioxidative enzymes, including superoxide dismutase, catalase, and glutathione peroxidase/reductase, convert reactive oxygen species into nonreactive oxygen molecules. Proteins showing antioxidant properties are listed in Table 14.1. Superoxide dismutase (SOD) converts superoxide anion into hydrogen peroxide and oxygen. There are two types of SOD: a magnesium-containing SOD and a copper–zinc-dependent SOD. Catalase is involved in cellular detoxification and can convert hydrogen peroxide into water and oxygen (Figure 14.3). Glutathione peroxidase is the most important hydrogen peroxide removing enzyme existing in the membrane. Glutathione disulfide reductase is a flavoprotein that permits the conversion of oxidized glutathione (GSSG) to reduced glutathione (GSH) by the oxidation of NADH to NAD^+ (Figure 14.3) [41].

14.4.2 Hydrogen-Donating Nutraceuticals

Antioxidative nutraceuticals, which can donate hydrogen atoms to free radicals, can scavenge free radicals and prevent lipid oxidation. Lipid oxidation in foods and in biological systems is a typical free-radical chain reaction of unsaturated fatty acids

TABLE 14.1
Antioxidative Enzymes

Proteins	Functions
Superoxide dismutase	Superoxide removal
Catalase	Hydroperoxide removal
Glutathione peroxidase	Hydroperoxide removal
Glutathione disulfide reductase	Oxidized glutathione reduction
Glutathione-*S*-transferase	Lipid hydroperoxide removal
Methionine sulfoxide reductase	Repair oxidized methionine residues
Peroxidase	Decomposition of hydrogen peroxide and lipid hydroperoxide

$2\cdot O_2^- + 2H^+ \rightarrow H_2O_2 + O_2$ (Superoxide dismutase)

$2H_2O_2 \rightarrow 2H_2O + O_2$ (Catalase)

$2GSH + H_2O_2 \rightarrow GSSG + 2H_2O$ (Glutathione peroxidase)
Reduced glutathione Oxidized glutathione
$GSSG + NADPH, H^+ \rightarrow 2GSH + NADP^+$ (Glutathione reductase)

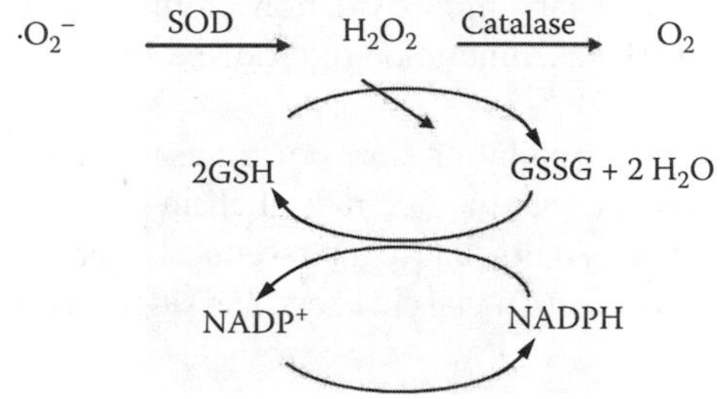

FIGURE 14.3 Antioxidative enzymes and their reaction mechanisms.

with initiation, propagation, and termination steps. The propagation step is a slow step in lipid oxidation and the concentration of peroxyl radicals is found to be the greatest of all fatty acid radicals [42]. Free-radical scavengers, which react with peroxyl radicals before the PUFAs react with peroxyl radicals, can prevent lipid oxidation. Chain-breaking antioxidants donate hydrogen atoms to peroxyl radicals and convert them to more stable and nonradical products (Table 14.2) [23,43]. Antioxidant radicals formed from hydrogen-donating antioxidants can react with alkyl, alkoxyl, and peroxyl radicals of PUFAs and generate nonradical stable compounds (Table 14.2).

Whether a compound acts as an antioxidant or a pro-oxidant can be determined by the standard one-electron reduction potential (Table 14.3). Standard one-electron reduction potentials of alkyl, peroxyl, and alkoxyl radicals of PUFAs are 600, 1000, and 1600 mV, respectively (Table 14.3) [31]. To work as an antioxidant and prevent lipid oxidation, the reduction potential of a free-radical scavenger should be lower than 600 mV, which is a reduction potential of PUFAs. For example, ascorbic acid

TABLE 14.2
Reaction of Hydrogen-Donating Antioxidants with Radicals

$R^\bullet$	+	AH	→	RH	+	$A^\bullet$
$RO^\bullet$	+	AH	→	ROH	+	$A^\bullet$
$ROO^\bullet$	+	AH	→	$ROOH^\bullet$	+	$A^\bullet$
$R^\bullet$	+	$A^\bullet$	→	RA		
$RO^\bullet$	+	$A^\bullet$	→	ROA		
$ROO^\bullet$	+	$A^\bullet$	→	ROOA		
Antioxidant	+	O_2	→	Oxidized antioxidant		

AH, antioxidant; $R^\bullet$, alkyl free radical; $RO^\bullet$, alkoxyl free radical; $ROO^\bullet$, peroxyl free radical.

TABLE 14.3
Standard One-Electron Reduction Potential (mV) at pH 7.0 for Selected Radical Couples

$HO^\bullet$, H^+/H_2O	2310
$RO^\bullet$, H^+/ROH	1600
$ROO^\bullet$, $H^+/ROOH$	1000
$GS^\bullet/GS^-$ (glutathione)	920
$PUFA^\bullet$, H^+/pUFA	600
$Catechol^\bullet$, H^+/catechol	530
α-$Tocopheroxyl^\bullet$, H^+/α-Tocopherol	480
H_2O_2, H^+/H_2O, $HO^\bullet$	320
$Ascorbate^{-\bullet}$, H^+/Ascorbate	282
$O_2/{}^\bullet O_2^-$	–330
$RSSR/RSSR^{-\bullet}$ (GSH)	–1500
H_2O/e^{-aq}	–2870

and tocopherol, which have lower standard one-electron reduction potential (282 and 480 mV, respectively) than PUFAs (600 mV), can donate a hydrogen atom to peroxyl radicals of PUFAs before PUFAs do [31].

The newly generated free radicals from antioxidative nutraceuticals should be stable enough not to participate in other lipid oxidation chain reactions. Radicals from phenolic compounds can be stabilized through resonance formation (Figure 14.4).

14.4.3 Metal-Chelating Nutraceuticals

Transition metals such as iron and copper play important roles in initiation and propagation steps of lipid oxidation. The initiation step of oxygen oxidation requires removal of a hydrogen atom. The presence of metal can accelerate the initiation step of lipid oxidation by the mechanism of $RH + M^{n+} \rightarrow R^\bullet + H^+ + M^{(n-1)+}$. Metals can decompose the hydroperoxide to form peroxyl radical and alkoxyl radical, and accelerate the lipid oxidation at the exponential rate [30,44]:

$$Fe^{3+} (Cu^{2+}) + ROOH \rightarrow Fe^{2+} (Cu^{+}) + ROO^\bullet + H^+$$

$$Fe^{2+} (Cu^{+}) + ROOH \rightarrow Fe^{3+}(Cu^{2+}) + RO^\bullet + OH^-$$

Metals are also involved in the formation of singlet oxygen:

$$Fe^{2+} + O_2 \rightarrow Fe^{3+} + O^-_2 \rightarrow {}^1O_2$$

Hydrogen peroxide can react with transition metal ions to form hydroxyl radical:

$$M^{n+} + H_2O_2 \rightarrow M^{(n+1)+} + {}^\bullet OH + OH^-$$

This reaction is dependent on the availability of transition metal ions such as copper and iron. The availability of metal ions is determined by the concentrations of metal-binding proteins, including ferritin, lactoferrin, and ceruloplasmin [23].

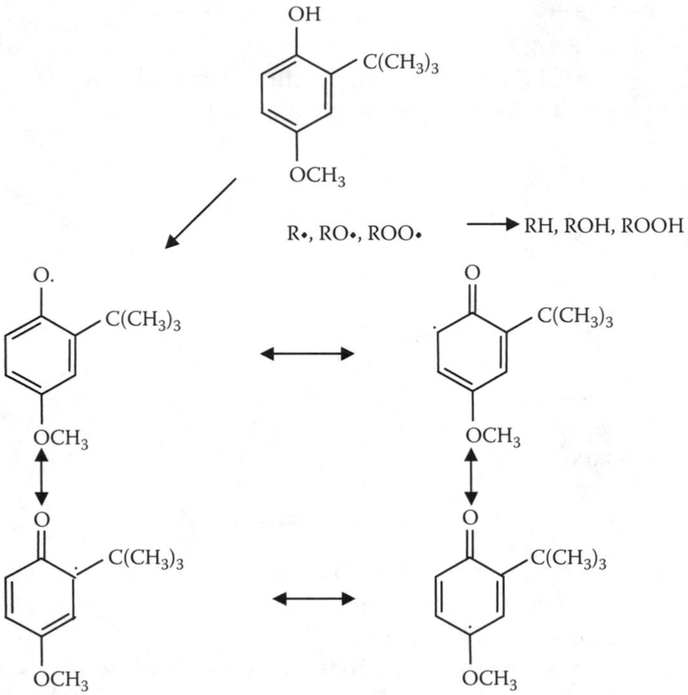

FIGURE 14.4 Resonance stabilization of phenolic antioxidant radicals. Antioxidant radicals are stabilized through resonance structures. $R^{\bullet}$, $RO^{\bullet}$, and $ROO^{\bullet}$: alky, alkoxyl, and peroxyl radicals, respectively.

Metal chelators, one type of antioxidative nutraceuticals, form complex ions or coordination compounds with metals by occupying all metal coordination sites and preventing metal redox cycling. Metal chelators can convert metal ions into insoluble metal complexes or generate steric hindrance, which can prevent the interactions between metals and lipid intermediates. Some reported metal chelating proteins are shown in Table 14.4. Metal chelators are phosphoric acid, citric acid, ascorbic acid,

TABLE 14.4
Metal-Chelating Proteins

Protein	Function
Ferritin	Iron storage
Transferrin	Iron storage
Lactoferrin	Iron storage
Haptoglobin	Hemoglobin sequestration
Ceruloplasmin	Copper storage
Albumin	Copper storage
Transferrin ferro-oxidase	Iron transport
Hemopexin	Stabilization of heme

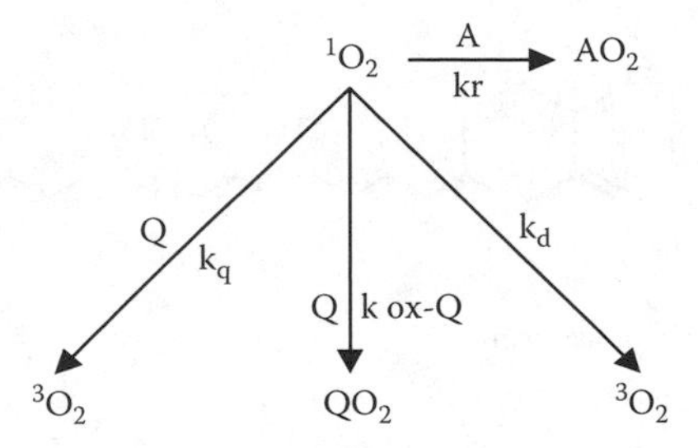

FIGURE 14.5 Singlet oxygen quenching mechanisms.

polyphenols like quercetin, carnosine, some amino acids, peptides, and proteins such as transferrin and ovotransferrin [45,46].

14.4.4 Singlet Oxygen Quenching Nutraceuticals

Singlet oxygen is highly reactive toward any molecules with π electrons or lone pairs of low ionization energy. There are two types of singlet oxygen quenching mechanisms: physical and chemical quenchings. Physical quenching converts singlet oxygen into triplet oxygen by either energy transfer or charge transfer without generating any other intermediates. Chemical quenching is involved with the generation of intermediates, such as oxidized products. Singlet oxygen reactions with compound (A) to form oxidized products (AO_2) are shown in Figure 14.5. Chemical quenching between singlet oxygen ($^1O_2$) and quencher (Q) involves the generation of an oxidized product "QO_2." Physical quenching converts singlet oxygen ($^1O_2$) to triplet oxygen ($^3O_2$) without production of oxidized product "QO_2." Detailed information on the singlet oxygen quenching mechanisms can be found in an excellent review by Boff and Min [7]. Singlet oxygen quenchers should have electron-rich structures such as double bonds in the molecules to react with singlet oxygen. Carotenoids, which have many double bonds, are well-known singlet oxygen quenchers [7]. Uric acid and ascorbic acid are also powerful quenchers of singlet oxygen [24]. Thioredoxin has been reported as a singlet oxygen quencher and a hydroxyl radical scavenger, which acts independently of the redox potential. [47]

14.5 ANTIOXIDATIVE NUTRACEUTICALS IN FOODS

14.5.1 Tocopherols and Tocotrienols

Tocopherols consist of a chroman ring and a long, saturated phytyl chain. Tocols are 2-methyl-2(4′,8′,12′-trimethyltridecyl)chroman-6-ols and tocotrienols have 3 double bonds at positions 3′, 7′, and 11′ of the side chain in tocols (Figure 14.6). The α-, β-, γ-, and σ-tocopherols and tocotrienols differ in the number and position of methyl groups attached to the 5, 7, and 8 of the ring structure [48]. Tocopherols and tocotrienols are very nonpolar and exist in lipid phase. Tocopherols are natural constituents of biological membranes. Tocotrienols are found mainly in palm oil, cereal grains, and kale [49].

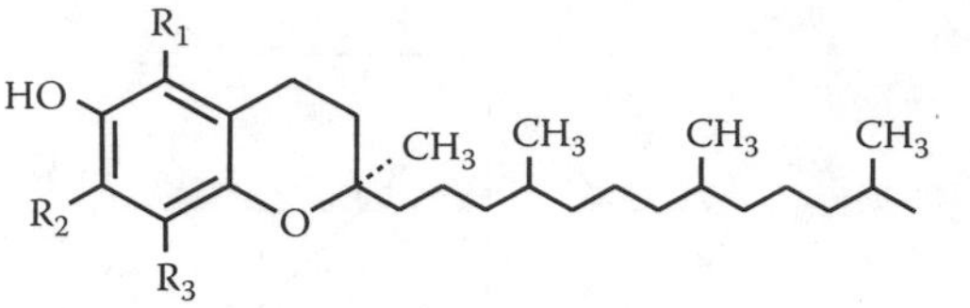

Trivial name	Chemical name	R_1	R_2	R_3
α-Tocopherol	5,7,8-Trimethyltocopherol	CH_3	CH_3	CH_3
β-Tocopherol	5,8-Dimethyltocopherol	CH_3	H	CH_3
γ-Tocopherol	7,8-Dimethyltocopherol	H	CH_3	CH_3
σ-Tocopherol	8-Methyltocopherol	H	H	CH_3

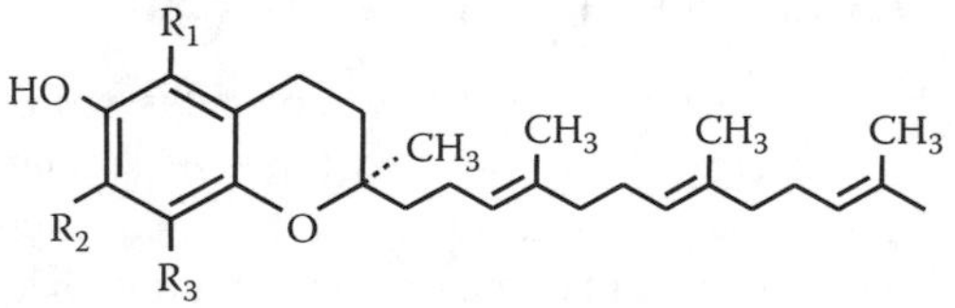

Trivial name	Chemical name	R_1	R_2	R_3
α-Tocotrienol	5,7,8-Trimethyltocotrienol	CH_3	CH_3	CH_3
β-Tocotrienol	5,8-Dimethyltocotrienol	CH_3	H	CH_3
γ-Tocotrienol	7,8-Dimethyltocotrienol	H	CH_3	CH_3
σ-Tocotrienol	8-Methyltocotrienol	H	H	CH_3

FIGURE 14.6 Structures of tocopherols and tocotrienols.

Tocopherols are typical and important antioxidants in humans. α-Tocopherol, which is present at the ratio of 1 to 1000 lipid molecules, is the most abundant among tocopherols. Tocopherols can protect PUFA within the membrane and LDL, and inhibit smooth muscle cell proliferation and protein kinase C activity. Tocopherol has been associated with the reduction of heart disease, delay of Alzheimer's disease, and prevention of cancer. γ-Tocopherols can reduce the concentration of nitrogen dioxide (NO_2) better than other tocopherols. Nitrogen dioxide is involved in carcinogenesis, arthritis, and neurological diseases. Tocotrienols have been shown to have anticancer activity and cholesterol lowering ability. Some *in vitro* studies showed that tocotrienols inhibited LDL oxidation better than tocopherols [49].

Antioxidant mechanisms of tocopherols involve transfer of a hydrogen atom at 6-hydroxyl group on the chroman ring, and scavenging of singlet oxygen and other reactive species. Tocopherols are regenerated in the presence of ascorbic acids. Phytyl chain in tocopherols can fit in the membrane bilayer while active chroman ring is closely positioned to the surface. This unique structure enables tocopherols to act as effective antioxidants and to be regenerated through reaction with other antioxidants like ascorbic acid [41]. α-Tocopherol has higher vitamin E activity and singlet oxygen quenching ability than β-, γ-, and δ- tocopherols while γ-tocopherol has better nitrogen dioxide and peroxynitrite radical scavenging ability than α-tocopherols [48]. Efficiency of scavenging hydroxyl, alkoxyl, and peroxyl radicals by α-tocopherol is approximately 10^{10}, 10^8, and 10^6 (M^{-1} s^{-1}), respectively [50].

Tocotrienols are potential nutraceuticals and the antioxidant mechanisms of tocotrienols are the same as that of tocopherols. Tocotrienols are more mobile within the biological membrane than tocopherols and have more recycling ability and more inhibition of liver oxidation [49].

14.5.2 Ascorbic Acid

L-Ascorbic acid is a 6-carbon lactone ring structure with 2,3-enediol moiety. The antioxidant activity of ascorbic acid comes from 2,3-enediol. L-Ascorbic acid first changes to semi-dehydroascorbic acid through donating one hydrogen atom and electron, and then L-dehydroascorbic acid by donating a second hydrogen atom and electron (Figure 14.7). Both L-ascorbic acid and L-dehydroascorbic acid retain the vitamin C activity. Ascorbic acid is highly susceptible to oxidation in the presence of metal ions such as Cu^{2+} and Fe^{3+}. Oxidation of ascorbic acid is also influenced by heat, light exposure, pH, oxygen concentration, and water activity [48].

Ascorbic acid may be associated with the prevention of some cancers, heart disease, and the common cold. Ascorbic acid and tocopherol supplementation can substantially reduce oxidative damage. The effects are greater in nonsmokers than smokers. Smoking induces oxidative stresses from numerous free-radical compounds in the gas phases and ascorbic acid radical could be pro-oxidant in smokers [12].

The antioxidant mechanisms of ascorbic acid are based on hydrogen atom donation to lipid radicals, quenching of singlet oxygen, and removal of molecular oxygen. Scavenging aqueous radicals and regeneration of α-tocopherol from the tocopheroxyl radical species are also well-known antioxidant mechanisms of ascorbic acid. Ascorbic acid is an excellent electron donor due to the low standard one-electron reduction potential (282 mV), the generation of relatively stable semidehydroascorbic acid, and the easy conversion of dehydroascorbic acid to ascorbic acid [51]. The reaction rate constants of ascorbic acid with other radicals are shown in Table 14.5. The kinetics of electron or hydrogen atom transfer reactions is rapid, with the resulting ascorbic acid as an excellent antioxidant. However, ascorbic acid can act as a pro-oxidant under certain conditions, including reducing ferric iron to more active ferrous iron.

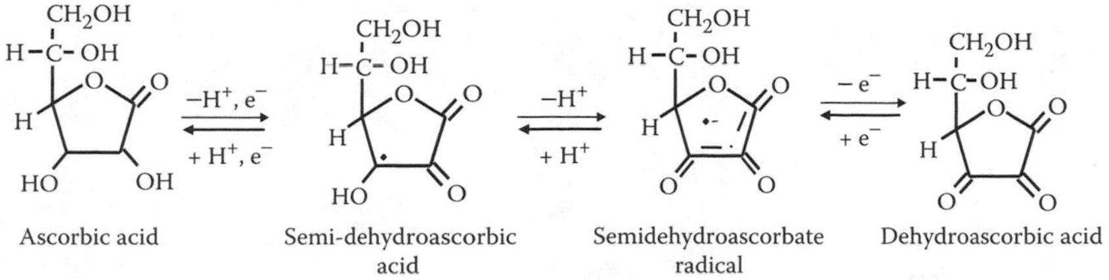

FIGURE 14.7 Sequential one-electron oxidations of L-ascorbic acid.

TABLE 14.5
Rate Constants for Reaction of Equilibrium Mixture of Ascorbic Acid/Semidehydroascorbic Acid/Dehydroascorbic Acid at pH 7.4

Radical	k (M^{-1} s^{-1})
$HO^{\bullet}$	1.1×10^{10}
$RO^{\bullet}$ (*tert*-butyl alkoxyl radical)	1.6×10^{9}
$ROO^{\bullet}$ (alkyl peroxyl radical)	$1–2 \times 10^{6}$
$GS^{\bullet}$ (glutathionyl radical)	6×10^{6} (pH 5.6)
Tocopheroxyl radical	2×10^{5}
Ascorbate$^{-\bullet}$ (dismutation)	2×10^{5}
$O_2^{-\bullet}/H_2O^{\bullet}$	1×10^{5}

Regeneration of tocopherol radicals to tocopherols by ascorbic acid has been known since the 1940s. Ascorbic acid can donate hydrogen atom to tocopheroxyl radical at the rate of 2×10^5 M^{-1} s^{-1} due to the difference of one-electron reduction potential between ascorbic acid (282 mV) and (480 mV). The phenol group of tocopherol is located near the interface of a biological membrane-water phase, and ascorbic acid can access easily to the antioxidant active site of tocopherols and regenerate tocopherols from tocopherol radicals (Figure 14.8) [52].

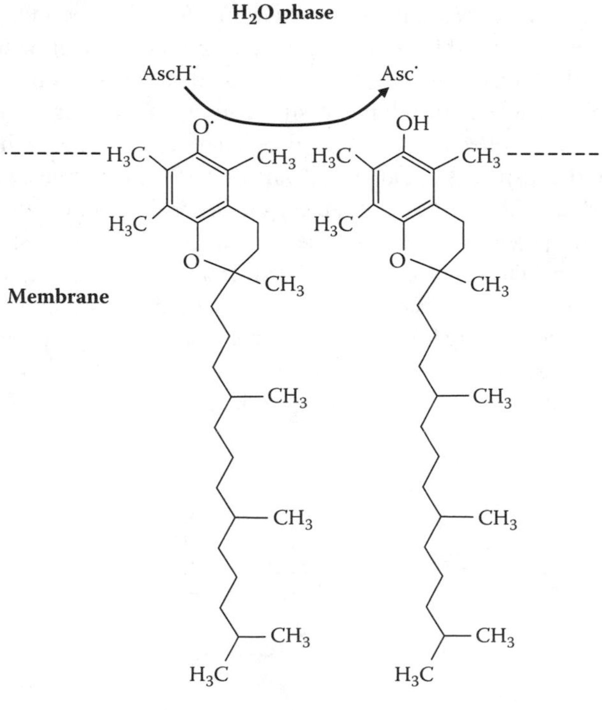

FIGURE 14.8 Regeneration of tocopherols by ascorbic acid.

14.5.3 Carotenoids

Carotenoids are a group of tetraterpenoids. The basic carotenoid structural backbone consists of isoprenoid units formed either by head-to-tail or by tail-to-tail biosynthesis. There are primarily two classes of carotenoids: carotenes and xanthophylls. Carotenes are hydrocarbon carotenoids and xanthophylls contain oxygen in the form of hydroxyl, methoxyl, carboxyl, keto, or epoxy groups. Lycopene and β-carotenes are typical carotenes while lutein in green leaves and zeaxanthin in corn are typical xanthophylls. The structures of carotenoids are acyclic, monocyclic, or bicyclic. For example, lycopene is acyclic, γ-carotene is monocyclic, and α- and β-carotenes are bicyclic carotenoids [53]. Double bonds in carotenoids are conjugated forms and usually the all *trans* forms of carotenoids are found in plant tissues (Figure 14.9).

The antioxidant potentials of carotenoids have been reported for the prevention of free-radical-initiated diseases, including atherosclerosis, cataracts, age-related

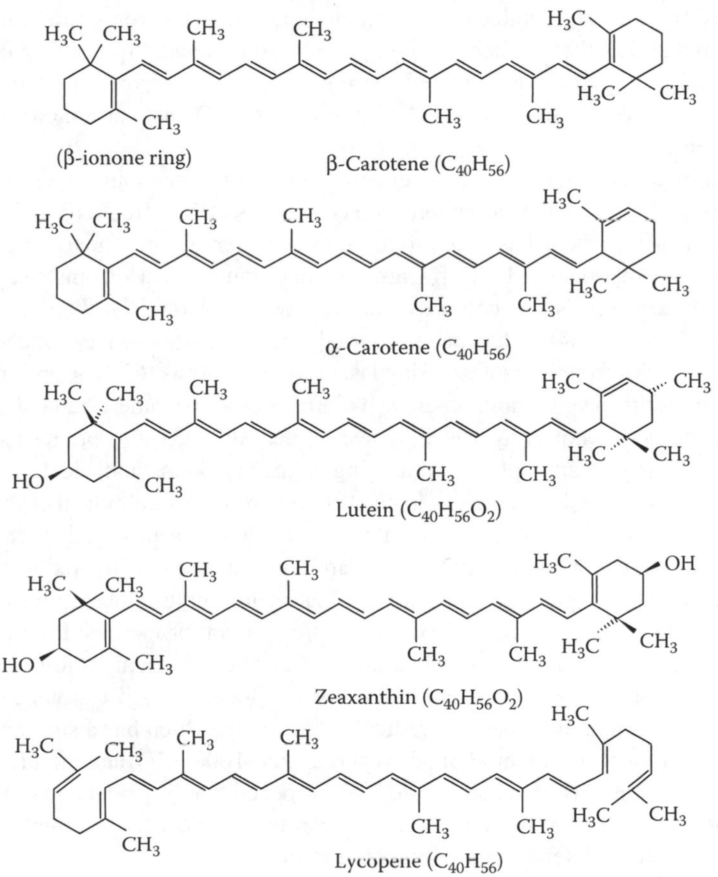

(β-ionone ring)

β-Carotene ($C_{40}H_{56}$)

α-Carotene ($C_{40}H_{56}$)

Lutein ($C_{40}H_{56}O_2$)

Zeaxanthin ($C_{40}H_{56}O_2$)

Lycopene ($C_{40}H_{56}$)

FIGURE 14.9 Structures of carotenoids. Lycopene and α- and β-carotenes are carotenes and lutein and zeaxanthin are xanthophylls.

muscular degeneration, and multiple sclerosis. Consumption of fresh tomatoes, tomato sauce, and pizza is significantly related to a low incidence of prostate cancer [54]. Lycopene, which is the main carotenoid of tomatoes and tomato products, has several health benefits including decreasing the development of cervical, colon, prostate, rectal, stomach, and other types of cancers [55,56]. Tomato juice with 40 mg of lycopene can reduce the endogenous levels of lymphocyte DNA breakage in a group of male smokers [57]. Carotenoids including lycopene and β-carotene inhibit the formation of oxidized products of LDL cholesterol, which are associated with coronary heart disease [58]. β-Carotene is involved in the protection of the skin against deleterious effects of sunlight. UV rays initiate free radicals in the epidermis by inducing lipid oxidation, which results in premature aging of the skin.

Carotenoids are the most efficient singlet oxygen quenchers in biological systems. One mole of β-carotene can quench 250 to 1000 molecules of singlet oxygen at a rate of 1.3×10^{10} M^{-1} s^{-1} [59]. The rate of singlet oxygen quenching by carotenoids is dependent on the number of conjugated double bonds, and the type and number of functional groups on the ring structure of the molecules [60]. To act as an effective singlet oxygen quencher, at least seven conjugated bonds are required and as the number of conjugated bond increases, quenching efficiency increases [7]. Singlet oxygen quenching mechanisms by carotenoids are physical quenching without generating oxidizing products ($^1O_2$ + ^{1}carotenoid $\rightarrow$ $^3O_2$ + ^{3}carotenoid).

Contrary to the singlet oxygen quenching ability of carotenoids, hydrogen donating antioxidant activities of carotenoids are controversial. The free-radical scavenging mechanism of β-carotene has been proposed to be different from the hydrogen donating phenolic compounds [61,62]. β-Carotene may donate electrons instead of hydrogen atom to free radicals, and become β-carotene radical cation ($R^\bullet$ + β-carotene $\rightarrow$ R^-+ β-carotene$^{\bullet+}$) [61,63,64]. It has been reported through the presence of near infrared absorption species from β-carotene using laser flash photolysis [65]. β-Carotene radical cation can absorb near infrared energy. β-Carotene can become a radical cation by donating an electron not by hydrogen. However, near infrared absorption species were not observed from xanthophylls containing hydroxyl, keto, and aldehyde groups, which may donate hydrogen atoms instead of electrons to free radicals [66].

β-Carotene in high oxygen concentration can act as a pro-oxidant rather than an antioxidant. Antioxidant activity of β-carotene increases at low oxygen concentration. Not only oxygen concentration but also carotenoid concentration plays an important role in determining antioxidant or pro-oxidant properties. Relatively high standard one-electron reduction potential of β-carotene radical cation (1060 mV) [67] could explain the pro-oxidant property of β-carotene. β-Carotene may not donate a hydrogen atom to peroxyl radicals effectively, which has a similar standard one-electron reduction potential of peroxyl radicals (1000 mV), and therefore cannot act as an antioxidant. Burton and Ingold [68] proposed that β-carotene may react with free radicals by addition mechanism and β-carotene molecules become resonance-stabilized, carbon-centered, and conjugated radicals.

Depending on the redox potentials of free radicals and chemical structures of carotenoids, especially the presence of oxygen-containing functional groups, either hydrogen or electron may transfer from carotenoids to free radicals [66]. β-Carotene can scavenge a superoxide anion with following equation [66]:

$$^{\bullet}O_2^- + CAR + 2H^+ \rightleftarrows CAR^+ + H_2O_2$$

14.5.4 Polyphenols

Phenolic compounds or polyphenols are ubiquitous in plants with more than 8000 structures reported [69]. The classes of phenolic compounds are shown in Table 14.6. Flavonoids, the most important single polyphenol group, are glycosides with a benzopyrone nucleus. The flavonoids including flavones, flavonols, flavanones, flavanonols, and anthocyanins are based on the common structures of carbon skeletons (Figure 14.10). The flavones have a double bond between C2–C3, while flavones have the saturated C2–C3. Flavanonols have an additional hydroxyl group at C3 position and flavonols are saturated between C2 and C3 with a hydroxyl group at C3 position. The most ubiquitous flavonoid is quercetin, 3,5,7,3′,4′-pentahydroxy flavone. Each flavonoid group is different depending on the number of hydroxyl, methoxyl, and other substituents on the two benzene rings.

Isoflavones, which do not have the common flavonoid structures, are chemically related with flavonoids. Soybeans contain significantly high isoflavone levels and are the major dietary source of isoflavones in humans. Isoflavones found in soybeans are aglycone forms, including genistein, daidzein, and glycitein, and their glycoside, malonyl glucoside, and acetyl glucoside derivatives (Figure 14.11). Genistein and its derivatives are found in the highest content in soybeans, followed by daidzein and its derivatives, and glycitein and its derivatives [70].

It has been reported that phenolic compounds have antioxidant, antimutagenic, and free-radical scavenging activities. Epidemiologic studies showed that increased consumption of phenolic compounds reduces the risk of cardiovascular disease and certain types of cancer. Moderate consumption of red wine, which contains a high content of polyphenols, is associated with a low risk of coronary heart disease [69,71].

Consumption of soy and soy products are related with biological effects, including anticarcinogenic, antiatherosclerotic, and antihemolytic. The bioactive components are isoflavones [70]. Soybean is the unique source of isoflavones with 1 to 3 mg/g raw soybean and with 0.025 to 3 mg/g soy products [72]. Antioxidant activities of isoflavones, especially genistein, were reported *in vivo* and *in vitro* [73], and in a simple lipid system like liposomes and a more complex system like lipoproteins [74]. Addition of purified forms of isoflavones inhibited copper-dependent LDL oxidation [75]. Oral intake of the isoflavone genistein is associated with an increased resistance of LDL oxidation and inhibition of plasma lipid oxidation products [76].

Antioxidant mechanisms of polyphenolic compounds are based on hydrogen donation abilities and chelating of metal ions [69]. After donating a hydrogen atom, phenolic compounds become resonance-stabilized radicals, which do not easily participate in other radical reactions. However, phenolic compounds act as prooxidants under certain conditions, such as high concentrations of phenolic compounds or metal ions, and high pH. Chemical structures also affect the antioxidant activities.

TABLE 14.6
Classes of Phenolic Compounds

Class	Basic Skeleton	Basic Structure
Simple phenols	C_6	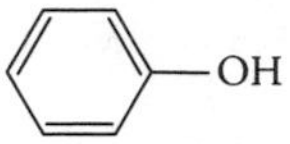 $-OH$
Benzoquinones	C_6	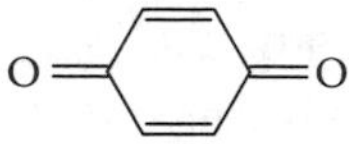 O= =O
Phenoic acids	C_6–C_1	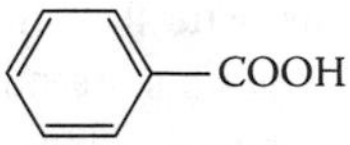 $-COOH$
Acetophenones	C_6–C_2	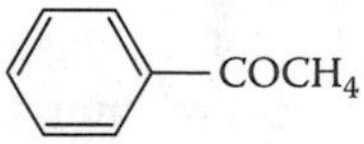 $-COCH_4$
Phenylacetic acids	C_6–C_2	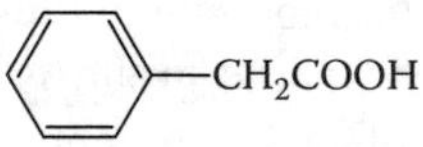 $-CH_2COOH$
Hydroxycinnamic acids	C_6–C_3	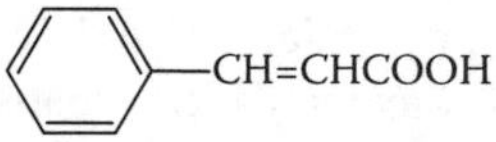 $-CH{=}CHCOOH$
Phenylpropenes	C_6–C_3	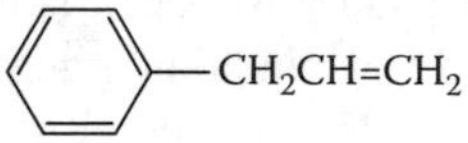 $-CH_2CH{=}CH_2$
Coumarins	C_6–C_3	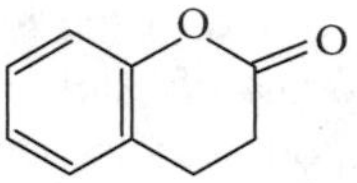 O, O
Chromones	C_6–C_3	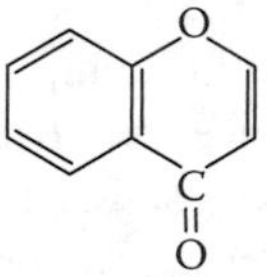 O, C=O
Anthraquinones	C_6–C_2–C_6	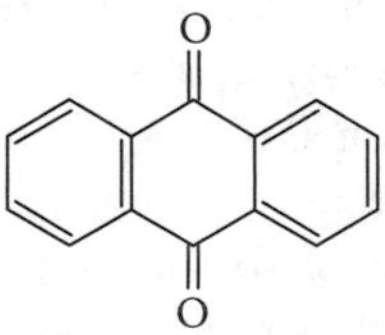 O, O
Flavonoids	C_6–C_3–C_6	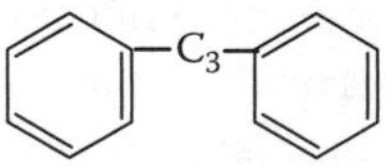 C_3

Flavonoids have the most potent antioxidant activities due to the chemical structures with the *o*-diphenolic group, a 2–3 double bond conjugated with the 4-oxo function, and hydroxyl groups in positions 3 and 5. Antioxidant activities of flavonoids are influenced by hydroxylation and the presence of sugar moiety [69].

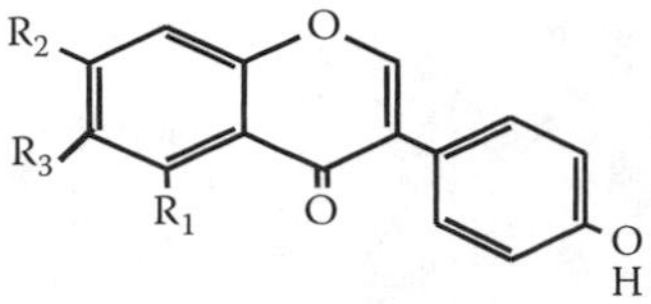

Aglycones	R_1	R_2	R_3
Genistein	OH	OH	H
Daidzein	H	OH	H
Glycitein	H	OH	OCH_3

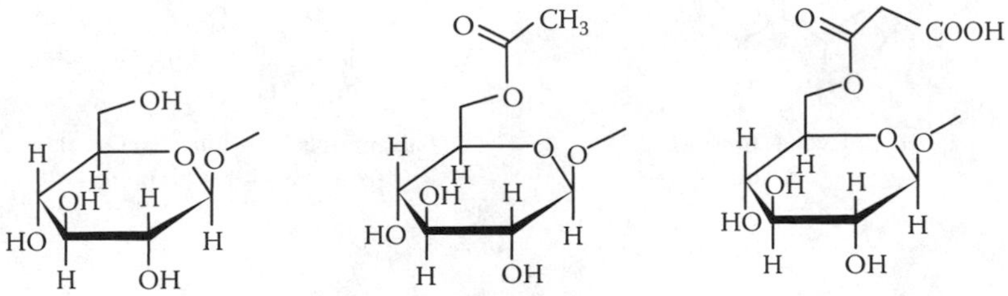

β-Glucoside form at R_2 position

Acetyl glucoside form at R_2 position

Malonyl glucoside form at R_2 position

FIGURE 14.11 Isoflavone structures. Genistein, daidzein, and glycitein are aglycones, and β-glucoside, acetyl glucoside, and malonyl glucosides of each aglycone are found in soybean.

14.6 CONCLUSIONS

Aging is a complex multifactorial process in which free-radical oxidative damage on lipids plays a very important role, but free-radical oxidative damage on lipids may not be the exclusive mechanism in aging. Antioxidant defense mechanisms in humans, such as antioxidative enzymes, tocopherol, and ascorbic acid, are linked to each other and balance with reactive oxygen species. The increased concentration of dietary foods containing antioxidative nutraceuticals with three to five servings from the vegetable group and two to four servings from the fruit group [77] can help humans reduce the deleterious reactive oxygen species and free radicals, and balance the oxidative stress, to slow down the aging process.

REFERENCES

1. Harman. D., Aging: overview, *Annu. New York Acad. Sci.*, 928, 1, 2000.
2. Spiteller, G., Lipid oxidation in aging and age-dependent disease, *Exp. Gerontol.*, 36, 1425, 2001.
3. Ashok, B. and Ali, R., The aging paradox: free radical theory of aging, *Exp. Gerontol.*, 34, 293, 1999.

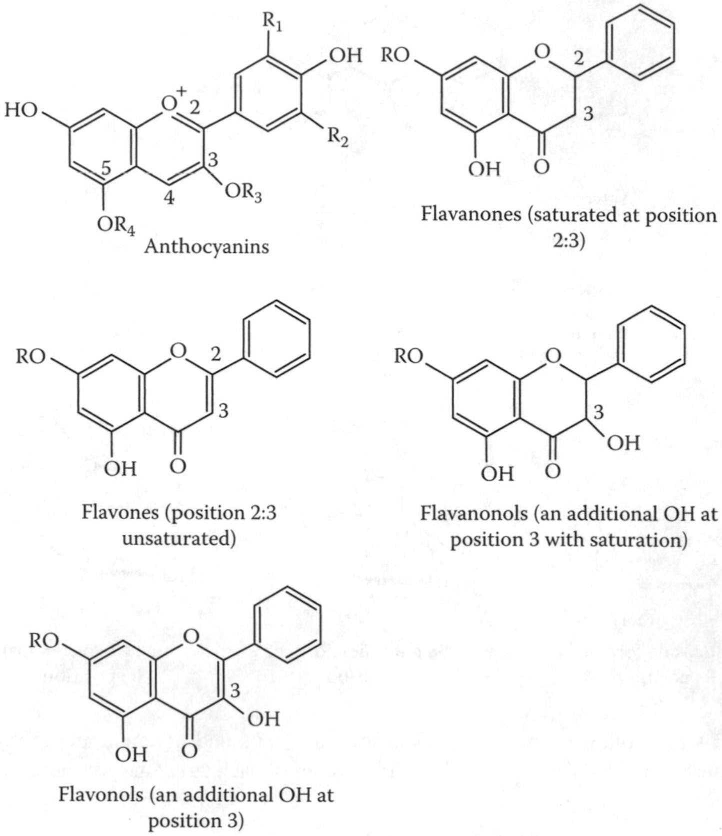

Anthocyanins

Flavanones (saturated at position 2:3)

Flavones (position 2:3 unsaturated)

Flavanonols (an additional OH at position 3 with saturation)

Flavonols (an additional OH at position 3)

FIGURE 14.10 Structures of flavonoids.

Flavonoids are effective hydroxyl radical and peroxyl radical scavengers. Flavonoids can form complexes with metals and inhibit metal initiating lipid oxidation [70].

The antioxidant mechanisms of isoflavones are not clearly understood and have been suggested to be different from conventional antioxidants. The structural similarities of genistein and daidzein to naturally occurring estrogens suggest that these compounds may protect against hormone-dependent cancers (i.e., prostate and mammary) by modulating the activity of estrogen [70]. Antioxidant activities of isoflavone on lipoxygenase catalyzed lipid oxidation were dependent on the concentrations and structures of isoflavones [73]. Glucose linkage to aglycone reduced the antioxidant activities of isoflavone. Isoflavones are not consumed during lipid oxidation and show synergic antioxidant effects with ascorbic acid [74]. Patel et al. [74] suggested antioxidant mechanisms of isoflavone as analogous of tocopherol-mediated peroxidation. Hwang et al. [75] suggested that isoflavones may prevent lipid oxidation through stabilizing LDL structures instead of being involved in the lipid oxidation chain reaction.

4. Ames, B.N., Shigenaga, M.K. and Hagen, T.M., Oxidants, antioxidants, and the degenerative diseases of aging, *Proc. Natl. Acad. Sci. USA,* 90, 7915, 1993.
5. Stief, T.W., The physiology and pharmacology of singlet oxygen, *Med. Hypotheses,* 60, 567, 2003.
6. Halliwell, B., Antioxidants and human disease: a general introduction, *Nutr. Rev.,* 55, S44, 1997.
7. Boff, J. and Min, D.B., Chemistry and reaction of singlet oxygen in foods, *Compr. Rev. Food Sci. Food Saf.,* 1, 58, 2002.
8. Gutteridge, J.M.C. and Halliwell, B., Transition metal ions and antioxidant proteins in extracellular fluids, *Atmos. Oxid. Antioxid.,* 3, 71, 1993.
9. Thomas, M.J., The role of free radicals and antioxidants: how do we know that they are working?, *Crit. Rev. Food Sci. Nutr.,* 35, 21, 1995.
10. Wickens, A.P., Aging and the free radical theory, *Resp. Physiol.,* 128, 379, 2001.
11. Rikans, L.E. and Hornbrook, K.R., Lipid peroxidation, antioxidant protection and aging, *Biochim. Biophys. Acta,* 1362, 116, 1997.
12. Kaur, C. and Kapoor, H.C., Antioxidants in fruits and vegetables — the millennium's health, *Int. J. Food Sci. Technol.,* 36, 703, 2001.
13. Halliwell, B. et al., Free radicals and antioxidants in food and *in vivo:* what they do and how they work, *Crit. Rev. Food Sci. Nutr.,* 35, 7, 1995.
14. Packer, L. and Weber, S.U., The role of vitamin E in the emerging field of nutraceuticals, in *Nutraceuticals in Health and Disease Prevention,* Kramer, K., Hoppe, P.P. and Packer, L., Eds., Marcel Dekker, New York, 2001, pp. 27, 15.
15. Fang, Y.Z., Yang, S., and Wu, G., Free radicals, antioxidants, and nutrition, *Nutrition,* 18, 872, 2002.
16. Korycka-Dahl, M.B. and Richardson, T., Activated oxygen species and oxidation of food constituents, *Crit. Rev. Food Sci. Nutr.,* 10, 209, 1978.
17. Padmaja, S. and Madison, S.A., Hydroxyl radical-induced oxidation of azo-dyes: a pulse radiolysis study, *J. Phys. Org. Chem.,* 12, 221, 1999.
18. Stief, T.W., The blood fibrinolysis/deep-sea analogy: a hypothesis on the cell signals singlet oxygen/photons as natural antithrombotics, *Thromb. Res.,* 99, 1, 2000.
19. Takayama, F., Egashira, T., and Yamanaka, Y., Singlet oxygen generation from phosphatidylcholine hydroperoxide in the presence of copper, *Life Sci.,* 68, 1807, 2001.
20. Girotti, A.W. and Korytowski, W., Cholesterol as a singlet oxygen detector in biological systems, in *Methods in Enzymology,* Packer, L. and Sies, H., Eds., Academic Press, New York, 2000, p. 85.
21. Foote, C.S., Definition of type I and type II photosensitized oxidation, *Photochem. Photobiol.,* 54, 659, 1991.
22. Yamazaki, S. et al., Photogeneration of 3β-hydroxy-5α-cholest-6-ene-5-hydroperoxide in rat skin: evidence for occurrence of singlet oxygen *in vivo, Free Radical Biol. Med.,* 27, 301, 1999.
23. Decker, E.A., Antioxidant mechanisms, in *Food Lipids, Chemistry, Nutrition, and Biotechnology,* Akoh, C.C. and Min, D.B., Eds., Marcel Dekker, New York, 1998, p. 397.
24. Halliwell, B., Uric acid: an example of antioxidant evaluation, in *Handbook of Antioxidants,* Cadenas, E. and Packer, L., Eds., Marcel Dekker, New York, 1996, p. 243.
25. Noguchi, N. and Niki, E., Chemistry of active oxygen species and antioxidants, in *Antioxidant Status, Diet, Nutrition, and Health,* Papas, A.M., Ed., CRC Press, Boca Raton, FL, 1999, p. 3.
26. Papas, A.M., Determinants of antioxidant status in humans, in *Antioxidant Status, Diet, Nutrition, and Health,* Papas, A.M., Ed., CRC Press, Boca Raton, FL, 1999, p. 21.

27. McVean, M., Kramer-Stickland, K., and Liebler, D.C., Oxidants and antioxidants in ultraviolet-induced nonmelanoma skin cancer, in *Antioxidant Status, Diet, Nutrition, and Health,* Papas, A.M., Ed., CRC Press, Boca Raton, FL, 1999, p. 401.
28. Babior, B.M., NADPH-oxidase: an update, *Blood,* 93, 1464, 1999.
29. Stuehr, D.J., Kwon, N.S., and Nathan, C.F., FAD and GSH participate in macrophage synthesis of nitric oxide, *Biochem. Biophys. Res. Commun.,* 168, 558, 1990.
30. Min, D.B., Lipid oxidation of edible oil, in *Food Lipids: Chemistry, Nutrition, and Biotechnology,* Akoh, C.C. and Min, D.B., Eds., Marcel Dekker, New York, 1998, p. 283.
31. Buettner, G.R., The pecking order of free radicals and antioxidants: lipid peroxidation, a-tocopherol, and ascorbate, *Arch. Biochem. Biophys.,* 300, 535, 1993.
32. Doleiden, F.H. et al., Reactivity of cholesterol and some fatty acids toward singlet oxygen, *Photochem. Photobiol.,* 20, 519, 1974.
33. Frei, B., Cardiovascular disease and nutrient antioxidants: role of low-density lipoprotein oxidation, *Crit. Rev. Food Sci. Nutr.,* 35, 83, 1995.
34. Nedeljkovic, Z.S., Gokce, N., and Loscalzo, J., Mechanisms of oxidative stress and vascular dysfunction, *Postgrad. Med.,* 79, 195, 2003.
35. Wildman, R.E.C., Nutraceuticals: a brief review of historical and teleological aspects, in *Handbook of Nutraceuticals and Functional Foods,* Wildman, R.E.C., Ed., CRC Press, Boca Raton, FL, 2001a, p. 1.
36. Dillard, C.J. and German, J.B., Phytochemicals: nutraceuticals and human health, *J. Sci. Food Agric.,* 80, 1744, 2000.
37. Wildman, R.E.C., Classifying nutraceuticals, in *Handbook of Nutraceuticals and Functional Foods,* Wildman, R.E.C., Ed., CRC Press, Boca Raton, FL, 2001, p. 13.
38. Duthie, S.J. et al., Antioxidant supplementation decreases oxidative DNA damage in human lymphocytes, *Cancer Res.,* 56, 1291, 1996.
39. Teoh, C.Y. and Davies, K.J.A., The broad spectrum of antioxidant defense and oxidant repair mechanisms, *Free Radical Res.,* 36, 8, 2002.
40. Wilson, R. et al., Dietary hydroxy fatty acids are absorbed in humans: implications for the measurement of "oxidative stress" *in vivo, Free Radical Biol. Med.,* 32, 62, 2002.
41. Papas, A.M., Other antioxidants, in *Antioxidant Status, Diet, Nutrition, and Health,* Papas, A.M., Ed., CRC Press, Boca Raton, FL, 1999, p. 231.
42. Frankel, E.N., Chemistry of autoxidation: mechanism, products and flavor significance, *AOCS Monogr.,* 15, 1, 1985.
43. Decker, M. et al., Analysing the antioxidant activity of food products: processing and matrix effects, *Toxicol. in Vitro,* 13, 797, 1999.
44. Papas, A.M., Diet and antioxidant status, in *Antioxidant Status, Diet, Nutrition, and Health,* Papas, A.M., Ed., CRC Press, Boca Raton, FL, 1999, p. 89.
45. Decker, E.A., The role of phenolics, conjugated linoleic acid, carnosine, and pyrroloquinoline quinone as nonessential dietary antioxidants, *Nutr. Rev.,* 53, 49, 1995.
46. Ramon, R. and Gonzalo, R., Renal damage mediated by oxidative stress: a hypothesis of protective effects of red wine, *Free Radical Bio. Med.,* 33, 409, 2002.
47. Kumuda, D.C. and Chandan, D.K., Thioredoxin, a Singlet oxygen quencher and hydroxyl radical scavenger: redox independent function, *Biochem. Biophys. Res. Commun.,* 277, 443, 2000.
48. Gregory, J.F. III, Vitamins, in *Food Chemistry,* 3rd ed., Fennema, O.R., Ed., Marcel Dekker, New York, 1996, p. 431.
49. Watkins, T.R., Bierenbaum, M.L., and Giampalolo, A., Tocotrienols: biological and health effects, in *Antioxidant Status, Diet, Nutrition, and Health,* Papas, A.M., Ed., CRC Press, Boca Raton, FL, 1999, p. 479.

50. Niki, E., α-Tocopherol, in *Handbook of Antioxidants,* Cadenas, E. and Packer, L. Eds., Marcel Dekker, New York, 1996, p. 3.
51. Rumsey, S.C., Wang, Y., and Levine, M., Vitamin C, in *Antioxidant Status, Diet, Nutrition, and Health,* Papas, A.M., Ed., CRC Press, Boca Raton, FL, 1999, p. 479.
52. Buettner, G.R. and Jurkiewicz, B.A., Chemistry and biochemistry of ascorbic acid, in *Handbook of Antioxidants,* Cadenas, E. and Packer, L. Eds., Marcel Dekker, New York, 1996, p. 91.
53. deMan, J.M., *Principles of Food Chemistry,* An Aspen publication, Maryland, 1999, p. 239.
54. Giovannucci, E. et al., Intake of carotenoids and retinol in relation to risk of prostate cancer, *J. Natl. Cancer Inst.,* 87, 1767, 1995.
55. Giovannucci, E., Tomatoes, tomato-based products, lycopene, and cancer: review of the epidemiologic literature, *J. Natl. Cancer Inst.,* 91, 317, 1999.
56. Giovannucci, E. and Clinton, S.K., Tomatoes, lycopene, and prostate cancer, *Prostate Cancer P. Soc. Exp. Biol. Med.,* 218, 129, 1998.
57. Pool-Zobel, B.L. et al., Consumption of vegetables reduces genetic damage in humans: first results of a human intervention trial with carotenoid-rich foods, *Carcinogenesis,* 18, 1847, 1997.
58. Weisburger, J.H., Mechanisms of action of antioxidants as exemplified in vegetables, tomatoes and tea, *Food Chem. Toxicol.,* 37, 943, 1999.
59. Foote, C.S., Photosensitized oxidation and singlet oxygen: consequences in biological systems, in *Free Radicals in Biology,* Pryor, W.A., Ed., Academic Press, New York, 1976, p. 85.
60. Beutner, S. et al., Synthetic singlet oxygen quenchers, *Method. Enzymol.,* 319, 226, 2000.
61. Liebler, D.C., Antioxidant reactions of carotenoids, *Ann. New York Acad. Sci.,* 691, 20, 1993.
62. Haila, K.M. et al., Carotenoid reaction with free radicals in acetone and toluene at different oxygen partial pressures, *Z. Lebensm. Unters Forsch. A.,* 204, 81, 1997.
63. Mortensen, A., Skibsted, L.H., and Truscott, T.G., The interaction of dietary carotenoids with radical species, *Arch. Biochem. Biophys.,* 385, 13, 2001.
64. Lee, J.H., Ozcelik, B., and Min, D.B., Electron donation mechanisms of β-carotene as a free radical scavenger, *J. Food Sci.,* 68, 861, 2003.
65. Mortensen, A. and Skibsted, L.H., Reactivity of β-carotene towards peroxyl radicals studied by laser flash and steady-state photolysis, *FEBS Lett.,* 426, 392, 1998.
66. Edge, R., Garvey, M.C., and Truscott, T.G., The carotenoids as antioxidants — a review, *J. Photochem. Photobiol. B: Biol.,* 41,189, 1997.
67. Edge, R. et al., The reduction potential of the β-carotene$^{\bullet+}$/β-carotene couple in an aqueous micro-heterogeneous environment, *FEBS Lett.,* 471, 125, 2000.
68. Burton, G.W. and Ingold, K.U., β-Carotene: an unusual type of lipid antioxidant, *Science,* 224, 569, 1984.
69. Bravo, L., Polyphenols: chemistry, dietary sources, metabolism, and nutritional significance, *Nutr. Rev.,* 56, 317, 1998.
70. Hendrich, S. et al., Isoflavone metabolism and bioavailability, in *Antioxidant Status, Diet, Nutrition, and Health,* Papas, A.M., Ed., CRC Press, Boca Raton, FL, 1999, p. 211.
71. German, J.B. and Walzem, R.L., The health benefits of wine, *Annu. Rev. Nutr.,* 20, 561, 2000.
72. Wang, H.J. and Murphy, P.A., Isoflavone content in commercial soybean foods, *J. Agric. Food Chem.,* 42, 1666, 1994.
73. Naim, M. et al., Antioxidative and antihemolytic activities of soybean isoflavones, *J. Agric. Food Chem.,* 24, 1174, 1976.

74. Patel, R.P. et al., Antioxidant mechanisms of isoflavones in lipid systems: paradoxical effects of peroxyl radical scavenging, *Free Radical Biol. Med.,* 31, 1570, 2001.
75. Hwang, J. et al., Synergistic inhibition of LDL oxidation by phytoestrogens and ascorbic acid, *Free Radical Biol. Med.,* 29, 79, 2000.
76. Wiseman, H. et al., Isoflavone phytoestrogens consumed in soy decrease F2-isoprostane concentrations and increase resistance of low-density lipoprotein to oxidation in humans, *Am. J. Clin. Nutr.,* 72, 395, 2000.
77. USDA/CNPP (U.S. Department of Agriculture, Center on Nutrition Policy and Promotion), Dietary guidelines for Americans, 2000. http://www.usda.gov/cnpp/Pubs/DG2000.

15 Dietary Lipids: Metabolism and Physiological Functions

Armand B. Christophe

CONTENTS

15.1 DIETARY LIPIDS

Dietary lipids can be defined as the material that can be extracted from foods with organic solvents. Depending on the food analyzed, different extraction conditions and solvents can be used. In general, dietary fats consist of a considerable number of different lipid classes and several types of molecules can make up each lipid class. Each of them, in mutual combination or in combination with other dietary components, may have effects that have an impact on health.

The major components of usual dietary fats are triglycerides (triacylglycerols [TAGs]). TAGs differ from each other in the fatty acids they contain and in the combination and/or stereochemical implantation of different fatty acids over the three glycerol positions. Other components with distinct nutritional effects that can be present in a usual food fat are phospholipids, sterols, fat-soluble vitamins, antioxidants and other fat-soluble components. Physical processing of natural fats such as refining can affect the level of minor lipid constituents. Natural food fats can also be modified chemically by processes such as hydrogenation and interesterification. Hydrogenation of oils may result in the formation of *trans* and *cis*-positional double bonds in their fatty acid moiety and one of the effects of interesterification is a change in the implantation of the different fatty acids at the different stereospecific positions of the TAGs. "Unintended" changes in minor components that can be very important from a pathophysiological point of view may also occur (see below).

By semisynthesis, minor natural fat components such as monoglycerides (monoacylglycerols [MAGs]), diglycerides (diacylglycerols [DAGs]), and phytosterol esters can be produced in large quantities. Lipids not present in nature, such as sucrose polyesters (SPEs), structured lipids (TAGs with medium-chain and long-chain fatty acids at specified positions; SL), and salatrims (TAGs with a mixture of acetic, propionic, butyric, and stearic acids), can also be synthesized. These components have special nutritional properties.

15.2 FAT DIGESTION AND ABSORPTION

In order to be a candidate for absorption, the material has to be able to reach the absorptive surface. For fats, this is mainly the surface of the microvilli of the jejunum. For anatomic–geometric reasons, only small particles can penetrate in the spaces between the microvilli. The rate-limiting step for fat absorption is diffusion through an "unstirred" water layer that lines the microvilli and is a few molecules thick. The size of emulsion droplets is too large for entering the spaces between the microvilli; water-soluble molecules such as medium-chain fatty acid anions and micelles can. In the stomach and intestines, food fats are usually present as emulsions. The conversion of fat emulsion droplets into micelles or water-soluble molecules is a requisite for fat absorption and is accomplished by fat digestion.

15.2.1 Fat Digestion

An average Western diet contains about 100 to 150 g food fat, about 97% of which are TAGs. The remainder is made up by phospholipids (4 to 8 g/d, mainly phosphatidylcholine), cholesterol, other sterols, and minor lipid components. When TAGs and phospholipids are present together in proportions as found in the diet, the latter act as emulsifiers by coating the TAG droplets, thus forming more stable emulsions. In practical situations, fats are always consumed as part of a meal, which also contains nonfat constituents such as carbohydrates, proteins, fiber, and minerals. These components may affect phenomena related to fat digestion and absorption. Proteins and their peptic digestion products stabilize TAG emulsions in the stomach and in the intestinal lumen [1]. Stabilized emulsion droplets are substrates for gastric lipase [2].

15.2.2 Gastric Digestion

Gastric lipase shows positional fatty acid and substrate specificity. It preferentially splits primary ester bonds [3]. With TAGs as substrate it catalyzes the hydrolysis of the fatty acid at the *sn*-3 position preferentially to that at *sn*-1 position [4] and prefers TAGs with short- and medium-chain lengths. Thus, TAGs with such fatty acids in the *sn*-3 position as are present in bovine milk and its products are preferential substrates. In principle, such TAGs can be synthesized. Gastric lipase is inhibited by free fatty acids. As a result of product inhibition by free fatty acids, the main digestion products of gastric lipase are free fatty acids and *sn*-1, -2-DAGs when TAGs are the substrate [5]. About 10% of the dietary fatty acids are liberated by gastric lipase from a test meal that contains TAGs with long-chain fatty acids but considerably higher amounts from a test meal that contains TAGs with short- or medium-chain fatty acids [6]. Even for long-chain fatty acid containing TAGs, this corresponds to the breakdown of about one third of the TAGs into DAGs and free fatty acids in the stomach. Furthermore there is evidence that gastric lipase is further active in the upper small intestine. It has been demonstrated that *sn*-1,3 DAGs are also substrates for gastric lipase, but with these substrates, the enantioselectivity for the *sn*-3 position no longer exists [7]. Gastric lipase does not act on phospholipids or on cholesteryl esters, its activity is not inhibited by phospholipids [8], and it is not active against SPEs.

The importance of gastric digestion is the generation of hydrolysis products which affect the excretion of some hormones involved in fat digestion, stabilize the lipid emulsion, and make it a better substrate for pancreatic lipase [9]. There is a compensatory increase in gastric lipase when pancreatic lipase is low. Nonetheless, in these conditions the capacity for fat absorption remains limited when usual food fats are fed. It can be expected that in these conditions the lipids of choice to enable feeding more fat and provide for energy and essential fatty acids would be semi-synthetic triglycerides with a short-chain fatty acid at the *sn*-3 position and long-chain fatty acids among which are essential fatty acids at the other positions.

15.2.3 Gastric Emptying

The composition of the fat that is emptied in the upper small intestine depends on the rate of gastric digestion and on the time the dietary fat was retained in the stomach. The physicochemical and chemical forms in which dietary fats are consumed affect the rate of gastric emptying. Fats that are ingested as emulsion are partitioned into an aqueous phase and an oil phase whereas fats consumed as nonemulsions form an oil phase. Fats in the aqueous phase are readily emptied from the stomach whereas fats in an oil phase are emptied much slower [10]. The chemical composition of the oil phase also influences the rate of emptying. The delaying effect depends on the saturation [11] and the chain length [12] of the fatty acids. In healthy individuals, the rate of gastric emptying can be rate limiting for fat absorption [13].

15.2.4 Reactions in the Small Intestines

The lipids that are emptied from the stomach into the small intestine are diluted by lipids from endogenous origin (~10 to 25 g/d when fasting). The two major sources of endogenous lipids originate from slaughtered intestinal cells and from biliary secretions. The amount of phospholipids from biliary origin (7 to 22 g/d; mainly phosphatidylcholine) is considerably larger than from dietary origin (3 to 8 g/d) [14]. The same is also often true for cholesterol (~1 g/d of endogenous origin). In bile, bile salt–phospholipid mixed micelles exist in cylindrical arrangement [15]. In the intestinal lumen, bile salts and phospholipid molecules are likely to partly adsorb to the lipid emulsion droplets.

Several chemical and physiochemical reactions occur simultaneously or quasi-simultaneously in the small intestine. The main reactions are digestion of the glycerides in the emulsion phase, isomerization of partial glycerides in the emulsion phase, transfer of the digestion products from the emulsion phase into micellar phases, digestion and isomerization in the micellar phases, and uptake of digestion products mainly from micellar phases. Thus the relative amounts of the different digestion products that are taken up by the intestinal cells depend on the relative rates of these reactions.

The major enzymes involved in fat digestion in the intestinal lumen are pancreatic lipase, pancreatic phospholipase A_2, and pancreatic cholesterol esterase, the latter also named bile-salt-dependent lipase. These enzymes work in concert with each other to yield the final reaction products.

Lipases act on the lipid/water interface. At the start and during fat digestion the lipid/water interface is rich in water and the digestion products are continuously removed from the interface favoring lipolysis. Human pancreatic lipase catalyzes the equilibrium reaction between ester formation and hydrolysis. It is specific for primary alcohols and esters of such alcohols. It splits off fatty acids from the *sn*-1 and *sn*-3 position of TAGs. It is more active against *sn*-1,2 and *sn*-2,3 DAGs thus formed than against TAGs [16]. Thus it can break down the TAGs into *sn*-1,2 and *sn*-2,3 DAGs which can be further hydrolyzed into *sn*-2 MAGs. These are no longer substrates for pancreatic lipase. Pancreatic lipase is less active against polyunsaturated fatty acids with a double bond close to the ester bond, in particular, when these are located in the *sn*-3 position [17], and clearly prefers the *sn*-1 position of both TAGs and DAGs [7]. Thus, if there would be neither isomerization nor absorption of DAGs and MAGs, the ultimate digestion products of TAGs would be a 2:1 mixture of free fatty acids and *sn*-2 MAGs. However, both reactions do occur. "Linear" partial glycerides are thermodynamically favored over "branched" partial glycerides. MAGs with equilibrium isomeric composition consist of about 45% *sn*-1, 45% *sn*-3, and 10% *sn*-2 isomers [18]. For DAGs, the isomer ratio is about 70:30 [19] for *sn*-1,3 and *sn*-1,2 plus 2,3 DAGs, respectively. Isomerization of partial glycerides generated from TAGs by pancreatic lipase yields products with primary ester bonds. These are substrates for pancreatic lipase and thus result in more complete digestion with the liberation of glycerol. As MAGs are readily absorbed, the rate of isomerization relative to that of uptake is important in order to be further digested. Isomerization rate increases with desaturation and decreases with chain length of the fatty acids they contain [20]. In any event, spontaneous isomerization of long-chain fatty acids containing *sn*-2 MAGs is low compared with their intestinal uptake [21]. About 15 to 20% of *sn*-2 MAGs would isomerize in the intestinal lumen [22].

Phospholipase A_2 hydrolyzes the fatty acid esterified at the *sn*-2 position in phosphatidylcholines to yield lysophosphatidylcholines and free fatty acids [23].

This reaction is of importance for the digestion and uptake of neutral glycerides as the presence in the intestinal lumen of intact phospholipids retards TAG digestion [24], affects the partitioning of lipid digestion products between the oil and micellar phase [25], and slows down the uptake of lipid digestion products from the micellar phase [26]. Lysophosphatidylcholines on the contrary favor micellar solubilization and uptake of fat digestion products [27] and other dietary lipophilic substances [28]. Dietary sphingolipids are broken down by sphyngomyelinase and the resulting ceramides by ceramidase.

Carboxyl ester lipase hydrolyzes sterol esters. It prefers substrates present in the micellar state but has low activity against monoglycerides [29]. Cholesterol is absorbed as free sterol. Phytostanols and phytosterols are set free by carboxyl ester lipase from their respective esters when foods are fortified with these substances. In the free form, plant sterols interfere with the absorption of cholesterol, at least in part by reducing the cholesterol content of the micelles that are present in the small intestine.

15.2.5 Uptake of Fat Digestion Products by the Intestinal Cells

Water-soluble components or components present in micelles can collide or come in close contact with the surface of the intestinal cells. The interior of the MAG/free fatty acid/conjugated bile salt mixed micelle is a lipid environment in which lipophilic substances such as cholesterol, carotenes, and fat-soluble vitamins also can dissolve. It is in this way that they can reach the absorptive surface. Depending on the chemical nature of the components this may, or may not, lead to absorption. Bile salts and phytosterols, for instance, are not absorbed by the jejunal cells, cholesterol is partially absorbed, and monoglycerides, free fatty acids, and lysophosphatidylcholines are almost completely absorbed.

When novel food fats prepared by industry are consumed, it is not always clear how they will be absorbed. *sn*-1,3 DAGs, for instance, which are the main components of commercial DAGs used in food products, can be completely hydrolyzed by pancreatic lipase but whether digestion will go to completion *in vivo* is not clear, as intermediary products such as *sn*-1 and *sn*-3 MAGs are likely to be absorbed, at least partially, as such. The same holds for structured lipids, which may yield α- or β-long-chain MAGs depending on their structure. Medium-chain triglycerides are known to be almost completely digested and absorbed as free fatty acids and glycerol whereas SPEs are known not to be digested or absorbed. Plant stanol/sterol esters are digested, their fatty acids are absorbed, and their sterol/stanol moiety is not.

15.2.6 Intestinal Metabolism of Fat Digestion Products

There are several enzymes in the cells of the small intestine that can act on free but not on esterified fatty acids. Free fatty acids are diluted with fatty acids originating from the plasma nonesterified fatty acid pool [30]. After activation they can be oxidized, elongated, chain desaturated, and converted into complex lipids [30,31]. The relative rates of the different metabolic routes depend on the nature of the fatty acids and on the presence of other components in the intestinal cells [32]. Conversion of saturated fatty acids in monounsaturates after entering the intestinal cells in free form (i.e., when they were present in the outer position of the dietary TAGs) could contribute to their lower hypercholesterolemic effect than when originally present at the inner position. Free fatty acids also affect intestinal gene expression [33] and the production of apolipoproteins and these effects are fatty acid dependent [34]. When *sn*-2 MAGs are present simultaneously with free fatty acids in the intestinal cells as is the case after TAG feeding, fatty acids and *sn*-2 MAGs are mainly converted by a multienzyme complex in TAGs which are excreted as chylomicrons. The excess of free fatty acids reacts with *sn*-3 glycerophosphate to form lysophosphatidic acids, which are further converted into phosphatidic acids. Dephosphorylation results in *sn*-1,2 DAGs, which are precursors of both TAGs and phospholipids. This metabolic route results in the formation of intestinal lipoproteins of smaller size [35].

Lysophosphatidylcholine affects intestinal metabolism [36] and can be converted into phosphatidylcholine again. There is evidence that phosphatidylcholine can be taken up intact and excreted as part of intestinal lipoproteins in the lymph [37].

TAG digestion products that are not or only absorbed in small quantities after "usual" fat feeding may be taken up by the intestinal cells after consuming novel food fats.

Intake of medium-chain triglycerides and structured lipids results in influx of medium-chain fatty acids in the intestinal cells. These are mainly excreted by the intestinal cells as such and transported to the liver by the portal vein [38].

What would happen if *sn*-1(3) MAGs would enter intestinal cells is not clear. A series of possible reactions must be considered such as complete digestion, acylation, transacylation, phosphorylation, and intact excretion out of the intestinal cells. Lipases that hydrolyze MAGs are present in the intestinal cells but their action seems to be reduced by other intestinal cell constituents [39]. Some of them show outspoken fatty acid specificity [40]. Both *sn*-1 and *sn*-3 MAGs can be acylated to form *sn*-1,3 DAGs. This transformation is fatty acid and species dependent [41]. Anyway, *sn*-1,3 DAGs are poor substrates for intestinal TAG synthesis [41,42]. *sn*-1 but not *sn*-3 MAGs can be phosphorylated into lysophosphatidic acid [43] and possibly be metabolized further as described above. If MAGs would escape intestinal metabolization and could leave the cell intact, they could be transported to the liver by portal blood bound to albumin [44]. Portal absorption of MAGs with medium-chain fatty acids has been documented [45]. It has been demonstrated that cultured liver cells convert *sn*-1 glycerol monoethers (*sn*-1 MAG analogues) into phospholipids and triglycerides [46]. There is some indirect evidence that after MAG feeding, intact *sn*-1 MAGs or their phosphorylated derivatives reach the liver with retention of the *sn*-1 ester bond [47].

15.2.7 Postprandial Effects

When TAGs are formed in the intestinal cells, as after feeding "usual" food fats, they are incorporated in intestinal lipoproteins with the size of chylomicrons. These are secreted in the lymphatics and transported via the lymphatic route. When they enter the blood-stream, they cause an increase in chylomicronemia, triglyceridemia, and change in the fatty acid composition of plasma TAGs toward that of the fat fed [48]. Chylomicron TAGs are hydrolyzed by lipoprotein lipase resulting in the formation of remnants and liberation of free fatty acids. As a consequence, the composition of the nonesterified fatty acids in plasma also changes toward that of the fat fed. Chylomicron remnants are taken up by the liver.

After feeding structured lipids, their long-chain fatty acids are dealt with similarly as the long-chain fatty acids after TAG feeding whereas their medium-chain fatty acids are transported mainly by the portal vein. However, part of their medium-chain fatty acids appear in "mixed chain" TAGs [49].

Postprandial (see below) and long-term effects are different when DAGs or TAGs are fed as food fat. The former result in a smaller increase in chylomicronemia and in hypertiglyceridemia than the latter [50].

15.3 PHYSIOLOGICAL EFFECTS

Several factors can have pronounced effects on the nutritional properties of "usual" and "novel" food fats. A major criterion that is often used in dietetics to evaluate whether a "usual" food fat is to be limited in the diet or not is its fatty acid composition. This is an oversimplification, as the stereospecific positions of the fatty acids and their combination in TAGs as well as the presence of minor components can be of considerable importance. These can be changed by industrial processing of the natural fats.

Interesterification of peanut oil, for instance, has been claimed to reduce its atherogenic potential in rats [51]. Initially it was thought that this was due to the rearrangement of fatty acids in TAGs. Later it was suggested that natural peanut oil contains small amounts of lectins, which caused the atherogenicity, and that these were destroyed by interesterification [52]. Differences in cholesterolemic effects of interesterified milk fat produced by chemical [53] or enzymatic processes [54] may be due to different unintended changes in minor constituents. Reducing plant sterols by physical refining may increase the cholesterolemic effects of oils.

It cannot be assumed that concepts that are valid for "usual" food fats also hold for "novel" food fats. Indeed, the latter are often used because they have properties deviating from those of "usual" food fats.

15.3.1 ENERGY VALUE

The heat of combustion of a component can easily be determined by bomb calorimetry. For all practical purposes, the heat of combustion in the body (37°C, no constant volume) is the same as in a calorimeter (25°C, constant volume) provided the end products are the same. (The heat of combustion is dependent on temperature and is different when determined under constant volume or not. Differences with heat of combustion in the body are small, however, and can be neglected.) This is the case for TAGs, of which the end products are CO_2 and water in both situations. To determine how much energy a fat would provide in the body, the heat of combustion is corrected for incomplete absorption. The heat of combustion of a TAG depends somewhat on the nature of the fatty acids it contains. So does the degree of absorption. For a "usual" food fat the energy value is close to 9 kcal/g (37 kJ/g in Europe which results in slightly lower energy values) and these values are conventionally used. These conventional values can no longer be applied when the end products of a food fat is different from CO_2 and water (e.g., cholesterol), when the chain lengths of fatty acids deviate considerably from those in a "usual" fat or when the fraction of the fat that is absorbed deviates considerably from that conventionally taken for a "usual" food fat (95%). For instance, the energy content of medium-chain triglycerides is about 7 to 8 kcal/g, that of salatrims (shorter chain lengths) and caprenin (its behenic acid poorly absorbed) about 5 kcal/g, and that of SPEs (not absorbed) 0 kcal/g. (In the United States, salatrim is recognized as having 5 kcal/g whereas in the European Union 6 kcal/g has to be taken as energy value [55].)

15.3.2 Effects Related to Fatty Acids

There was a time that available methodology allowed only to estimate groups of fatty acids such as saturated, monounsaturated, polyunsaturated, and *trans* fatty acids. Many physiological effects of fatty acids have been and still are described in terms of these groups. Even nutritional recommendations in many countries are given in terms of these groups. With the advancement of lipid methodology it was demonstrated that different fatty acids of each group have different effects. Not all saturated fatty acids increase serum cholesterol but some are neutral in this respect; not all polyunsaturated fatty acids reduce serum cholesterol but some reduce the level of serum triglycerides. *trans* Fatty acids of dairy origin are considered less detrimental than those formed by hydrogenation of oils. Different omega-3 fatty acids have some effects in common whereas others differ. In general, it can be said that each fatty acid has its own physiological effects. Reviewing these effects is beyond the scope of this work.

15.3.3 Effects Related to the Position of Fatty Acids in Triglycerides

When acyl groups are bound to the outer positions of glycerides, they can be set free by the action of lipases. In contrast to ester-bound fatty acids, free saturated fatty acids can form insoluble soaps with divalent ions that lead to reduced absorption of both minerals and these fatty acids. Compared with saturated fatty acids at the *sn*-2 position, the ones at the outer positions are absorbed less and can be transformed into monounsaturated fatty acids in the intestinal cell. Both these effects result in lower influx of saturated fatty acids in the general circulation and in the tissues. This may explain why the hypercholesterolemic effect of saturated fatty acids is lower when they are esterified at the outer positions than at the inner position of TAGs. The reverse is true for the hypocholesterolemic effect of linoleic acid. It has been suggested that a larger fraction of the fatty acids present at the inner position than in the outer positions of TAGs end up in the liver and that this could explain the positional effects of fatty acids not only on cholesterolemia but also on lipemia, platelet aggregation, and percentage conversion of linoleic acid into its higher metabolites [56]. Early postprandial insulinemia has been related to the positional implantation of fatty acids in TAGs as well [57]. Chylomicron clearing is retarded by saturated fatty acids in the *sn*-2 position compared with unsaturated fatty acids [58]. More generally, plasma clearing of chylomicrons depends on the specific arrangement of acyl chains of the constituting TAGs and not necessarily on their overall saturation [59]. Clear differences between feeding seal oil, with long-chain omega-3 fatty acids mainly at the outer TAG positions, or fish or squid oil with their omega-3 fatty acids mainly at the inner position, were obtained on eicosanoid production [60], reduction of triglyceridemia [61], and serum phospholipid arachidonic acid content [62]. In general, the positive effects of these fatty acids seem to be more pronounced when they are esterified at the outer position of dietary of triglycerides. Moreover, fatty acid effects are not only dependent on the nature of the fatty acid itself and its positional incorporation in glycerides but also on coingestion with other fats [63] and other dietary macronutrients [64]. The combination of fatty acids in the TAGs may also be of importance. Thus physical and interesterified mixtures of MCT and long-chain TAGs have different properties.

15.3.4 Effects Related to the Glyceride Composition

TAGs and partial glycerides with the same fatty acids have different postprandial and long-term effects. For instance, for fats with the same fatty acid composition, long-term DAG feeding in humans results in lower adiposity, insulinemia, and leptinemia than feeding TAGs [19], and long-term and postprandial effects of MAG intake are different from those of TAG feeding [35].

15.3.5 Effects Related to Novel Combinations

When fatty acids are combined with sucrose, SPEs are formed. These are not digested, remain in the intestinal lumen as an emulsion phase, and, thus, affect the partitioning of fat-soluble components between the micellar and emulsion phase. This results in lower absorption and lower status of some fat-soluble vitamins. As they are not absorbed, SPEs can have a role in body weight maintenance [65]. The effect of the ingestion of SPE-based foods on energy balance and feeding behavior in humans has been reviewed [66].

Phytosterol/stanol esters are digested and the free sterols/stanols influence micellar solubilization of cholesterol but also of some lipophilic substances. Absorption of β-carotene is reduced but a healthy diet rich in carotenoids is effective in maintaining normal blood carotenoid levels [67]. Metabolic effects of sterols and stanols have been reviewed recently [68].

REFERENCES

1. Meyer, J.H., Stevenson, E.A., and Watts, H.D. The potential role of protein in the absorption of fat, *Gastroenterology* 70, 232, 1997.
2. Armand, M. et al. Digestion and absorption of 2 fat emulsions with different droplet sizes in the human digestive tract, *Am. J. Clin. Nutr.* 70, 1096, 1999.
3. Tiruppathi, C. and Balasubramanian, K.A. Purification and properties of an acid lipase from human gastric juice, *Biochim. Biophys. Acta* 712, 692, 1982.
4. Rogalska, E., Ransac, S., and Verger, R. Stereoselectivity of lipases. II. Stereoselective hydrolysis of triglycerides by gastric and pancreatic lipases, *J. Biol. Chem.* 265, 20271, 1990.
5. Pafumi, Y. et al. Mechanisms of inhibition of triacylglycerol hydrolysis by human gastric lipase, *J. Biol. Chem.* 27, 28070, 2002.

6. Carriere, F. et al. Secretion and contribution of lipolysis by gastric and pancreatic lipases during a test meal in humans, *Gastroenterology* 105, 876, 1993.
7. Carriere, F. et al. In vivo and in vitro studies on the stereoselective hydrolysis of triglycerides and diglycerides by gastric and pancreatic lipases, *Bioorg. Med. Chem.* 5, 429, 1997.
8. Jensen, R.G. et al. Fatty acid and positional selectivities of gastric lipase from premature human infants: in vitro studies, *Lipids* 29, 433, 1994.
9. Dahim, M. and Brockman, H. How colipase–fatty acid interactions mediate absorption of pancreatic lipase to interfaces, *Biochemistry* 37, 8396, 1998.
10. Vian, R. et al. Gastric emptying and intragastric distribution of lipids in man. A new scintigraphic method of study, *Dig. Dis. Sci.* 27, 705, 1982.
11. Maes, B.D. et al. Gastric emptying of liquid, solid and oil phase of a meal in normal volunteers and patients with Bilroth II gastrojejustomy, *Eur. J. Clin. Invest.* 28, 197, 1996.
12. Hunt, J.N. and Knox, M.T. A relation between the chain length of fatty acids and the slowing of gastric emptying, *J. Physiol.* 194, 327, 1968.
13. Maes, B.D. et al. Relation between gastric emptying rate and the rate of intraluminal lipolysis, *Gut* 38, 23, 1996.
14. Northfield, T.C. and Hofman, A.M. Biliary lipid output during three meals and an overnight fast, *Gut* 16, 1, 1975.
15. Hjelm, R.P. et al. Probing self assembly in biological mixed colloids by SANS, deuteration, and molecular manipulation, *J. Phys. Chem.* 99, 16395, 1995.
16. Lykidis, A., Mougios, V., and Arzoglou, P. Kinetics of the two-step hydrolysis of triacylglycerol by pancreatic lipases, *Eur. J. Biochem.* 230, 892, 1995.
17. Bottino, N.R., Vandenburg, G.A., and Reriser, R. Resistance of certain long-chain polyunsaturated fatty acids of marine oils to pancreatic lipase hydrolysis, *Lipids* 2, 489, 1967.
18. Christophe, A. and Verdonk, G. Effects of substituting monoglycerides for natural fats in the diet on fasting serum lipids in man, *Biochem. Soc. Trans.* 5, 1041, 1977.
19. Murase, T. et al. Dietary diacylglycerol suppresses high fat and high-sucrose diet-induced body fat accumulation in C57BL/6J mice, *J. Lipid Res.* 42, 372, 2001.
20. Boswinkel, G. et al. Kinetics of acyl migration in monoglycerides and dependence on acyl chain length, *J. Am. Oil Chem. Soc.* 73, 707, 1996.
21. Lyubachesskaya, G. and Boyle-Rode, E. Kinetics of 2 monoacylglycerol acyl migration in model chylomicra, *Lipids* 35:1353, 2000.
22. Skipski, V.P., Morehous, M.G., and Duel, H.J. The absorption in the rat of a dioleoyl-2-deuteriostearoyl glyceride C^{14}, *Arch. Biochem. Biophys.* 81, 93,1959.
23. Nalbone, G. et al. Pancreatic phospholipase A2 hydrolysis of phosphatidylcholines in various physicochemical states, *Biochim. Biophys. Acta* 620, 612, 1980.
24. Wickham, M., Wilde, P., and Fillery-Travis, A. A physicochemical investigation of two phosphatidylcholine/bile salt interfaces: implications for lipase activation, *Biochim. Biophys. Acta* 1580, 110, 2002.
25. Saunders, D.R. and Sillery, J. Lecithin inhibits fatty acid and bile salt absorption from rat small intestine *in vivo*, *Lipids* 11, 830, 1976.
26. Surpuriya, V. and Higuchi, W.I. Enhancing effect of calcium ions on transport of cholesterol from aqueous sodium taurocholate-lecithin micellar phase to oil phase, *J. Pharm. Sci.* 63:1325, 1994.
27. Rampone, A.J. and Long, L.R. The effect of phosphatidylcholine and lysophosphatidylcholine on the absorption and mucosal metabolism of oleic acid and cholesterol in vitro, *Biochim. Biophys. Acta* 486, 500, 1977.
28. Sugawara, T. et al. Lysophosphatidylcholine enhanced carotenoid uptake from mixed micelles by Caco-2 human intestinal cells, *J. Nutr.* 131, 2921, 2001.

29. Lindstrom, M.B., Sterny, B., and Borgstrom, B. Concerted action of carboxyl ester lipase and pancreatic lipase during lipid digestion in vitro: importance of the physicochemical state of the substrate, *Biochim. Biophys. Acta* 959, 178, 1988.
30. Nilsson, A. and Becker, B. Uptake and interconversion of plasma unesterified n-3 polyunsaturated fatty acids by the GI tract of rats, *Am. J. Physiol.* 268, G732, 1995.
31. Huang, Y.S., Lin, J.W., Koba, K., and Anderson, S.N. N-3 and n-6 fatty acid metabolism in undifferentiated and differentiated human intestine cell line (Caco-2), *Mol. Cell. Biochem.* 18, 121, 1995.
32. Chen, Q. and Nillson, A. Interconversion of alpha-linolenic acid in rat intestinal mucosa: studies in vivo and in isolated villus and crypt cells, *J. Lipid Res.* 35, 601, 1994.
33. Murase, T. et al. Abundant expression of uncoupling protein-2 in the small intestine: up-regulation by dietary fish oil and fibrates, *Biochim. Biophys. Acta* 1530,15, 2001.
34. Sanderson, I.R. and Naik, S. Dietary regulation of intestinal gene expression, *Annu. Rev. Nutr.* 20, 311, 2000.
35. Christophe, A. and Verdonk, G. Postprandial changes in light scattering intensity of serum lipoproteins of different size after single fat feedings in man, in *Protides of the Biological Fluids, Proceedings of the 25th Colloquium,* Peeters, H. Ed., Pergamon Press, New York, 1978, p. 367.
36. O'Doherty, P.J.A., Yousef, I.M., and Kuksis, A. Effect of phosphatidylcholine on triacylglycerol synthesis in rat intestinal mucosa, *Can. J. Biochem.* 52, 726, 1974.
37. Zierenberg, O., Odenthal, J., and Betzing, H. Incorporation of PPC into serum lipoproteins after oral or i.v. administration, *Atherosclerosis* 34, 259, 1979.
38. Mu, H. and Hoy, CE Effect of different medium-chain fatty acids on intestinal absorption of structured triacylglycerols, *Lipids* 35, 83, 2000.
39. De Jong, B.J. and Hulsmann, B.C. Monoacylglycerol hydrolase activity of isolated rat small intestinal cells, *Biochim. Biophys. Acta* 27, 36, 1978.
40. Hanel, A.M. and Gelb, M.H. Multiple enzymatic activities of the human cytosolic 85-kDa phospholipase A2: hydrolytic reactions and acyl transfer to glycerol, *Biochemistry* 20, 7807, 1995.
41. Bierbach, H. Triacylglycerol biosynthesis in human small intestinal mucosal Ayl-CoA: monoglyceride acyltransferase, *Digestion* 28, 138, 1983.
42. Lehner, R., Kuksis, A., and Itabashi, Y. Stereospecificity of monoacylglycerol and diacylglycerol acyltransferases from rat intestine as determined by chiral phase high performance liquid chromatography, *Lipids* 28, 29, 1993.
43. Paris, R. and Clement, G. Biosynthesis of lysophosphatidic acid from 1-monoolein and ATP by subcellular particles from intestinal mucosa, *Proc. Soc. Exp. Biol. Med.* 131:363, 1969.
44. Duff, S.M., Kalambur, S., and Boyle-Roden, E. Serum albumin binds beta- and alpha-monoolein in vivo, *J. Nutr.* 131, 774, 2001.
45. Tso, P., Karlstad, M.D., Bistrian, B.R., and DeMichelle, S.J. Intestinal digestion, absorption, and transport of structured triglycerides or cholesterol in rats, *Am. J. Physiol.* 268, G568, 1995.
46. Mackenzie, C.G. et al. Metabolism of glycerol monoethers in cultured liver cells and implications for monoglyceride pathways, *J. Lipid Res.* 36, 125, 1976.
47. Christophe, A. and Verdonk, G. Feeding monoglycerides as food fat results in an unusual serum triglyceride carbon number composition compatible with the idea of portal monoglyceride absorption with structure retention, *Arch. Intern. Physiol. Biochim.* 88, B221, 1980.
48. Sakr, S.W. et al. Fatty acid composition of an oral load affects chylomicron size in human subjects, *Br. J. Nutr.* 77, 19, 1997.

49. Christophe, A. et al. Fatty acid chain length combinations in ascitic fluid triglycerides containing lymphatic absorbed medium chain fatty acids, *Lipids* 17, 759, 1982.
50. Tagushi, H. et al. Double-blind controlled study on the effects of dietary diacylglycerol on postprandial serum and chylomicron triacylglycerol responses in healthy humans, *J. Am. Coll. Nutr.* 19, 789, 2000.
51. Kritchevski, D. et al. Cholesterol vehicle in experimental atherosclerosis. Part 13. Randomized peanut oil, *Atherosclerosis* 17, 225, 1973.
52. Kritchevski, D., Tepper, S.A., and Klurfeld, D.M. Lectin may contribute to the atherogenicity of peanut oil, *Lipids* 33, 821, 1998.
53. Christophe, A. et al. Nutritional studies with randomized butter. Cholesterolemic effects of butter oil in man, *Arch. Intern. Physiol. Biochem.* 86, 414, 1978.
54. Christophe, A.B. et al. Substituting enzymatically interesterified butter for native butter has no effect on lipemia or lipoproteinemia in man, *Ann. Nutr. Metab.* 44, 61, 2000.
55. Anonymous. Salatrim is approved by the EC, *Inform* 15, 105, 2000.
56. Renaud, S.C., Ruf, J.C., and Petithory, D. The positional distribution of fatty acids in palm oil and lard influences their biological effects in rats, *J. Nutr.* 125, 229, 1995.
57. Yli-Jokipii, K. et al. Effects of palm oil and transesterified palm oil on chylomicron and VLDL triacylglycerol structures and postprandial lipid response, *J. Lipid Res.* 42, 1618, 2001.
58. Ikeda, I. et al. Effects of long-term feeding of marine oils with different positional distribution of eicosapentaenoic and docosahexaenoic acid on lipid metabolism, eicosanoid production and platelet aggregation in hypercholesterolemic rats, *Lipids* 33, 897, 1998.
59. Mortimer, B.C. et al. Plasma clearance of model lipoproteins containing saturated and polyunsaturated monoacylglycerols injected intravenously in the rat, *Biochim. Biophys. Acta* 1127, 67, 1992.
60. Mortimer, B.C. et al. Effects of triacylglycerol-saturated acyl chains on the clearance of chylomicron-like emulsions from the plasma of rats, *Biochim. Biophys. Acta* 1211, 171, 1994.
61. Yoshida, H. et al. Effect of dietary seal and fish oils on triacylglycerol metabolism in rats, *J. Nutr. Sci. Vitaminol.* 45, 411, 1999.
62. Yoshida, H. et al. Effect of dietary seal and fish oils on lipid metabolism in hamsters, *J. Nutr. Sci. Vitaminol.* 47, 242, 2001.
63. Lawson, L.D. and Hughes, B.G. Absorption of eicosapentaenoic acid and docosahexaenoic acid from fish oil triacylglycerols or fish oil ethyl esters co-ingested with a high fat meal, *Biochem. Biophys. Res. Commun.* 156, 960, 1988.
64. Kaku, S. et al. Interaction of dietary fats and proteins on fatty acid composition of immune cells and LTB4 production by peritoneal exudate of rats, *Biosci. Biotechnol. Biochem.* 65, 315, 2001.
65. Eldridge, A.L., Cooper, D.A., and Peters, J.C. A role for olestra in body weight management, *Obes. Rev.* 3, 17, 2002.
66. Stubbs, R.J. The effect of ingesting olestra-based foods on feeding behavior and energy balance in humans, *Crit. Rev. Food Sci. Nutr.* 41, 363, 2001.
67. Ntanios, F.Y. and Duchateau, D.S. A healthy diet rich in carotenoids is effective in maintaining normal blood carotenoid levels during the daily use of plant sterol-enriched spreads, *Int. J. Vitam. Nutr. Res.* 72, 32, 2002.
68. de Jong, A., Plat, J., and Mensink, R.P. Metabolic effects of plant sterols and stanols, *J. Nutr. Biochem.* 362, 2003.

16 CLA in Human Nutrition and Health: Human Studies

Yong Li and Bruce A. Watkins

CONTENTS

16.1 INTRODUCTION

The relationship between dietary fat and human nutrition continues as a chief research emphasis to advance the understanding of health and make recommendations to reduce the incidence of chronic diseases. The 2002 report from the Institute of Medicine of the National Academies on dietary reference intakes (DRIs) revised the recommended levels for fat and individual fatty acids; however, no guidelines were provided for conjugated linoleic acid (CLA). The DRIs establish an adequate intake (AI) of linoleic acid level at 17 g/d for young men and 12 g/d for young women in the United States [1,2]. Though no DRIs are recommended for saturated fat and *trans* fatty acids because of their perceived adverse effects on health, the tolerable upper intake levels were not set for these fatty acids because of practical issues. At this time CLA isomers are a part of the fat component of a normal diet and are perceived as having a health benefit.

The family of fatty acids known collectively as CLA is a group of positional and geometric fatty acid isomers of octadecadienoic acid. The CLA isomers are purported to possess antioxidant properties, inhibit carcinogen–DNA adduct formation, induce apoptosis and cytotoxic activity, modulate tissue fatty acid composition and prostanoid formation, and affect the expression and action of cytokines and growth factors [3]. Though numerous biological actions of CLA have been reported, the most consistent findings include anticancer effects in rodents and cancer cells and reduction of body fat in growing animals. In some cases the biological responses observed from CLA isomers in animal models were influenced by the amounts of dietary n-6 and n-3 polyunsaturated fatty acids (PUFAs) present [4–6]. The purpose of this chapter is to present the current findings describing the actions of CLA in human subjects.

16.2 HEALTH ASPECTS OF CLA

A search result of two databases (Medline and Current Contents) revealed that 34 original journal articles (in English) were published from January 2000 to February 2004 reporting the effects of dietary CLA or epidemiologic evidence in human subjects. These studies describe the actions of CLA isomers on body weight and fat mass control (Table 16.1), blood lipids, lipid peroxidation (Table 16.2), insulin resistance (Table 16.3), lipid metabolism (Table 16.4), immune function (Table 16.5), as well as epidemiologic data on the correlation between CLA intake and breast cancer risk (Table 16.6). Of the human studies examined herein all data, conflicting or inconsistent, are presented without exception.

16.2.1 Body Weight and Fat Control

Though several investigations indicated a fat loss or redistribution when CLA isomers were given to animals [7–9], when they were tested in human subjects, the results were variable in original studies [10] and in a recent review [10–12]. Some studies reported a decline in body fat when CLA was administered to human subjects. Out of the 15 publications that looked at the effect of CLA on body weight and/or body fat, 6 reported a positive effect on reducing body fat, 1 indicated an increase in lean body mass, and none found a significant reduction in body weight (Table 16.1). The dose of CLA supplementation ranged from 0.4 to 6.8 g/d and the duration of treatments was from 4 weeks to 6 months. Of the subjects involved in these investigations, eight studies used healthy subjects, five used overweight or obese individuals, and one used subjects with type 2 diabetes mellitus. In all of these studies, a mixture of CLA isomers was used for the dietary treatment and most of these preparations had equal amounts of c9,t11 and t10,c12, which are the two major isomers present in many commercially available CLA supplements. In general, CLA seemed to be effective in both genders.

In a double-blind, placebo-controlled study, 21 patients with type, 2 diabetes were given either 8.0 g/d of CLA oil (76% total CLA, 37% c9,t11 + 39% t10,c12 isomer, in free fatty acid form) or 8.0 g/d of safflower oil as the control [13]. After 8 weeks of intervention, the plasma concentrations of the t10,c12 CLA isomer were inversely correlated with body weight changes as well as serum leptin levels, indicating

TABLE 16.1
Reported Findings on the Effects of CLA Isomers on Body Weight and Fat Control in Human Studies

Effect	CLA Dose[a]	Duration	CLA Isomeric Composition	Subject and Design	Indicators	References
Lowering morning appetite	1.8 or 3.6 g/d	13 weeks	Tonalin™ 75% CLA	26 men and 28 women overweight subjects (average age 37.8 ± 7.7 yr), double-blind, placebo-controlled randomized	Body weight maintenance after weight loss	Kamphuis et al. 2003 [10]
Reduced body fat but not body weight	3.8 g/d	6 months	8.3% c9,t11, 7.9% t10,c12, 6.0% t8,c10, 7.1% c11,t13, 4.7% c,c, and 17.7% t,t isomers	14 healthy resistant training subjects (5 male and 9 female), nonplacebo controlled	Body composition and endocrine parameters	von Loeffelholz et al. 2003 [11]
CLA t10,c12 inversely related to body weight	6.0 g/d	8 weeks	37% c9,t11 and 39% t10,c12	21 subjects (gender not specified) with type 2 diabetes mellitus, double-blind, randomized block design	Body weight, leptin	Belury et al. 2003 [13]
Reduced body fat but not body weight	1.8 g/d	12 weeks	Equal amount of c9,t11 and t10,c12 isomers	10 male and 10 female healthy subjects (18 to 30 yr) of normal body weight, random-double-blind, placebo controlled	Body fat and weight	Thom et al. 2001 [14]
Reduced percentage body fat, but not body weight	4.2 g/d	12 weeks	Equal amount of c9,t11 and t10,c12	53 healthy subjects (both genders) aged 23 to 63 yr randomized, double-blind placebo controlled	Anthropometric and metabolic variables, fatty acid metabolism	Smedman et al. 2001 [15]

(continued)

TABLE 16.1 (Continued)
Reported Findings on the Effects of CLA Isomers on Body Weight and Fat Control in Human Studies

Effect	CLA Dose[a]	Duration	CLA Isomeric Composition	Subject and Design	Indicators	References
Reduction of body fat and increased lean body mass (6.8 g/d)	1.7–6.8 g/d	12 weeks	Equal amounts of c9,t11 and t10,c12 isomers	60 overweight or obese volunteers (17 male and 35 female at completion), randomized, double-blind, placebo controlled	Body composition	Blankson et al. 2000 [16]
Reduced body fat mass and reduced HDL-cholesterol	0.7 g/d followed by 1.4 g/d	4 weeks followed by another 4 weeks	Equal amounts of c9,t11 and t10,c12 isomers	14 volunteers (gender not specified, aged 19 to 24 yr), double-blind, placebo controlled	Body fat, serum lipid profile, creatine kinase, and cortisol	Mougios et al. 2001 [17]
Decreased abdominal fat	4.2 g/d	4 weeks	36.9% c9,t11, 37% t10,c12, 1.4% t,t isomer, and 2.2% c,c isomer	25 middle-aged abdominally obese men, double-blind, randomized placebo controlled	Sagittal abdominal diameter, serum lipid profile, glucose and insulin	Riserus et al. 2001 [18]
Enhancing the effect of herbal anticellulite pills	0.4 or 0.8 g/d	60 d	CLA capsule (no detailed information)	Females with visible cellulite, randomized	Thigh circumference, skin appearance	Birnbaum et al. 2001 [19]
No effect on body weight regain	1.8 or 3.6 g/d	13 weeks	Tonalin™ 75% CLA	26 men and 28 women overweight subjects (average age 37.8 ± 7.7 yr), double-blind, placebo-controlled randomized	Body weight maintenance after weight loss	Kamphuis et al. 2003 [20]

Lowered leptin, no effect on fat mass	3 g/d	94 d (64 d CLA treatment	17.6% c9,t11, 22.6% t10,c12, 23.6% c11,t13, 16.6% t8,c10, 7.7% t,t, and 11.9% other isomers	24 healthy women, double-blind, placebo-controlled randomized	Circulating leptin concentrations and appetite	Medina et al. 2000 [21]
No significant differences were found in body fat	2.1 g/d	45 d	Equal amount of c9,t11 and t10,c12 isomers	17 young healthy nonobese sedentary women (aged 19 to 24 yr), randomized double-blind crossover, placebo controlled	Body fat, serum leptin, TAG, total cholesterol, HDL-cholesterol, and alanine aminotransferase	Petridou et al. 2003 [22]
No significant effect on body fat and energy expenditure	3 g/d	94 d (64 d CLA treatment)	17.6% c9,t11, 22.6% t10,c12, 23.6% c11,t13, 16.6% t8,c10, 7.7% t,t, and 11.9% other isomers	17 healthy, adult women, randomized, blind, placebo controlled	Body composition and energy expenditure	Zambell et al. 2000 [23]
No significant differences were detected	6 g/d	28 d	17.6% c9,t11, 22.6% t10,c12, 23.6% c11,t13, 16.6% t8,c10, 7.7% t,t, and 11.9% other isomers	23 experienced, resistance-trained subjects	Body composition, blood chemistry, and immunity parameters	Kreider et al. 2002 [24]
No reduction in body weight and no adverse effects	3.4 g/d	12 weeks	Tonalin™	60 overweight or obese volunteers (30 male and 17 female at completion, >18 yr), randomized, placebo controlled, and double blind	Body composition and blood parameters	Berven et al. 2000 [25]

[a] Whenever possible, CLA dose was given as the amount of pure CLA when the CLA source was a mixture of CLA and other fatty acids.

TABLE 16.2
Reported Findings on the Effects of CLA Isomers on Lipid Peroxidation in Human Studies

Effect	CLA Dose[a]	Duration	CLA Isomeric Composition	Subject and Design	Indicators	References
The t10,c12 isomer increased insulin resistance, lowered HDL cholesterol, and elevated oxidative stress	3.4 g/d	12 weeks	Purified t10,c12 (76.5% t10, c12) and CLA isomer mixture (35.4% c9,t11 and 35.9% t10,c12)	60 obese men with metabolic syndrome, double-blind, placebo-controlled trial	Plasma lipids and insulin resistance	Riserus et al. 2002 [26]
Inducing lipid peroxidation	4.2 g/d	12 weeks	Equal amount of c9,t11 and t10,c12	53 healthy men and women (aged 23 to 63 yr), double blind, randomized, placebo controlled	Circulating and urinary 15-keto-dihydro-$PGF_{2\alpha}$ and 8-iso-$PGF_{2\alpha}$	Basu et al. 2000 [27]
The t10,c12 isomer was more potent in inducing enzymatic and nonenzymatic lipid peroxidation	3.5 g/d CLA mix or 4.0 g/d CLA 1012	4 weeks	*CLA mix*: 3.5 g CLA/d, containing 38% t10,c12 and 36.5% (c9,t11 + t8,c10) *CLA 1012*: 85.1% t10,c12 and 9.1% c9,t11	60 healthy subjects (25 men and 35 women, aged 21 to 63 yr), randomized	Urinary 15-keto-dihydro-$PGF_{2\alpha}$ and 8-iso-$PGF_{2\alpha}$	Smedman et al. 2004 [28]
Inducing lipid peroxidation	4.2 g/d	1 month	Equal amounts of c9,t11 and t10,c12	24 middle-aged men with abdominal obesity, randomized controlled	Urinary 15-keto-dihydro-$PGF_{2\alpha}$ and 8-iso-$PGF_{2\alpha}$	Basu et al. 2000 [29]

[a] Whenever possible, CLA dose was given as the amount of pure CLA when the CLA source was a mixture of CLA and other fatty acids.

TABLE 16.3
Reported Findings on the Effects of CLA Isomers on Insulin-Resistance and Diabetes in Human Studies

Effect	CLA Dose[a]	Duration	CLA Isomeric Composition	Subjectand Design	Indicators	References
CLA t10,c12 inversely related to serum leptin	6.0 g/d	8 weeks	37% c9,t11 and 39% t10,c12	21 subjects (gender not specified) with type II diabetes mellitus, double-blind, randomized block design	Body weight, leptin	Belury et al. 2003 [13]
Induced insulin resistance and reduced HDL cholesterol	3.4 g/d	12 weeks	*CLA mixture*: 35.4% c9,t11 and 35.9% t10,c12. *t10,c12 CLA:* 76.5% t10,c12 and 2.9% c9,t11.	60 Caucasian obese men (aged 35 to 65 yr) with metabolic syndrome, randomized, double blind, placebo controlled	Insulin sensitivity	Riserus et al. 2002 [30]

[a] Whenever possible, CLA dose was given as the amount of pure CLA when the CLA source was a mixture of CLA and other fatty acids.

TABLE 16.4
Reported Findings on the Effects of CLA Isomers on Lipid Metabolism in Human Studies

Effect	CLA Dose[a]	Duration	CLA Isomeric Composition	Subject and Design	Indicators	References
50:50 CLA mixture reduced blood TAG and VLDL cholesterol	3 g/d	8 weeks	c9,t11:t10,c12 ratios of 50:50 or 80:20	51 normolipidemic subjects (18 male and 33 female, 31.6 ± 10.0 yr), randomized, double blind, placebo controlled	Body fat; blood lipids, glucose, and insulin	Noone et al. 2002 [31]
Reduced the fat content of milk but not total milk	1.5 g/d	17 d	Tonalin, 36.4% c9,t11, 37.2% t10,c12, 1.3% c9,c11, and 1.1% c10,t12	10 healthy lactating women (1 to 12 months postpartum), randomized, double blind, crossover, placebo controlled	Milk fat content and total milk output	Masters et al. 2002 [32]
No effect	3.9 g/d	94 d (64 d CLA treatment)	11.4% c9,t11, 10.8% t8,c10, 15.3% c11,t13, and 14.7% t10,c12	17 healthy normolipidemic female subjects, randomized	Blood lipids, and tissue fatty acid composition	Benito et al. 2001 [33]
No effect on fatty acid or glycerol metabolism	3.9 g/d	4 weeks	17.6% c9,t11, 22.6% t10,c12, 23.6% c11,t13, 16.6% t8,c10, 7.7% t9,t11 and t10,t12, and 11.9% other isomers	6 healthy, weight-stable, adult women (aged 24 to 41 yr), randomized, blind, and placebo controlled	Staple isotope infusion on fatty acid and glycerol metabolism	Zambell et al. 2001 [34]

[a] Whenever possible, CLA dose was given as the amount of pure CLA when the CLA source was a mixture of CLA and other fatty acids.

TABLE 16.5
Reported Findings on the Effects of CLA Isomers on Immune Function in Human Studies

Effect	CLA Dose[a]	Duration	CLA Isomeric Composition	Subject and Design	Indicators	References
The t10,c12 enhanced response to hepatitis B vaccination	1.7 or 1.6 g/d	12 weeks	50:50 and 80:20 mixtures of c9,t11/t10,c12	81 healthy Caucasian males (31 to 69 yr), stratified, double-blind randomized	Antibody levels to hepatitis B vaccination	Albers et al. 2003 [35]
No effect was observed on the function of peripheral blood mononuclear cells	3.9 g/d	93 d (63 d CLA treatment)	22.6% t10,c12, 23.6% c11,t13, 17.6% c9,t11, 16.6% t8,c10, and 19.6% other isomers	17 healthy young women (aged 20 to 41 yr), randomized and placebo controlled	Function of peripheral blood mononuclear cells	Kelley et al. 2001 [36]
CLA did not alter any of the indices of immune status tested	3.9 g/d	93 d (63 d CLA treatment)	22.6% t10,c12, 23.6% c11,t13, 17.6% c9,t11, 16.6% t8,c10, and 19.6% other isomers	17 healthy young women (aged 20 to 41 yr), randomized and placebo controlled	Fatty acid composition and indices of immune status	Kelley et al. 2000 [37]
CLA did not alter the blood-clotting parameters	3.9 g/d	93 d (63 d CLA treatment)	22.6% t10,c12, 23.6% c11,t13, 17.6% c9,t11, 16.6% t8,c10, and 19.6% other isomers	17 healthy young women (aged 20 to 41 yr), randomized and placebo controlled	Blood coagulation and platelet function	Benito et al. 2001 [38]

[a] Whenever possible, CLA dose was given as the amount of pure CLA when the CLA source was a mixture of CLA and other fatty acids.

TABLE 16.6
Reported Findings on the Effects of CLA Isomers on Breast Cancer Risk in Epidemiologic Studies

Effect	Duration	CLA Isomeric Composition	Subject and Design	Indicators	References
A weak, but positive correlation between CLA (c9,t11 or t9,c11) intake and breast cancer incidence	6.3 yr	Energy-adjusted CLA intake: 0.2 ± 0.1 g/d for both cancer cases (941) and subcohort members (1598)	Cohort of 62,573 women subjects, aged 55 to 69 yr with 941 incident cases of breast cancer, cohort study, the Netherlands	Dietary CLA intake and cancer incidence	Voorrips et al. 2002 [39]
Not related	Cross-sectional	c9,t11 was the major isomer found in breast adipose tissue, total CLA was 0.44% of total fatty acids in cases and 0.43% in controls	329 female patients (241 invasive breast cancer cases and 88 benign breast pathologies as control), case-control study, France	CLA content in breast adipose tissue and breast cancer risk	Chajes et al. 2002 [40]
No association	Median follow-up 7.5 yr	c9,t11 was the major isomer present in breast adipose tissue, the sum of CLA isomers was at 0.44% of total fatty acids	209 female breast cancer patients aged 24.2 to 82.9 yr (mean 55 yr), 43% premenopausal and 57% postmenopausal, 7.5 yr cohort study, central France	Breast biopsy tissue CLA content and cancer metastasis	Chajes et al. 2003 [41]
Low intake of CLA (esp. cheese) was associated with increased cancer risk	Cross-sectional	CLA intake: 142.3 ± 60.6 and 131.9 ± 58.9 mg/d for precases and controls, respectively, and 126.8 ± 63.9 and 132.8 ± 65.2 mg/d for postcases and controls, respectively	195 female breast cancer cases (68 premenopausal and 127 postmenopausal) and 208 control subjects (75 premenopausal and 133 postmenopausal), case-control study, Finland	CLA intake and breast cancer risk	Aro et al. 2000 [42]

a possible role for this CLA isomer in body weight regulation in these patients. CLA (1.8 g/d) (Tonalin™, Natural Lmt, ASA; containing equal amounts of c9,t11 and t10,c12 isomers) supplementation to healthy exercising human subjects (10 males and 10 females, aged 18 to 30 years (yr)) with normal body weight for 12 weeks in a double-blind and placebo (hydrogel) controlled study reduced body fat but not body weight as measured using near infrared light [14]. In another double-blind, placebo-controlled study, 53 healthy subjects (both genders) aged 23 to 63 years were randomly assigned to CLA supplementation (4.2 g/d, equal amounts of c9,t11 and t10,c12 isomers) or the same amount of olive oil for 12 weeks [15]. At the end of the study, the percentage of body fat decreased by 3.8% ($p < 0.001$) within the CLA-treated group and this reduction was also significant when compared with the placebo control ($p = .05$), while body weight, serum lipids, and glucose metabolism were not affected by CLA.

Blankson et al. [16] investigated the dose–response relationship of CLA (Tonalin, with equal amounts of c9,t11 and t10,c12 isomers, Natural Lipids, Norway) to body fat mass in 60 overweight or obese volunteers (body mass index ranged from 25 to 35 kg/m^2). The subjects were randomly divided into five groups and were given the respective treatments (9 g olive oil and 1.7, 3.4, 5.1, or 6.8 g/d CLA) for 12 weeks. At the end of the study, a significantly higher reduction in body fat mass was found in the CLA groups compared with the placebo group ($p = .03$). The reduction in body fat within the CLA subjects was significant for the groups given 3.4 and 6.8 g CLA when compared with the baseline level within the group (a reduction of 5.7% $p = .05$ and 3.7% $p = .02$, respectively). Other than the reduction in body fat, no significant differences were observed among the groups for lean body mass and body mass index. However, within group, the 6.8 g/d CLA treatment significantly increased the lean body mass when compared with the baseline.

Mougious et al. [17] examined the effect of dietary CLA supplementation (0.7 g of CLA for 4 weeks and 1.4 g of CLA for the next 4 weeks) in 24 volunteers (14 men and 10 women, aged 19 to 24 years) in a double-blind, placebo-controlled design and found that fat mass was significantly reduced in the CLA group during the second period ($p < .004$) but not over the full duration of the study. Subjects were administered capsules (CLA-70 in capsules manufactured by TrofoCell, Hamburg, Germany) that contained equal amounts of the c9,t11 and t10,c12 isomers. The serum lipid profiles were affected by CLA as serum HDL-cholesterol decreased significantly ($p < .001$) and triacylglycerols (TAGs) as well as total cholesterol tended to decrease in the CLA group during the first period.

The short-term metabolic effects of CLA in abdominally obese men were investigated by Riserus et al. [18]. In this study, 25 abdominally obese men (body mass index 32 ± 2.7 kg/m^2) participated in a double-blind, randomized controlled trial for 4 weeks, during which the CLA group received 4.2 g/d CLA (36.9% c9,t11, 37% t10,c12, 1.4% t,t isomers, and 2.2% c,c isomers) and the placebo-controlled group was given the same amount of olive oil. After 4 weeks, there was a significant decrease in sagittal abdominal diameter in the CLA group compared with the placebo ($p = .04$).

The fat reduction property of CLA isomers was also found to be useful in cosmetic applications. CLA (the isomeric composition information was not available in the publication) enhanced the effect of herbal anticellulite pills when the two were used concomitantly in females with visible cellulite [19].

Kamphuis et al. [10,20] evaluated the effect of CLA on body weight maintenance after weight loss. In this study, overweight subjects (26 men and 28 women, average age 37.8 ± 7.7 years, average body mass index 27.8 ± 1.5 kg/m^2) were first subjected to a low calorie diet for 3 weeks and then administered 1.8 g/d or 3.6 g/d CLA (Tonalin 75% TG, Hovdebygda, Norway) or given a placebo over a 13-week intervention period. Though body weight was not affected by any of the treatments, appetite (hunger, satiety, and fullness) was favorably decreased by CLA.

Medina et al. [21] conducted a study with 24 healthy women (aged 20 to 41 years) given CLA (3 g/d, 17.6% c9,t11, 22.6% t10,c12, 23.6% c11,t13, 16.6% t8,c10, 7.7% t9,t11 and t10,t12, and 11.9% other isomers) for 64. At the end of the study, CLA treatment in subjects lowered circulating leptin, but did not affect appetite nor body fat compared with the placebo (sunflower oil) control group. In another study, 17 young healthy nonobese sedentary women (aged 19 to 24 years) were given 2.1 g/d CLA (equal amounts of c9,t11 and t10,c12 isomers) or placebo for 45 d in a randomized, double-blind, crossover design [22]. At the conclusion of the study, no significant differences were observed in energy, carbohydrate, lipid, or protein intake between the CLA and placebo groups. No significant differences were found in body fat or lipid metabolism parameters, such as serum leptin, TAG, total cholesterol, HDL-cholesterol, and alanine aminotransferase between the CLA and the placebo groups. Therefore, although daily supplementation with CLA increased blood levels of the dietary isomers, CLA had no effect on body composition or the lipidemic profile in nonobese women. An important observation in this study was that the 2-week washout period after the end of CLA supplementation was sufficient to deplete CLA isomers in the serum lipids.

In healthy resistance-training subjects, beginner or advanced athletes, 6 months of CLA supplementation significantly reduced body fat compared with the baseline level [11]. In this 6-month period of dietary supplementation, 7 g/d CLA-oil (containing 54% CLA, EuroChem Feinchemie GmbH, Groebenzell, Germany, of the total fatty acids, 8.3% c9,t11, 7.9% t10,c12, 6.0% t8,c10, 7.1% c11,t13, 4.7% c,c, and 17.7 t,t) was given to two groups of male and female resistance-training athletes (seven beginners, three female/four male, and seven advanced, two female/five male). In this study, CLA did not affect body weight or selected endocrine parameters (serum leptin, soluble leptin receptor, and IGF-I) [11].

In another study, 17 healthy adult women were given 3 g/d CLA (17.6% c9,t11, 22.6% t10,c12, 23.6% c11,t13, 16.6% t8,c10, 7.7% [t9,t11 + t10,t12], and 11.9% other isomers) or sunflower oil as a placebo and confined to a metabolic suite for 94 d (30 d baseline and 64 d CLA treatment). CLA had no significant effect on energy expenditure, fat oxidation, or respiratory exchange ratio at rest or during exercise [23]. Kreider et al. [24] administered either 9 g/d of an olive oil placebo or 6 g/d of CLA (Tonalin, PharmaNutrients, with 17.6% c9,t11, 22.6% t10,c12, 23.6% c11,t13, 16.6% t8,c10, 7.7% t9,t11 and t10,t12, and 11.9% other isomers) with 3 g/d of other fatty acids to 23 experienced, resistance-trained subjects for 28 d. At the end of the study, no significant differences were detected in total body mass, fat-free mass, fat mass, percentage of body fat, bone mass, bone strength, clinical blood chemistries (total protein, albumin, globulin, glucose, electrolytes, liver enzymes, lipid profiles, total bilirubin, hemoglobin, hematocrit, red blood cells, and white blood cells), or general markers of catabolism (creatinine, blood urea nitrogen, creatine kinase, and lactate dehydrogenase) and immunity (neutrophil/lymphocyte ratio) during training for the CLA group compared with the placebo group.

In a study that examined the safety of CLA supplementation, no significant effect of 3.4 g/d CLA (Tonalin) on body weight and body mass index (BMI) was observed [25]. After the 12-week period of fatty acid treatment, no reduction in body weight was found in overweight or obese volunteers (60 subjects), though there appeared to be no clinically significant adverse effects [25].

16.2.2 Lipid Peroxidation, Insulin Resistance, and Lipid Metabolism

Although limited, the effects of CLA on plasma lipids and insulin resistance suggest specific actions of individual isomers that are detrimental to health (Tables 16.2 and 16.3). Riserus et al. [26] reported findings from a double-blind, placebo-controlled trial in which 60 obese men with metabolic syndrome were randomized to one of three groups receiving 3.4 g/d of purified t10,c12 isomer (76.5% t10,c12 CLA), a CLA isomer mixture (35.4% c9,t11 and 35.9% t10,c12), or placebo (olive oil) for 12 weeks (Table 16.2). At the end of the study, supplementation with the t10,c12 isomer of CLA significantly increased insulin resistance, lowered HDL cholesterol, and elevated oxidative stress and inflammatory biomarkers in obese men. In another 3-month study, the use of a CLA supplement (consisting of equal amounts of c9,t11 and t10,c12 isomers) clearly induced both nonenzymatic and enzymatic lipid peroxidation in 53 healthy men and women (aged 23 to 63 years) subjects [27]. In a recent publication [28], two major CLA isomers (CLA mix: 3.5 g/d CLA, containing 38% t10,c12, 36.5% c9,t11 + t8,c10; CLA 1012: 4.0 g/d CLA, or consisting of 85.1% t10,c12 and 9.1% of c9,t11; Natural Ltd, Oslo, Norway) were compared during a 6-week study involving 60 healthy male and female subjects (Table 16.2). The investigators found that the t10,c12 isomer was twice as potent in inducing lipid peroxidation compared with the c9,t11 isomer [28]. CLA isomers also increased urinary 8-iso-$PGF_{2\alpha}$, a biomarker of nonenzymatic free-radical-induced lipid peroxidation in subjects when compared before and after CLA treatments were administered. A short-term, 1-month study showed similar results in 24 middle-aged obese men (mean age of 53 years) that the urinary levels of 8-iso-$PGF_{2\alpha}$ and 15-keto-dihydro-$PGF_{2\alpha}$ (a major metabolite of $PGF_{2\alpha}$ as an indicator of enzymatic oxidation of arachidonic acid) were significantly increased by daily intake of 4.2 g of a CLA isomeric mixture (equal amounts of c9,t11 and t10,c12, Natural Ltd, Oslo, Norway) compared with the placebo control (olive oil) group [29]. These findings raise some legitimate concerns (increased risk of oxidative stress) for the effectiveness and safety of CLA supplements in humans that must be addressed in future research.

Riserus et al. [30] investigated the effect of the t10,c12 CLA isomer and a commercial CLA mixture on insulin sensitivity, lipid metabolism, and body composition in 60 Caucasian obese men (aged 35 to 65 years) with symptoms of metabolic syndrome in a randomized, double-blind, controlled trial with 3.4 g/d CLA (isomeric mixture (35.4% c9,t11 and 35.9% t10,c12), purified t10,c12 CLA (76.5% t10,c12 and 2.9% c9,t11) both as oil supplement capsules, Natural Lipids, Hovebygda,

Norway), or a placebo (content not specified in the original publication) for 12 weeks (Table 16.3). At the end of the study, the individuals consuming the t10,c12 CLA product demonstrated increased insulin resistance (19%; $p < .01$) and glycemia (4%; $p < .001$), and reduced high-density lipoprotein (HDL) cholesterol (–4%; $p < .01$) compared with the values in the placebo group. The t10,c12 CLA isomer was also shown to be inversely related to the body weight and serum leptin in subjects with type 2 diabetes [13].

Noone et al. [31] reported that a specific isomeric mixture of a CLA supplement had beneficial effects on circulating lipid profiles compared with those given linoleic acid in a study of 51 normolipidemic subjects (18 males and 33 females, mean age 31.6 ± 10.03 years) (Table 16.4). Administering 3 g/d of the CLA mixture of a 50:50 blend of c9,t11:t10,c12 for 8 weeks lowered fasting triacylglycerols and very low density lipoprotein (VLDL) cholesterol levels compared with those who consumed the 80:20 blend of similar CLA isomeric mixture or linoleic acid (all supplements were free fatty acids). In a double-blind study of 53 healthy male and female subjects randomly assigned to either supplementation with CLA (4.2 g/d, equal amounts of c9,t11 and t10,c12) or the same level of olive oil for 12 weeks [15], those in the CLA group showed higher proportions of stearic, docosatetraenoic, and docosapentaenoic acids in serum lipids and thrombocytes, while proportions of palmitic, oleic, and dihomo γ–linolenic acids were decreased, indicating a decrease in the activities of the Δ6 and Δ9 desaturases and an increase in the activity of Δ5 desaturase. The inhibitory effect of CLA isomers on lipid metabolism was also evident in reducing the fat content of milk in healthy breast-feeding women (1 to 12 months postpartum) when supplemented with 1.5 g of CLA/d (Tonalin, PharmaNutrients Inc., Lake Bluff, IL), though the total amount of milk output was not affected [32]. The authors recommended not taking a commercial CLA supplement while lactating.

In a study of 17 healthy female volunteers [33], the effects of dietary CLA (11.4% c9,t11, 10.8% t8,c10, 15.3% c11,t13, and 14.7% t10,c12) on blood lipids, lipoproteins, and tissue fatty acid composition were investigated (Table 16.4). After the 94-d treatment period (30-d stabilization period + 64-d treatment, subjects were given 3.9 g/d CLA for the intervention group or an equivalent amount of sunflower oil for the control group), CLA did not change the levels of plasma cholesterol, low-density lipoprotein cholesterol, high-density lipoprotein cholesterol, or TAG, even though the plasma CLA isomer levels were significantly elevated by CLA treatment. In a similar study [34] with six healthy, weight-stable, adult women (aged 24 to 41 years), 4 weeks of supplementation with 3.9 g/d CLA (17.6% c9,t11, 22.6% t10,c12, 23.6% c11,t13, 16.6% t8,c10, 7.7% t9,t11 and t10,t12, and 11.9% other isomers) did not significantly affect fatty acid or glycerol metabolism either at rest or during exercise (Table 16.4).

16.2.3 Immune Function

Albers et al. [35] showed that by giving 1.7 g/d of a CLA mixture containing a 50:50 blend of c9,t11:t10,c12, the seroprotective rates in 81 healthy subjects that received hepatitis B vaccination were boosted by more than twofold compared with subjects that received about the same amount of either a different CLA mix (1.6 g/d, 80:20 c9,t11:t10,c12) or safflower oil (Table 16.5). It seems that the t10,c12 isomer was the effective factor in the CLA mixture because of its relative abundance that resulted in enhanced immune response.

A series of studies were performed to examine the effects of CLA isomers on fatty acid composition and function of peripheral blood mononuclear cells [36], indices of immune status [37], and blood coagulation and platelet function [38] (Table 16.4). In one of these studies [36], 17 healthy young women (aged 20 to 41 years) were recruited to participate in a 93-d study and given CLA (Tonalin) at 3.9 g/d (mixture of CLA isomers [t10,c12, 22.6%; c11,t13, 23.6%; c9,t11, 17.6%; t8,c10, 16.6%; other isomers 19.6%]) or placebo (sunflower oil) for 63 d. After CLA supplementation, no effect was observed on the function of peripheral blood mononuclear cells as evidenced by lack of changes for *in vitro* secretion of prostaglandin E_2, leukotriene B_4, interleukin-1β, or tumor necrosis factor α (TNFα) stimulated with lipopolysaccharide and the secretion of interleukin (IL)-2 stimulated with phytohemagglutinin. CLA treatment also failed to alter the percentages of T cells producing IL-2 and interferon-γ and monocytes producing TNFα after these cells were activated/labeled *in vitro* and measured by flow cytometry. In the test on immune response after immunization with an influenza vaccine [37], CLA supplementation (3.9 g/d of an isomeric mixture) to young healthy women (n = 10 in CLA group with average age of 27.0 ± 1.8 years, n = 7 in control group with average age of 29.3 ± 2.6 years) failed to alter any of the indices of immune status tested (number of circulating white blood cells, granulocytes, monocytes, lymphocytes, and their subsets; lymphocyte proliferation in response to phytohemagglutinin; and influenza vaccine, serum influenza antibody titers, and delayed-type hypersensitivity [DTH] response) compared with the placebo control (sunflower oil). These data indicate that CLA at the given doses and durations did not offer any additional benefit to the immune status of these subjects. In the same study, when compared with sunflower oil, adding 3.9 g/d dietary CLA to a typical Western diet did not alter the blood-clotting parameters in healthy adult females [38].

16.2.4 Correlation between Habitual CLA Dietary Intake and Breast Cancer Risk

The consistent anticancer findings with CLA are from animal and cell culture studies that utilized mixtures of CLA isomers. However, their application to humans is lacking. A recent 6.3-year cohort study in the Netherlands [39] reported a weak, but positive, correlation between CLA (c9,t11 or t9,c11) intake and breast cancer incidence that was similar to the intake of total *trans* fatty acids and saturated fatty acids (Table 16.6). Most of the epidemiologic studies did not show a positive effect of CLA (mean CLA level was 0.44% as measured in adipose tissue obtained during surgery and the main isomer was the c9,t11 18:2 [40,41]) on breast cancer prevention [39–41], though one study suggested that a low intake of CLA (energy adjusted) as measured by food frequency questionnaire was associated with increased breast cancer risk in female subjects in a case-control Kuopio Breast Cancer Study (Finland) [42]. The serum CLA level in breast cancer patients was also lower ($p < .01$) compared with the age and residency matched controls [42]. At present, these human studies suggest an increased risk or no association with cancer and the intake of CLA from dietary sources (Table 16.6).

16.2.5 Metabolism of CLA Isomers

Emken et al. [43] investigated the absorption, incorporation in complex lipids, and elongation/desaturation of dietary CLA isomers (t10,c12 and c9,t11) in comparison to oleic acid (18:1n-9) and linoleic acid (18:2n-6) in six healthy Caucasian adult women (aged 23 to 41 years) who had consumed normal diets supplemented with 6 g/d of sunflower oil or 3.9 g/d of CLA for 63 d (Table 16.7). This study showed that CLA isomers had a lower absorption (20 to 25% less) relative to oleic and linoleic acids by using deuterated fatty acids (stable isotopes) and the relative amounts measured by gas chromatography mass spectrometry (GC-MS). The distribution pattern of the two major CLA isomers in esterified lipids was also different in that the incorporation of the t10,c12 isomer was fourfold higher than the c9,t11 isomer in 1-acyl phosphatidylcholine, while the incorporation of the c9,t11 isomer was two- to threefold higher than the t10,c12 isomer in cholesterol esters. Elongated and desaturated CLA metabolites were also detected in this study, though the amount was very low and found only in the triacylglycerol fraction.

Lucchi et al. [44] studied the distribution of CLA in chronic renal failure patients, end-stage chronic renal failure patients in conservative treatment, hemodialysis patients, and healthy controls (Table 16.7). They found that the incorporation of CLA in red blood cells was significantly and gradually reduced as the state of renal failure became more severe; however, CLA in the plasma and adipose tissue was found to be the highest in the end-stage chronic renal failure patients indicating either a reduced metabolism or an altered site distribution of CLA isomers (accumulation in the adipose tissue).

TABLE 16.7
Reported Findings on CLA Isomer Metabolism in Human Studies

Effect	CLA Dose[a]	Duration	Subject and Design	Indicators	References
CLA isomers had a lower absorption (20–25% less) relative to oleic and linoleic acids	3.9 g/d	63 d	Six healthy Caucasian adult women (aged 23 to 41 yr)	Absorption, incorporation, and elongation/desaturation of dietary CLA isomers (c9,t11 and t10,c12)	Emken et al. 2002 [43]
CLA tissue distribution was altered during renal failure			74 chronic renal failure patients and 30 controls (gender not specified)	CLA tissue distribution	Lucchi et al. 2000 [44]

[a] Whenever possible, CLA dose was given as the amount of pure CLA when the CLA source was a mixture of CLA and other fatty acids.

16.3 BIOCHEMICAL AND MOLECULAR ASPECTS

CLA was reported to alter leptin concentration or expression in animal and cell culture studies [45–48]. As a regulator of appetite and lipid metabolism and a proposed neuroendocrine regulator of bone mass, leptin is a likely target of CLA action in adipose tissue for regulating fat mass, and possibly in bone metabolism to maintain mineral content [49,50].

Specific effects of CLA isomers on activity and expression of enzymes associated with fatty acid desaturation have been reported. For example, CLA isomers (other than c9,t11) were observed to decrease the mRNA level of the stearoyl-CoA desaturase in both liver tissue and hepatocyte cultures [51]. On the catabolic side, dietary administration of the t10,c12 isomer in mice reduced body fat and inhibited the uptake of fatty acids in cultured 3T3-L1 adipocytes purportedly by down regulating lipoprotein lipase and stearoyl-CoA desaturase [52]. A similar finding was reported by Choi et al. [53] in mouse 3T3-L1 preadipocyte cultures with the t10,c12 isomer [53]. Dietary CLA (approximately equal amounts of the c9,t11 [43%] and t10,c12 [45%] isomers) reduced adipose depot weight but not the total body weight in Sprague-Dawley rats and C57Bl/6 mice; though the possible mechanism for this reduction in body fat could be attributed to enhanced β-oxidation in muscle for the mice as evidence by increased expression of uncoupling protein, this was not the case for the observation in rats [54]. Rahman et al. [55] showed that feeding CLA, either as a free or triacylglycerol (33.2% c9,t11/t9,c11 and 34.2% t10,c12) produced a 5.2% decrease in body weight compared with the control rats [55]. The mechanism responsible for the reported actions of CLA remains unresolved; however, evidence is mounting to suggest that these are specific for the isomers and animal species.

CLA lowered the production of pro-inflammatory products in macrophages through activation of peroxisome proliferators activated receptor (PPARγ) [56] and lowered basal and lipopolysaccharide (LPS) stimulated IL-6 and basal TNF production in rat resident peritoneal macrophages [4]. Through activation of PPARγ, CLA decreased interferon-γ induced mRNA expression of cyclooxygenase (COX)-2, inducible nitric oxide synthase (iNOS), TNFα, and cytokines (IL-1β and IL-6) in RAW macrophage cultures [56]. It is well recognized that some PUFAs are ligands of PPARs which can potentially alter COX-2 protein levels and mRNA expression [57]. We reported that dietary CLA isomers reduced *ex vivo* prostaglandin E_2 (PGE_2) production in rat bone organ cultures [5] and similar effects of CLA on PGE_2 production in various biological systems are reported [58–60]. In a recent publication, CLA isomers were not effective in reducing bone lost in the ovariectormized rat [61]. Recognizing the biochemical and molecular relationships between these actions of CLA isomers, the targets that integrate their potential biological impact include PPARs, COX enzymes, and genes to influence health status.

16.4 PRINCIPAL BIOLOGICAL ACTIONS OF CLA

CLA is the only known antioxidant and anticarcinogen associated with foods originating from animal sources. Ha and co-workers [62] were one of the first groups to report that CLA from beef was protective against chemically induced cancer. In this example, CLA isolated from extracts of grilled ground beef were found to reduce skin tumors in mice treated with 7,12-dimethylbenz[α]anthracene (DMBA), a known carcinogen [62]. Since then, many investigators have reported biological actions of CLA isomers. The research on CLA isomeric mixtures has relied almost entirely on animal models and various cell culture systems. The properties of CLA include anticarcinogenic [62–72] and antiatherosclerotic [73,74]. Other CLA effects include antioxidative [64,65,73,74] and immunomodulative [4,75–78]. More recently, preliminary data suggest that CLA may have a role in controlling obesity [52,79,80], reducing the risk of diabetes [81], and modulating bone metabolism [5,6].

Extensive investigations into the anticarcinogenic properties of CLA showed that CLA isomers (chiefly mixtures [c9,t11 and t10,c12] of variable purity) reduced chemically induced tumorigenesis in rat mammary gland and colon [63,65–69, 72,82] and modulated chemically induced carcinogenesis mouse skin [62,83] and forestomach [64]. Sources of CLA also inhibited the growth of human tumor cell lines in culture [71,84,85] and in severe combined immunodeficient (SCID) mice [70,86]. The cytotoxic effects of CLA isomers on growth of various human and animal-derived cancer cells seem to be linked to decreased expression of the gene transcription factor Bcl-2 family whose members inhibit apoptotic cell death and/or induce caspase-dependent apoptosis [87–91]. CLA also prevented basic fibroblast growth factor-induced angiogenesis [92], a critical process for growth and metastasis of cancers. Evidence for the anticarcinogenic effects of the c9,t11 isomer suggests a modifying role in PPARα action [93].

Research in growing animals indicates that CLA reduced body fat accumulation [94–96] but not all CLA isomers appear to work equally as reported in mice [97,98]. In an experiment the t10,c12 isomer, but not the c9,t11 isomer, induced body fat loss and adipocyte apoptosis in mice [94]. Others reported an observed fat loss or redistribution when CLA isomers were given to chickens (CLA mixture, animals weighed less in CLA group) [7], rats (reduce body fat, need isomer info) [8], and pigs (fat repartitioning, equal amounts of c9,t11 and t10,c12 with small percentages of c,c and t,t isomers) [9], but when tested in human subjects the response was not as obvious [12]. The aforementioned studies suggest that close attention be paid to endocrine parameters and gender and age effects on energy metabolism before designing new investigations to examine the actions of CLA on weight loss in the human.

16.5 IMPLICATIONS FOR HUMAN NUTRITION

Habitual dietary intake of CLA for humans has been estimated to be in the range of 0.1 to 0.4 g/d. Intake of the c9,t11 CLA isomer was determined to be 94.9 ± 40.6 mg/d for 22 free-living Canadians by analyzing two 7-d diet records taken 6 months apart [99]. This level of CLA was significantly correlated to the intake of saturated fat when expressed as per unit of energy consumed [99]. Total CLA (c9,t11 and t10,c12 isomers, at a ratio of about 1.32:1) intake was estimated to be 0.212 ± 0.014 g/d for men and 0.151 ± 0.014 g/d for women (in the western United States) in a survey using food duplicate methodology [100]. In another study of young female college students in Germany, CLA (c9,t11) intake was calculated to be 0.246 g/d by food frequency questionnaires or 0.323 g/d by a 7-d estimated record [101]. In addition,

Fritsche and co-workers [102] estimated the CLA intake to be 0.36 g/d for women and 0.44 g/d for men in Germany.

The published data on CLA and human subjects has increased; however, the evidence for any benefit to health is inconclusive. Moreover, some investigators indicate that the consumption of CLA isomers by humans should not be recommended at present [103]. The possible reasons for the inconsistency among the human studies reviewed herein are likely, but not limited to the following: first, the different CLA isomeric compositions of the supplements; second, the age, gender, and health status of the subjects; third, the duration of the treatments; and fourth, some of the studies were conducted in controlled environments while others utilized free living subjects. In contrast, more convincing evidence of CLAs positive effects on various biological targets in animal and cell culture studies supports the need to continue the work in human subjects. New research should limit the scope, but instead, focus on specific actions and biological targets for successful applications in the human.

The marketing of commercial CLA products has been largely directed toward body weight reduction to address the obesity problem in the United States. However, it is premature to recommend CLA as a means to control obesity based on the limited published research. The future application of CLA and other nutraceutical fatty acids as supplements and in functional food formulations continues to be an important area of public health research [104] and a major effort of product development for the food industry [105].

16.6 SUMMARY

The published human studies described in this chapter were performed on very diverse subjects and tested a variety of CLA products that contained mixtures of various isomers. The inconsistencies in the observed responses are likely attributed to these differences as well as the duration of investigation. The studies using animal models avoid many of these variables, some of which can be controlled and others are unavoidable or inherent to human investigations. Hence, it is difficult to draw definitive conclusions from the human studies.

CLA isomers are a group of unusual unsaturated fatty acids, present in dairy and red meat products and certain vegetable oil supplements that appear to be involved in a variety of biological functions. The biological and physiological effects of CLA isomers reviewed on various health conditions in the human included body weight control, lipid metabolism, diabetes and insulin resistance, immune function, and epidemiologic findings. Although these conditions may at first seem too diverse for a common investigative approach, in many cases a general biological target can be studied to advance the understanding of CLA isomers. An integrative approach to understanding fat accumulation, insulin actions, immune function, and bone biology would involve PPAR, COX, and other specific transcription factors. Therefore, recognizing the potential diverse actions of these nutrients provides an opportunity for collaborative scientific inquiry to systematically study the functions of CLA isomers for improving health. Since CLA isomers, as well as other PUFAs, are recognized as natural PPAR ligands, investigating how these isomers modulate transcription factors and their target genes could lead to elucidating the potential benefits of CLA in humans. Future investigations with CLA should take into account the recognized actions and potential applications of these isomers to specific physiological states and candidate targets for these nutrients. Until this approach is taken in understanding the role of CLA in human biology, there is not enough evidence to pursue human studies to elucidate a mechanism of action.

ACKNOWLEDGMENT

This is supported by a grant from the 21st Century Research and Technology Fund and the Center for Enhancing Foods to Protect Health (www.efph.purdue.edu).

REFERENCES

1. Dietary fats: total fat and fatty acids. Dietary reference intakes — Energy, carbohydrate, fiber, fat, fatty acids, cholesterol, protein, and amino acids. [8], (8-1)-(8-97) (2002). The National Academies Press, Washington, DC.
2. Dietary fats: total fat and fatty acids. Dietary reference intakes — Energy, carbohydrate, fiber, fat, fatty acids, cholesterol, protein, and amino acids. [11], (11-1)-(11-87) (2002). The National Academies Press, Washington, DC.
3. Watkins, B.A. and Li, Y. (2001) Conjugated linoleic acid: the present state of knowledge, in *Handbook of nutraceuticals and functional foods* (Wildman, R.E.C., Ed.), pp. 445–476, CRC Press, Boca Raton, FL.
4. Turek, J.J., Li, Y., Schoenlein, I.A., Allen, K.G.D., and Watkins, B.A. (1998) Modulation of macrophage cytokine production by conjugated linoleic acids is influenced by the dietary n-6:n-3 fatty acid ratio, *J. Nutr. Biochem. 9,* 258–266.
5. Li, Y. and Watkins, B.A. (1998) Conjugated linoleic acids alter bone fatty acid composition and reduce ex vivo prostaglandin E_2 biosynthesis in rats fed n-6 or n-3 fatty acids, *Lipids 33,* 417–425.
6. Li, Y., Seifert, M.F., Ney, D.M., Grahn, M., Grant, A.L., Allen, K.G., and Watkins, B.A. (1999) Dietary conjugated linoleic acids alter serum IGF-I and IGF binding protein concentrations and reduce bone formation in rats fed (n-6) or (n-3) fatty acids, *J. Bone Miner. Res. 14,* 1153–1162.
7. Badinga, L., Selberg, K.T., Dinges, A.C., Corner, C.W., and Miles, R.D. (2003) Dietary conjugated linoleic acid alters hepatic lipid content and fatty acid composition in broiler chickens, *Poult. Sci. 82,* 111–116.
8. Yamasaki, M., Ikeda, A., Oji, M., Tanaka, Y., Hirao, A., Kasai, M., Iwata, T., Tachibana, H., and Yamada, K. (2003) Modulation of body fat and serum leptin levels by dietary conjugated linoleic acid in Sprague Dawley rats fed various fat-level diets, *Nutrition 19,* 30–35.
9. Meadus, W.J., MacInnis, R., and Dugan, M.E. (2002) Prolonged dietary treatment with conjugated linoleic acid stimulates porcine muscle peroxisome proliferator activated receptor gamma and glutamine-fructose aminotransferase gene expression *in vivo, J. Mol. Endocrinol. 28,* 79–86.
10. Kamphuis, M.M.J.W., Lejeune, M.P.G.M., Saris, W.H.M., and Westerterp-Plantenga, M.S. (2003) The effect of conjugated linoleic acid supplementation after weight loss on body weight regain, body composition, and resting metabolic rate in overweight subjects, *Int. J. Obes. 27,* 840–847.

11. von Loeffelholz, C., Kratzsch, J., and Jahreis, G. (2003) Influence of conjugated linoleic acids on body composition and selected serum and endocrine parameters in resistance-trained athletes, *Eur. J. Lipid Sci. Technol. 105,* 251–259.
12. Riserus, U., Smedman, A., Basu, S., and Vessby, B. (2003) CLA and body weight regulation in humans, *Lipids 38,* 133–137.
13. Belury, M.A., Mahon, A., and Banni, S. (2003) The conjugated linoleic acid (CLA) isomer, t10c12-CLA, is inversely associated with changes in body weight and serum leptin in subjects with type 2 diabetes mellitus, *J. Nutr. 133,* 257S-260S.
14. Thom, E., Wadstein, J., and Gudmundsen, O. (2001) Conjugated linoleic acid reduces body fat in healthy exercising humans, *J. Int. Med. Res. 29,* 392–396.
15. Smedman, A. and Vessby, B. (2001) Conjugated linoleic acid supplementation in humans — metabolic effects, *Lipids 36,* 773–781.
16. Blankson, H., Stakkestad, J.A., Fagertun, H., Thom, E., Wadstein, J., and Gudmundsen, O. (2000) Conjugated linoleic acid reduces body fat mass in overweight and obese humans, *J. Nutr. 130,* 2943–2948.
17. Mougios, V., Matsakas, A., Petridou, A., Ring, S., Sagredos, A., Melissopoulou, A., Tsigilis, N., and Nikolaidis, M. (2001) Effect of supplementation with conjugated linoleic acid on human serum lipids and body fat, *J. Nutr. Biochem. 12,* 585–594.
18. Riserus, U., Berglund, L., and Vessby, B. (2001) Conjugated linoleic acid (CLA) reduced abdominal adipose tissue in obese middle-aged men with signs of the metabolic syndrome: a randomised controlled trial, *Int. J. Obes. Relat. Metab. Disord. 25,* 1129–1135.
19. Birnbaum, L. (2001) Addition of conjugated linoleic acid to a herbal anticellulite pill, *Adv. Ther. 18,* 225–229.
20. Kamphuis, M.M., Lejeune, M.P., Saris, W.H., and Westerterp-Plantenga, M.S. (2003) Effect of conjugated linoleic acid supplementation after weight loss on appetite and food intake in overweight subjects, *Eur. J. Clin. Nutr. 57,* 1268–1274.
21. Medina, E.A., Horn, W.F., Keim, N.L., Havel, P.J., Benito, P., Kelley, D.S., Nelson, G.J., and Erickson, K.L. (2000) Conjugated linoleic acid supplementation in humans: effects on circulating leptin concentrations and appetite, *Lipids 35,* 783–788.
22. Petridou, A., Mougios, V., and Sagredos, A. (2003) Supplementation with CLA: isomer incorporation into serum lipids and effect on body fat of women, *Lipids 38,* 805–811.
23. Zambell, K.L., Keim, N.L., Van Loan, M.D., Gale, B., Benito, P., Kelley, D.S., and Nelson, G.J., (2000) Conjugated linoleic acid supplementation in humans: effects on body composition and energy expenditure, *Lipids 35,* 777–782.
24. Kreider, R.B., Ferreira, M.P., Greenwood, M., Wilson, M., and Almada, A.L. (2002) Effects of conjugated linoleic acid supplementation during resistance training on body composition, bone density, strength, and selected hematological markers, *J. Strength Cond. Res. 16,* 325–334.
25. Berven, G., Bye, A., Hals, O., Blankson, H., Fagertun, H., Thom, E., Wadstein, J., and Gudmundsen, O. (2000) Safety of conjugated linoleic acid (CLA) in overweight or obese human volunteers, *Eur. J. Lipid Sci. Technol. 102,* 455–462.
26. Riserus, U., Basu, S., Jovinge, S., Fredrikson, G.N., Arnlov, J., and Vessby, B. (2002) Supplementation with conjugated linoleic acid causes isomer-dependent oxidative stress and elevated C-reactive protein — A potential link to fatty acid-induced insulin resistance, *Circulation 106,* 1925–1929.
27. Basu, S., Smedman, A., and Vessby, B. (2000) Conjugated linoleic acid induces lipid peroxidation in humans, *FEBS Lett. 468,* 33–36.
28. Smedman, A., Vessby, B., and Basu, S. (2004) Isomer-specific effects of conjugated linoleic acid on lipid peroxidation in humans: regulation by alpha-tocopherol and cyclo-oxygenase-2 inhibitor, *Clin. Sci. 106,* 67–73.

29. Basu, S., Riserus, U., Turpeinen, A., and Vessby, B. (2000) Conjugated linoleic acid induces lipid peroxidation in men with abdominal obesity, *Clin. Sci. 99,* 511–516.
30. Riserus, U., Arner, P., Brismar, K., and Vessby, B. (2002) Treatment with dietary trans10cis12 conjugated linoleic acid causes isomer-specific insulin resistance in obese men with the metabolic syndrome, *Diabetes Care 25,* 1516–1521.
31. Noone, E.J., Roche, H.M., Nugent, A.P., and Gibney, M.J. (2002) The effect of dietary supplementation using isomeric blends of conjugated linoleic acid on lipid metabolism in healthy human subjects, *Br. J. Nutr. 88,* 243–251.
32. Masters, N., McGuire, M.A., Beerman, K.A., Dasgupta, N., and McGuire, M.K. (2002) Maternal supplementation with CLA decreases milk fat in humans, *Lipids 37,* 133–138.
33. Benito, P., Nelson, G.J., Kelley, D.S., Bartolini, G., Schmidt, P.C., and Simon, V. (2001) The effect of conjugated linoleic acid on plasma lipoproteins and tissue fatty acid composition in humans, *Lipids 36,* 229–236.
34. Zambell, K.L., Horn, W.F., and Keim, N.L. (2001) Conjugated linoleic acid supplementation in humans: effects on fatty acid and glycerol kinetics, *Lipids 36,* 767–772.
35. Albers, R., van der Wielen, R.P.J., Brink, E.J., Hendriks, H.F.J., Dorovska-Taran, V.N., and Mohede, I.C.M. (2003) Effects of cis-9, trans-11 and trans-10, cis-12 conjugated linoleic acid (CLA) isomers on immune function in healthy men, *Eur. J. Clin. Nutr. 57,* 595–603.
36. Kelley, D.S., Simon, V.A., Taylor, P.C., Rudolph, I.L., Benito, P., Nelson, G.J., Mackey, B.E., and Erickson, K.L. (2001) Dietary supplementation with conjugated linoleic acid increased its concentration in human peripheral blood mononuclear cells, but did not alter their function, *Lipids 36,* 669–674.
37. Kelley, D.S., Taylor, P.C., Rudolph, I.L., Benito, P., Nelson, G.J., Mackey, B.E., and Erickson, K.L. (2000) Dietary conjugated linoleic acid did not alter immune status in young healthy women, *Lipids 35,* 1065–1071.
38. Benito, P., Nelson, G.J., Kelley, D.S., Bartolini, G., Schmidt, P.C., and Simon, V. (2001) The effect of conjugated linoleic acid on platelet function, platelet fatty acid composition, and blood coagulation in humans, *Lipids 36,* 221–227.
39. Voorrips, L.E., Brants, H.A., Kardinaal, A.F., Hiddink, G.J., van den Brandt, P.A., and Goldbohm, R.A. (2002) Intake of conjugated linoleic acid, fat, and other fatty acids in relation to postmenopausal breast cancer: the Netherlands Cohort Study on Diet and Cancer, *Am. J. Clin. Nutr. 76,* 873–882.
40. Chajes, V., Lavillonniere, F., Ferrari, P., Jourdan, M.L., Pinault, M., Maillard, V., Sebedio, J.L., and Bougnoux, P. (2002) Conjugated linoleic acid content in breast adipose tissue is not associated with the relative risk of breast cancer in a population of French patients, *Cancer Epidemiol. Biomarkers Prev. 11,* 672–673.
41. Chajes, V., Lavillonniere, F., Maillard, V., Giraudeau, B., Jourdan, M.L., Sebedio, J.L., and Bougnoux, P. (2003) Conjugated linoleic acid content in breast adipose tissue of breast cancer patients and the risk of metastasis, *Nutr. Cancer 45,* 17–23.
42. Aro, A., Mannisto, S., Salminen, I., Ovaskainen, M.L., Kataja, V., and Uusitupa, M. (2000) Inverse association between dietary and serum conjugated linoleic acid and risk of breast cancer in postmenopausal women, *Nutr. Cancer 38,* 151–157.
43. Emken, E.A., Adlof, R.O., Duval, S., Nelson, G., and Benito, P. (2002) Effect of dietary conjugated linoleic acid (CLA) on metabolism of isotope-labeled oleic, linoleic, and CLA isomers in women, *Lipids 37,* 741–750.
44. Lucchi, L., Banni, S., Melis, M.P., Angioni, E., Carta, G., Casu, V., Rapana, R., Ciuffreda, A., Corongiu, F.P., and Albertazzi, A. (2000) Changes in conjugated linoleic acid and its metabolites in patients with chronic renal failure, *Kidney Int. 58,* 1695–1702.

45. Corino, C., Mourot, J., Magni, S., Pastorelli, G., and Rosi, F. (2002) Influence of dietary conjugated linoleic acid on growth, meat quality, lipogenesis, plasma leptin and physiological variables of lipid metabolism in rabbits, *J. Anim. Sci. 80,* 1020–1028.
46. Rodriguez, E., Ribot, J., and Palou, A. (2002) Trans-10, cis-12, but not cis-9, trans-11 CLA isomer, inhibits brown adipocyte thermogenic capacity, *Am. J. Physiol.— Regul. Integr. Comp. Physiol. 282,* R1789-R1797.
47. Kang, K. and Pariza, M.W. (2001) trans-10, cis-12-Conjugated linoleic acid reduces leptin secretion from 3T3-L1 adipocytes, *Biochem. Biophys. Res. Commun. 287,* 377–382.
48. Watkins, B.A., Li, Y., Romsos, D.R., and Seifert, M.F. (2003) CLA and bone modeling in rats, in *Advances in Conjugated Linoleic Acid Research* (Sebedio, J.L., Christie, W.W., and Adlof, R., Eds.), pp. 218–250, AOCS Press, Champaign, IL.
49. Reseland, J.E. and Gordeladze, J.O. (2002) Role of leptin in bone growth: central player or peripheral supporter? *FEBS Lett. 528,* 40–42.
50. Thomas, T. and Burguera, B. (2002) Is leptin the link between fat and bone mass?, *J. Bone Miner. Res. 17,* 1563–1569.
51. Lee, K.N., Pariza, M.W., and Ntambi, J.M. (1998) Conjugated linoleic acid decreases hepatic stearoyl-CoA desaturase mRNA expression, *Biochem. Biophys. Res. Commun. 248,* 817–821.
52. Park, Y., Storkson, J.M., Albright, K.J., Liu, W., and Pariza, M.W. (1999) Evidence that the *trans*-10, *cis*-12 isomer of conjugated linoleic acid induces body composition changes in mice, *Lipids 34,* 235–241.
53. Choi, Y.J., Kim, Y.C., Han, Y.B., Park, Y., Pariza, M.W., and Ntambi, J.M. (2000) The trans-10, cis-12 isomer of conjugated linoleic acid downregulates stearoyl-CoA desaturase 1 gene expression in 3T3-L1 adipocytes, *J. Nutr. 130,* 1920–1924.
54. Ealey, K.N., El Sohemy, A., and Archer, M.C. (2002) Effects of dietary conjugated linoleic acid on the expression of uncoupling proteins in mice and rats, *Lipids 37,* 853–861.
55. Rahman, S.M., Wang, Y., Yotsumoto, H., Cha, J., Han, S., Inoue, S., and Yanagita, T. (2001) Effects of conjugated linoleic acid on serum leptin concentration, body-fat accumulation, and beta-oxidation of fatty acid in OLETF rats, *Nutrition 17,* 385–390.
56. Yu, Y., Correll, P.H., and Vanden Heuvel, J.P. (2002) Conjugated linoleic acid decreases production of pro-inflammatory products in macrophages: evidence for a PPAR gamma-dependent mechanism, *Biochim. Biophys. Acta 1581,* 89–99.
57. Meade, E.A., McIntyre, T.M., Zimmerman, G.A., and Prescott, S.M. (1999) Peroxisome proliferators enhance cyclooxygenase-2 expression in epithelial cells, *J. Biol. Chem. 274,* 8328–8334.
58. Sugano, M., Tsujita, A., Yamasaki, M., Yamada, K., Ikeda, I., and Kritchevsky, D. (1997) Lymphatic recovery, tissue distribution, and metabolic effects of conjugated linoleic acid in rats, *J. Nutr. Biochem. 8,* 38–43.
59. Bulgarella, J.A., Patton, D., and Bull, A.W. (2001) Modulation of prostaglandin H synthase activity by conjugated linoleic acid (CLA) and specific CLA isomers, *Lipids 36,* 407–412.
60. Liu, K.L. and Belury, M.A. (1998) Conjugated linoleic acid reduces arachidonic acid content and PGE_2 synthesis in murine keratinocytes, *Cancer Lett. 127,* 15–22.
61. Watkins, B.A., Li, Y., Lippman, H.E., Reinwald, S., and Seifert, M.F. (2004) A test of Ockham's razor: implications of conjugated linoleic acid in bone biology, *Am. J. Clin. Nutr. 79,* 1175S-1185S.
62. Ha, Y.L., Grimm, N.K., and Pariza, M.W. (1987) Anticarcinogens from fried ground beef: heat-altered derivatives of linoleic acid, *Carcinogenesis 8,* 1881–1887.

63. Ip, C., Singh, M., Thompson, H.J., and Scimeca, J.A. (1994) Conjugated linoleic acid suppresses mammary carcinogenesis and proliferative activity of the mammary gland in the rat, *Cancer Res. 54,* 1212–1215.
64. Ha, Y.L., Storkson, J., and Pariza, M.W. (1990) Inhibition of benzo(a)pyrene-induced mouse forestomach neoplasia by conjugated dienoic derivatives of linoleic acid, *Cancer Res. 50,* 1097–1101.
65. Ip, C., Chin, S.F., Scimeca, J.A., and Pariza, M.W. (1991) Mammary cancer prevention by conjugated dienoic derivative of linoleic acid, *Cancer Res. 51,* 6118–6124.
66. Ip, C., Briggs, S.P., Haegele, A.D., Thompson, H.J., Storkson, J., and Scimeca, J.A. (1996) The efficacy of conjugated linoleic acid in mammary cancer prevention is independent of the level or type of fat in the diet, *Carcinogenesis 17,* 1045–1050.
67. Ip, C., Jiang, C., Thompson, H.J., and Scimeca, J.A. (1997) Retention of conjugated linoleic acid in the mammary gland is associated with tumor inhibition during the post-initiation phase of carcinogenesis, *Carcinogenesis 18,* 755–759.
68. Liew, C., Schut, H.A., Chin, S.F., Pariza, M.W., and Dashwood, R.H. (1995) Protection of conjugated linoleic acids against 2-amino-3-methylimidazo[4,5-f]quinoline-induced colon carcinogenesis in the F344 rat: a study of inhibitory mechanisms, *Carcinogenesis 16,* 3037–3043.
69. Banni, S., Angioni, E., Casu, V., Melis, M.P., Carta, G., Corongiu, F.P., Thompson, H., and Ip, C. (1999) Decrease in linoleic acid metabolites as a potential mechanism in cancer risk reduction by conjugated linoleic acid, *Carcinogenesis 20,* 1019–1024.
70. Cesano, A., Visonneau, S., Scimeca, J.A., Kritchevsky, D., and Santoli, D. (1998) Opposite effects of linoleic acid and conjugated linoleic acid on human prostatic cancer in SCID mice, *Anticancer Res. 18,* 1429–1434.
71. Cunningham, D.C., Harrison, L.Y., and Shultz, T.D. (1997) Proliferative responses of normal human mammary and MCF-7 breast cancer cells to linoleic acid, conjugated linoleic acid and eicosanoid synthesis inhibitors in culture, *Anticancer Res. 17,* 197–203.
72. Ip, M.M., Masso-Welch, P.A., Shoemaker, S.F., Shea-Eaton, W.K., and Ip, C. (1999) Conjugated linoleic acid inhibits proliferation and induces apoptosis of normal rat mammary epithelial cells in primary culture, *Exp. Cell Res. 250,* 22–34.
73. Lee, K.N., Kritchevsky, D., and Pariza, M.W. (1994) Conjugated linoleic acid and atherosclerosis in rabbits, *Atherosclerosis 108,* 19–25.
74. Nicolosi, R.J., Rogers, E.J., Kritchevsky, D., Scimeca, J.A., and Huth, P.J. (1997) Dietary conjugated linoleic acid reduces plasma lipoproteins and early aortic atherosclerosis in hypercholesterolemic hamsters, *Artery 22,* 266–277.
75. Chin, S.F., Storkson, J.M., Liu, W., Albright, K.J., and Pariza, M.W. (1994) Conjugated linoleic acid (9,11- and 10,12-octadecadienoic acid) is produced in conventional but not germ-free rats fed linoleic acid, *J. Nutr. 124,* 694–701.
76. Cook, M.E., Miller, C.C., Park, Y., and Pariza, M. (1993) Immune modulation by altered nutrient metabolism: nutritional control of immune-induced growth depression, *Poult. Sci. 72,* 1301–1305.
77. Miller, C.C., Park, Y., Pariza, M.W., and Cook, M.E. (1994) Feeding conjugated linoleic acid to animals partially overcomes catabolic responses due to endotoxin injection, *Biochem. Biophys. Res. Commun. 198,* 1107–1112.
78. Sugano, M., Tsujita, A., Yamasaki, M., Noguchi, M., and Yamada, K. (1998) Conjugated linoleic acid modulates tissue levels of chemical mediators and immunoglobulins in rats, *Lipids 33,* 521–527.
79. DeLany, J.P., Blohm, F., Truett, A.A., Scimeca, J.A., and West, D.B. (1999) Conjugated linoleic acid rapidly reduces body fat content in mice without affecting energy intake, *Am. J. Physiol.—Regul. Integr. Comp. Physiol. 45,* R1172-R1179.

80. Yamasaki, M., Mansho, K., Mishima, H., Kasai, M., Sugano, M., Tachibana, H., and Yamada, K. (1999) Dietary effect of conjugated linoleic acid on lipid levels in white adipose tissue of Sprague-Dawley rats, *Biosci. Biotechnol. Biochem. 63,* 1104–1106.
81. Houseknecht, K.L., Heuvel, J.P.V., Moya-Camerena, S.Y., Portocarrero, C.P., Nickel, K.P., and Belury, M.A. (1998) Dietary conjugated linoleic acid normalizes impaired glucose tolerance in the Zucker diabetic fatty fa/fa rat, *Biochem. Biophys. Res. Commun. 244,* 678–682.
82. Thompson, H., Zhu, Z., Banni, S., Darcy, K., Loftus, T., and Ip, C. (1997) Morphological and biochemical status of the mammary gland as influenced by conjugated linoleic acid: implication for a reduction in mammary cancer risk, *Cancer Res. 57,* 5067–5072.
83. Belury, M.A., Nickel, K.P., Bird, C.E., and Wu, Y. (1996) Dietary conjugated linoleic acid modulation of phorbol ester skin tumor promotion, *Nutr. Cancer 26,* 149–157.
84. Schonberg, S. and Krokan, H.E. (1995) The inhibitory effect of conjugated dienoic derivatives (CLA) of linoleic acid on the growth of human tumor cell lines is in part due to increased lipid peroxidation, *Anticancer Res. 15,* 1241–1246.
85. Shultz, T.D., Chew, B.P., and Seaman, W.R. (1992) Differential stimulatory and inhibitory responses of human MCF-7 breast cancer cells to linoleic acid and conjugated linoleic acid in culture, *Anticancer Res. 12,* 2143–2145.
86. Visonneau, S., Cesano, A., Tepper, S.A., Scimeca, J.A., Santoli, D., and Kritchevsky, D. (1997) Conjugated linoleic acid suppresses the growth of human breast adenocarcinoma cells in SCID mice, *Anticancer Res. 17,* 969–973.
87. Liu, J.R., Chen, B.Q., Yang, Y.M., Wang, X.L., Xue, Y.B., Zheng, Y.M., and Liu, R.H. (2002) Effect of apoptosis on gastric adenocarcinoma cell line SGC-7901 induced by cis-9, trans-11-conjugated linoleic acid, *World J. Gastroenterol. 8,* 999–1004.
88. Miller, A., Stanton, C., and Devery, R. (2002) Cis 9, trans 11- and trans 10, cis 12-conjugated linoleic acid isomers induce apoptosis in cultured SW480 cells, *Anticancer Res. 22,* 3879–3887.
89. Kim, E.J., Jun, J.G., Park, H.S., Kim, S.M., Ha, Y.L., and Park, J.H. (2002) Conjugated linoleic acid (CLA) inhibits growth of Caco-2 colon cancer cells: possible mediation by oleamide, *Anticancer Res. 22,* 2193–2197.
90. Palombo, J.D., Ganguly, A., Bistrian, B.R., and Menard, M.P. (2002) The antiproliferative effects of biologically active isomers of conjugated linoleic acid on human colorectal and prostatic cancer cells, *Cancer Lett. 177,* 163–172.
91. Yamasaki, M., Chujo, H., Koga, Y., Oishi, A., Rikimaru, T., Shimada, M., Sugimachi, K., Tachibana, H., and Yamada, K. (2002) Potent cytotoxic effect of the trans10, cis12 isomer of conjugated linoleic acid on rat hepatoma dRLh-84 cells, *Cancer Lett. 188,* 171–180.
92. Moon, E.J., Lee, Y.M., and Kim, K.W. (2003) Anti-angiogenic activity of conjugated linoleic acid on basic fibroblast growth factor-induced angiogenesis, *Oncol. Rep. 10,* 617–621.
93. Thuillier, P., Anchiraico, G.J., Nickel, K.P., Maldve, R.E., Gimenez-Conti, I., Muga, SJ, Liu, K.L., Fischer, S.M., and Belury, M.A. (2000) Activators of peroxisome proliferator-activated receptor-alpha partially inhibit mouse skin tumor promotion, *Mol. Carcinog. 29,* 134–142.
94. Hargrave, K.M., Li, C.L., Meyer, B.J., Kachman, S.D., Hartzell, D.L., Della-Fera, M.A., Miner, J.L., and Baile, C.A. (2002) Adipose depletion and apoptosis induced by trans-10, cis-12 conjugated linoleic acid in mice, *Obes. Res. 10,* 1284–1290.
95. Tsuboyama-Kasaoka, N., Takahashi, M., Tanemura, K., Kim, H.J., Tange, T., Okuyama, H., Kasai, M., Ikemoto, S., and Ezaki, O. (2000) Conjugated linoleic acid supplementation reduces adipose tissue by apoptosis and develops lipodystrophy in mice, *Diabetes 49,* 1534–1542.

96. Akahoshi, A., Goto, Y., Murao, K., Miyazaki, T., Yamasaki, M., Nonaka, M., Yamada, K., and Sugano, M. (2002) Conjugated linoleic acid reduces body fats and cytokine levels of mice, *Biosci. Biotechnol. Biochem. 66,* 916–920.
97. Takahashi, Y., Kushiro, M., Shinohara, K., and Ide, T. (2003) Activity and mRNA levels of enzymes involved in hepatic fatty acid synthesis and oxidation in mice fed conjugated linoleic acid, *Biochim. Biophys. Acta 1631,* 265–273.
98. Warren, J.M., Simon, V.A., Bartolini, G., Erickson, K.L., Mackey, B.E., and Kelley, D.S. (2003) Trans-10, cis-12 CLA increases liver and decreases adipose tissue lipids in mice: possible roles of specific lipid metabolism genes, *Lipids 38,* 497–504.
99. Ens, J.G., Ma, D.W.L., Cole, K.S., Field, C.J., and Clandinin, M.T. (2001) An assessment of c9,t11 linoleic acid intake in a small group of young Canadians, *Nutr. Res. 21,* 955–960.
100. Ritzenthaler, K.L., McGuire, M.K., Falen, R., Shultz, T.D., Dasgupta, N., and McGuire, M.A. (2001) Estimation of conjugated linoleic acid intake by written dietary assessment methodologies underestimates actual intake evaluated by food duplicate methodology, *J. Nutr. 131,* 1548–1554.
101. Fremann, D., Linseisen, J., and Wolfram, G. (2002) Dietary conjugated linoleic acid (CLA) intake assessment and possible biomarkers of CLA intake in young women, *Public Health Nutr. 5,* 73–80.
102. Fritsche, J., Rickert, R., Steinhart, H., Yurawecz, M.P., Mossoba, M.M., Sehat, N., Roach, J.A.G., Kramer, J.K.G., and Ku, Y. (1999) Conjugated linoleic acid (CLA) isomers: formation, analysis, amounts in foods, and dietary intake, *Fett-Lipid 101,* 272–276.
103. Kelley, D.S. and Erickson, K.L. (2003) Modulation of body composition and immune cell functions by conjugated linoleic acid in humans and animal models: benefits vs. risks, *Lipids 38,* 377–386.
104. Exploring a vision: integrating knowledge for food and health. Rouse, T.I. and Davis, D.P. (2004). A workshop summary, National Research Council of the National Academies, National Academy Press, Washington, DC.
105. Camire, M.E., Childs, N., Hasler, C.M., Pike, L.M., Shahidi, F., and Watkins, B.A. Nutraceuticals for health promotion and disease prevention. Issue Paper Number 24, 1–16 (2003). Council for Agricultural Science and Technology.

17 Fortified Foods: A Way to Correct Low Intakes of EPA and DHA

Reto Muggli

CONTENTS

17.1 INTRODUCTION

In the opinion of the general public and also that of some experts, the less fat you eat and the less fat you have on your body, the better. While it is certainly true that the excessive intake of dietary fat contributes to the development of chronic diseases such as obesity, atherosclerosis, and diabetes, it is equally true that without fat, or lipids, for that matter, life would be impossible. Like all nutrients, fat is beneficial in appropriate quantities — and it is harmful to ingest either too little or too much of it.

Dietary fats and oils consist largely of saturated, monounsaturated, and polyunsaturated fatty acids (PUFAs). PUFAs are broadly subdivided into two key families: the omega-6 and the omega-3 family. The key members are linoleic acid and α-linolenic acid, both of which are essential — that is, they are fatty acids that the body is unable to manufacture itself. Inadequate intake of either of these fatty acids can result in deficiency symptoms. The essentiality of the omega-6 family has been recognized for decades, but for the omega-3 family it has been a matter of debate for some time. However, today there is no longer doubt that omega-3 fatty acids are essential. The focus of interest is currently on the omega-3 long-chain PUFAs

(LC-PUFAs), eicosapentaenoic acid (EPA) and docosahexaenoic acid (DHA), which have been reported to have many nutritional benefits, including protection against cardiovascular disease and anti-inflammatory effects. DHA is found in high concentrations within the cells of the brain, retina, and reproductive organs and plays an essential role in the visual and neural development of infants [1].

During the last several hundred years, the human diet has changed considerably in terms of quantity and quality. In particular, the intake of calories in the form of fats has risen dramatically. With the introduction of refined vegetable oils, the daily intake of saturated fatty acids, linoleic acids, and *trans*-fatty acids has risen severalfold, concomitant with lower intakes of oils rich in α-linolenic acid and of omega-3 LC-PUFAs (i.e., EPA and DHA, notably from fish). Today it is generally accepted that there is a growing deficiency in these two essential fatty acids, and it is believed that these dietary changes are partially responsible for the expansion of some of the most prevalent chronic diseases [2].

The most compelling evidence, however, comes from animal and epidemiologic studies and clinical intervention trials, which show that fish oils protect the heart from lethal arrhythmias in the wake of a myocardial infarction. In one recently published study, mortality over a 3.5-year period was down significantly, declining by more than 15%, in patients who had received a daily dose of fish oil containing approximately 1 g of EPA/DHA following myocardial infarction [3]. Because the evidence for the cardioprotective effect of EPA/DHA is so strong and consistent, various authorities and professional societies recommend higher intakes of EPA/DHA, either through fatty fish, fortified foods, or dietary supplements. The British Nutrition Foundation, for example, has recommended a daily intake of 1.25 g of EPA/DHA for adults of normal weight [4]. In 1999, a National Institutes of Health workshop on the essentiality of recommended intakes of omega-6 and omega-3 fatty acids defined the adequate intake for EPA/DHA to be 650 mg daily [5]. In 2002, the American Heart Association (AHA) concluded that omega-3 fatty acids benefit the hearts of healthy people, and those at high risk of — or who have — cardiovascular disease; the AHA recommends eating fish (particularly fatty fish) at least two times a week [6]. A proposal for an unconditional health claim for omega-3 LC-PUFAs and EPA/DHA has been submitted to the Food and Drug Administration.

17.2 WAYS TO CORRECT FOR LOW EPA/DHA INTAKES

The obvious approach to correct low EPA/DHA intakes is the increase of fish consumption. However, fish consumption is notoriously low in Western societies, and increasing the consumption to the recommended levels would involve major dietary changes. In addition, there are large variations in the EPA/DHA content in the various fish species, fresh fish is not always available and often expensive, and many people do not like to eat fish. As a consequence, dietary supplements or fortified foods with EPA/DHA should be considered as dietary alternatives to eating fish. The regular consumption of encapsulated EPA/DHA oils is difficult to maintain over a prolonged period and many people dislike capsules as pharmaceutical application forms. A more natural way to increase intake of omega-3 LC-PUFAs is the enrichment with EPA/DHA of foods that are regularly consumed by the majority of the population.

This approach coincides with an increasing awareness of a relationship among health, diet, and functional foods or nutraceuticals. This group of food products is designed to enhance the nutritional value of general food products while maintaining their sensorial (taste and smell) properties, thus allowing consumers to balance their diet without changing their eating habits. Using food as a vehicle to increase the intake of certain nutrients is not a totally new concept: iodine fortification of salt and iron fortification of flour have been common in many countries for many years.

17.3 PROBLEMS OF FOOD FORTIFICATION

In general, food companies have been relatively slow to respond to the call from the dietary sciences for developing food products rich in EPA/DHA to help correct the low intakes of EPA/DHA. This reluctance is explained in part by the pressure from consumers to reduce fat in general and by the technical difficulties often encountered in the manufacture of EPA/DHA-rich food products. However, with the realization that the quality of dietary fats is as important as quantity, the concept of adding "good" fats to foods is gaining acceptance. Furthermore, with the development of novel refining techniques by the fats and oil industry, EPA/DHA oils of much better quality can be produced than before with considerable less fish taste and odor, making it easier to add these oils to a range of foods without affecting the flavor profile of the product.

Nevertheless, the fortification of food products with EPA/DHA poses certain difficulties and has delayed the introduction of such products in the mass markets. Working with EPA/DHA requires special knowledge and attention, as these fatty acids have intrinsically poor stability. The many double bonds in these fatty acids make them extremely susceptible to oxidation and therefore extremely sensitive to oxygen, heat, and light. Consequently they go rancid very quickly, which can be a barrier to their use as functional nutritional ingredients, particularly when they have been emulsified into various food systems. The problem is aggravated with omega-3 LC-PUFAs, as high secondary oxidation levels are responsible for the fishy taste and smell. Classical tests to assess oil quality, which include acidity, peroxide value, *p*-anisidine value, 2-thiobarbituric acid value, conjugated dienes, and many other physical and chemical measurements, are of limited value in the assessment of the quality of an omega-3 PUFA oil for food fortification purposes. Macfarlane [7] has shown that none of the measured molecules are useful for correlation with a trained panel taste response, and Jacobsen [8] has indicated that there is no correlation between peroxide value and taste panel response in mayonnaise that incorporates LC-PUFAs.

When a headspace of omega-3 PUFAs is created, just about all types of chemistry are possible. Oxidation of omega-3 PUFAs leads to volatile primary oxidation products, which, though odorless and tasteless, are precursors of a wide range of strong oxidation products with a range of off-flavors present at parts per billion levels that classical wet chemical testing and electronic noses fail to detect [7,9]. Among the molecules formed are those that give rise to a fishy smell and taste. Humans are extremely sensitive to small changes in levels of oxidation in EPA/DHA oils. Until now, sensory evaluation of EPA/DHA oils or the food product containing the oil has been the only reliable method of deciding whether the oil is possibly fit for use; but to recruit, train, and maintain an expert taste panel is no easy task and expensive. Equipment of adequate sensitivity for measuring the headspace has become available and techniques developed [10]. Volatiles of static or dynamic (purge and trap) headspace are analyzed by gas chromatography/mass spectrometry against a known internal standard and either specific components or total volatiles are compared with the flavor scores [7]. These techniques offer a way of quantifying the molecules responsible for off-flavors and correlating them with human sensory evaluations.

17.4 SOLUTIONS TO THE PROBLEMS

Fishy smell and taste are the single most important criteria for consumers' nonacceptance of any food item fortified with EPA/DHA, as most people find the flavor and smell of oily fish off-putting and unpalatable. Nutritionists and food developers must realize that these foods should not only be healthy but also taste as good as similar existing products. With the necessary experience, due care, and proper precautions taken, however, it is possible to produce a surprising variety of foodstuffs enriched with EPA/DHA. How can the sensorial properties and oxidative stability of EPA/DHA oils be improved and preserved to make them more widely available for human consumption?

The solution is fourfold: (1) work with high-quality raw materials; (2) remove components that could give rise to a fishy taste and smell; (3) incorporate a powerful antioxidant system into the purified oil and the food products to slow down oxidation and to prevent the build-up of new peroxides; and (4) protect oils and foods from oxygen, heat, and light at all times during storage and manufacture. Coupled with the techniques described above to monitor the sensory quality of the resulting products, these factors provide an avenue for successful inclusion of EPA and DHA in functional foods.

17.4.1 RAW MATERIALS

There are different groups of raw materials that may be used to fortify foods with EPA/DHA and each has its own advantages and limitations. EPA and DHA are found naturally in fish, particularly oily fish, making marine oils a cheap and easily available source of EPA/DHA. However, most current crude fish oils are of very low quality since they are by-products of fishmeal production. Extraction, winterization, and storage of crude marine oils result in extensive oxidative damage. Crude fish oils vary extensively in composition and quality and have high levels of free fatty acids, pigments, polymers, and primary and secondary oxidation products, all of which require extensive reduction by refining. In recent years, the development of algal strains rich in DHA has opened up the possibility to produce DHA oils by large-scale fermentation. A third group of raw material source comprises animal products enriched in EPA/DHA by raising domestic animals with feed high in omega-3 PUFAs [11]. Eggs rich in EPA/DHA are now widely available, and egg yolk or the oil extracted from it can be used to fortify foods. Similarly, it has been shown that with special feeding regimes it is, in principle, possible to produce milk,

beef, and pork high in EPA/DHA. Whether such foods represent economical ingredient sources for the fortification of foods with EPA/DHA remains to be seen. It goes without saying that whatever the EPA/DHA source used for the manufacture of EPA/DHA foods, raw material quality is of key importance. A special word of caution is appropriate regarding the choice of nonmarine oils on the assumption that EPA/DHA sources not derived from fish are more stable and devoid from fishy smell and taste. Oxidative instability is an intrinsic property of all omega-3 oils, and fishy smell and taste molecules are obligatory volatile secondary oxidation products of EPA and DHA, regardless of the origin (be it from fermentation or from fish), as has been clearly shown by direct comparison of a marine oil with a DHA algal oil [12].

17.4.2 Refining

Whatever the raw material source, it is imperative to remove, as far as possible, any contaminants that are known or suspected to pose a safety problem and components that could give rise to fishy smell and taste molecules. A clear maintenance of quality issues is of paramount importance. In 2002, the Omega-3 Working Group of the Council for Responsible Nutrition [13] developed a voluntary monograph that applies to EPA and DHA obtained from fish, plant, and microbial sources of marine algae. The monograph sets forth rigorous and validated methodology for the assay methods measuring the amount of EPA and DHA, as well as quality standards and limits on environmental contaminants. The conventional refining of these oils results in substantial reduction of undesirable components such as heavy metals and pesticide residues. However, the standard oil refining techniques of degumming, neutralizing, bleaching, and deodorization still do not produce oils of acceptable smell and taste. Treatment with silica, molecular distillation, and other more specialized techniques may be required to remove objectionable fishy smell and taste molecules or their precursors. With respect to the latter, residual peroxide values (PV) seem to be of key importance. Let et al. [9] concluded that the initial PV of oils is a critical determinant for the oxidation rates. A PV of 0.1 meq/kg seems to be an upper limit for food-grade omega-3 oils of high quality.

17.4.3 Stabilization

The sensory stability of unprotected omega-3 LC-PUFA oils is very poor. Since the reaction of EPA and DHA produce particularly unpleasant off-flavors, even at very low degrees of oxidation, efficient measures to prevent oxidation is a *conditio sine qua non* for the manufacture of bland and stable EPA/DHA food products.

The stability of omega-3 LC-PUFA oils can be prolonged substantially by the addition of suitable antioxidants. Most commonly used is α-tocopherol (vitamin E) in combination with ascorbyl palmitate. Vitamin E is the most important lipid-soluble antioxidant in nature and appears to be the major free-radical scavenger inside human membranes and in plasma lipids. Tocopherols inhibit lipid peroxidation because they scavenge lipid peroxyl radicals much faster than these can react with adjacent fatty acid side chains or with membrane proteins. The hydroxyl group of the tocopherol gives up its hydrogen atom to the peroxyl radical, converting it to lipid peroxide.

By leaving an unpaired electron on the oxygen this produces a tocopheryl radical. Hence, after breaking the chain reaction of lipid peroxidation, α-tocopherol is itself converted to a radical and consumed during the process. The α-tocopherol radical can be regenerated to tocopherol by ascorbate. However, studies have shown those conventional chain-breaking antioxidants systems, including tocopherols and ascorbic acid, may not provide optimal protection [14,15]. Recently, other natural antioxidants, such as rosemary, thyme, oregano, and sage extract, have been proposed in addition to the tocopherols and ascorbates to improve stability.

The combination of heavy metal ions with traces of hydroperoxides catalyze both the initiation of lipid peroxidation and the decomposition of hydroperoxides. Omega-3 hydroperoxides, in particular, seem susceptible to trace metal-mediated degradation into undesirable volatile components [16]. Metal chelators such as citric acid and ethylenediaminetetraacetic acid (EDTA) have been shown to prevent this cycle and to be very effective inhibitors of oxidative deterioration of food products [17,18].

Microencapsulation or spray drying EPA/DHA oils into powders together with starches or proteins is another way to protect EPA/DHA oils from being rapidly oxidized and from the various forms of stresses imposed during the processing of foods [19]. When correctly formulated, powders can be dispersible in aqueous media and exceptionally stable. These properties, together with a neutral taste, make them attractive for enriching foods of low fat content or foods where a high dispersibility is required, such as in low-fat milk drinks or functional beverages.

17.5 FOOD MANUFACTURE

Despite today's availability of high-quality EPA/DHA oils and powders for various food applications, the manufacture of foodstuffs enriched with EPA/DHA with no off-flavors is not a trivial task. In any food manufacturing process there are a number of factors that need to be addressed when the addition of EPA/DHA oils is considered. As discussed above, oxygen, heavy metals ions, elevated temperatures, and light greatly accelerate PUFA oxidation. As a general rule, exposure to these factors should be kept as low as possible during all steps of food product manufacture, transport, and storage. The observance of the following general dos and don'ts will greatly improve the chances of success (Table 17.1). However, it must be realized that development of an EPA/DHA food must be considered on a case-by-case basis and on the particular product type with expert help from food technologists familiar with the chemistry of PUFAs. A demonstration of how success may depend critically on the combination of new technology and novel process control can be found in the form of pasteurized fresh milk launched in Ireland in the spring of 2004. Thanks to aseptic dosing of the heat-sensitive EPA/DHA to processed milk just before it is packaged in cartons, the fatty acids do not have to be included in the milk pasteurization process, which would destroy their stability. Previously, only ultrahigh temperature (UHT) milk could be fortified with omega-3 LC-PUFAs.

Since added water is a common ingredient of most foods, whenever possible this should be deionized and deaerated before coming in contact with PUFAs. The addition of a metal chelator such as citric acid will help overcome the pro-oxidant effect of residual heavy metal atoms, such as iron and copper. The removal of oxygen

TABLE 17.1
General Handling Rules for Maximizing the Stability and the Flavor Profile of EPA/DHA Oils and Food Products

Critical Factors	General Handling Rules
Oxygen	Work and package under vacuum
	Work and package under nitrogen
	Use airtight storage material
Heavy metals (Cu, Fe)	Choose high-quality raw materials low in heavy metals
	Minimize contact with metal equipments and containers
	Add chelating agents
Peroxides	Choose raw materials low in peroxides
	Remove peroxides with novel refining technology
	Use a combination of antioxidants
Temperature	Work and store at the lowest temperature and for shortest time possible
Light	Use opaque equipment and packaging material

is particularly important in low-fat products such as margarines and spreads containing emulsifiers, thickeners, and flavoring agents. In all cases, additional antioxidants will be required to protect the product and provide the desired shelf life. Typically tocopherols, ascorbyl palmitate, or sodium ascorbate are used. The exact nature and amounts required will depend upon the individual product concerned, the fat content and overall composition of the fat blend, and the shelf life needed. One has to be extremely careful in the choice and the final concentration of the antioxidant combination. Overdosing may turn the antioxidant into a pro-oxidant, and possible "oversynergistic" effects between added antioxidants and those naturally present in other food components (e.g., flavonoids, catechins, carotenes, and tocopherols) must be considered. On the other hand, as the consumption of EPA/DHA foods increases the physiological requirements of vitamin E as a function of the fortification levels, a compensation, preferably in the form of tocopheryl acetate must therefore be applied [20].

Foods with a high fat content, such as margarines, spreads, and salad oils, lend themselves most easily to the addition of EPA/DHA oils. Oily forms can be added to products just like any other fat, most preferably as close as possible to the end of the manufacturing process to limit the time of exposure of the oils to oxygen, heat, and light.

Another food group that contains appreciable amounts of fat is milks and yogurts. Milk has been shown to be an efficient carrier for EPA/DHA, most probably because of the highly dispersed milk lipid micelles into which omega-3 oils can diffuse. Milk varies considerably in composition depending on the forage of the dairy cattle and is beset with all the difficulties when working with emulsions. Milk trace metals, notably iron and copper, and components of the milk fat globule membrane promote oxidation of milk fat and development of off-flavors even in the absence of added omega-3 LC-PUFAs [21]. Conversely, casein, citric acid, and

TABLE 17.2
Intakes of EPA/DHA through Fortified Foods

		Fortification Levels with a Novel Refined Fish Oil of 25% EPA + DHA								
	Serving size	**0.2%**	**0.5%**	**1.0%**	**1.5%**	**2.0%**	**2.5%**	**3.0%**	**3.5%**	**4.0%**
Margarine	25 g	12 mg	31 mg	62 mg	94 mg	**125 mg**	**157 mg**	**187 mg**		
Bread	100 g	50 mg	**125 mg**							
Cake	50 g	25 mg	62 mg	**125 mg**						
Cereal bar	35 g	17 mg	44 mg	87 mg						
Cookies/crackers	30 g	15 mg	38 mg	75 mg						
Milk	300 ml	**150 mg**								
Yogurt	180 g	113 mg	**225 mg**							
Beverage (100% juice)	300 ml	**150 mg**								
Sauce, marinara	125 g	62 mg	**156 mg**							
Salad oil	30 g	15 mg	38 mg	75 mg						

Fortification levels that provide minimum intakes of 125 mg EPA/DHA per serving (about 20% of the NIH Workshop recommendation, 1999 [5]), are bolded. Shaded area (blank) represents fortification levels that are currently not feasible or not recommended.

lactoferrin are stabilizing milk components by virtue of their well-recognized efficient metal chelating properties [22,23].

The manufacture of food products that involve an emulsification step is particularly complex. Emulsified food systems with EPA/DHA show wide variations in oxidative stability mainly due to the large surface area of the oil droplets [24], the rheological stress incurred, and the likelihood of air being introduced during homogenization or emulsification. Whenever homogenization or emulsification are required, it is recommended that the process be carried out under either a vacuum or an inert gas blanket to minimize exposure to oxygen.

In multiphase food mixtures and lipid emulsions, antioxidants do not necessarily provide the same protection as in bulk oils (the so-called polarity paradox). What may work in oil may not necessarily work in an emulsion, as the partitioning between the aqueous and lipid phases, the physical properties of the components, and the nature of the interfaces are important factors that affect antioxidant activity and are difficult to reproduce [16,25,26]. In one example, DHA single-cell oil supplemented with 1937 ppm of mixed tocopherols lost its high stability completely in the corresponding emulsion [12]. On the other hand, EDTA seems to be particularly suited as an inhibitor of lipid oxidation in emulsion systems by chelation of iron [12,17].

In liquid applications, the low viscosity presents a particular challenge to food technologists for the incorporation of EPA/DHA oils, and EPA/DHA oils often need to be mixed with emulsifiers or density modifiers to prevent separation and ringing. Powders may also be easier to incorporate into beverages with heavy suspension, where the cloud hides any impact of the nutritional oil, but here the dispersibility and leakage of oil from the powder are critical. Well-packaged products with short shelf lives also lessen formulation challenges. EPA/DHA-fortified products should not be stored in transparent packing since light can cause them to oxidize and metal packing should also be avoided to minimize oxidation. However, even with all precautions taken, there is a very small window open between the amounts needed for a significant enhancement of the nutritional value and the achievable fortification levels without affecting the flavor and taste profile of the food product (Table 17.2).

17.6 PRODUCTS ON THE MARKET

Increasing consumer awareness of the health benefits of EPA and DHA has generated a wealth of opportunities for their use in functional foods. Many food and beverage companies with functional food programs consider EPA and DHA as important food ingredients. Food technology has now advanced to the point where, in theory, any food can be enriched with EPA and DHA [27]. At the forefront of developments are infant formulas and baby follow-on food in Europe and the Far East. Over the last couple of years, a surprisingly large number of EPA/DHA-fortified food products have entered the market, ranging from meat, eggs, and dairy products to cereals, cereal bars, and infant formulas (Table 17.3). Breads, low-fat spreads, and milk for the delivery of EPA/DHA have entered the mainstream in Europe, are rapidly gaining acceptance elsewhere, and are appearing on the shelves of supermarkets and groceries.

TABLE 17.3
A Selection of EPA/DHA Foods Marketed in Different Countries

	Bread and Bakery Products
Germany	Omega-3 bread and rolls (VK Mühlen, Hamburg); *Diamant Vital*, bread mix (Diamant Mühle); *Wellness Aktifit*, bread (Ruf Lebensmittelwerke)
Portugal	*Tostagrill Saude*, toast bread (Diatosta)
Spain	*Cuetara F-plus*, biscuits (Cuetara), *Galleta Omega 2*, biscuits (Grupo Siro)
Sweden	*Leva*, bread (Pågen)
United Kingdom	*Nimble Heartbeat*, bread (British Bakeries); *Heartwatch Omega*; sliced bread, sponge cakes, biscuit varieties (Functional Nutrition)
South Africa	*Richbake Omega Brown*, bread mix (Credin Bakery)
Australia	*Tip Top*, sliced bread (TipTop Bakeries)
Malaysia	*Sweetie*, bread (Today Bakery)
Vietnam	*Marie*, biscuits (Kinh Do Bakery)
Central America	*Club*, crackers (Pascual)
	Margarines and Spreads
Belgium	*Vitelma*, margarine (Vandemoortele)
Finland	*Lätt & Lagom*, margarine (Arla Foods)
France	*Primevère*, margarine (Céma)
Germany	*Vitaquell Omega-3*, fat spread (Vitaquell)
Scandinavia	*Gaio*, fat spread (MD Foods)
United Kingdom	*Heartwatch Omega*, reduced fat spread (Functional Nutrition)
United States	*Smart Balance OmegaPlus*, margarine (GFA)
Australia	*Sea Change*, margarine (Caines)
Chile	*Dos Alamos*, margarine (Grasco)
	Eggs and Egg Products
Finland	*Minicol Omega*, pasteurized eggs (Wammala Food)
Germany	*Eiplus*, eggs (Eifrisch)
Netherlands	*Columbus*, eggs (Belovo)
Italy	*Sereno*, eggs (Maia Agroalimentare); *Oro*, eggs (Unione Cascine Valpadana); *OvitoPiù*, eggs (Fattorie del Garda)
United States	*Gold Circle Farm*, eggs (OmegaTech & NutraSweet Kelco); *Liquid Eggs* (Burnbrae Farms); *Eggs Plus* (Pilgrim's Pride)
	Pasta
United Kingdom	*Heartwatch Omega*, pasta spirals and crests (Functional Nutrition)
	Milk and Dairy Products
France	*Candia Omega*, milk (Candia)
Ireland	*Dawn Omega Milk* (Dawn Dairies)
Italy	*Plus Omega-3*, UHT milk (Parmalat)
Portugal	*Especial Omega-3*, milk (Mimosa)
Spain	*Omega-3*, skimmed milk and yogurt (Puleva); *Candia aux Omega 3* and *Lauki Omega +*, UHT milk (Candia)

TABLE 17.3(Continued) A Selection of EPA/DHA Foods Marketed in Different Countries

Country	Products
Australia	*Farmers Best with Omega 3*, skimmed milk (Dairy Farmers); *Heart Plus*, skimmed milk (Peter & Brownes)
Brazil	*Vitalat*, UHT semi-skimmed milk (Parmalat)
Chile	*Margarina Belmont Omega-3*, margarine (Watt's Alimentos)
Colombia	*Omega-3*, milk (Parmalat)
Peru	*Gloria Omega-3:6*, milk (Gloria)
	Juice and Soft Drinks
Germany	*Riobella*, fruit drink (Kirberg)
United Kingdom	*Bertrams Exclusive Omega*, fruit juice (Bertrams)
	Meat, Poultry, and Seafood Products
Finland	*Atria Turkey Luncheon*, meat slices (Atria)
Spain	*Omega-3 Jamon Cocido*, cooked ham; *Terra y Mar*, turkey breast (Carnicas Serrano)
Israel	*Mega Off*, frozen chicken pieces (Off Tene)
United States	Line of franks, burgers, sausages based on salmon and tuna (AquaCuisine); surimi (Wegman's); seafood salad and crab surimi (Shining Ocean); crab surimi (Nichirei Trident); salmon chowder (Tabatchnik)
Philippines	Hotdog (Pure Foods)
Australia	Range of meat products (Hans Small Goods)
	Snacks
United States	*OmegaZone*, nutrition bar (Sears Labs); *PediaSure*, child nutrition bar (Abbott Laboratories)
	Term Infant Formula
Europe	*Aptamil First, Aptamil Extra* and *Aptamil Forward* (Milupa)
United States	*Isomil Advance* (Ross Products); *Enfamil Lipil* (Mead Johnson); *Good Star Supreme DHA and ARA* (Nestlé)
Asia Pacific	*Enfalac A+ with DHA/AA, Enfapro with DHA, Enfakid with DHA* (Bristol-Myers Squibb); *S26 Gold with DHA/AA, Promil Gold with DHA/AA, Progress Gold with DHA/AA* (Wyeth); *Gain Plus* (Abbott)
Indonesia	*SGM 1, 2, and 3 with DHA* (Sari Husada); *Chilmil and Chilkid* (Sanghiang Perkasa/Morinaga)
	Preterm Infant Formulas
United Kingdom	*Osterprem* (Farley's)
Belgium	*Premilon* (Nutricia)
United States	*Similac NeoSure Advance* and *Similac Natural Care Advance* (Ross Products); *Enfamil Premature Lipil* (Mead Johnson)
	Baby Food
United States	*First Advantage* (Beech-Nut)
Venezuela	Chicken soup with strained vegetables (Heinz)

17.7 DO EPA/DHA FOODS WORK?

The direct clinical evidence for a health benefit of EPA/DHA comes from studies with natural or concentrated fish or single-cell DHA oils. It may be legitimate to ask whether these effects can be reproduced with food products enriched with EPA/DHA. A number of studies have addressed this question and provided direct proof for a health beneficial effect of commercially available EPA/DHA food products. In healthy humans, skimmed milk supplemented with fish oil reduced the levels of total LDL cholesterol, homocysteine, and endothelial adhesion molecules — all accepted biochemical risk factors for cardiovascular disease — or conversely increased HDL cholesterol (the "healthy" form of cholesterol) concentrations [28,29]. Similarly, bread containing 0.5 g of omega-3 fatty acids in the form of fish oil lowered plasma triglyceride and increased HDL-cholesterol levels in subjects with hyperlipidemia [30]. It is unquestionable that EPA/DHA foods lead to a desirable modification of cardiovascular disease risk factors, which, hence, offer an alternative to supplements or a change of dietary habits. Despite the technical issues of incorporating EPA/DHA into foodstuffs, there is no doubt that this concept offers food processors the opportunity to introduce a new range of foods associated with definite health benefits that will enjoy the support of the scientific community as helping in disease prevention. Sooner or later, EPA/DHA will become important and widely accepted ingredients in foods.

REFERENCES

1. Connor, W.E., Importance of n-3 fatty acids in health and disease, *Am. J. Clin. Nutr.*, 71, 171S, 2000.
2. Simopoulos, A.P., Essential fatty acids in health and chronic disease, *Am. J. Clin. Nutr.*, 70, 560S, 1999.
3. GISSI-Prevenzione Investigators (Gruppo Italiano per lo Studio della Sopravvivenza nell'Infarto miocardico), Dietary supplementation with n-3 polyunsaturated fatty acids and vitamin E after myocardial infarction: results of the GISSI-Prevenzione trial, *Lancet*, 354, 447, 1999.
4. The British Nutrition Foundation, *Unsaturated fatty acids: nutritional and physiological significance*, Task Force Report, 1992.
5. Simopoulos, A.P., Leaf, A., and Salem, N., Workshop on the essentiality of and recommended dietary intakes for omega-6 and omega-3 fatty acids, *J. Am. Coll. Nutr.*, 18, 487, 1999.
6. American Heart Association, AHA Scientific Statement: fish consumption, fish oil, omega-3 fatty acids and cardiovascular disease, *Circulation, 106*, 2747, 2002.
7. Macfarlane, N., The FAST Index — a fishy scale, *Inform*, 12, 244, 2001.
8. Jacobsen, C., Sensory impact of lipid oxidation in complex food systems, *Fett/Lipid*, 12, 484, 1999.
9. Let, M.B. et al., Oxidative flavour deterioration of fish oil enriched milk, *Eur. J. Lipid Sci. Technol.*, 105, 518, 2003.
10. American Oil Chemists' Society, Correlation of oil volatiles with flavor scores of edible oils, AOCS Recommended Practice Cg 1-83 reapproved, 1997.

11. Jacobsen, K., Dietary modifications of animal fats: status and future perspectives, *Fett/Lipid,* 12, 475, 1999.
12. Frankel, E.N. et al., Oxidative stability of fish and algae oils containing long-chain polyunsaturated fatty acids in bulk and in oil-in-water emulsions, *J. Agric. Food Chem.,* 50, 2094, 2002.
13. CRN Omega-3 Working Group, Voluntary monograph on long-chain omega-3 EPA and DHA, 2002.
14. Jacobsen, C., Adler-Nissen, J, and Meyer, A.S., Effect of ascorbic acid on iron release from the emulsifier interface and on the oxidative flavor deterioration in fish oil enriched mayonnaise, *J. Agric. Food Chem.,* 47, 4917, 1999.
15. Jacobsen, C. et al., Oxidation in fish-oil-enriched mayonnaise. 2. Assessment of the efficacy of different tocopherol antioxidant systems by discriminant partial least squares regression analysis, *Eur. Food Res. Technol.,* 210, 242, 2000.
16. Frankel, E.N., *Lipid Oxidation,* The Oily Press, Dundee, UK, 1998.
17. Jacobsen, C. et al., Lipid oxidation in fish oil enriched mayonnaise: calcium disodium ethylenediaminetetraacetate, but not gallic acid, strongly inhibited oxidative deterioration, *J. Agric. Food Chem.,* 49, 1009, 2001.
18. Fomuso, L.B., Corredig, M., and Akoh, C.C., Metal-catalyzed oxidation of a structured lipid model emulsion, *J. Agric. Food Chem.,* 50, 7114, 2002.
19. Schrooyen, P.M., van der Mer, R., and De Kruif, C.G., Microencapsulation: its application in nutrition, *Proc. Nutr. Soc.,* 60, 475, 2001.
20. Muggli, R. Physiological requirements of vitamin E as a function of the amount and type of polyunsaturated fatty acid, *World Rev. Nutr.,* 75, 166, 1994.
21. King, R.L. and Dunkley, W.L., Relation of natural copper in milk to incidence of spontaneous oxidised flavour, *J. Dairy Res.,* 42, 420, 1959.
22. Allen, J.C. and Wrieden, W.L., Influence of milk proteins on lipid oxidation in aqueous emulsion. I. Casein, whey protein and α-lactalbumin, *J. Dairy Sci.,* 49, 239, 1982.
23. Fox, P.F. and McSweeney, P.L.H., *Dairy Chemistry and Biochemistry.* Blackie Academic & Professional, London, 1998.
24. Gothani, S. et al., Effect of droplet size on oxidation of docosahexaenoic acid in emulsion system, *J. Disper. Sci. Technol.,* 20, 1319, 1999.
25. McClements, D.J. and Decker, E.A., Lipid oxidation in oil-in-water emulsions: impact of molecular environment on chemical reaction in heterogeneous food systems, *J. Food Sci.,* 65, 1270, 2000.
26. Schwarz, K. et al., Activities of antioxidants are affected by colloidal properties of oil-in-water and water-in-oil emulsions and bulk oils, *J. Agric. Food Chem.,* 48, 4874, 2000.
27. Trautwein, E.A., n-3 Fatty acids — physiological and technical aspects for their use in food, *Eur. J. Lipid Sci. Technol.,* 103, 45, 2001.
28. Baró, L. et al., n-3 Fatty acids plus oleic acid and vitamin supplemented milk consumption reduces total and LDL cholesterol, homocysteine and levels of endothelial adhesion molecules in healthy humans, *Clin. Nutr.,* 22, 175, 2003.
29. Visioli, F. et al., Very low intakes of n-3 fatty acids incorporated into bovine milk reduce plasma triacylglycerol and increase HDL-cholesterol concentrations in healthy subjects, *Pharmacol. Res.,* 41, 571, 2000.
30. Liu, M., Wallin, R., and Saldeen, T., Effect of bread containing stable fish oil on plasma phospholipid fatty acids, triglycerides, HDL-cholesterol, and malondialdehyde in subjects with hyperlipidemia, *Nutr. Res.,* 21, 1403, 2001.

18 Phytosterols and Human Health

Vivienne V. Yankah

CONTENTS

18.1 INTRODUCTION

Increased plasma levels of low-density lipoprotein (LDL) cholesterol in humans constitute the main risk factor for cardiovascular disease (CVD), the leading cause of death in most developed countries. Dietary and pharmaceutical interventions have been used to control human cholesterol levels. The challenges encountered stem from the fact that the body has a natural tendency for the continual *de novo* synthesis of cholesterol.

Every cell in the body contains cholesterol and has a ready access to a large extracellular supply. In addition, normal diets can make major contributions to the body cholesterol content. Excess cholesterol predisposes people to severe pathology.

Dietary and nutritional guidelines for prevention of CVD are aimed at promoting the consumption of lower cholesterol diets and foods with natural hypocholesterolemic effects. There is currently intense research interest in the dietary sources of secondary plant metabolites because of their potential preventive effects on the chronic diseases of Western societies, especially CVD and cancer. Phytosterols (plant sterols) in the diet are poorly absorbed and interfere with cholesterol absorption. The effect of phytosterols in diet on modifying the risk of certain diseases has been known for centuries. Phytosterols in foods were identified as natural cholesterol-lowering agents in the 1950s [1]; however, the development and convenience of using pharmaceutical drugs in that era made drugs preferable as cholesterol-lowering therapies compared with phytosterols. Through scientific research, the understanding of the relationships among phytosterols, physiological function, and disease has been the focus of both food and pharmaceutical industries seeking to provide the consumer with the normal foods they enjoy eating daily, but with a health-promoting component.

18.2 DESCRIPTION OF PHYTOSTEROLS

Phytosterols encompass the entire class of sterols found in membrane extracts of the unsaponifiable fat-soluble fractions of plants, including algae. These include pure sterols, stanols, their phenolic acids, and conjugated glucosides. Common known sources are vegetables, wood, and vegetable oils. Phytosterols are members of the triterpene family. All triterpenes are synthesized via a pathway that starts with reduction of 3-hydroxy-3-methylglutaryl coenzyme A (HMG-CoA) (six carbons) to mevalonate (five carbons). Six mevalonate units are then assembled into two farnesyl diphosphate molecules, which are combined to make squalene (30 carbons or "three terpenes"). Enzymatic ring closure steps then form cycloartenol (also 30 carbons), and additional enzymatic reactions form common plant triterpenes such as phytosterols, triterpene alcohols, and brassinosteroids [2].

18.2.1 Differences and Similarities of Phytosterols to Cholesterol Structure And Biosynthesis

Phytosterols are C-28 or C-29 sterols, structurally very similar to cholesterol (C-27), but differ in their nucleus and/or side chain configuration or polar groups (Figure 18.1). Common substitutions include methyl, ethyl, or double bond at the C-24 position in the sterol side chain. β-Sitosterol contains an ethyl group at C-24, campesterol has a methyl group at C-24, and stigmasterol is considered to be an unsaturated phytosterol because of the double bond at C-22. Chemical saturation of the -5 double bond leads to the 5-α position of the hydrogen atom (i.e., cholestanol, campestanol, sitostanol), while enzymatic transformation by gut bacteria leads to

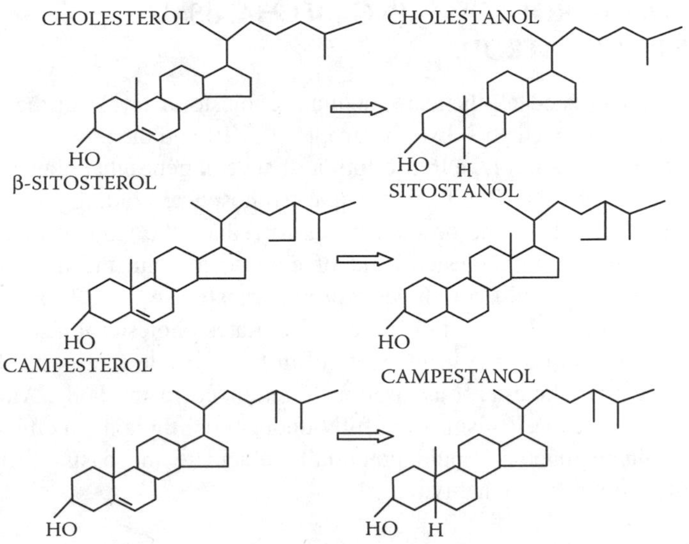

CHOLESTEROL
CHOLESTANOL
HO
HO H
β-SITOSTEROL
SITOSTANOL
HO
HO
CAMPESTEROL
CAMPESTANOL
HO
HO H

FIGURE 18.1 Structures of sterols and their saturated derivatives.

epimerization of the H atom at position 5 to the β configuration as represented by coprostanol and its derivatives [3].

Cholesterol is the predominant sterol in animals. Plant membranes contain little or no cholesterol, and they are composed of several types of phytosterols. Both cholesterol and phytosterols have structural functions. The function of phytosterols in plants is primarily related to their ability to affect membrane fluidity and water permeability [4]. Phytosterols are also thought to stabilize plant membranes and play a role in membrane rigidification based on existing sterol/phospholipid ratios. The primary role of cholesterol in the body appears to be as a constituent of membranes. It is also the precursor for synthesis of steroids and bile acids. Free cholesterol serves as a stabilizer for cell membranes and cholesteryl fatty acid esters are the storage/transport form, usually associated with triacylglycerols. Both cholesterol and phytosterol synthesis have been known to be controlled by the enzyme HMG-CoA reductase. However, recent findings on the Pojak and Rohmer shunts debate the essential role of HMG-CoA reductase as a rate-determining enzyme for phytosterol synthesis in plants as well as cholesterol biosynthesis in animals [5].

Individual phytosterols differ in their effect on membrane stability. Stigmasterol has been reported to cause disorder in membranes, and the molar ratio of stigmasterol to other phytosterols in the plasma membrane increases during senescence [6]. Phytosterols also act as plant hormones and hormonal precursors [7]. The most frequent phytosterols in nature, and thus in human diets, include sitosterol, campesterol, stigmasterol (4-desmethylsterols of the cholestane series), and dihydrobrassicasterol — the latter unique to the Brassica species (canola). The typical Western daily diet contains 100 to 300 mg phytosterols and 20 to 50 mg plant stanols [8].

18.3 CHOLESTEROL-LOWERING MECHANISM OF PHYTOSTEROLS

A series of collaborated mechanisms regulate cholesterol levels in the body. The two nuclear hormone receptors, liver X receptor (LXR) and the farnesoid X receptor (FXR), are known to control the transcription of several genes that play a major role in cholesterol metabolism [9,10]. Adenosine triphosphate-binding cassette (ABC) transporter is encoded by one of such genes to reduce the amount of cholesterol absorbed by transporting cholesterol out of enterocytes into the intestinal lumen [27,28]. The enzyme Acyl CoA:cholesterol acyltransferase (ACAT) also catalyzes the esterification of cholesterol in cells and facilitates cholesterol absorption [21]. The physicochemical presence of phytosterol in the diet interferes with the normal processes by which cholesterol absorption is regulated in the body. Although the hypocholesterolemic mechanism is not fully understood, three main effects of plant sterol metabolism, together, could potentially elucidate the basis of the proven hypocholesterolemic functionality:

- Phytosterol effects on cholesterol absorption
- Phytosterol effects in the intestinal lumen
- Phytosterol effects on intracellular cholesterol pool

18.3.1 Phytosterol Effects on Cholesterol Absorption

Due to their physical structure, chemical nature, and biological activities, phytosterols reduce cholesterol absorption and increase cholesterol excretion with no effects on high-density lipoprotein (HDL) or triglyceride (TG) levels. The serum cholesterol-lowering effects of phytosterols are believed to be associated with an inhibition of cholesterol absorption resulting from the higher affinity of phytosterols than of cholesterol for micelles [11]. Phytosterols appear to decrease the solubility of cholesterol in the oil and micellar phases (the aggregation of molecules with both lyophilic and hydrophilic groupings at the interphase), thereby displacing cholesterol from bile salt micelles and interfering with its absorption [12]. In situations where the concentration of β-sitosterol in mixed micellar solutions is maintained similar to that of cholesterol, phytosterols are absorbed at rates equal to cholesterol [13].

A cholesterol molecule with a methyl or ethyl group (phytosterols) results in poor intestinal absorption of phytosterols in humans. The longer the side chain, the less it is absorbed because of increased hydrophobicity [14] relative to cholesterol, the base molecule. This increases the ability of phytosterols to mix with micelles in competition against cholesterol, thereby blocking cholesterol absorption into the body. Studies have suggested that the mechanism that accounts for inhibition of cholesterol absorption is different from that controlling the absorption of phytosterols from mixed micelles. Specifically, sitosterol [12,13] and fucosterol [12] displace cholesterol from micelle solution, accounting for the inhibition of its absorption, but micellar solubilization does not ensure absorption since sitosterol in not absorbed even where fully micellar solubilized [15]. About 5% or less of phytosterol is absorbed from the human intestine compared with 30 to 60% for dietary cholesterol [7]. It has been reported that the different absorption rates of cholesterol vs. phytosterols may be due to differences in the rates of esterification before incorporation into chylomicrons [16]. About 70 to 80% of the absorbed cholesterol is esterified, compared with 12% for phytosterols [15]. Animal and human studies, as well as *in vitro* experiments, indicate an inverse relationship between the absorbability of plant sterols and their efficiency in inhibiting cholesterol absorption [17]. In animals, it has been shown that the absorption rate of campesterol is considerably higher than that of sitosterol and stigmasterol [18]. The carrier for absorbed phytosterols circulating in human plasma is LDL, and HDL is the carrier in rats [21].

The efficacy of phytosterols as cholesterol-lowering agents has been compared among phytosterols differing in saturation or as free and esterified forms. Hydrogenation of the 5-α nucleus double bond leads to a decrease in sterol absorbability, as has been demonstrated for cholesterol/cholestanol [19] and sitosterol/sitostanol [17]. The more saturated equivalents of phytosterols, namely, the phytostanols/plant stanols, are almost unabsorbable. Hence, it is proposed that phytostanols are more effective as cholesterol-lowering agents than their sterol counterparts, although this has been disputed by other studies.

About 20% of the absorbed β-sitosterol is converted to cholic and chenodeoxycholic acids. The remainder is excreted in bile as free sterol and this excretion is more rapid than that of cholesterol [20]. While studies with humans have supported the formation of cholic and chenodeoxycholic acid from labeled sitosterol, rat studies could not confirm the conversion of phytosterols to bile acids. Unabsorbed phytosterols may undergo bacterial transformation by intestinal microflora to produce metabolites as coprostanol and coprostanone detected in fecal excreta [21].

18.3.2 Phytosterol Effects in the Intestinal Lumen

The small intestine is believed to be the major site for cholesterol absorption. The phytosterol effects at the absorption sites of the intestinal lumen could therefore be the location for distinguishing between the absorbable and nonabsorbable phytosterols. Intestinal esterification of sterols by ACAT is reported to be more efficient for cholesterol than for phytosterols. Esterification rate is reported at 89, 79, and 34% for campesterol, sitosterol, and stigmasterol, respectively, compared with cholesterol [21]. Slower rate of transfer of sitosterol from the cell surface to intracellular site, compared with cholesterol, may also contribute to the lower absorption rate for plant sterol [21]. Cocrystallization of phytosterols and cholesterol in the gastrointestinal tract could lead to the reduction of intestinal cholesterol uptake. This mechanism is supported by reports on formation of mixed cholesterol:β-sitosterol crystals from a mixed solution of cholesterol and β-sitosterol and formation of solids/mixed crystals from mixed cholesterol:β-sitosterol after solvent evaporation [22].

Specific phytosterols may increase bile acid excretion [23] and hinder cholesterol esterification rate in the intestinal mucosa *in vitro* [24]. *In vivo* studies indicate that phytosterols may modify hepatic acetyl-CoA carboxylase [25] and cholesterol-7α-hydroxylase enzyme activities [26] in animals and humans. The expression of ABC in the small intestine has been reported to affect the efflux of phytosterols from enterocytes back out into the intestinal lumen, thereby limiting their absorption [27,28].

18.3.3 Effects on Intracellular Cholesterol Pool

Hepatic cholesterol content is the major regulator of the plasma cholesterol level. The hepatic availability of cholesterol is determined by intestinal absorption and by synthesis and degradation in the liver. Because all cells in the body have the ability to synthesize cholesterol *de novo*, this natural feature of cells could neutralize the phytosterol-induced cholesterol reduction by enhancing cholesterol synthesis. Several reports have discussed the extent to which cholesterogenesis offsets the hypocholesterolemic effects of phytosterols. Studies with the enzyme HMG-CoA reductase show that phytosterols have less influence on HMG-CoA activity compared with cholesterol [29]. Therefore, the hypocholesterolemic mechanism of phytosterols may not involve suppression of cholesterol synthesis. Different interpretations of the effect of phytosterols on cholesterogenesis have been reported. The principles underlying the methods used to assess cholesterogenesis vary, including methods using stable isotope tracers and endogenous pathway precursors; this could be a factor for the observed differences in the trend of cholesterol synthesis measured in the studies. β-sitosterol has been shown to inhibit cholesterol synthesis [30] in animals. A sitostanol-rich phytosterol mixture did not result in significant changes in endogenous cholesterol synthesis in hypercholesterolemic men [32]. The hypocholesterolemic mechanism may [7,31,57] or may not [32,64] result in elevation in endogenous cholesterol synthesis. Decreases in the serum concentration of cholesterol were dependent on the dose of ingested cholesterol, and the fractional synthesis rate of cholesterol was highest at the maximal dose of phytosterol feeding. Generally, the reported observations suggest that any compensatory cholesterol synthesis is offset by the hypocholesterolemic effects of phytosterols if the recommended dose of phytosterol is ingested.

18.4 NUTRITIONAL STUDIES WITH PHYTOSTEROLS

18.4.1 Effective Dietary Phytosterol Application Levels

The amounts of plant sterols present in ordinary American diets are not sufficient to block the absorption of cholesterol by competitive blockade of absorptive sites. On average, humans consume about 200 to 300 mg/d [33], with intakes in vegetarians and Japanese estimated to be higher, 300 to 500 mg/d [34]. Prehistoric diets had intakes of phytosterols at levels higher than 1 g/d.

Earlier reports on the therapeutic use of phytosterols for treatment of hypercholesterolemia suggested cholesterol-lowering efficacy only at high doses of 10 to 20 g/d; thus, the application of less absorbable sterols for therapeutic use was recommended [35]. In addition, a dose–response effect was reported, with the concentration of phytosterols [31] and phytostanols [36] plateauing after levels higher than 2.5 g were consumed. Recent nutritional studies have reported efficacy of phytosterols and their derivatives at lower levels of about 1.5 to 3.0 g/d administered as a capsule, in a suspension, or as a mixture with margarines or other oils. Reports from two studies [37,38] indicated that reduction in cholesterol absorption is less effective with low cholesterol (less than 200 mg/d) diets. In a study, 0.74 g/d soybean phytosterols mixed in butter (36% fat, 0.4 to 0.45 g/d) lowered total and LDL cholesterol by 10% and 15%, respectively, in 12 healthy males [39] after 4 weeks. In hypercholesterolemic males and females, 0.6 to 0.8 g/d of phytosterols and their derivatives have been shown to reduce total and LDL cholesterol by 1 to 4% and up to 7%, respectively, over a minimum of 9 weeks [40,75,76]. The health benefits of serum cholesterol concentration reduction also appear to be related to age. A 10% reduction in total serum cholesterol concentration produces a reduction in CVD of 50% at age 40, 40% at age 50, 30% at age 60, and 20% at age 70, with a greater part of the lower risk benefits realizable after 2 years, and the full benefit after 5 years [41].

Epidemiologic and experimental studies on membrane structure and function of tumor host tissue, apoptosis, immune function, and cholesterol metabolism [42] suggest that dietary phytosterols may offer protection from some common cancers in Western societies, including colon, breast, and prostate. In the past few years, issues about safety and efficacy have shifted the focus of most scientific studies from phytosterols to phytostanols and their derivatives. However, with more researchers working with phytostanols, it has been observed that the higher hypocholesterolemic effect reported with phytostanols is not consistent [14,44] and remains a major issue of controversy associated with phytosterols and their derivatives. The observed inconsistencies could be due to variation in the length of the nutritional study, differences in the data collection protocols, analytical methods used in the reports, and the physical form in which the phytosterol is dispersed in the diet [43]. A more probable conclusion that can be drawn from the observed trend suggests that a mixture of phytosterols [44] or phytosterols and stanols may be as effective as stanols alone [45], irrespective of the natural raw material source [46].

18.4.1.1 Effects of Phytosterols in LDL Cholesterol Reduction

Currently, the U.S. National Cholesterol Education Program (NCEP) targets low LDL cholesterol levels as a goal of therapy [47] to prevent atherosclerosis-related pathologies. Several studies with various phytosterol mixtures support the assertion that in normal, mildly, and overtly hypercholesterolemic humans, phytosterols are effective in reducing total cholesterol and LDL cholesterol levels to about 10 and 15%, respectively [19,23,48–51]. A minimum of 3.0 g phytosterol is needed a day for at least 2 weeks to realize these reductions. Coadministration of phytosterols with divalent calcium and magnesium cations further enhances cholesterol lowering [52]. When phytosterol feeding is discontinued, the LDL cholesterol levels return to their initial levels [51] within 6 weeks [31]. HDL cholesterol and triglyceride levels are not significantly affected by dietary phytosterols [31,50,51] although an increase in HDL cholesterol level has occasionally been reported [48].

18.4.1.2 Phytosterols in Combination with Low-Fat Diets

Most studies report a general trend of lower cholesterol and LDL cholesterol levels during the run-in period of clinical studies when subjects are consuming the control diet.

These pretreatment diets are usually low fat or are made up of polyunsaturated fats [46]. It is argued that feeding a prudent diet of low fat and phytosterols as control may lower the cholesterol-lowering efficacy of dietary phytosterols [53]. An American Heart Association advisory indicates that the cholesterol-lowering effect of plant sterols is consistent regardless of the fat level in the background diet [54]. Phytosterols have been highly efficacious in children [55] and adults [49,50] when used in combination with low-fat diets.

18.4.1.3 Phytosterols in Combination with Cholesterol-Lowering Medications

The two most important pharmaceutical approaches used to lower circulating cholesterol levels are (1) the bile-sequestering polymeric resins, cholestyramine (quaternary ammonium functional groups attached to a styrene-divinylbenzene copolymer) and colestipol (a copolymer of tetraethylenepentamine and epichlorohydrin in the ratio 2:5); and (2) 3-hydroxy-3-methylglutaryl coenzyme A reductase inhibitors or statins. Statins (e.g., Atorvastatin), inhibit HMG-CoA reductase, the rate-determining enzyme in cholesterol synthesis, and thereby reduce cholesterol synthesis [56]. Drugs are often used in combination to improve effectiveness. For example, a combination of statin plus resin inhibits cholesterol synthesis and enhances fecal elimination of cholesterol by bile acid binding.

Combination therapy using statins with phytosterols has proved efficacious for patients with hypercholesterolemia. Statins suppress the synthesis of cholesterol markedly more effectively in subjects with high rather than low baseline synthesis ($29.4 \pm 0.9\%$ vs. $25.6 \pm 0.9\%$, $p < .001$) [40]. Since most studies with phytosterols show no observable increase in HDL nor reductions in TG levels, such a combination of a natural food component with medication that enables prevention of cholesterol synthesis as well as absorption holds extensive benefits for patients.

18.4.2 Effects of Phytosterols on Fat-Soluble Vitamins

The mechanism by which phytosterols achieve lower LDL cholesterol levels via the inhibition of cholesterol absorption raises concern about the impact of phytosterols on the absorption of other fat-soluble compounds, such as vitamins and their precursors, in the intestine. Earlier studies have reported that phytosterol consumption may reduce the absorption of some fat-soluble vitamins; however, the results could be difficult to interpret. Lipoproteins are the major transport vehicles for newly absorbed fat-soluble vitamins. A decrease in lipoprotein levels may limit transport of fat-soluble vitamins or lower efficiency of absorption. Randomized trials have inconsistently shown a reduction in blood concentrations of β-carotene up to about 25%, α-carotene by 10%, lycopene by 10%, and vitamin E by 8% [36,57]. Decreases in the serum absolute concentrations of β-carotene and α-tocopherol concentrations are expected since both vitamins are transported in serum in lipoproteins, where concentrations decrease during experimental diet periods [49]. To facilitate understanding of the physiological significance of reduced fat-soluble vitamin levels associated with the cholesterol-lowering effects of plant sterols, the expression of fat-soluble vitamin levels relative to total or LDL cholesterol levels was adapted. When serum β-carotene concentrations and tocopherol are related to the serum total cholesterol concentrations, the decrease is insignificant [49].

Vitamin D and vitamin A (retinal) concentrations are on average unaffected by sterols and stanols [58]. Vitamin K-dependent clotting factors did not change in subjects fed stanols [46]. Eight patients were given stanol esters with no significant changes in prothrombin time and no major changes in coumarin dose, suggesting that vitamin K status was unchanged [58]. In a study of 185 volunteers, lipid-adjusted, fat-soluble vitamin concentrations remained unchanged; however, phytosterol intake reduced lipid-adjusted α- and β-carotene concentrations by 15 to 25% after 1 year, relative to control [59]. With the consistent exception of β-carotene where decreases are reported after data were normalized to total cholesterol level, phytosterols appear to have no significant effect on fat-soluble vitamin levels.

18.4.3 Safety Evaluation and Toxicity of Phytosterols

With the exception of the reported lower circulating levels of some carotenoids, studies on the potential toxicity of phytosterols have suggested few biochemical or clinical side effects. Reports showed no significant effect of 0.8 to 3.2 g of plant sterol esters on alkaline phosphatase, alanine transaminase, aspartate transaminase, and γ-glutamate transaminase levels [60]. Reported toxicity work on coagulation and fibrinolytic parameters indicates no significant differences in fibrinogen levels, factor VII coagulant activity, factor VII amidolytic activity, antithrombin III activity, plasminogen activator inhibitor type I activity, and tissue plasminogen activator among three dietary treatments [46]. Phytosterols had little or no effect on thyrotrophin, creatinine, glucose, or hemoglobin levels, white or red blood cell or platelet counts, or hematocrit [46].

Some concerns exist about potential estrogenic and endocrine effects of phytosterols consumed by growing children. Phytosterol consumption did not show any estrogenic or endocrine effects *in vivo* and *in vitro* [74]. A nominal phytosterol ester concentration of 8.1% fed to rats for 90 d, an equivalent to a dose of 6.6 g/kg body weight/d or 4.1 g/kg phytosterol, was considered to be at the no observed adverse effect level [61]. No toxicity was observed in a wide range of reproductive and developmental parameters in studies conducted over two generations of rats (81) [62]. In healthy adult males and females, a high intake of vegetable oil phytosterol esters (8.6 g vegetable oil phytosterol) increased the amount of neutral sterols in feces, but did not result in the increased formation of bile acids or sterol metabolites [44]. The products of bacterial metabolism of cholesterol and primary bile acids, such as secondary bile acids, deoxycholic acid, and litocholic acid, have been implicated in colon cancer development [63]. Since the large majority of phytosterols are not absorbed in the small intestine, such metabolites would accumulate for a period in the large intestine before excretion and could exert carcinogenic effects. Current animal and epidemiologic data, however, suggest that phytosterols may decrease colon cancer [42].

18.4.3.1 Hyperphytosterolemia

The incidence of high levels of phytosterols in the blood exists as hyperphytosterolemia or sitosterolemia, a very rare condition due to an inborn error of phytosterol metabolism. Recent data indicate that the genes ABCG5 and ABCG8 are mutually involved in the intestinal absorption of phytosterols and possibly also dietary cholesterol, as well as in the biliary secretion of sterols. Mutations in ABCG5 and ABCG8, the two half-transporters of the ABC transporter family involved in the absorption of phytosterols into the mucosal cells, have been identified as the cause of dietary sterol accumulation in patients with sitosterolemia. Patients also show elevated plasma cholesterol levels.

Some reports suggest that this condition could be cured by administration of phytosterols [64], such as sitostanol, which is also known to lower serum levels of other plant sterols [31] without being absorbed. However, this therapy has not been used for treatment of humans with this condition. Currently, severe cases of sitosterolemia are treated with bile acid resins such as cholestyramine or colestipol [45].

18.4.3.2 Oxidative Effects of Phytosterols

Plant sterols and other cholesterol-like unsaturated lipids are subjected to oxidation when exposed to air. Oxidation is enhanced by heating, ionizing radiation, exposure to light, or chemical catalysts. Sterol oxidation is a free-radical chain reaction and begins with formation of a hydroperoxide that may, in turn, decompose to various compounds. Similar to cholesterol, the main phytosterol oxidation products are hydroxy, keto, and epoxy compounds with the possible formation of triol compounds on hydration of the latter intermediates [65]. Many *in vitro* studies have shown that cholesterol oxides (oxysterols) have cyotoxic, mutagenic, and atherogenic properties; induce apoptosis; and are potent regulators of cholesterol metabolism. Owing to the structural similarities between plant sterols and cholesterol, the possible impact on health of oxidized plant sterols (oxyphytosterols) in food has been considered; however, unlike cholesterol, information on phytosterol oxidation products is limited. Oxysterols in foods have been measured to examine the stability. Small amounts of phytoxysterols have been detected in wheat flour, coffee, fried potatoes, and vegetable oils [66]. The major phytoxysterols detected were similar to those formed from cholesterol (i.e., 7-hydoxy, 7-keto, and 5,6-epoxy compounds).

Plant sterols may be oxidized even more easily than cholesterol [67]. A recommended daily intake of 2 to 4 g phytosterols could result in the ingestion of 2 to 4 mg phytosterol oxides [65]. Only a few studies have measured the biological effects of phytosterol oxides; they are similar to cholesterol oxides but show less severe effects. Studies have demonstrated that phytosterol oxides cause cellular damage in cultured macrophage-derived cells [68]. 7-Ketocholesterol inhibited cholesterogenesis by 20 to 30%, whereas 7-ketositosterol had no effect [69]. Thermally oxidized derivatives of β-sitosterol demonstrate similar biological effects as 7-β-OH in U937 cells, but at higher concentrations [70].

18.5 COMMERCIALIZATION OF PHYTOSTEROL-INCORPORATED PRODUCTS

Research-based sound scientific evidence has established the hypocholesterolemic impact of phytosterols on human health and heightened the appeal of health-conscious consumers for phytosterol-incorporating functional food products. Phytosterols are added in processed commercially available fat-based foods as spreads, yogurt, and snack bars. The phytosterols/phytostanols are commonly incorporated in the esterified form, free sterols having a solubility of about 2.55 to 3% in fat, whereas the solubility of sterol esters is roughly 30% [71,72]. The products have been successfully launched globally under legislation of the respective food regulatory agency [73] of the region.

18.5.1 Food and Pharmaceutical Companies Involved in Producing Phytosterol-Enriched Products

Cytellin, a preparation of predominantly β-sitosterol, is the first known commercial product marketed in the United States to treat hypercholesterolemia (from the 1950s to 1982). Eli Lilly Company introduced Cytellin as a drug to the market. Both soya oil (rich in campesterol and stigmasterol) and tall oil (from pine trees—93% sitosterol) were utilized as phytosterol sources for Cytellin.

Benecol has been manufactured in Finland since 1995. Benecol contains phytosterols from wood pulp as well as other sources. McNeil Consumer Healthcare, a part of Johnson & Johnson, has produced margarine containing phytostanol esters providing the consumer with 1.7 g of stanol esters per serving. Studies show that Benecol lowers total and LDL cholesterol levels by 10 to 15% in adults with high blood cholesterol levels [51,57]. Other phytostanol ester-containing products such as products containing Benecol, such as cream cheese spreads, milk, mayonnaise, yogurt, and snack bars, are available in Europe.

Lipton, a New Jersey-based unit of Unilever, produces Take Control, a margarine containing phytosterol esters obtained from soybean, in the United States. Three varieties of Take Control salad dressings—Italian, blue cheese, and ranch—were launched in September 1999. This is made by the Wish-Bone manufacturing unit of Lipton. A 2-tablespoon serving contains 1120 mg of natural soybean plant sterol extract.

Forbes Medi-Tech Inc., a pharmaceutical company, produces Phytrol (consumer-branded as Reducol), a mixture of unesterified phytosterols and stanols isolated from coniferous trees. Phytrol is exclusively licensed on a worldwide basis to Novartis Consumer Health Inc. for use as a functional food ingredient, dietary supplement, and inclusion in certain over-the-counter products. In 2000, Novartis announced a joint venture with Quaker Oats to form Altus Foods, to manufacture "healthy foods" containing Phytrol.

Meadow Lea manufactures Logicol spreads and milk products containing phytosterols available in Australia. Degussa Bioactives produces Cholestatin tablets containing free phytosterols.

REFERENCES

1. Peterson, D.W. Effect of soybean sterols in the diet of on plasma and liver cholesterol in chicks. *Proc Soc Biol Med* 1951;78:143–147.
2. Moreau, R.A., Whitaker, B.D., and Hicks, K.B. Phytosterols, phytostanols, and their conjugates in foods; structural diversity, quantitative analysis, and health promoting uses. *Prog Lipid Res* 2002;41:457–500.
3. Heinemann, T., Kullak-Ublick, G.A., Pietruck, B., and von Bergmann. K. Mechanisms of action of plant sterols on inhibition of cholesterol absorption. Comparison of sitosterol and sitostanol. *Eur J Clin Pharmacol* 1991;40 (suppl 1):S59–S63.
4. Piironen, V., Lindsay, D.G., Miettinen, T.A., Toivo, J., and Lampi, A.-M. Plant sterols: biosynthesis, biological function and their importance to human nutrition. *J Sci Food Agric* 2000;80:939–966.
5. Yankah, V.V. and Jones, P.J.H. Phytosterols and health implications—chemical nature and occurrence. *Inform* 2001;12:808–811.
6. Moreau R.A., Whitaker B.D., and Hicks K.B. Phytosterols, phytostanols, and their conjugates in foods; structural diversity, quantitative analysis, and health promoting uses. *Prog Lipid Res* 2002;41:457–500.
7. Grundy, S.M., Ahrens E.H. Jr., and Davignon, J. The interaction of cholesterol absorption and cholesterol synthesis in man. *J Lipid Res* 1969;10:304–315.
8. Czubayko, F., Beumers, B., Lammsfuss, S., Lutjohann, D., and von Bergmann, K. A simplified micro-method for quantification of fecal excretion of neutral and acidic sterols for outpatient studies in humans. *J Lipid Res* 1991;32:1861–1867.
9. Lu, T.T., Repa, J.J., and Mangelsdorf, D.J. Orphan nuclear receptors as eLiXirs and FiXeRs of sterol metabolism. *J Biol Chem* 2001;276:37735–37738.
10. Chawla, A., Repa, J.J., Evans, R.M., and Mangelsdorf, D.J. Nuclear receptors and lipid physiology: opening the X-files. *Science* 2001;1866–11870.
11. Armstrong, M.J. and Carey, M.C. Thermodynamic and molecular determination of sterol solubilities in bile salt micelles. *J Lipid Res* 1987;28:1144–1155.
12. Ikeda, I. and Sugano, M. Inhibition of cholesterol absorption by plant sterols for mass intervention. *Curr Opin Lipidol* 1998;9:527–531.
13. Slota, T., Kozlov, N.A., and Ammon, H.V. Comparison of cholesterol and beta-sitosterol: effects of jejunal fluid secretion induced by oleate, and absorption from mixed micellar solutions. *Gut* 1983;24:653–658.
14. Heinemann, T., Axtmann, G., and von Bergmann, K. Comparison of intestinal absorption of cholesterol with different plant sterols in man. *Eur J Clin Invest* 1993;23:827–831.
15. Ikeda, I., Tanaka, K., Sugano, M., Vahouny, G.V., and Gallo, L.L. Inhibition of cholesterol absorption in rats by plant sterols. J *Lipid Res* 1988;29:1573–1582.
16. Grundy, S.M. Absorption and metabolism of dietary cholesterol. *Ann Rev Nutr* 1983;3:71–96.
17. Heinemann, T., Kullak-Ublick, G.A., Pietruck, B., and von Bergmann. K. Mechanisms of action of plant sterols on inhibition of cholesterol absorption. Comparison of sitosterol and sitostanol. *Eur J Clin Pharmacol* 1991:40 (suppl 1):S59–S63.
18. Bhattacharyya, A. Uptake and esterification of plant sterols by rat small intestine. *Am J Physiol* 1981:240:G50–55.
19. Vahouny, G.V., Mayer R.M., and. Treadwell, C.R. Comparison of lymphatic absorption of dihydrocholesterol and cholesterol in rat. *Arch Biochem Biophys* 1960;86:215–218.
20. Salen, G., Ahrens, E.H. Jr., and Grundy, S.M. Metabolism of beta sitosterol in man. *J Clin Invest* 1970;49:952–967.

21. Moghadasian, M.H. Pharmacological properties of plant sterol in vivo and in vitro observations. *Life Sci* 2000;67:605–615.
22. Trautwein E.A., Duchateau, G.S., Lin, Y., Mel'nikov, S.M., Molhuizen, H.O., and Ntanois, F.Y. Proposed mechanisms of cholesterol-lowering action of plant sterols. *Eur J Lipid Technol* 2003;105:171–185.
23. Becker, M., Staab, D., and von Bergman, K. Treatment of severe familial hypercholesterolemia in childhood with sitosterol and sitostanol. *J Pediatr* 1993;122:292–296.
24. Child, P. and Kuksis, A. Critical role of ring structure in the differential uptake of cholesterol and plant sterols by membrane preparations in vitro. *J Lipid Res* 1983;24:1196–1209.
25. Laraki, L., Pelletier, X., Mourot, J., and Derby G. Effects of dietary phytosterols on liver lipids and lipid metabolism enzymes. *Ann Nutr Metab* 1993;37:129–133.
26. Shefer, S., Salen, G., Bullocks, J., Nguyen, L.B., Ness, G.C., Vhao, Z., Belamavich, P.F., Chowdhary, I., Lerner, S., Batta, A.K., and Tint, G.S. The effect of increased hepatic sitosterol on the regulation of 3-hydroxy-3-methylglutaryl-coenzyme A reductase and cholesterol 7α-hydroxylase in the rat and sitosterolemic homozygotes. *Hepatology* 1994;20:213–219.
27. Lee, M.H., Lu, K., Hazard, S., Yu, H., Shulenin, S., Hidaka, H., and Kojima, H. Identification of a genem ABCG5, important in the regulation of dietary cholesterol absorption. *Nat Genet* 2001;27:79–83.
28. Berge, K.E., Tian, H., Graf, G.A., Yu, L., Grishin, N.V., Schultz, J., and Kwiterovich, P. Accumulation of dietary cholesterol in sitosterolemia caused by mutations in adjacent ABC transporters. *Science* 2000;290:1771–1775.
29. Nguyen, L.B., Salen G., Shefer, S., Tint, G.S., and Ruiz, F. Macrophage 3-hydroxy-3-methylglutaryl coenzyme a reductase activity in sitosterolemia: effects of increased cellular cholesterol and sitosterol concentrations. *Metabolism* 2001; 50:1224–1229.
30. Kakis, G. and Kuksis, A. Effect of intravenous infusion of intralipid, cholesterol, and plant sterols on hepatic cholesterogenesis. *Can J Biochem Cell Biol* 1984;62:1–10.
31. Nguyen, T.T., Dale, L.C., von Bergman, K., and Croghan, I.T. Cholesterol-lowering effect of stanol ester in a US population of mildly hypercholesterolemic men and women: a randomized controlled trial. *Mayo Clin Proc* 1999;74:1198–1206.
32. Jones, P.J.H., Ntanois, F.Y., Raeini-Sarjaz, M., and Vanstone, C.A. Cholesterol-lowering efficacy of a sitostanol-containing phytosterol mixture with a prudent diet in hyperlipidemic men. *Am J Clin Nutr* 1999;69:1144–1150.
33. Morton, G.M., Lee, S.M., Buss, H., David, H., and Lawrence, P. Intakes and major dietary sources of cholesterol and phytosterols in the British diet. *J Hum Nutr Diet* 1995;429–440.
34. Ling, W.H. and Jones, P.J.H. Dietary phytosterols: a review of metabolism, benefits and side effects. *Life Sci.* 1995;57:195–206.
35. Lees, R.S. and Lees, A.M. Effects of sitosterol on plasma lipid and lipoprotein concentration. In Greten, H., Ed., *Lipoprotein Metabolism.* Springer-Verlag, Berlin, 1976, pp 119–130.
36. Hallikainen, M.A., Sarkkinen, F.S., and Uusitupa, M.I.J. Plant stanol esters affect serum cholesterol concentrations of hypercholesterolemic men and women in a dose-dependent manner. *J Nutr* 2000;130:767–776.
37. Briones, E.R., Steiger, D., Palumbo, P.J., and Kottke, B.A. Primary hypercholesterolemia: effect of treatment on serum lipids, lipoprotein fractions, cholesterol absorption, sterol balance, and platelet aggregation. *Mayo Clin Proc* 1984;59: 251–257.

38. Denke, M.A. Lack of efficacy of low-dose sitostanol therapy as an adjunct to a cholesterol-lowering diet in men with moderate hypercholesterolemia. *Am J Clin Nutr* 1995;61:392–396.
39. Pelletier, X., Belbraouet, S., Mirabel, D., Mordret, F., Perrin, J.L., Pages, X., and Debry, G. A diet moderately enriched in phytosterols lowers plasma cholesterol concentrations in normocholesterolemic humans. *Ann Nutr Metab* 1995;39:291–295.
40. Miettinen, T.A., Strandberg, T.E., and Gylling, H. Noncholesterol sterols and cholesterol lowering by long-term simvastatin treatment in coronary patients-relation to basal serum cholestanol. *Atheroscler Thromb Vasc Biol* 2000;20:1340–1352.
41. Law, M.R., Wald, M.J., and Thompson, S.G. By how much and how quickly does a reduction in serum cholesterol concentrations lower risk of ischaemic heart disease? *Br Med J* 1994;308:367–372.
42. Awad, A.B. and Fink, C.S. Phytosterols as anticancer dietary components: evidence and mechanism of action. *J Nutr* 2000;130:2127–2130.
43. Yankah, V.V. and Jones, P.J.H. Phytosterols and health implications — efficacy and nutritional aspects. *Inform* 2001;12:899–903.
44. Weststrate, J.A., Ayesh, R., Bauer-Plank, C., and Drewitt, P.N. Safety evaluation of phytosterol esters. Part 4. Faecal concentrations of bile acids and neutral sterols in healthy normolipidaemic volunteers consuming a controlled diet either with or without a phytosterol ester-enriched margarine. *Food Chem Toxicol* 1999;37:1063–1071.
45. Moghadasian, M.H. and Frohlich, J.J. Effects of dietary phytosterols on cholesterol metabolism and atherosclerosis: clinical an experimental evidence. *Am J Med* 1999;107:588–594.
46. Plat, J. and Mensink, R.P. Vegetable based versus wood based stanol ester mixtures: effects on serum lipids and hemostatic factors in non-hypercholesterolemic subjects. *Atherosclerosis* 2000;148:101–112.
47. National Cholesterol Education Program. Second report of the Expert Panel on Detection, Evaluation, and Treatment of high blood cholesterol in adults (Adult Treatment Panel II). *Circulation* 1994;89:1329–1445.
48. Gylling, H. and Miettinen, T.A. Serum cholesterol and cholesterol and lipoprotein metabolism in hypercholesterolaemic NIDDM patients before and during sitostanol ester-margarine treatment. *Diabetologia* 1994;37:773–780.
49. Hallikainen, M.A. and Uusitupa, M.I.J. Effects of 2 low-fat stanol ester-containing margarines on serum cholesterol concentrations as part of a low-fat diet in hypercholesterolemic subjects. *Am J Clin Nutr* 1999;69:403–410.
50. Jones, P.J.H., Ntanois, F.Y., Raeini-Sarjaz, M., and Vanstone, C.A. Cholesterol-lowering efficacy of a sitostanol-containing phytosterol mixture with a prudent diet in hyperlipidemic men. *Am J Clin Nutr* 1999;69:1144–1150.
51. Miettinen, T.A., Puska, P., Gylling, H., Vanhanen, H., and Vantiainen, E. Reduction of serum cholesterol with sitostanol-ester margarine in a mildly hypercholesterolemic population. *N Engl J Med* 1995;333:1308–1312.
52. Vaskonen, T., Mervaala, E., Seppanen-Laakso, T., and Karppanen H. Diet enrichment with calcium and magnesium enhances the cholesterol lowing effect of plant sterols in obese Zucker rats. *Nutr Metab Cardiovasc Dis* 2001;11:158–167.
53. Miettinen, T.A. and Vanhanen, H. Dietary sitostanol related to absorption, synthesis and serum level of cholesterol in different apolipoprotein E phenotype. *Atherosclerosis* 1994;105:217–226.
54. Lichtenstein, A.H. and Deckelbaum, R.J. Stanol/sterol ester-containing foods and blood cholesterol levels. *Circulation* 2001;103:1177–1183.

55. Gylling, H., Siimes, M.A., and Miettinen, T.A. Sitostanol ester margarine in dietary treatment of children with familial hypercholesterolemia. *J Lipid Res* 1995;36:1807–1812.
56. Vanhanen, H. Cholesterol malabsorption caused by sitostanol ester feeding and neomycin in pravastatin-treated hypercholesterolaemic patients. *Eur J Clin Pharmacol* 1994;47:169–176.
57. Gylling, H. and Miettinen, T.A. Cholesterol reduction by different plant stanol mixtures and with variable fat intake. *Metabolism* 1999;48:575–580.
58. Katan, M.B., Grundy, S.M. Jones, P., Law, M., Miettinen, T., Paoletti, R., and Stresa Workshop Participants. The efficacy of plant stanols and sterols in the management of blood cholesterol levels. *Mayo Clin Proc* 2003;78(8):965–978.
59. Hendricks, H.J.F., Brink, E.J., Meijer, G.W., Princen, H.M.G., and Ntanois, F.Y. Safety of long-term consumption of plant sterol ester-enriched spread. *Eur J Clin Nutr* 2003;57:681–692.
60. Hendricks, H.F.J., Westrate, J.A., van Vliet, T., and Meijer, G.W. Spreads enriched with three different levels of vegetable oil sterols and the degree of cholesterol lowering in normocholesterolaemic and mildly hypercholesterolaemic subjects. *Eur J Clin Nutr* 1999;53:319–327.
61. Hepburn, P.A., Horner, S.A., and Smith, M. Safety evaluation of phytosterols esters. Part 2. Subchronic 90-day Oral Toxicity Study on phytosterol esters — A novel functional food. *Food Chem Toxicol* 1999;37:521–532.
62. Waalkens-Berendsen, D.H., Wolterbeek, A.P.M., Wijnands, M.V.W., Richold, M., and Hepburn, P.A. Safety evaluation of phytosterol esters. Part 3. Two generation reproduction study in rats with phytosterol esters — a novel functional food. *Food Chem Toxicol* 1999;37:683–696.
63. Wilpart, M. Co-mutagenicity of bile acid: structure–activity relations. *Eur Cancer Prev* 1991;1 (suppl. 2):45–48.
64. Lutjohann, D., Bjorkhem, I., Beil, U.F., and von Bergmann, K. Sterol absorption and sterol balance in phytosterolemia evaluated by deuterium-labeled sterol: effect of sitostanol treatment. *J Lipid Res* 1995;36:1763–1773.
65. Lampi, A.-M., Juntunen, L., Toivo, J., and Piironen, V. Determination of thermo-oxidation products of plant sterols. *J Chromatogr B* 2002;777:83–92.
66. Dutta, P.C. Studies on phytosterol oxides. II Content in some vegetable oils and French fries prepared in these oils. *J Am Oil Chem Soc* 1997;74:659–666.
67. Li, W. and Przybylski, R.. Oxidation products formed from phytosterols. *Inform* 1995;6:499–500.
68. Adcox, C., Boyd, L., Oehrl, L., Allen, J., and Fenner, G. Comparative effects of phytosteroloxides and cholesterol oxides in cultured macrophage-derived cell lines. *J Agric Food Chem* 2001;49:2090–2095.
69. Kakis, G., Kuksis, A., and Myher, J.J. Injected 7-oxycholesterol and plant sterol derivatives and hepatic cholesterogenesis. *Adv Exp Med Biol* 1977;82:297–299.
70. Maguire, L., Konoplyannikov, M., Ford, A., Maguire, A.R., and O'Brien, N.M. Comparison of the cytotoxic effects of beta-sitosterol oxides and a cholesterol oxide, 7-beta-hydroxycholesterol, in cultured mammalian cells. *Br J Nutr* 2003;90:767–775.
71. Sierksma, A., Westrate, J.A., and Meijer, G.W. Spreads enriched with plant sterols, either esterified 4,4-dimethylsterols or free 4-desmethylsterols, and plasma total- and LDL-cholesterol concentrations. *Br J Nutr* 1999;82:273–282.
72. Mattson, F.H., Grundy, S.M., and Crouse, J.R. Optimizing the effect of plant sterols in cholesterol absorption in man. *Am J Clin Nutr*1982;35:697–700.

73. Yankah, V.V. and Jones, P.J.H. Phytosterols and health implications — commercial products and their regulation. *Inform* 2001;12:1011–1016.
74. Baker, V.A., Hepburn, P.A., Kennedy, S.J., Jones, P.A., Lea, L.J., Sumpter J.P., and Ashby, J. Safety evaluation of phytosterol esters. Part 1. Assessment of oestrogenicity using a combination of in vivo and in vitro assays. *Food Chem Toxicol* 1999;37:13–22.
75. Vanhanen, H.T., Kajander, J., Lehtovirta, H., and Miettinen, T.A. Serum levels, absorption efficiency, faecal elimination and synthesis of cholesterol during increasing doses of dietary sitosterol esters in hypercholesterolemic subjects. *Clin Sci (Lond.)* 1994;87:61–67.
76. Vanhanen, H.T. and Miettinen, T.A. Effects of unsaturated and saturated dietary plant sterols on their serum contents. *Clin Chim Acta* 1992;205:97–107.

19 Clinical Studies with Structured Triacylglycerols

Trine Porsgaard

CONTENTS

19.1 INTRODUCTION

Structured triacylglycerols (STs) can be defined in several ways, but in this chapter the term will cover triacylglycerols (TAGs) that have been modified or restructured from natural oils, fats, and fatty acids. The STs are synthesized for nutritional use or specific purposes, such as human milk-fat substitutes, low-calorie fats, oils enriched in essential fatty acids (linoleic and α-linolenic acid) or long-chain n-3 polyunsaturated fatty acids (eicosapentaenoic [EPA] and docosahexaenoic acid [DHA]), and TAGs containing both long-chain polyunsaturated fatty acids and medium-chain fatty acids (MCFAs) for specific nutritional or medical applications.

Early work with STs focused on the needs of hospitalized patients. The interest was primarily on STs containing MCFAs and long-chain fatty acids (LCFAs). With these fatty acids on the same glycerol molecule, MCFAs could supply rapidly available energy and LCFAs could provide essential fatty acids to patients with malabsorption or with high demands for energy.

During the last 25 years, a range of animal studies has been conducted using STs and constitutes the background for many of the clinical studies performed subsequently. These animal studies have included basic investigations of the absorption characteristics of STs, as well as animal models of human disease. In the following, results from a few of these animal studies will be included to support the results from the clinical studies. However, the main focus of this chapter will be the many clinical studies that have been conducted investigating the effects of STs.

19.2 SYNTHESIS OF STRUCTURED TRIACYLGLYCEROLS

STs can be produced by different methods. Chemical interesterification with a catalyst like sodium methoxide results in TAGs with random distribution of the fatty acids within the glycerol molecule. This randomization distributes the fatty acids equally in all three positions of the TAG. Most of the clinical studies referred in this chapter were performed with randomized TAGs.

More specific structures can be achieved by enzymatic interesterification using *sn*-1,3 specific lipases, which results in specific replacement of the fatty acids in the *sn*-1,3 positions of the TAG. The outcome of this process depends primarily on the substrates and the ratio between them, the choice of enzyme, and the process technology [1,2]. The cost of producing these specific structures is severalfold higher than producing the randomized products primarily due to the high price of the enzymes and has therefore discouraged commercialization of specific STs. New lipases with lower cost are now available and this will probably influence the production of specific STs.

19.3 ABSORPTION OF STRUCTURED TRIACYLGLYCEROLS

The preduodenal lipases (lingual and gastric lipases) perform the first step in the hydrolysis of dietary TAGs. These lipases possess high activity toward fatty acids located in the *sn*-3 position of the TAG, especially short-chain fatty acids (SCFAs). In some disease states, like cystic fibrosis and chronic alcoholism with pancreatic insufficiency, and in preterm infants, these lipases are the major lipolytic agents [3].

The pancreatic lipase catalyzes the further intestinal hydrolysis of ingested TAGs with preference toward the *sn*-1,3 positions resulting in *sn*-2 monoacylglycerols (2-MAGs) and free fatty acids. The nature of the fatty acids in the TAG molecule determines the hydrolysis rate with faster hydrolysis of MCFAs [4] and slow hydrolysis of the highly unsaturated LCFAs, in particular, EPA and DHA [5]. The major route of absorption for the LCFAs and 2-MAGs is through the enterocytes with conservation of approximately 75% of the fatty acids located in the *sn*-2 position of the dietary fat [6]. The specificity and preference of the pancreatic lipase are important issues when considering the possible advantages of tailor-making fat with particular TAG structures and maintenance of fatty acids in specific positions following absorption. In the enterocytes, the 2-MAGs are reesterified with fatty acids of exogenous and endogenous origin to form a new population of TAGs, and these are secreted to the lymph packed into chylomicrons. Because of the high polarity of the MCFAs, the low affinity for the cytosolic fatty acid binding proteins, and the low activation to CoA esters, these fatty acids are primarily absorbed via the portal vein for oxidation in the liver [7], although some MCFAs are detected in lymph as well [8].

The absorption characteristics of STs have been investigated in studies with lymph-cannulated rats, and these studies have emphasized the preferential absorption of the fatty acid in the *sn*-2 position. Jensen et al. [9] compared the intestinal absorption of fatty acids from a randomized oil with the absorption of fatty acids from a specific ST of the medium-chain/long-chain/medium-chain fatty acids (MLM) type. The absorption of MCFAs was highest from the randomized oil, where approximately 33% of the MCFAs were in the *sn*-2 position, and the absorption of LCFAs was highest from the structured oil, with 100% LCFAs in the *sn*-2 position, emphasizing the enhanced absorption of fatty acids in the *sn*-2 position. Similar results were obtained by Christensen et al. [10] comparing the absorption of an MLM TAG and a randomized oil based on MCFAs and EPA/DHA, and by Ikeda et al. [11] comparing the absorption of fatty acids from a long-chain triacylglycerol (LCT), a medium-chain triacylglycerol (MCT), and STs of the MLM and long-chain/medium-chain/long-chain fatty acids (LML) types. These studies concluded that STs of the MLM type would be a better fat supplement than MCT and LCT in the treatment of fat malabsorption because of the rapid hydrolysis and absorption of this structure together with the important supply of essential fatty acids.

The influence of STs on the metabolism of chylomicrons has not yet been examined in as much detail as the influence on the absorption process. Swift et al. [12] showed that MCFAs to some extent were incorporated into chylomicrons following intake of MCT in humans. Hultin et al. [13] compared the clearance of LCT, MCT/LCT, and MLM emulsions after bolus injection to rats. MLM and MCT/LCT emulsions were cleared more rapidly from the blood than the LCT emulsion. We have shown that the size and number of chylomicrons were affected by the structure of TAGs during absorption in rats after administration of MCT, LCT, MLM, and LML, indicating that the clearance of the chylomicrons also may be affected by the structure of the ingested TAG (Porsgaard et al., unpublished results).

19.4 CLINICAL STUDIES WITH STRUCTURED TRIACYLGLYCEROLS

19.4.1 In Disease States

After major surgery or in critically ill patients, most people become catabolic and lose net protein mass, resulting in loss of muscle and organ tissue and thereby function. Improvement of the nitrogen balance is therefore important for the conservation of muscle mass and organ function. Mixed fuel systems that incorporate intravenous lipid emulsions are increasingly utilized in total parenteral nutrition (TPN). For the supply of lipids in TPN, different concepts have been in use, and for many years lipid emulsions based on LCT derived from soybean or safflower oils were the first choice. LCTs supply essential fatty acids and fatty acids for energy. Therefore, they inhibit the hydrolysis of protein to amino acids for energy and the body can use amino acids as protein and not as a calorie source. However, the use of pure LCT emulsions is connected with slow clearance from the bloodstream and impairment of the reticuloendothelial system (RES) [14]. RES functions in the phagocytosis of microorganisms and foreign material and in the secretion of chemical mediators of the inflammatory response. Animal studies showed that parenteral administration of LCT emulsions led to RES overload, thereby suggesting a role for RES in lipid clearance as well, and RES overload led to interference with the clearance of bacteria [15,16].

The potential detrimental effects of infusing LCT emulsions have prompted a search for alternative lipid sources. MCTs derived from palm-kernel or coconut oils offer potential advantages that include rapid clearance from the bloodstream, rapid utilization with mitochondrial uptake and metabolism independently of the carnitine shuttle, and high oxidation rate that results in a reduction in net lipid storage in comparison to an equivalent amount of LCT [17,18]. Pure MCT emulsions unfortunately do not provide any essential fatty acids and may have side effects, such as metabolic acidosis; therefore, the use of mixed LCT/MCT emulsions have attracted increased attention. Results from animal studies suggested that mixed LCT/MCT or pure MCT emulsions in TPN did not depress RES function and may better support host bactericidal capacity than similar emulsions composed of pure LCTs [15,16]. Infusion of an MCT/LCT emulsion in traumatized rats led to less atrophy of the gut mucosa, a major drawback of TPN, during 7 d of TPN compared with infusion of an LCT emulsion [19]. RES function during LCT or LCT/MCT infusion was examined in a clinical study and showed that provision of LCT in a mixture with MCT did not result in RES dysfunction in patients and may be of particular benefit in critically ill or septic patients [20]. Furthermore, the study showed that the mode of LCT administration was of particular importance regarding RES function. Continuous mode of LCT administration in contrast to intermittent infusion enabled adequate metabolism of the lipid emulsion without impairment of RES function.

The positive results in TPN with mixed MCT/LCT emulsions led to the use of STs in clinical experiments. A range of studies was published in which STs were tested in animal models of human disease. These created the background for many of the human clinical studies with STs. Numerous investigations have demonstrated that STs have different metabolic effects compared with the natural oils and physical mixtures of TAGs.

In rat models of malabsorption in which the common bile and pancreatic duct was cannulated, thereby creating a malabsorption state, the absorption of STs, both randomized and specific, was compared with the absorption of LCT and physical mixtures of MCT and LCT. Administration of specific STs of the MLM structure with LCFAs from soybean oil [21] and from rapeseed oil [22] showed higher absorption of the LCFAs compared with the other oils. These studies indicated that STs of the MLM type may provide a means to increase the absorption of essential fatty acids in diseases of fat malabsorption, probably due to improved hydrolysis and absorption when MCFAs are present in the outer positions of the TAG molecule. Patients with fat malabsorption and with cystic fibrosis and pancreatic insufficiency are at risk of developing essential fatty acid deficiency [23,24] and could also be at risk of low fat-soluble vitamin status. In patients with cystic fibrosis, STs (Captex products) were absorbed faster than safflower oil, and patients receiving Captex 810D with 40% linoleic acid and MCFAs had a higher increase in plasma linoleic acid compared with controls [25,26]. Tso et al. [27] showed that a randomized oil produced from MCT and fish oil improved the absorption of fat-soluble vitamins compared with the physical mixture of the two fats in a rat model of fat malabsorption. In Caco-2 cells rendered deficient in essential fatty acids, the administration of STs produced a marked increase in the cellular levels of linoleic and arachidonic acids, thereby correcting the deficiency [28]. In fat malabsorbing patients with Crohn's disease or short bowel syndrome on home parenteral nutrition, the effect of long-term treatment with STs (Structolipid 20%) was compared with the effect of an LCT emulsion (Intralipid 20%) in a crossover study [29]. No differences were observed between the groups with respect to clinical safety, but the use of STs on a long-term basis may be associated with possible reductions in liver dysfunction since two patients receiving the LCT emulsion developed abnormal liver function, which resolved after switching to the ST emulsion. The results from these studies indicate that the use of STs with MCFAs and LCFAs may have some beneficial effects regarding increased absorption of fat in cases of malabsorption and reversal of essential fatty acid deficiency.

A collection of studies published in the 1980s and early 1990s in various animal models of burn injury [30–33], endotoxic shock [34], and cancer [35,36] demonstrated that enteral or parenteral feeding with randomized STs may diminish the catabolic response to injury and improve nitrogen balance compared with natural fats and physical mixtures of oils with similar fatty acid compositions.

In humans, several studies were performed comparing the effects of STs in patients requiring TPN postoperatively with the effects of other fat emulsions. Sandström et al. [37] conducted a safety and tolerance study in humans in which a ST emulsion (fat emulsion 73403) was compared with an LCT emulsion (Intralipid 20%) in postoperative patients requiring TPN after major surgery. Safety and tolerance variables were similar in the two groups, and the measured physiological and biochemical variables suggested that the STs were rapidly cleared and metabolized. A similar study was performed by Bellantone et al. [38] comparing the safety, tolerance, and efficacy of the same lipid emulsions as used by Sandström et al., but this time in colorectal surgical patients. The results showed that the STs were safe, well tolerated, and as efficacious as the LCT emulsion in creating a positive nitrogen balance postoperatively in the patients. A nitrogen sparing effect of STs was observed by Lindgren et al. [39] in critically ill patients and by Chambrier et al. [40] in postoperative patients. Kruimel et al. [41] observed improved nitrogen balance after ST (Structolipid) infusion compared with infusion of a physical mixture of MCT and LCT (Lipofundin MCT/LCT 20%) in catabolic patients requiring TPN after placement of an aortic prosthesis. Serum TAGs and plasma MCFAs increased two times more after infusion of the physical mixture and this suggested faster clearance from the blood after ST infusion. A study in postoperative patients comparing the administration of ST (FE 73403) and LCT (Intralipid 20%) demonstrated also a faster clearance of the STs as well as an increased whole body fat oxidation as determined by indirect calorimetry [42]. These results suggested that STs were hydrolyzed and oxidized faster than the other lipid emulsions.

In postsurgical cancer patients, the effect of enteral feeding of a fish oil-based, randomized ST was investigated [43]. The rationale for incorporating n-3 fatty acids into STs was to change the eicosanoid and cytokine production, yielding improved immunocompetence and reduced inflammatory response to injury and disease.

The results showed that the STs were well tolerated and that the number of infections and gastrointestinal complications were reduced. In addition, improved renal and liver functions were suggested through modulation of urinary prostaglandin levels. In a follow-up study, the presumed mechanism for these observed effects was investigated and a reduction in eicosanoid production from peripheral blood mononuclear cells was observed [44].

Overall, the results from all these studies in which STs have been compared with LCT or physical mixtures in patients suggest that STs have a future in TPN regimens as well as in enteral nutrition. The experiments show important improvements in the nitrogen balance and increases in body fat oxidation, factors that are important for the conservation of muscle mass and organ function during disease states and after major surgery. Reasons for not using STs in lipid emulsions for TPN may be the higher price compared with LCT or physical mixtures, but a lower cost of lipases for use in production of STs in the future may help improve their use in enteral and parenteral nutrition.

19.4.2 In Human Milk-Fat Substitutes

Human milk contains 3 to 5% total lipid, and the fatty acid composition of this lipid is highly dependent on the maternal dietary habit. Saturated and monounsaturated fatty acids predominate, each accounting for approximately 40% of total fatty acids. The most abundant saturated fatty acid in human milk fat is palmitic acid, and approximately 70% of this fatty acid is positioned at the *sn*-2 position of the glycerol backbone [45] and is absorbed as 2-MAG [46]. Fat for infant formulas is normally produced from vegetable oils in which palmitic acid is esterified to the outer *sn*-1 and *sn*-3 positions. Lard is an exception, with almost 70% palmitic acid in the *sn*-2 position, but its use in infant formulae offers some limitations in many countries due to religious reasons. This means that palmitic acid in vegetable oil-based formulae is hydrolyzed from the glycerol backbone by the action of the lipases in the stomach and in the intestine. Free long-chain saturated fatty acids, when present in the intestine, tend to form insoluble soaps with calcium that are excreted with feces, resulting in unnecessary loss of dietary energy and calcium [47]. Furthermore, there appears to be a link between high levels of insoluble calcium soaps and an increase in constipation and stool hardness [48], overall factors that are undesirable for the growth and well-being of the infant.

Betapol is an enzymatically interesterified lipid from Loders Croklaan produced as an alternative fat for use in human infant formulae. To produce Betapol, tripalmitin is reacted with unsaturated fatty acids from vegetable oils to give TAGs with up to 60% of the palmitic acid at the *sn*-2 position, thereby to a higher degree than conventional formulae mimicking the structure of human breast milk fat. Animal studies by Lien et al. [49,50] confirmed that palmitic acid can be efficiently absorbed, avoiding fatty soap formation if it is present in the *sn*-2 position.

In the mid- and late 1990s, a range of clinical studies was published in which preterm and term infants were fed Betapol or comparable TAGs with similar overall fatty acid composition but with different structures containing less palmitic acid in the *sn*-2 position. In a study by Carnielli et al. [51], preterm infants were randomized to be fed 1 week with Betapol and 1 week with a comparable TAG with 10% palmitic acid in the *sn*-2 position in a crossover design. The fatty acid composition of blood showed that the palmitic acid content in sterol esters, TAG, and free fatty acids was higher after Betapol feeding compared with feeding the control TAG, indicating increased absorption of palmitic acid when present in high amounts in the *sn*-2 position. Calcium excretion was lower in Betapol-fed preterm infants, indicating that this formula improved mineral balance in comparison with a conventional formula [52]. Lucas et al. [53] observed improved absorption of palmitic acid and calcium as well as reduced formation of insoluble calcium soaps in the stool of preterm infants fed Betapol with 74% palmitic acid in the *sn*-2 position, compared with infants fed formulas with 8 or 28% palmitic acid in the *sn*-2 position.

In term infants fed formulas with 66% (Betapol), 39 or 13% palmitic acid in the *sn*-2 position from birth, significantly higher absorption of fat (97.6, 93.2, and 90.4% in the three groups, respectively) and calcium (53.1, 35.4, and 32.5% in the three groups, respectively) was observed after Betapol feeding compared with the other formulas [54]. The effect on stool hardness and skeletal mineral deposition was investigated in term infants receiving Betapol with 50% *sn*-2 palmitic acid, in breast-fed infants, and in infants receiving a formula with 12% *sn*-2 palmitic acid [55]. Feeding Betapol as well as breast-feeding resulted in reduced stool soap fatty acids, softer stools, and greater bone mass compared with the infants fed the control formula. From the referred studies, it is obvious that feeding an infant formula like Betapol, which resembles human milk to a higher degree than conventional formulae (not only in fatty acid composition but also in overall TAG structure due to the interesterification process), offers some advantages in fat and calcium absorption compared with other formulae in preterm and term infants.

19.4.3 In Low-Energy Fats

Conventional fats and oils contain 9 kcal/g compared with the 4 kcal/g energy content of carbohydrates and proteins. In Western countries, the daily intake of fat is far beyond the recommendations that no more than 30% of the dietary calories should be derived from fat. The increasing problem with obesity and obesity-related diseases has increased consumers' and producers' interests in low-fat products and, thereby, also in the development of low-energy fats. The approaches to reduce the energy content of fat have included the use of SCFAs and MCFAs with lower energy content compared with LCFAs and the use of saturated LCFAs with reduced absorption efficiencies. The best-known representatives of these low-energy fats are salatrim and caprenin.

Danisco Cultor manufactures and markets Benefat, the trade name for salatrim, originally developed by Nabisco Inc. and introduced in the United States, Japan, and Taiwan in the mid-1990s. Salatrim is a mixture of TAGs composed of short- and long-chain saturated fatty acids randomly esterified to glycerol. Stearic acid is the principal LCFA in salatrim, originating from fully hydrogenated canola, soybean, cottonseed, or sunflower oil, and the SCFAs used are acetic, propionic, and butyric acids [56]. Due to the relatively low absorption coefficient of stearic acid [57,58] and the low caloric value of SCFAs, salatrim provides only 5 kcal/g (U.S. regulation) or 6 kcal/g (European Union regulation) compared with the 9 kcal/g of typical fats, thereby creating a reduced-calorie fat. Benefat is used in chocolate confectionery products and in baked goods in the United States and was approved for use in Europe in December 2003. Through changes in the number of LCFAs and the number and chain lengths of the SCFAs in the TAGs, the different preparations can be varied to be used for cocoa butter substitutes, shortenings, and fats suitable for filled milk and other dairy products.

In 1994, the results were published from five clinical studies with salatrim in which salatrim was fed in a variety of food items, such as chocolate bars, cookies, ice cream, and yogurt, to healthy subjects with intakes between 0 and 60 g/d in single-exposure and long-term studies [59,60]. The studies concluded that salatrim ingestion produced no significant clinical effects when consumed at the anticipated maximum use level of 30 g/d. Excretion of fecal fat and stearic acid increased during periods of salatrim feeding compared with periods with control fat (coconut oil) feeding. This was expected due to the relatively poor absorption of stearic acid. High doses of salatrim (60 g/d) led to gastrointestinal discomfort in several of the subjects, but this was not reported at the expected level of maximum daily intake. The effect on plasma lipids was investigated in hypercholesterolemic subjects fed margarines rich in either salatrim or a palmitic acid–rich test fat for 5 weeks each in a crossover design [61]. The results showed that plasma total and low-density lipoprotein (LDL) cholesterol during the high-fat test diets did not differ from concentrations measured during a period of low-fat intake, indicating that salatrim did not adversely affect plasma lipoprotein cholesterol levels.

Caprenin is a randomized TAG containing behenic acid and two MCFAs. It is produced by Procter & Gamble Company. and has functional and organoleptic properties comparable to cocoa butter. Because of the incomplete absorption of the LCFA [62] and the low-energy content of the MCFAs, caprenin has reduced energy value and provides only 5 kcal/g. A feeding study in hypercholesterolemic men consuming a caprenin-rich diet for 6 weeks showed decreased high-density lipoprotein (HDL) cholesterol in serum and increased ratio of total cholesterol to HDL cholesterol compared with a baseline diet enriched in palm oil/palm-kernel oil [63]. This study indicated that one or more of the fatty acids in caprenin can contribute to hypercholesterolemia in men. Caprenin was used as a cocoa butter substitute in the mid-1990s in the United States but its use has been taken over by other reduced fat alternatives.

19.4.4 Effects on Blood Lipid Parameters

Fats with the same fatty acid composition, but with different distribution of the fatty acids on the TAG molecule, may exhibit different effects on blood lipid parameters. A range of studies has been performed in healthy subjects investigating the effects on blood lipid parameters of STs in comparison with other fats. Nordenström et al. [64] investigated the short-term effect of infusing equimolar doses of a randomized TAG (FE 73403) and LCT (Intralipid 20%) emulsions into healthy subjects. The increase in serum TAG concentrations was greater during LCT infusion than during ST infusion. Plasma fatty acid compositions during fat infusions were similar to the fatty acid compositions of the infused emulsions, indicating that the capacity of healthy subjects to hydrolyze ST was at least as high as that to hydrolyze LCT.

Depending on the position of individual fatty acids in the TAG molecule, they may influence the lipidemic response differently. In a study by Zampelas et al. [65], Betapol was used to examine the influence of the positional distribution of fatty acids on the postprandial response after a liquid test meal. The effect of Betapol was compared with a TAG with similar fatty acid composition, but the saturated fatty acids were predominantly located in the *sn*-1,3 positions. The results demonstrated that the positional distribution of fatty acids was not an important determinant of postprandial lipaemia in healthy men.

Due to the absorption mode of dietary TAG, interesterification could be an advantage if it moved the cholesterol-raising fatty acids out of the *sn*-2 position and neutral fatty acids in. Christophe et al. investigated the effects on plasma lipid and lipoprotein levels in healthy men when native butter was substituted by either chemically [66] or enzymatically [67] interesterified butter. Enzymatic interesterification of butter fat resulted in considerable decreases in lauric, myristic, and palmitic acid with a concomitant increase of stearic and oleic acid at the *sn*-2 position of the TAG [68]. When chemically interesterified butter fat replaced native butter fat, a decrease in serum cholesterol was observed, but when native butter was replaced by enzymatically interesterified butter, for unknown reasons it did not affect plasma lipid and lipoprotein levels. Mascioli et al. [69] investigated the long-term effect of feeding butter or an interesterified mixture of butter, MCT, and safflower oil, which increased the content of MCFAs in *sn*-2 on the lipidemic response in hypercholesterolemic adults. The interesterified mixture had no appreciable effect on plasma cholesterol concentrations, but was associated with a modest rise in plasma TAG. This indicated that MCFAs in the *sn*-2 position had the same potential as the longer chain saturated fatty acids, such as myristic and palmitic acid, to increase plasma cholesterol concentrations. The effect on plasma lipids of interesterifying a high-palm oil blend was investigated in mildly hypercholesterolemic men [70]. Interesterification transferred substantial proportions of palmitic acid into the *sn*-2 position of TAG and the unsaturated fatty acids into the *sn*-1,3 positions, but the ingestion of the interesterified margarine did not result in higher plasma cholesterol than the original high-palm oil margarine, although the presence of palmitic acid in the *sn*-2 position would expectedly result in a cholesterol-raising effect. Zock et al. [71] used palm oil and enzymatically modified oil (Betapol) to investigate the effect of the positional distribution of fatty acids within dietary TAG on serum lipoproteins in healthy humans. In the palm oil diet, 18% of palmitic acid was attached to the *sn*-2 position, while in the enzymatically modified oil 65% was in the *sn*-2 position. Despite these large differences in fatty acid configuration, only very small differences were observed in lipoprotein concentrations.

To summarize, the effects of STs on blood cholesterol and lipoprotein concentrations seem to be minimal compared with native dietary fats with similar fatty acid compositions but with different TAG structures, indicating that dietary intake of STs does not lead to any deleterious effects on these blood parameters. In view of the more modest differences in positional distribution that can be achieved in everyday diets, the observed results after intake of STs are very small.

19.4.5 Effects in Healthy Subjects

Since dietary fats are metabolized differently in the human body, they may also influence energy expenditure, substrate oxidation, appetite regulation, and body weight regulation differently. The effect on these parameters of rapeseed oil in comparison with three modified fats (lipase interesterified fat, chemically interesterified fat, and a physically mixture; all produced from rapeseed oil and octanoic acid) was investigated in a short-term study in healthy men [72]. No differences were observed in appetite sensation or *ad libitum* energy intake among the four fats. The three modified fats resulted in higher postprandial energy expenditure and fat oxidation than the rapeseed oil, but no differences among these three fats were observed. This means that the modified fats will promote negative energy and fat balance, overall beneficial effects with regard to body weight regulation. Similarly, Matsuo et al. [73] compared the short-term effect of soybean oil with a randomized TAG composed of 80% LCFAs and 20% MCFAs produced by transesterification of MCT and rapeseed oil in healthy young women. The postingestive energy expenditure and thermic effect were higher after ingestion of the ST compared with LCT ingestion. These results suggested that long-term substitution of ST for LCT would result in body fat loss if the energy intake remained constant. The group also investigated the effect on body fat accumulation in healthy men [74]. They compared the long-term effect of a liquid-formula diet supplement containing ST (10% MCFAs, 90% LCFAs) with that of LCT. Following 12 weeks supplement with the test fats, the body fat accumulation was lower in the subjects receiving the ST than in those receiving LCT. Kasai et al. [75] investigated the effect of an ST-containing diet of the MLM type on body fat accumulation in healthy humans in comparison with an LCT diet. The subjects consumed the test fat, in bread for breakfast for 12 weeks. Significant decreases in body weight, amount of body fat, and serum total cholesterol were observed in the ST group compared with the LCT group. These studies indicate that ST may have some beneficial effects regarding energy expenditure and body fat accumulation in healthy subjects.

One study has been published in which the effect of an enzymatically interesterified TAG with MLM structure in combination with glucose on exercise performance was tested against glucose alone [76]. Well-trained cyclists consumed the test solutions during prolonged submaximal exercise followed by a time trial. Performance during the time trial was similar after intake of the two test solutions. MCFAs were not detected in plasma TAGs, fatty acids, and phospholipids, suggesting that the lack in time trial improvement after intake of MCFAs might be due to no available MCFAs in the systemic circulation.

19.5 FUTURE APPLICATIONS

In the pharmaceutical industry, the oral absorption of lipophilic compounds may be limited due to their physicochemical properties. This has led to a growing effort to develop pharmaceutical formulations enhancing the oral bioavailability of these lipophilic compounds. An approach to overcome the low bioavailability is to incorporate the active lipophilic component into lipid vehicles and ST may be an example of such a vehicle. Studies by Holm et al. in rat [77] and canine [78] models using the lipophilic antimalaria drug halofantrine have indicated that ST may be a possible lipid vehicle in humans in the future to improve the absorption of lipophilic drugs.

Another area of great economic potential is the use of STs in animal feed, thereby influencing human consumption indirectly. Studies in postweaning piglets in which they were fed different test fats in the diet for 3 weeks showed increased fat digestibility and nitrogen retention after feeding interesterified fats produced from MCFAs and rapeseed oil compared with pure rapeseed oil [79]. These results indicate that feeding STs may lead to faster growth of domestic animals compared with feeding native fats.

In the future, work with structured lipids other than STs, such as structured diacylglycerols and phospholipids, will definitely be performed. The intake of diacylglycerols has been shown to result in lipid-lowering and antiobesity effects, but previous work has concentrated only on the glyceride backbone and not on the fatty acid composition of the diacylglycerols. Phospholipids are emulsifying agents with widespread use, and only a few studies have investigated the effects of changing fatty acid composition. Thus, many experiments investigating the effects of different structured lipids are waiting to be performed.

ACKNOWLEDGMENT

This work was supported by the Danish Technological Research Council.

REFERENCES

1. Xu, X., Production of specific-structured triacylglycerols by lipase-catalyzed reactions: a review, *Eur. J. Lipid Sci. Technol.,* 102, 287, 2000.
2. Xu, X., Engineering of enzymatic reactions and reactors for lipid modification and synthesis, *Eur. J. Lipid Sci. Technol.,* 105, 289, 2003.
3. Hamosh, M., Preduodenal fat digestion, in *Fat Digestion and Absorption,* 1st ed., Christophe, A. and Vriese, S. De, Eds., AOCS Press, Champaign, IL, 2000, chap. 1.
4. Jandacek, R.J. et al., The rapid hydrolysis and efficient absorption of triglycerides with octanoic acid in the 1 and 3 positions and long-chain fatty acid in the 2 position, *Am. J. Clin. Nutr.,* 45, 940, 1987.
5. Bottino, N.R., Vandenburg, G.A., and Reiser, R., Resistance of certain long-chain polyunsaturated fatty acids of marine oils to pancreatic lipase hydrolysis, *Lipids,* 2, 489, 1967.
6. Åkesson, B. et al., Absorption of synthetic, stereochemically defined acylglycerols in the rat, *Lipids,* 13, 338, 1978.
7. Bernard, A. and Carlier, H., Absorption and intestinal catabolism of fatty acids in the rat: effect of chain length and unsaturation, *Exp. Physiol.,* 76, 445, 1991.
8. Mu, H. and Høy, C.-E., Effects of different medium-chain fatty acids on intestinal absorption of structured triacylglycerols, *Lipids,* 35, 83, 2000.
9. Jensen, M.M., Christensen, M.S., and Høy, C.-E., Intestinal absorption of octanoic, decanoic, and linoleic acids: effect of triglyceride structure, *Ann. Nutr. Metab.,* 38, 104, 1994.

10. Christensen, M.S. et al., Intestinal absorption and lymphatic transport of eicosapentaenoic (EPA), docosahexanoic (DHA), and decanoic acids: dependence on intramolecular triacylglycerol structure, *Am. J. Clin. Nutr.,* 61, 56, 1995.
11. Ikeda, I. et al., Lymphatic absorption of structured glycerolipids containing medium-chain fatty acids and linoleic acid, and their effect on cholesterol absorption in rats, *Lipids,* 26, 369, 1991.
12. Swift, L.L. et al., Medium-chain fatty acids: evidence for incorporation into chylomicron triglycerides in humans, *Am. J. Clin. Nutr.,* 52, 834, 1990.
13. Hultin, M. et al., Metabolism of emulsions containing medium- and long-chain triglycerides or interesterified triglycerides, *J. Lipid Res.,* 35, 1850, 1994.
14. Seidner, D.L. et al., Effects of long-chain triglyceride emulsions on reticuloendothelial system function in humans, *J. Parenteral Enterol. Nutr.,* 13, 614, 1989.
15. Sobrado, J. et al., Lipid emulsions and reticuloendothelial system function in healthy and burned guinea pigs, *Am. J. Clin. Nutr.,* 42, 855, 1985.
16. Hamawy, K.J. et al., The effect of lipid emulsions on reticuloendothelial system function in the injured animal, *J. Parenteral Enterol. Nutr.,* 9, 559, 1985.
17. Bach, A.C., Frey, A., and Lutz, O., Clinical and experimental effects of medium-chain-triglyceride-based fat emulsions-a review, *Clin. Nutr.,* 8, 223, 1989.
18. Ulrich, H. et al., Parenteral use of medium-chain triglycerides: a reappraisal, *Nutrition,* 12, 231, 1996.
19. Linseisen, J. and Wolfram, G., Efficacy of different triglycerides in total parenteral nutrition for preventing atrophy of the gut in traumatized rats, *J. Parenteral Enterol. Nutr.,* 21, 21, 1997.
20. Jensen, G.L. et al., Parenteral infusion of long- and medium-chain triglycerides and reticuloendothelial system function in man, *J. Parenteral Enterol. Nutr.,* 14, 467, 1990.
21. Christensen, M.S., Müllertz, A., and Høy, C.-E., Absorption of triglycerides with defined or random structure by rats with biliary and pancreatic diversion, *Lipids,* 30, 521, 1995.
22. Straarup, E.M. and Høy, C.-E., Structured lipids improve fat absorption in normal and malabsorbing rats, *J. Nutr.,* 130, 2802, 2000.
23. Jeppesen, P.B., Høy, C.-E., and Mortensen, P.B., Essential fatty acid deficiency in patients receiving home parenteral nutrition, *Am. J. Clin. Nutr.,* 68, 126, 1998.
24. Lloyd-Still, J.D., Essential fatty acid deficiency and nutritional supplementation in cystic fibrosis, *J. Pediatr.,* 141, 157, 2002.
25. McKenna, M.C., Hubbard, V.S., and Bieri, J.G., Linoleic acid absorption from lipid supplements in patients with cystic fibrosis with pancreatic insufficiency and in control subjects, *J. Pediatr. Gastroenterol. Nutr.,* 4, 45, 1985.
26. Hubbard, V.S. and McKenna, M.C., Absorption of safflower oil and structured lipid preparations in patients with cystic fibrosis, *Lipids,* 22, 424, 1987.
27. Tso, P., Lee, T. and DeMichele, S.J., Randomized structured triglycerides increase lymphatic absorption of tocopherol and retinol compared with the equivalent physical mixture in a rat model of fat malabsorption, *J. Nutr.,* 131, 2157, 2001.
28. Spalinger, J.H. et al., Uptake and metabolism of structured triglyceride by Caco-2 cells: reversal of essential fatty acid deficiency, *Am. J. Physiol.,* 275, G652, 1988.
29. Rubin, M. et al., Structured triacylglycerol emulsion, containing both medium- and long-chain fatty acids, in long-term home parenteral nutrition: a double-blind randomized cross-over study, *Nutrition,* 16, 95, 2000.
30. Teo, T.C. et al., Administration of structured lipid composed of MCT and fish oil reduces net protein catabolism in enterally fed burned rats, *Ann. Surg.,* 210, 100, 1989.
31. Maiz, A. et al., Protein metabolism during total parenteral nutrition (TPN) in injured rats using medium-chain triglycerides, *Metabolism,* 33, 901, 1984.

32. Mok, K.T. et al., Structured medium-chain and long-chain triglyceride emulsions are superior to physical mixtures in sparing body protein in the burned rat, *Metabolism,* 33, 910, 1984.
33. Swenson, E.S. et al., Persistence of metabolic effects after long-term oral feeding of a structured triglyceride derived from medium-chain triglyceride and fish oil in burned and normal rats, *Metabolism,* 40, 484, 1991.
34. Teo, T.C. et al., Long-term feeding with structured lipid composed of medium-chain and n-3 fatty acids ameliorates endotoxic shock in guinea pigs, *Metabolism,* 40, 1152, 1991.
35. Mendez, B. et al., Effects of different lipid sources in total parenteral nutrition on whole body protein kinetics and tumor growth, *J. Parenteral Enterol. Nutr.,* 16, 545, 1992.
36. Ling, P.R. et al., Structured lipid made from fish oil and medium-chain triglycerides alters tumor and host metabolism in Yoshida-sarcoma-bearing rats, *Am. J. Clin. Nutr.,* 53, 1177, 1991.
37. Sandström, R. et al., Structured triglycerides to postoperative patients: a safety and tolerance study, *J. Parenteral Enterol. Nutr.,* 17, 153, 1993.
38. Bellantone, R. et al., Structured versus long-chain triglycerides: a safety, tolerance, and efficacy randomized study in colorectal surgical patients, *J. Parenteral Enterol. Nutr.,* 23, 123, 1999.
39. Lindgren, B.F. et al., Nitrogen sparing effect of structured triglycerides containing both medium- and long-chain fatty acids in critically ill patients; a double blind randomized controlled trial, *Clin. Nutr.,* 20, 43, 2001.
40. Chambrier, C. et al., Medium- and long-chain triacylglycerols in postoperative patients: structured lipids versus a physical mixture, *Nutrition,* 15, 274, 1999.
41. Kruimel, J.W. et al., Parenteral structured triglyceride emulsion improves nitrogen balance and is cleared faster from the blood in moderately catabolic patients, *J. Parenteral Enterol. Nutr.,* 25, 237, 2001.
42. Sandström, R. et al., Structured triglycerides were well tolerated and induced increased whole body fat oxidation compared with long-chain triglycerides in postoperative patients, *J. Parenteral Enterol. Nutr.,* 19, 381, 1995.
43. Kenler, A.S. et al., Early enteral feeding in postsurgical cancer patients. Fish oil structured lipid-based polymeric formula *versus* a standard polymeric formula, *Ann. Surg.,* 223, 316, 1996.
44. Swails, W.S. et al., Effect of a fish oil structured lipid-based diet on prostaglandin release from mononuclear cells in cancer patients after surgery, *J. Parenteral Enterol. Nutr.,* 21, 266, 1997.
45. Christie, W.W. and Clapperton, J.L., Structures of the triglycerides of cows' milk, fortified milks (including infant formulae), and human milk, *J. Soc. Dairy Tech.,* 35, 22, 1982.
46. Innis, S.M., Dyer, R., and Nelson, C.M., Evidence that palmitic acid is absorbed as *sn*-2 monoacylglycerol from human milk by breast-fed infants, *Lipids,* 29, 541, 1994.
47. Chappell, J.E. et al., Fatty acid balance studies in premature infants fed human milk or formula: effect of calcium supplementation, *J. Pediatr.,* 108, 439, 1986.
48. Quinlan, P.T. et al., The relationship between stool hardness and stool composition in breast- and formula-fed infants, *J. Pediatr. Gastroenterol. Nutr.,* 20, 81, 1995.
49. Lien, E.L. et al., Corandomization of fats improves absorption in rats, *J. Nutr.,* 123, 1859, 1993.
50. Lien, E.L. et al., The effect of triglyceride positional distribution on fatty acid absorption in rats, *J. Pediatr. Gastroenterol. Nutr.,* 25, 167, 1997.
51. Carnielli, V.P. et al., Effect of dietary triacylglycerol fatty acid positional distribution on plasma lipid classes and their fatty acid composition in preterm infants, *Am. J. Clin. Nutr.,* 62, 776, 1995.

52. Carnielli, V.P. et al., Feeding premature newborn infants palmitic acid in amounts and stereoisomeric position similar to that of human milk: effects on fat and mineral balance, *Am. J. Clin. Nutr.,* 61, 1037, 1995.
53. Lucas, A. et al., Randomised controlled trial of a synthetic triglyceride milk formula for preterm infants, *Arch. Dis. Child.,* 77, F178, 1997.
54. Carnielli, V.P. et al., Structural position and amount of palmitic acid in infant formulas: effects on fat, fatty acid, and mineral balance, *J. Pediatr. Gastroenterol. Nutr.,* 23, 553, 1996.
55. Kennedy, K. et al., Double-blind, randomized trial of a synthetic triacylglycerol in formula-fed term infants: effects on stool biochemistry, stool characteristics, and bone mineralization, *Am. J. Clin. Nutr.,* 70, 920, 1999.
56. Softly, B.J. et al., Composition of representative SALATRIM fat preparations, *J. Agric. Food Chem.,* 42, 461, 1994.
57. Klemann, L.P., Finley, J.W., and Leveille, G.A., Estimation of the absorption coefficient of stearic acid in SALATRIM fats, *J. Agric. Food Chem.,* 42, 484, 1994.
58. Finley, J.W. et al., Caloric availability of SALATRIM in rats and humans, *J. Agric. Food Chem.,* 42, 495, 1994.
59. Finley, J.W. et al., Clinical assessment of SALATRIM, a reduced-calorie triacylglycerol, *J. Agric. Food Chem.,* 42, 581, 1994.
60. Finley, J.W. et al., Clinical study of the effects of exposure of various SALATRIM preparations to subjects in a free-living environment, *J. Agric. Food Chem.,* 42, 597, 1994.
61. Nestel, P.J. et al., Effect of a stearic acid-rich, structured triacylglycerol on plasma lipid concentrations, *Am. J. Clin. Nutr.,* 68, 1196, 1998.
62. Peters, J.C. et al., Caprenin 3. Absorption and caloric value in adult humans, *J. Am. Coll. Toxicol.,* 10, 357, 1991.
63. Wardlaw, G.M. et al., Relative effects on serum lipids and apolipoproteins of a caprenin-rich diet compared with diets rich in palm oil/palm-kernel oil or butter, *Am. J. Clin. Nutr.,* 61, 535, 1995.
64. Nordenström, J., Thörne, A., and Olivecrona, T., Metabolic effects of infusion of a structured-triglyceride emulsion in healthy subjects, *Nutrition,* 11, 269, 1995.
65. Zampelas, A. et al., The effect of triacylglycerol fatty acid positional distribution on postprandial plasma metabolite and hormone responses in normal adult men, *Br. J. Nutr.,* 71, 401, 1994.
66. Christophe, A. et al., Nutritional studies with randomized butter. Cholesterolemic effects of butter-oil and randomized butter-oil in man, *Arch. Intern. Biophys. Biochim.,* 86, 413, 1978.
67. Christophe, A. et al., Substituting enzymatically interesterified butter for native butter has no effect on lipemia or lipoproteinemia in man, *Ann. Nutr. Metab.,* 44, 61, 2000.
68. Pabai, F., Kermasha, S., and Morin, A., Lipase from *Pseudomonas fragi* CRDA 323: partial purification, characterization and interesterification of butter fat, *Appl. Microbiol. Biotechnol.,* 43, 42, 1995.
69. Mascioli, E.A. et al., Lipidemic effects of an interesterified mixture of butter, medium-chain triacylglycerol and safflower oils, *Lipids,* 34, 889, 1999.
70. Nestel, P.J. et al., Effect on plasma lipids of interesterifying a mix of edible oils, *Am. J. Clin. Nutr.,* 62, 950, 1995.
71. Zock, P.L. et al., Positional distribution of fatty acids in dietary triglycerides: effects on fasting blood lipoprotein concentrations in humans, *Am. J. Clin. Nutr.,* 61, 48, 1995.
72. Bendixen, H. et al., Effect of 3 modified fats and a conventional fat on appetite, energy intake, energy expenditure, and substrate oxidation in healthy men, *Am. J. Clin. Nutr.,* 75, 47, 2002.

73. Matsuo, T. et al., The thermic effect is greater for structured medium- and long-chain triacylglycerols versus long-chain triacylglycerols in healthy young women, *Metabolism,* 50, 125, 2001.
74. Matsuo, T. et al., Effects of a liquid diet supplement containing structured medium- and long-chain triacylglycerols on bodyfat accumulation in healthy young subjects, *Asia Pacific J. Clin. Nutr.,* 10, 46, 2001.
75. Kasai, M. et al., Effect of dietary medium- and long-chain triacylglycerols (MLCT) on accumulation of body fat in healthy humans, *Asia Pacific J. Clin. Nutr.,* 12, 151, 2003.
76. Vistisen, B. et al., Minor amounts of plasma medium-chain fatty acids and no improved time trial performance after consuming lipids, *J. Appl. Physiol.,* 95, 2434, 2003.
77. Holm, R. et al., Structured triglyceride vehicles for oral delivery of halofantrine: examination of intestinal lymphatic transport and bioavailability in conscious rats, *Pharm. Res.,* 19, 1354, 2002.
78. Holm, R. et al., Examination of oral absorption and lymphatic transport of halofantrine in a triple-cannulated canine model after administration in self-microemulsifying drug delivery systems (SMEDDS) containing structured triglycerides, *Eur. J. Pharm. Sci.,* 20, 91, 2003.
79. Straarup, E.M. et al., Fat digestibility, nitrogen retention, and fatty acid profiles in blood and tissues of post-weaning piglets fed interesterified fats, *J. Animal Feed Sci.,* 12, 539, 2003.

Part IV

Role of Biotechnology and Market Potential for Functional Lipids

20 Enzymatic Modification of Lipids for Functional Foods and Nutraceuticals

Yuji Shimada

CONTENTS

20.1 INTRODUCTION

Lipases have been referred to as three-measure digestive enzymes, together with amylases and proteases, since the hydrolysis activity of oils and fats was discovered in pancreatic homogenate by Bernard in 1856. However, basic and application studies on lipases were delayed significantly, compared with those on amylases and proteases, because the substrates, oils and fats, do not dissolve in water, and because determination of structures of triacylglycerols (TAGs) in oils and fats are very difficult in small amounts. Studies on lipase accelerated in the 1980s. Lipases were reported to be active and stable in water-immiscible organic solvents and started to be used as catalysts in many organic reactions. A lipase was also used as an ingredient in detergents, and an immobilized 1,3-position specific lipase was applied for industrial production of cocoa butter substitute with a fixed-bed bioreactor. In addition, cloning of lipase genes advanced and x-ray structures of several lipases were determined, leading to the clarification of oil–water interface activation of lipases, which can be explained by a helix (lid) covering the catalytic site [1,2].

Based on these basic and application studies, the use of lipases has been increasing in the oil and fat industry. This chapter deals with recent achievements of lipase studies on the modification of edible oils and fats.

20.2 CHARACTERISTICS OF LIPASES

20.2.1 Lipase-Catalyzed Reactions and Their Substrate Specificities

Lipases are defined as enzymes that hydrolyze ester bonds of long-chain fatty acids and alcohols and also catalyze esterification and transesterification (acidolysis, alcoholysis, and interesterification). In general, hydrolysis occurs preferentially in a system containing a large amount of water, and esterification proceeds effectively in a system containing only a small amount of water. Transesterification is catalyzed efficiently in a mixture without water using an immobilized enzyme.

Lipases have the following substrate specificities: fatty acid specificity, alcohol specificity, positional specificity, TAG specificity that lipases recognize the overall structure of TAG [3], and acylglycerol specificity that they have generally strong activity on acylglycerols in the order of monoacylglycerol (MAG) > diacylglycerol (DAG) > TAG [4,5]. Among these specificities, fatty acid and positional specificities are especially important for modification of lipids. In general, lipases act strongly on C_8 to C_{24} of medium- and long-chain fatty acids, but weakly on polyunsaturated fatty acids (PUFAs) of $\geq C_{18}$ with ≥ 3 double bonds (except for α-linolenic acid). Lipases are classified into 1,3-position specific enzymes, which act on primary alcohol esters, and nonspecific enzymes, which act on primary and secondary alcohol esters. *Geotrichum candidum* lipases III and IV act preferentially on ester bonds at the 2-position of TAGs [6,7], but 2-position specific lipase has not been found. Isolation of the 2-position specific lipase is strongly desired because the lipase can be used as a tool for precise processing of oils and fats.

20.2.2 Classification of Industrial Lipases

With the development of gene technology, many lipase genes have been cloned, and microbial lipases can be classified based on the homologies of their primary structure deduced from the nucleotide sequences. Homology of the primary structure shows similarity of tertiary structure, and lipases with similar tertiary structure are assumed to have similar properties. We thus classify industrial lipases into five groups based on the homologies of their primary structures (Table 20.1).

TABLE 20.1
Classification of Industrial Lipases Based on Their Primary Structures

Group	Microorganism	Property
Bacteria		
Group 1	*Burkholderia cepacia* *B. glumae* *Pseudomonas aeruginosa*	Positional specificity: nonspecific or 1,3-position preferential Acts somewhat on PUFAs
Group 2	*P. fluorescens* *Serratia marcescens*	Positional specificity: nonspecific or 1,3-position preferential Acts somewhat on PUFAs
Yeasts		
Group 3	*Candida rugosa* *Geotrichum candidum*	Positional specificity: nonspecific Acts very weakly on C_{20} fatty acids and PUFAs Fatty acid specificity: relatively strict Acts on sterol and L-menthol Hydrolysis activity: strong
Group 4	*C. antarctica*	Positional specificity: 1,3-position preferential, 1,3-position specific in a reaction Acts strongly on PUFAs and short-chain alcohols
Fungi		
Group 5	*Rhizomucor miehei* *Rhizopus oryzae* *Thermomyces lanuginose* *Fusarium hetersporum*	Positional specificity: 1,3-position specific Acts weakly on PUFAs Acts strongly on C_8 to C_{24} saturated and monoenoic fatty acids
	Penicillium camembertii	Positional specificity: 1,3-position specific Does not act on TAGs

20.3 APPLICATIONS OF LIPASES TO OIL PROCESSING

Application studies of lipases have been actively ongoing, and the use of lipases in the oil and fat industry is increasingly accelerating. Present applications of lipases are summarized in Table 20.2, and their new applications reported recently are summarized in Table 20.3. Among them, production of functional lipids with lipases in the food industry is classified from the viewpoint of the reactions (hydrolysis, esterification, and transesterification) and is described below.

20.3.1 Production of Oil Rich in PUFAs by Selective Hydrolysis

Lipases generally act weakly on PUFAs; thus, hydrolysis of PUFA-containing oils enriches PUFAs in undigested acylglycerols [8]. *Candida rugosa* lipase is suitable for this purpose because, among industrially available lipases, the lipase acts most weakly on PUFAs. When tuna oil containing ca. 25% docosahexaenoic

TABLE 20.2
Industrial Applications of Lipases

Application	Origin of Lipase	Appendix
Detergent	*Thermomyces lanuginosa*	
Optical resolution	*C. rugosa*	Positionally nonspecific enzyme
	C. antarctica	
	Serratia marcescens	
	Pseudomonas	
	Burkholderia, etc.	
Medicine (digestive)	*Aspergillus niger*	
Diagnostic reagent	*Pseudomonas aeruginosa*	Assay of TAGs and cholesterol esters in blood
Production of unsaturated fatty acids by hydrolysis of oils	*C. rugosa*	Linoleic, α-linolenic, and ricinoleic acids
	Pseudomonas	PUFAs
Removal of lipids in egg white	*Mucor javanicus*	Heat labile enzyme
		Degradation of TAGs
Paper making	Closed	Degradation of TAGs in woods
Leather making	Closed	Removal of fats
Addition of flavor (production of flavor)	*C. rugosa*	Butter flavor
	Penicillium roqueforti	Cheese flavor
	Rhizopus oryzae	Bread flavor
	A. niger	
Bread making	*Fusarium oxysporum*	Partial digestion of TAGs
		Partial acylglycerols act as surfactants
		Production of functional oils and fats
Cocoa butter substitute and human milk fat substitute	*Rhizomucor miehei*	Immobilized 1,3-position specific lipase
	Rhizopus oryzae	
	R. niveus, etc.	
PUFA-rich oil	*C. rugosa*	Selective hydrolysis
Diacylglycerols	Closed	Esterification with immobilized1,3-position specific lipase
TAGs containing medium- and long-chain fatty acids	Closed	Transesterification with lipase
Production of steryl esters	*C. rugosa*	

acid (DHA) was hydrolyzed with the lipase in the presence of 50% water, the content of DHA in acylglycerols increased to 50% at 70% hydrolysis [9]. The content of DHA increased further by repeating hydrolysis of acylglycerols obtained from a single reaction. Several repetitions of the selective hydrolysis produced oils containing nearly 70% DHA. Short-path distillation or *n*-hexane fractionation is adopted industrially to recover acylglycerols from the oil layer [9,10]. Oil containing high concentrations of DHA has been on the market in Japan as a nutraceutical since 1994.

Selective hydrolysis of borage oil containing γ-linolenic acid (GLA) increased the content of GLA from 22 to 45% at 60% hydrolysis using *C. rugosa* lipase as a

TABLE 20.3
Application of Lipases That Was Recently Reported

Application	Origin of Lipase	Appendix
Production of functional lipids		
Highly absorbable structured TAGs		
	Rhizomucor miehei	Immobilized 1,3-position specific lipase
	Thermomyces lanuginosa	
	Rhizopus niveus	
	R. oryzae, etc	
Monoacylglycerols	Pseudomonas	
	Penicillium camembertii	
	R. oryzae	
	C. rugosa	
	C. antarctica	
Synthesis of useful esters		
L-Menthyl esters	*C. rugosa*	
Steryl esters	*C. rugosa*	
Ascorbic acid esters	*C. antarctica*	
Sugar esters	*C. antarctica*	
Capsaicins	*C. antarctica*	
Purification of useful compounds		
PUFAs	*R. oryzae*	Selective esterification and selective alcoholysis
	Rhizomucor miehei	
	C. rugosa	
Tocopherols, sterols, steryl esters, and astaxanthin	*C. rugosa*	
	Pseudomonas aeruginosa	
Production of biodiesel fuel (fatty acid methyl esters)	*C. antarctica*	Stepwise methanolysis of TAGs
Production of fatty acid ethyl esters	*C. antarctica*	Stepwise ethanolysis of TAGs Ethyl esterification of fatty acids

catalyst [10,11]. An oil containing 57% arachidonic acid (AA) was also produced by hydrolyzing a single-cell oil containing 40% AA with the same lipase [12].

20.3.2 Production of Acylglycerols by Esterification

It has been reported that lipase-catalyzed esterification synthesizes various esters, such as fatty acid esters of primary and secondary alcohols (wax esters) [13], lactones [14], estolides [15,16], L-methyl esters [17], steryl esters [18], sugar esters [19,20], and ascorbic acid esters [21,22]. In this section, production of acylglycerols is described from the viewpoint of lipid modification.

20.3.2.1 Production of TAG

TAGs with oxidatively unstable fatty acids can be synthesized from glycerol and the free fatty acids (FFAs) by lipase-catalyzed esterification. Because this reaction is conducted in a system without the addition of water, the use of immobilized lipase is effective. When the esterification is conducted over 50°C, spontaneous acyl migration occurs easily. Hence, 1,3-position specific lipases can be available as well as nonspecific lipases. TAGs were synthesized efficiently even in an organic solvent-free system by mixing completely 1 mol of glycerol and 3 mol of FFA with immobilized *Candida antarctica* lipase (nonspecific; 1,3-position preferential) or *Rhizomucor miehei* lipase (1,3-position specific) [23,24]. The degree of esterification increased by evaporating generated water under reduced pressure with a vacuum pump, and the content of TAG reached nearly 90%. TAGs with PUFAs were also synthesized using *C. antarctica* lipase, which acted strongly on PUFAs (Table 20.4) [23,24].

20.3.2.2 Production of DAG

During digestion, TAGs are hydrolyzed normally to 2-MAGs and fatty acids by pancreatic lipase and incorporated into intestinal mucosa. The 2-MAGs and fatty acids are resynthesized to TAGs in epithelial cells, and a part of TAGs is accumulated in adipose tissue. Meanwhile, DAGs seem to be resynthesized with difficultly to TAGs after incorporation of the digested components (mainly 1-MAGs and FFAs) into intestinal mucosa, resulting in reduction of body fat [25].

DAGs are synthesized basically by esterification of glycerol with 2 mol of fatty acids using immobilized 1,3-position specific lipase. In this reaction, removal of generated water increases the yield of DAGs. DAGs produced by an enzymatic process have been on the market in Japan as a "food for specified health uses" since 1999.

20.3.2.3 Production of MAG

MAGs are very good emulsifiers and are widely used as food additives. MAGs with saturated and monoenoic fatty acids are produced industrially by chemical

TABLE 20.4
Synthesis of TAGs with PUFAs Using *C. antarctica* Lipase

	Composition (wt%)				
PUFA	TAG	1,3-DAG	1,2-DAG	MAG	FFA
GLA	89.1	4.0	1.5	1.1	4.3
AA	88.8	3.4	1.3	1.9	4.6
EPA	88.4	3.8	1.8	1.4	4.6
DHA	82.8	5.8	1.0	2.9	7.5

A mixture of PUFA/glycerol (1:3, mol/mol) and 5% immobilized *C. antarctica* lipase was agitated for 24 h at 40°C and 15 mmHg.

alcoholysis of oils and fats with 2 mol of glycerol at high temperatures of 210 to 240°C [26,27]. But the process cannot be applied to synthesize MAGs with unstable fatty acids. Meanwhile, an enzymatic process is available for production of MAGs with unstable fatty acids for nutritional application because it proceeds efficiently under mild conditions.

Enzymatic synthesis of MAGs has been actively studied since about 1990. Many of the reports were syntheses of MAGs with saturated and monoenoic acids by hydrolysis of TAGs, esterification of fatty acids with glycerol, glycerolysis of TAGs, and ethanolysis of TAGs in organic solvent systems [28]. An organic solvent-free system is preferable from the viewpoint of industrial production, and several systems have been proposed; glycerolysis of TAGs with *Pseudomonas* lipase [29,30] and esterification of fatty acids and glycerol with *Penicillium camembertii* mono- and diacylglycerol lipase [31]. An important fact can be drawn from these reports: the yield of MAG increases in the reaction at low temperatures. This conclusion is explained by the melting point of MAG. Its melting point is the highest among components (FFA, MAG, DAG, and TAG) in the reaction mixture if the constituent fatty acids are the same. Because lipases act very weakly on solid-state substrate, MAG solidified at low reaction temperatures and does not participate in the reaction. A high yield of MAG was therefore achieved [29,30,32].

In an industrial production of MAG, short-path distillation (molecular distillation) is presently adopted for purification of MAG from the reaction mixture. It is, however, difficult to separate MAG and FFA by distillation. Hence, the content of contaminating FFA after the reaction should be less than 5%. In light of this matter, we are studying the production of MAG with FFA for nutritional applications. While the study has not yet been completed, we developed two processes for production of MAG with conjugated linoleic acid (CLA). The processes are described below.

20.3.2.3.1 Two-Step In Situ *Reaction System*

CLA is a group of C_{18} fatty acids containing a pair of conjugated double bonds in either *cis* or *trans* configuration. A typical commercial product (referred to as FFA-CLA) contains almost equal amounts of 9*cis*, 11*trans* (9*c*,11*t*)-CLA, and 10*t*,12*c*-CLA. The FFA-CLA has been reported to have various physiological activities, such as reduction of the incidence of cancer [33,34], decrease in body fat content [35,36], beneficial effects on atherosclerosis [37,38], and improvement of immune function [39]. These activities attracted a great deal of attention, and synthesis of MAG from FFA-CLA and glycerol was attempted because the first product in the industrial process is the FFA mixture containing CLA.

Monoacylglycerol lipase synthesizes MAG from FFA and glycerol, and may not catalyze conversion of MAG to DAG. However, since there was no monoacylglycerol lipase available as an industrial enzyme, *Penicillium camembertii* mono- and diacylglycerol lipase (referred to as lipase) was selected and used as a catalyst for production of MAG with CLA by a two-step *in situ* reaction system [40].

The first step was esterification of FFA-CLA with 5 mol of glycerol. This reaction required 2% water by weight of the reaction mixture for maximal expression of

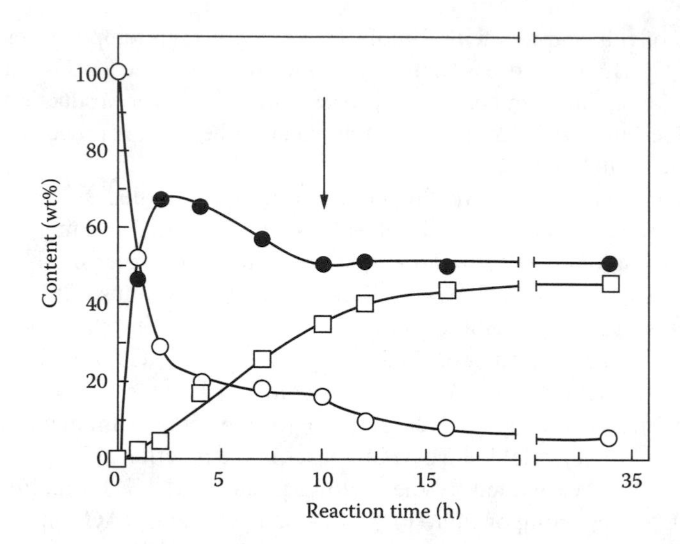

FIGURE 20.1 Esterification of FFA-CLA with glycerol with *P. camembertii* lipase. The reaction was conducted at 30°C in a mixture of 294 g FFA-CLA/glycerol (1:5, mol/mol) and 6 ml (60,000 U) *P. camembertii* lipase, with agitation at 250 rpm. At 20 h, as indicated with the arrow, the reactor was connected to a vacuum pump, and the reaction was continued with dehydration at 5 mmHg. Open circles, FFA-CLA; closed circles, MAG-CLA; squares, DAG-CLA.

activity of *P. camembertii* lipase. However, this water and the water generated by esterification hindered the achievement of a high degree of esterification, and the reaction reached a steady state at 80% esterification. Hence, after the reaction reached a steady state (10 h), the water was removed continuously by evacuation at 5 mmHg with a vacuum pump. The degree of esterification reached 95% after 24 h (34 h in total), and the contents of MAG and DAG were 49 and 46%, respectively (Figure 20.1) [40].

The second step is glycerolysis of DAG. The reaction mixture in the first-step esterification was solidified by vigorous agitation on ice. When the solidified mixture was allowed to stand on ice for 2 weeks, glycerolysis of DAG proceeded successfully, and the content of MAG in the reaction mixture increased to 89%. Hydrolysis did not occur during the glycerolysis, and the content of FFA decreased slightly from 5 to 4% [40].

20.3.2.3.2 Esterification at Low Temperature

MAG-CLA was efficiently produced by the two-step *in situ* reaction described in the previous section. But this process included a drawback that >2 weeks was necessary for completion of the reaction. If *P. camembertii* lipase does not recognize MAG synthesized by esterification of FFA-CLA with glycerol, the yield of MAG will increase because the MAG will not convert to DAG. We thus attempted esterification of FFA-CLA with glycerol at low temperatures.

When FFA-CLA was esterified at 5°C with 5 mol of glycerol using *P. camembertii* lipase in the presence of 2% water, the degree of esterification reached 90%

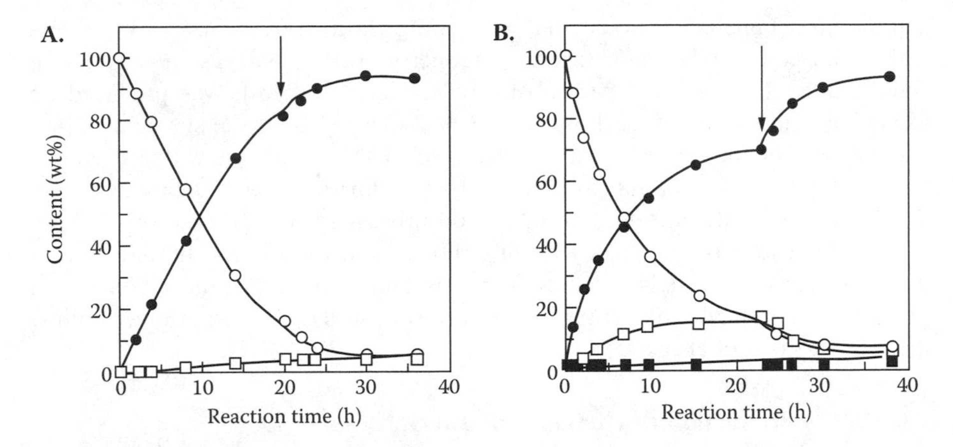

FIGURE 20.2 Production of MAG-CLA by esterification lipase at low temperature followed by dehydration. A 300-g mixture of FFA-CLA/glycerol (1:5, mol/mol), 60,000 U *P. camembertii* or *R. oryzae* lipase, and 2% of water originating from lipase solution was agitated at 5°C and 250 rpm. (A) Reaction with *P. camembertii* lipase. At 20 h, the reactor was connected to a vacuum pump, and the reaction was continued with dehydration at 5 mmHg. (B) Reaction with *R. oryzae* lipase. At 24 h, dehydration was started. Arrows indicate the time when dehydration was started. Open circles, FFA-CLA; closed circles, MAG-CLA; open squares, DAG-CLA; closed squares, TAG-CLA.

after 45 h and the contents of MAGs and DAGs were 87 and 5%, respectively [41]. TAGs were not synthesized in this *P. camembertii* lipase-catalyzed esterification. After the esterification was conducted for 20 h (the degree of esterification, 81%), dehydration was started by evaporation at 5 mmHg with a vacuum pump. The degree of esterification increased concomitantly with dehydration and reached 95% after 16 h (36 h in total). The contents of MAGs and DAGs were 93 and 3%, respectively (Figure 20.2A) [41]. When esterification system at low temperature was adopted, other triacylglycerols lipases (from *Rhizopus oryzae* and *C. rugosa*) were also effective for production of MAG-CLA. When *C. rugosa* lipase was used instead of *P. camembertii* lipase and the reaction was conducted under similar conditions, the degree of esterification reached 95% and the content of MAGs was 93% (Figure 20.2B) [32]. These results show that solidification of MAG at low reaction temperatures is effective for an increase in the reaction yield.

20.3.3 Production of Modified TAG by Transesterification

20.3.3.1 Production of Cocoa Butter Substitute

The main component of cocoa butter is TAG with palmitic and stearic acids at the 1,3-positions and oleic acid at the 2-position. The physical property, which has a sharp melting point around body temperature, is due to the specific structure of TAG, and the fat is important as a material for making chocolate. An enzymatic process was attempted to produce a fat that has a property (melting point) similar to that of cocoa butter. The modified fat, 1,3-stearoyl-2-oleoyl glycerol, was produced by exchange of fatty acids at the 1,3-positions of 2-oleoyl TAGs with stearic acid [42] and has been on the market in Japan since 1983. The new process with a fixed-bed bioreactor packed with an immobilized 1,3-specific lipase attracted attention in those days and significantly affected subsequent oil processing with lipases.

The bioreactor is operated by feeding substrates dissolved in *n*-hexane; thus, a modified fat having a higher melting point can be produced by similar process. At present, 1,3-behenoyl-2-oleoyl glycerol is also produced and is used for controlling the melting point of chocolate.

20.3.3.2 Production of Human Milk-Fat Substitute

Human milk fat contains 20 to 25% palmitic acid (PA), and about 70% of the fatty acid is esterified to the 2-position of TAGs [43,44]. In addition, the main component of the milk dienoic TAGs is 1,3-oleoyl-2-palmitoyl glycerol. Gastric and pancreatic lipases in infants hydrolyze dietary fat to 2-MAGs and fatty acids, and the absorption efficiency of free PA is relatively low compared with that of free unsaturated fatty acids [45]. Hence, the fat absorption is higher in infants fed fats with PA at the 2-position of TAGs than the 1,3-positions [46]. It has been hypothesized from these facts that the high absorption efficiency of human milk fat is the result of specific positioning of PA at the 2-position of the TAG moiety [46,47].

Based on these facts, 1,3-oleoyl-2-palmitoyl glycerol was produced as human milk-fat substitute. The fat was produced by similar reaction to that for production of cocoa butter substitute: transesterification of fatty acids at 2-palmitoyl TAGs with oleic acid (or oleic acid ethyl ester) using immobilized 1,3-position specific lipase [48,49]. The fat substitute is presently used as an ingredient in infant formula.

Human milk contains AA, which accelerates the growth of preterm infants, as does DHA [50,51]. Therefore, 1,3-arachidonoyl-2-palmitoyl glycerol was synthesized by acidolysis of tripalmitin with AA in an organic solvent-free system using immobilized *R. oryzae* lipase [52].

Because lipases act very weakly on solid-state substrates, high reaction temperatures are required when a high melting point fat is used as a substrate in an organic solvent-free system. Thermostable immobilized lipases are preferable for this purpose. It was reported that the thermostability is significantly increased by selecting a suitable carrier for the immobilization [53]. In addition, thermostable lipases were created by gene technology; *Fusarium heterosporum* lipase with the C-terminal prosequence [54-56] and *R. oryzae* lipase with the N-terminal prosequence [56,57] are thermostable.

20.3.3.3 Production of TAGs Containing Medium- and Long-Chain Fatty Acids

TAGs in various vegetable oils, such as soybean and rapeseed oils, are mainly composed of C_{18} long-chain fatty acids. As described in Section 20.3.2.2, TAGs are hydrolyzed to 2-MAGs and fatty acids by pancreatic lipase and incorporated into intestinal mucosa. The 2-MAGs and fatty acids are resynthesized to TAGs in epithelial cells, and a part of TAGs is accumulated in adipose tissue. Meanwhile, ester bonds of medium-chain fatty acids, such as C_8 and C_{10}, are hydrolyzed not only by pancreatic lipase, but also by gastric lipase, and the fatty acids undergo rapid β-oxidation in liver after their adsorption into intestinal mucosa. Hence, TAGs containing medium-chain fatty acids do not accumulate in adipose tissue. Attention was focused on this property of medium-chain fatty acids, and TAGs with long- and medium-chain fatty acids were produced industrially. The oil was produced by interesterification of rapeseed oil and medium-chain TAG, and has been on the market in Japan as a "Food for specified health uses" since 2003. A reaction with powdered *Alcaligenes* lipase is reported to be effective for production of this type of transesterified oil [58].

20.3.3.4 Production of Highly Absorbable Structured Lipids

In general, natural oils and fats contain saturated or monoenoic fatty acids at the 1,3-positions and highly unsaturated fatty acids at the 2-position. However, the distribution of fatty acids along the glycerol backbone is not specified. Meanwhile, TAGs having particular fatty acids at a specific position of glycerol are referred to as structured TAGs. TAGs with medium-chain fatty acid at the 1,3-positions and long-chain fatty acid at the 2-position (MLM-type) are hydrolyzed to 2-MAGs and fatty acids faster than natural oils and fats with long-chain fatty acids (LLL-type), resulting in efficient absorption into intestinal mucosa [59,60]. Because PUFAs play a role in the prevention of a number of human diseases, MLM containing PUFAs are expected as nutrition for patients with maldigestion and malabsorption of lipids and as high-value added nutraceuticals for the elderly. Studies on enzymatic production of structured TAGs have been conducted at many laboratories since 1995, and many processes have been proposed. Typical processes for the production are described below, although all of them do not adopt transesterification.

20.3.3.4.1 Production by Acidolysis or Inter esterification of Natural Oils

A typical production process of MLM-type structured TAGs is the same as that of production cocoa butter substitute and human milk-fat substitute. The structured TAGs can be produced by acidolysis of natural oils with medium-chain fatty acid or by their interesterification with medium-chain fatty acid ethyl esters using immobilized 1,3-position specific lipases (e.g., lipases from *Rhizopus oryzae, R. miehei*, and *Thermomyces lanuginosa*) [9,61–63]. MLM-type structured TAGs containing DHA were produced using tuna oil as a starting material [64]. MLM-type TAGs with GLA [65] and AA [66] were produced using borage oil and a single-cell oil containing high content of AA, respectively. In addition, MLM-type TAGs containing linoleic and α-linolenic acids were produced from safflower and linseed oils, respectively [67]. Purification of structured TAGs is efficiently achieved by short-path distillation [65]. Hexane extraction under alkaline conditions is also available for their purification from the reaction mixture obtained by acidolysis. Spontaneous acyl migration can be repressed by restricting the amount of water in the reaction mixture and by conducting at low temperatures (<30°C). We believe acidolysis or interesterification of natural oils is suitable for industrial production of MLM-type TAGs.

The content of medium-chain fatty acid in TAGs was increased to 40 to 50 mol% by single reaction. If all fatty acids at the 1,3-positions are exchanged with medium-chain fatty acid, the content will reach 66.7 mol%. Hence, the content of medium-chain fatty acid in TAGs showed that all of the TAGs in the reaction mixture were not MLM-type. To increase the content of MLM, repeated reaction was effective. Safflower and linseed oils underwent acidolysis with caprylic acid (CA) using immobilized *R. oryzae* lipase (Table 20.5). The contents of CA in TAGs were 49 and 47 mol%, respectively. After the reaction, their acylglycerol fractions were recovered and were then allowed to react again, resulting in 60 mol% of CA content. Three-times repetitions of acidolysis of the two oils reached 67 mol% of CA content, showing that all fatty acids at the 1,3-positions were exchanged with CA. Actually, high-performance liquid chromatography analysis indicated that TAGs changed almost completely to MLM-type [66,67].

Repeated reaction was effective for complete exchange of fatty acids at the 1,3-positions of natural oils with CA. However, because 1,3-position specific lipases act weakly on PUFAs, especially on DHA, tuna oil was not converted

TABLE 20.5
Production of Structured TAGs Rich in MLM from Safflower and Linseed oils by Repeated Reaction

Oil	Treatment	Fatty Acid Composition (mol%)						Acidolysis (%)
		8:0	16:0	18:0	18:1	18:2	18:3	
Safflower	None	n.d.	7.7	2.5	13.5	74.3	n.d.	—
	First	48.9	1.4	0.5	7.2	41.9	n.d.	73.3
	Second	59.8	0.5	n.d.	5.7	41.9	n.d.	89.7
	Third	67.4	n.d.	n.d.	5.2	27.4	n.d.	101.2
Linseed	None	n.d.	6.0	2.9	16.7	15.4	57.6	—
	First	46.6	1.3	0.6	10.0	10.5	30.9	69.9
	Second	60.8	0.4	n.d.	7.6	8.4	22.7	91.2
	Third	66.7	n.d.	n.d.	6.3	7.3	19.7	100.0

Safflower and linseed oils underwent acidolysis at 30°C for 48 h with two weight parts of CA using 4% immobilized *Rhizopus oryzae* lipase. After the single acidolysis, acylglycerols were recovered and then allowed to react again under similar conditions. The acidolysis was repeated three times in total. n.d., not detected.

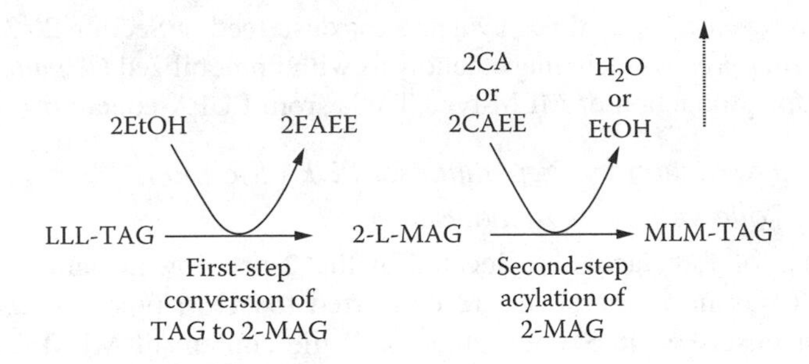

FIGURE 20.3 Production of MLM-type structured TAG by a process comprising ethanolysis of natural oil and acylation of the resulting 2-MAG. LLL-TAG, natural oil; 2-L-MAG, MAG with long-chain fatty acids at the 2-position.

completely to MLM-type TAGs, even though the acidolysis was repeated three times [66]. This problem could be solved using a process described in the next section.

20.3.3.4.2 Production by Preparation of 2-MAG Followed by Its Acylation

This process does not include transesterification reaction, which is the subject in this section, but is described for reference because of an interesting process for production of MLM-type structured TAGs.

MLM-type TAGs were produced by preparation of 2-MAG from natural oil, followed by acylation of the resulting 2-MAGs (Figure 20.3). In a process, 2-MAGs were prepared by ethanolysis of triolein or trilinolein with immobilized *R. miehei* lipase in a system containing methyl *t*-butyl ether (yield, 72%). The resulting 2-MAGs were esterified with CA using immobilized *R. miehei* or *R. oryzae* lipase in a mixture containing *n*-hexane (esterification, 90%). The synthesized TAGs contained >90% CA at the 1,3-positions and 99% long-chain fatty acids at the 2-position [68].

In another process, bonito oil underwent ethanolysis with excess amounts of ethanol (EtOH) (>40 mol/mol of TAGs) using immobilized *C. antarctica* lipase. This lipase acted strongly on C_8 to C_{24} of saturated and unsaturated fatty acids, and also acted on only fatty acids at the 1,3-positions in the presence of excess amount of EtOH. Hence, 2-MAGs were efficiently prepared (yield, 93%). Acylation of the 2-MAGs was then performed with CA ethyl ester (CAEE) using immobilized *R. miehei* lipase. Because this reaction by-produced EtOH, it was shifted to the right side by removing EtOH under reduced pressure. MLM-type TAGs were produced at 71% yield through this two-step process [69].

Natural oils containing PUFAs do not carry all PUFAs at the 2-position, and a part of PUFAs is located at the 1,3-positions. When natural oils containing PUFAs are converted to MLM-type structured TAGs using a 1,3-position specific lipase, it is difficult to convert all of the oils to MLM. Indeed, DHA at the 1,3-positions of tuna oil were converted with CA at only 30%, even though acidolysis of tuna oil with CA was repeated three times by a process described in Section 20.3.3.4.1 [66]. Hence, the latter process including ethanolysis with immobilized *C. antarctica* lipase is effective for production of MLM-type TAGs from PUFA-containing oils.

20.3.3.4.3 Production by Preparation of PUFA-Rich Oil Followed by Its Acidolysis

Several kinds of fatty acids are located at the 2-position in natural oils. Even though all TAGs in a natural oil are converted to MLM-type through the two-step process described in Section 20.3.3.4.2, the content of MLM containing a desired fatty acid is determined by the content of the fatty acid esterified at the 2-position of the original oil. Including a pretreatment for increasing the content of a desired fatty acid at the 2-position, MLM-type TAGs with that fatty acid are produced in high yield. This section deals with production of structured TAGs rich in 1,3-capryloyl-2-γ-linolenoyl glycerol (CGC) by a process comprising selective hydrolysis of borage oil and repeated acidolysis of the resulting acylglycerols with CA.

The content of GLA at the 2-position of borage oil was 49 mol% (Table 20.6). Hence, even though all TAGs were converted to MLM, the content should not exceed 49 mol%. The content of GLA at the 2-position will increase by previous removal of TAGs with fatty acids except for GLA at the 2-position. Selective hydrolysis described in Section 20.3 was effective for this purpose. An oil containing 45% GLA (GLA45 oil) was produced by selective hydrolysis of borage oil with *C. rugosa* lipase, and the content of GLA at the 2-position of GLA45 TAGs increased to 85 mol% (Table 20.6). Acidolysis of TAG45 TAGs with CA using immobilized *R. oryzae* lipase produced structured TAGs containing 45 mol% CGC, and the content of CGC was increased to 61 mol% by repeating the acidolysis three times [70]. Because three-time repetition of acidolysis of borage oil produced structured TAGs containing only 35 mol% CGC, the process including selective hydrolysis was shown to be effective for increasing the content of MLM containing a desired PUFA.

This two-step process was adopted for production of structured TAGs rich in 1,3-capryloyl-2-arachidonoyl glycerol (CAC) from a single-cell oil containing 40% AA. While acidolysis of the single-cell oil was repeated three times, the content of CAC

TABLE 20.6
Fatty Acid Composition at the 1,3- and 2-Positions of Borage Oil and GLA45 TAGs

		Fatty Acid Composition (mol%)							
Oil	**Position**	**16:0**	**18:0**	**18:1**	**18:2**	**18:3**	**20:1**	**22:1**	**24:1**
Borage	1,3	16.1	5.8	17.8	36.6	9.3	5.5	3.3	2.1
	2	0.6	n.d.	12.3	34.8	48.9	n.d.	0.3	n.d.
GLA45 TAGs	1,3	9.1	5.4	15.1	28.5	25.3	7.7	4.5	2.4
	2	n.d.	0.3	2.4	10.8	84.8	0.5	0.3	0.2

increased to only 36 mol%. The process including selective hydrolysis of the oil, however, increased the content of CAC to 51 mol% [71], confirming that this two-step process is valuable for increasing the content of a desired MLM.

20.3.3.4.4 Production by Synthesis of Simple TAG Followed by Its Acidolysis

MLM with only one fatty acid at the 2-position will not be synthesized as far as natural oils are used as starting materials, even though the reaction efficiency is raised. A high-purity MLM can be synthesized by preparation of a simple TAG, followed by exchange of fatty acids at the 1,3-positions with medium-chain fatty acid.

Chemical reaction is effective for synthesis of a simple TAG with saturated or monoenoic acid, but is not suitable for synthesis of a TAG with unstable PUFA. For this purpose, an enzymatic reaction is desirable. Simple TAGs with PUFAs were synthesized by esterification of 1 mol of glycerol with 3 mol of FFA using immobilized *C. antarctica* lipases as shown in Table 20.4. High-purity MLM with PUFA at the 2-position can be synthesized by acidolysis of TAG-PUFA with CA or by interesterification of the TAG with CAEE using immobilized 1,3-position specific lipases.

According to this strategy, 1,3-capryloyl-2-eicosapentaenoyl glycerol (CEC) and CAC were synthesized. Trieicosapentaenoin was first synthesized from 1 mol of glycerol and 3 mol of eicosapentaenoic acid (EPA) using immobilized *C. antarctica* lipase, and then underwent interesterification with 100 mol of CAEE using immobilized *R. miehei* lipase (yield of CEC, 88%) [72]. In addition, when acidolysis of triarachidonin and triesicosapentaenoin with 13 mol of CA using immobilized *R. oryzae* lipase was repeated three times, CEC and CAC were synthesized in 87 and 86% yields, respectively [24].

As described in Section 20.3.3.4.2, transesterification with 1,3-position specific lipases includes a drawback: the enzymes act weakly on PUFA. Tridocasahexaenoin actually underwent acidolysis with CA using immobilized *R. oryzae* lipase. While the reaction was repeated three times, the yield of 1,3-capryloyl-2-docosahexaenoyl glycerol (CDC) was only 19% [24]. High-yield synthesis of CDC was achieved through a two-step process comprising ethanolysis of tridocasahexaenoin using immobilized *C. antarctica* lipase for preparing 2-monodocosahexaenoin and acylation of the 2-MAG with CAEE using immobilized *R. miehei* lipase. The yield of CDC was 85% [73].

20.4 CONCLUSIONS

Modification of lipids with lipases has been described in this chapter. Lipases, however, are very useful for production of useful esters, such as wax esters, estolides, steryl esters, methyl esters, and capsicines (vanillyl amine fatty acid amides). The enzyme reactions are also effective for purification of PUFAs and oil- and fat-related compounds, such as tocopherols, sterols, steryl esters, and astaxanthin. In addition, lecithin can be converted to phosphatidylserine by phospholipase D. Because enzyme reactions have advantages that by-reactions scarcely occur and create only a little waste material, it is hoped that lipases will be used more and more in the oil and fat industry.

REFERENCES

1. Brzozowski, A.M. et al., A model for interfacial activation in lipases from the structure of a fungal lipase-inhibitor complex, *Nature,* 351, 491, 1991.
2. Tilbeurgh, H. et al., Interfacial activation of the lipase-procolipase complex by mixed micelles revealed by x-ray crystallography, *Nature,* 362, 814, 1993.
3. Tanaka, Y. et al., Triglyceride specificity of *Candida cylindracea* lipase: effect of docosahexaenoic acid on resistance of triglyceride to lipase, *J. Am. Oil Chem. Soc.,* 70, 1031, 1993.
4. Shimada, Y. et al., Selective hydrolysis of polyunsaturated fatty acid-containing oil with *Geotrichum candidum* lipase, *J. Am. Oil Chem. Soc.,* 72, 1577, 1995.
5. Watanabe, Y. et al., Enzymatic conversion of waste edible oil to biodiesel fuel in a fixed-bed bioreactor, *J. Am. Oil Chem. Soc.,* 78, 703, 2001.
6. Sugihara, A. et al., A novel *Geotrichum candidum* lipase with some preference for the 2-position on a triglyceride molecule, *Appl. Microbiol. Biotechnol.,* 35, 738, 1991.
7. Sugihara, A. et al., Positional and fatty acid specificities of *Geotrichum candidum* lipases, *Prot. Eng.,* 7, 585, 1994.
8. Hoshino, T., Yamane, T., and Shimizu, S., Selective hydrolysis of fish oil by lipase to concentrate n-3 polyunsaturated fatty acids, *Agric. Biol. Chem.,* 54, 1459, 1990.
9. Shimada, Y., Sugihara, A., and Tominaga, Y., Production of functional lipids containing polyunsaturated fatty acids with lipase, in *Enzymes in Lipid Modification,* Bornscheuer, U.T., Ed., Wiley-VCH, Weinheim, Germany, 2000, p. 128.
10. Shimada, Y. et al., Selective hydrolysis of borage oil with *Candida rugosa* lipase: two factors affecting the reaction, *J. Am. Oil Chem. Soc.,* 75, 1581, 1998.
11. Syed Rahmatullah, M.S.K., Shukla, V.K.S., and Mukherjee, K.D., Enrichment of γ-linolenic acid from evening primrose oil and borage oil via lipase-catalyzed hydrolysis, *J. Am. Oil Chem. Soc.,* 71, 569, 1994.
12. Shimada, Y. et al., Enzymatic purification of n-6 polyunsaturated fatty acids, *Kagaku to Kogyo,* 73, 125, 1999 (in Japanese).
13. Okumura S., Iwai, M., and Tsujisaka, Y., Synthesis of various kinds of esters by four microbial lipases, *Biochim. Biophys. Acta,* 575, 156, 1979.
14. Makita, A., Nihira, T., and Yamada, Y., Lipase catalyzed synthesis of macrocyclic lactones in organic solvents, *Tetrahedron Lett.,* 28, 805, 1987.
15. Okumura, S., Iwai, M., and Tsujisaka, Y., Synthesis of estolides during hydrolysis of castor oil by *Geotrichum candidum* lipase, *J. Jpn. Oil Chem. Soc.,* 32, 271, 1983.
16. Yoshida, Y. et al., Synthesis of estolides with immobilized lipase, *J. Jpn. Oil Chem. Soc.,* 44, 328, 1995.
17. Shimada, Y. et al., Enzymatic synthesis of L-methyl esters in organic solvent-free system, *J. Am. Oil Chem. Soc.,* 76, 1139, 1999.
18. Shimada, Y. et al., Enzymatic synthesis of steryl esters of polyunsaturated fatty acids, *J. Am. Oil Chem. Soc.,* 76, 713, 1999.
19. Zhang, X. et al., Lipase-catalyzed synthesis of monolauroyl maltose through condensation of maltose and lauric acid, *Food Sci. Technol. Res.,* 9, 110, 2003.
20. Kobayashi, T., Adachi, S., and R. Matsuno, Kinetic analysis of the immobilized-lipase-catalyzed synthesis of octanoyl octyl glucoside in acetonitrile, *Biochem. Eng. J.,* 16, 323, 2003.
21. Kuwabara, K. et al., Synthesis of 6-O-unsaturated acyl L-ascorbates by immobilized lipase in acetone in the presence of molecular sieve, *Biochem. Eng. J.,* 16, 17, 2003.

22. Kuwabara, K. et al., Continuous production of acyl L-ascorbates using a packed-bed reactor with immobilized lipase, *J. Am. Oil Chem. Soc.,* 80, 895, 2003.
23. Haraldsson, G.G., Gudmundsson, B.O., and Almarsson, O., The preparation of homogeneous triglycerides of eicosapentaenoic acid and docosahexaenoic acid by lipase, *Tetrahedron Lett.,* 34, 5791, 1993.
24. Kawashima, A. et al., Enzymatic synthesis of high-purity structured lipids with caprylic acid at 1,3-positions and polyunsaturated fatty acid at 2-position, *J. Am. Oil Chem. Soc.,* 78, 611, 2001.
25. Matsuo, N. and Tokimitsu, I., Methabolic characteristics of diacylglycerol: an edible oil that is less likely to become body fat, *INFORM,* 12, 1098, 2001.
26. Sonntag, N.O.V., New developments in the fatty acid industry in America, *J. Am. Oil Chem. Soc.,* 61, 229, 1984.
27. Lauridsen, J.B., Food emulsifiers: surface activity, edibility, manufacture, composition, and application, *J. Am. Oil Chem. Soc.,* 53, 400, 1976.
28. Aha, B. et al., Lipase-catalyzed synthesis of regioisomerically pure mono- and diglycerides, in *Enzymes in Lipid Modification,* Bornscheuer, U.T., Ed., Wiley-VCH, Weinheim Germany, 2000, p. 100.
29. McNeil, G..P., Shimizu, S., and Yamane, T., High-yield enzymatic glycerolysis of fats and oils, *J. Am. Oil Chem. Soc.,* 68, 1, 1991.
30. McNeil, G..P. and Yamane, T., Further improvements in the yield of monoglycerides during enzymatic glycerolysis of fats and oils, *J. Am. Oil Chem. Soc.,* 68, 6, 1991.
31. Yamaguchi, S. and Mase, T., High-yield synthesis of monoglyceride by mono- and diacylglycerol lipase from *Penicillium camembertii* U-150, *J. Ferment. Bioeng.,* 72, 162, 1991.
32. Watanabe, Y. et al., Production of MAG of CLA in a solvent-free system at low temperature with *Candida rugosa* lipase, *J. Am. Oil Chem. Soc.,* 80, 909, 2003.
33. Pariza M.W., CLA, a new cancer inhibitor in dairy products, *Bull. Int. Dairy Fed.,* 257, 29, 1991.
34. Ip, C. et al., Mammary cancer prevention by conjugated dienoic derivatives of linoleic acid, *Cancer Res.,* 51, 6118, 1991.
35. Park, Y. et al., Effect of conjugated linoleic acid on body composition in mice, *Lipids,* 32, 853, 1997.
36. Ostrowska, E. et al., Dietary conjugated linoleic acids increase lean tissue and decrease fat deposition in growing pigs, *J. Nutr.,* 129, 2037, 1999.
37. Lee, K.N., Kritchevsky, D., and Pariza M.W., Conjugated linoleic acid and atherosclerosis in rabbits, *Atherosclerosis,* 108, 19, 1994.
38. Nicolosi, R.J. et al., Dietary conjugated linoleic acid reduces plasma lipoproteins and early aortic atherosclerosis in hypercholesterolemic hamsters, *Artery,* 22, 266, 1997.
39. Sugano, M. et al., Conjugated linoleic acid modulates tissue levels of chemical mediators and immunoglobulins in rats, *Lipids,* 33, 521, 1998.
40. Watanabe, Y. et al., Synthesis of MAG of CLA with *Penicillium camembertii* lipase, *J. Am. Oil Chem. Soc.,* 79, 891, 2002.
41. Watanabe, Y. et al., Production of monoacylglycerol of conjugated linoleic acid by esterification followed by dehydration at low temperature using *Penicillium camembertii* lipase, *J. Mol. Catal. B: Enzymol.,* 27, 249, 2004.
42. Yokozeki, K. et al., Application of immobilized lipase to region-specific interesterification of triglyceride in organic solvent, *Eur. J. Appl. Microbiol.,* 14, 1, 1982.
43. Breckenridge, W.C., Marai, L., and Kuksis, A., Triglyceride structure of human milk fat, *Can. J. Biochem,* 47, 761, 1969.

44. Martin, J.C. et al., Triacylglycerol structure of human colostrum and mature milk, *Lipids,* 28, 637, 1993.
45. Jensen, C., Buist, N.R., and Wilson, T., Absorption of individual fatty acids from long chain and medium chain triglycerides in very small infants, *Am. J. Clin. Nutr.,* 43, 745, 1986.
46. Innis, S.M., Dyer, R., and Nelson, C.M., Evidence that palmitic acid is absorbed as *sn*-2-monoacylglycerol from human milk by breast-fed infants, *Lipids,* 29, 541, 1994.
47. Fomon, S.J. et al., Excretion of fat by normal full-term infants fed various milks and formulas, *Am. J. Clin. Nutr.,* 23, 1299, 1970.
48. Akoh, C.C. and Xu, X., Enzymatic production of Betapol and other specialty fats, in *Lipid Biotechnology,* Kuo, T.M. and Gardner, H.W., Eds., Marcel Dekker, New York, 2002, 461.
49. Nagao, T. et al., Use of thermostable *Fusarium heterosporum* lipase for production of structured lipid containing oleic and palmitic acids in organic solvent-free system, *J. Am. Oil Chem. Soc.,* 78,167, 2001.
50. Carlson, S.E. et al., Arachidonic acid status correlates with first year growth in preterm infants, *Proc. Natl. Acad. Sci. USA,* 90, 1073, 1993.
51. Lanting, C.I. et al., Neurological differences between 9-year-oil children fed breast-milk as babies, *Lancet,* 344, 1319, 1994.
52. Shimada, Y. et al., Enzymatic synthesis of structured lipid containing arachidonic and palmitic acids, *J. Am. Oil Chem. Soc.,* 77, 89, 2000.
53. Shimada, Y. et al., Immobilization of *Rhizopus delemar* lipase, *Kagaku to Kogyo,* 74, 117, 2000 (in Japanese).
54. Nagao, T. et al., C-Terminal peptide of *Fusarium heterosporum* lipase is necessary for its increasing thermostability, *J. Biochem.,* 124, 1124, 1998.
55. Nagao, T. et al., Amino acid residues contributing to stabilization of *Fusarium heterosporum* lipase, *J. Biosci. Bioeng.,* 89, 446, 2000.
56. Nagao, T. et al., Increase in stability of *Fusarium heterosporum* lipase, *J. Mol. Catal. B: Enzymol.,* 17, 125, 2002.
57. Beer H.D. et al., The folding and activity of the extracellular lipase of *Rhizopus oryzae* are modulated by a prosequence, *Biochem. J.,* 319, 351, 1996.
58. Negishi, S. et al., Activation of powdered lipase by cluster water and the use of lipase powders for commercial esterification of food oils, *Enzmol. Microb. Technol.,* 32, 66, 2003.
59. Ikeda, I. et al., Lymphatic absorption of structured glycerolipids containing medium-chain fatty acids and linoleic acid, and their effect on cholesterol absorption in rats, *Lipids,* 26, 369, 1991.
60. Christensen, M.S. et al., Intestinal absorption and lymphatic transport of eicosapentaenoic (EPA), docosahexaenoic (DHA), and decanoic acids: dependence on intramolecular triacylglycerols structure, *Am. J. Clin. Nutr.,* 61, 56, 1995.
61. Xu, X, Enzymatic production of structured lipids: process reactions and acyl migration, *Inform,* 11, 1121, 2000.
62. Iwasaki, Y. and Yamane T., Lipase-catalyzed synthesis of structured triacylglycerols containing polyunsaturated fatty acids: monitoring the reaction and increasing the yield, in *Enzymes in Lipid Modification,* Bornscheuer, U.T., Ed., Wiley-VCH, Weinheim, Germany, 2000, p. 148.
63. Akoh, C.C. et al., Enzymatic synthesis of structured lipids, in *Lipid Biotechnology,* Kuo, T.M. and Gardner, H.W., Eds., Marcel Dekker, New York, 2002, p. 433.
64. Shimada, Y. et al., Production of structured lipid containing docosahexaenoic and caprylic acids using immobilized *Rhizopus delemar* lipase, *J. Ferment. Bioeng.,* 81, 299, 1996.

65. Shimada, Y. et al., Continuous production of structured lipid containing γ-linolenic and caprylic acids by immobilized *Rhizopus delemar* lipase, *J. Am. Oil Chem. Soc.,* 76, 189, 1999.
66. Shimada, Y. et al., Fatty acid specificity of *Rhizopus delemar* lipase in acidolysis, *J. Ferment. Bioeng.,* 83, 321, 1997.
67. Shimada, Y. et al., Production of structured lipids containing essential fatty acids by immobilized *Rhizopus delemar* lipase, *J. Am. Oil Chem. Soc.,* 73, 1415, 1996.
68. Soumanou, M.M., Bornscheuer, U.T., and Schmid, R.D., Two-step enzymatic reaction for synthesis of pure structured triacylglycerides, *J. Am. Oil Chem. Soc.,* 75, 703, 1998.
69. Irimescu, R. et al., Two-step enzymatic synthesis of docosahexaenoic acid-rich symmetrically structured triacylglycerols *via* 2-monoacylglycerols, *J. Am. Oil Chem. Soc.,* 78, 743, 2001.
70. Kawashima, A. et al., Production of structured TAG rich in 1,3-dicapryloyl-2-γ-linolenoyl glycerol from borage oil, *J. Am. Oil Chem. Soc.,* 79, 871, 2002.
71. Nagao, T. et al., Production of structured TAG rich in 1,3-capryloyl-2-arachidonoyl glycerol from *Mortierella* single-cell oil, *J. Am. Oil Chem. Soc.,* 80, 867, 2003.
72. Irimescu, R. et al., Enzymatic synthesis of 1,3-dicapryloyl-2-eicosapentaenoyl glycerol, *J. Am. Oil Chem. Soc.,* 77, 501, 2000.
73. Irimescu, R. et al., Utilization of reaction medium-dependent regiospecificity of *Candida antarctica* lipase (Novozym 435) for the synthesis of 1,3-dicapryloyl-2- docosahexaenoyl (or eicosapentaenoyl) glycerol, *J. Am. Oil Chem. Soc.,* 78, 285, 2001.

21 Application of Structured Lipids in Emulsions

Hannah T. Osborn-Barnes and Casimir C. Akoh

CONTENTS

21.1 INTRODUCTION

Structured lipids (SLs) are a new generation of fats and oils with improved nutritional or physical properties because of modifications to incorporate new fatty acids or to change the position of existing fatty acids on the glycerol backbone by chemically and enzymatically catalyzed reactions and/or genetic engineering. Comprehensive reviews on the rationale, production, analysis, commercial outlook, and future prospects of SLs are available [1–3]. With the ability to combine the beneficial characteristics of component fatty acids into one triacylglycerol (TAG) molecule, lipid modification enhances the role fats and oils play in food, nutrition, and health applications. Many of these potential applications exist either partly or wholly as emulsions. Examples include nutritional beverages, infant formulae, mayonnaise, margarine, and salad dressings.

An emulsion consists of two immiscible liquids (usually oil and water), one dispersed in the other in the form of small spherical droplets. A system that consists of oil droplets dispersed in an aqueous phase is known as an oil-in-water (o/w) emulsion, whereas a system that consists of water droplets dispersed in an oil phase is known as a water-in-oil (w/o) emulsion. Emulsions are thermodynamically unstable systems because of the positive free energy required to increase the surface area between the oil and water phases [4]. With time, emulsions tend to separate into a system that consists of a layer of oil on the top of a layer of water. To form emulsions that are kinetically stable for a reasonable period of time, emulsifiers are required. Emulsifiers are surface-active molecules that adsorb to the surface of freshly formed droplets during homogenization, forming a protective "membrane" that prevents the droplets from aggregating [5]. Emulsifiers are often amphiphilic molecules that have both polar and nonpolar regions on the same molecule. Once emulsifiers have been added to the oil and water, forming drops is easy, but breaking them up into small droplets requires additional energy. Homogenizers provide the inertial forces needed to deform and break up emulsion droplets, which are produced by the rapid, intensive pressure fluctuations occurring in turbulent flow [6]. O/w emulsions can be divided into three distinct regions: the interior of the droplet (oil), the continuous phase (water), and the interfacial region (emulsifier).

Emulsified foods may contain a variety of ingredients, including water, lipids, proteins, carbohydrates, minerals, sugars, small-molecule surfactants, transition metals, and antioxidants. Food emulsions undergo diverse treatments, such as storage at various conditions, before they reach the consumer's plate. These treatments induce emulsion modifications, including chemical reactions such as lipid oxidation, which can produce undesirable off-flavors and potentially toxic reaction compounds that make the products no longer acceptable for human consumption. Therefore, lipid oxidation is of great concern to the food industry and manufacturers must develop methods of preventing or retarding oxidation in foods [5].

In addition to altering the nutritional and physical properties, changing the fatty acids on a TAG will also affect its oxidation properties. SLs often contain both long-chain polyunsaturated fatty acids and saturated medium-chain fatty acids (MCFAs) on the same glycerol backbone. Because of the inherent structural differences between SLs and unmodified oils, the oxidation properties of SL-based emulsions cannot be predicted based on current knowledge of conventional lipids in emulsions alone.

This chapter covers information on emulsions formulated with SLs with improved functional and nutritional properties. Potential food and medical applications for SL-based emulsions and the factors affecting the oxidation of SL-based emulsions are discussed.

21.2 APPLICATIONS OF STRUCTURED LIPID-BASED EMULSIONS

21.2.1 Structured Lipids with Improved Nutritional Properties

The relationship between stereospecific fatty acid location and lipid nutrition suggests that the process of interesterification or acidolysis could be used to improve the nutrition profile of certain TAGs. Manufacturers of specialty food ingredients for infant formulae, parenteral supplements, and nutritional beverages should design fats with saturated fatty acids at the *sn*-2 position to provide increased caloric intake [7].

21.2.1.1 Infant Formulas

The fat component of infant formulae should ideally contain the fatty acids, such as MCFAs, linoleic acid, linolenic acid, and polyunsaturated fatty acids (PUFAs), in the same position and amount as those found in human milk. Human milk is composed of 20 to 30% palmitic acid, with 33% at the *sn*-2 position [8]. The fat in most infant formulae is of vegetable origin and tends to have unsaturated fatty acids in the *sn*-2 position. Innis et al. [9] found that infants fed human milk had 26% palmitic acid in their plasma TAGs compared with 7.4% in infants fed vegetable oil–based infant formula with the same total concentration of palmitic acid, but not at the *sn*-2 position. When rats were fed a coconut oil and palm olein SL, absorption was increased due to the increased proportion of long-chain saturates at the *sn*-2 position [10]. Although infant formula emulsions have not been prepared with SLs and administered to infants, evidence exists that SLs with high proportions of palmitic acid at the *sn*-2 position would provide a fat with improved absorption capability in infants [8]. Such studies are merited, but care must be taken with regard to the concentrations of the saturates at the *sn*-2 position, because palmitic acid is the only saturate that has been studied extensively and other long-chain saturates may have hypercholesterolemic effects [11].

21.2.1.2 Parenteral Nutrition

While medium chain triacylglycerols (MCTs) have many advantages, humans still require a minimum amount of long-chain triacylglycerols (LCTs) to provide essential fatty acids (EFAs). Physical mixtures of MCTs and LCTs have proved useful in the past for parenteral (intravenous feeding) nutrition. More recently, structured TAGs composed of long-chain fatty acids (LCFAs) and MCFAs have emerged as the preferred alternative to physical mixtures for treatment of patients, although both products provide identical fat contents. Structured lipids composed of both LCFAs and MCFAs are designed to provide simultaneous delivery of the fatty acids and a slower, more controlled release of the LCFA into the bloodstream [12]. The advantages of parenterally fed SLs may well relate to differences in absorption and processing. Structured TAGs that contain MCFA may provide a vehicle for rapid hydrolysis and absorption due to their smaller molecular size and greater water solubility in comparison to long-chain TAGs. [13] Structured lipids offer several advantages over native oils and physical mixtures, including improved immune function, decreased cancer risk, thrombosis prevention, cholesterol lowering, improved nitrogen balance, and no risk of reticuloendothelial system impairment [14]. Increases in protein synthetic rates in both skeletal muscle and liver have also been demonstrated in patients receiving SL [15].

The TAGs in total parenteral nutrition (TPN) are normally administered as an emulsion. These emulsions are suspected of suppressing the immune function because pneumonia and wound infection often occur in patients treated with TPN. However, the influence of various lipid emulsions on leukocyte function is still unclear.

Kruimel et al. [16] attempted to explain this phenomenon through *in vitro* studies of emulsions containing fatty acids with different chain lengths on the production of radicals by polymorphonuclear leukocytes. Emulsions made from LCT, a physical mixture of LCT and MCT, and an SL were compared in this study. The results indicated physical mixtures caused higher peak levels and faster production of oxygen radicals, compared with LCT and SL emulsions. Chambrier et al. [17] conducted a similar study comparing the effect of physical mixtures and SL on postoperative patients. They did not see the hepatic function disturbances in patients given the SL, which are often observed with TPN. The plasma TAG levels remained normal in patients given SL, whereas they significantly increased with the physical mixture. Bellantone et al. [18] gave lipid emulsions to patients after colorectal surgery. The differences between the SL and physical mixture groups were less marked in this study.

SLs containing MCFAs and n-3 PUFAs could be a therapeutic or medical lipid source and may be useful in parenteral nutrition as well. Fish oil and MCTs were used to synthesize structured lipids, which were administered to patients undergoing surgery for upper gastrointestinal malignancies. The control diet used for comparison in the study differed only in its fat source. The SL diet was tolerated significantly better, led to improved hepatic and renal function, and reduced the number of infections per patient [19].

Lee et al. [20] fed female mice diets supplemented with an SL containing n-3 PUFAs and caprylic acid or soybean oil for 21 d. The effect of the diet on serum lipids and glucose concentrations were determined at the end of the feeding period. In spite of the higher content of caprylic acid in the SL, the fatty acid was not detected in the livers of the SL fed mice. High amounts of n-3 PUFAs were found in the livers of the SL fed mice. These findings suggest that the caprylic acid was metabolized quickly for energy and that different fatty acids in the diet may eventually lead to change in fatty acid composition of the liver. This SL could decrease serum cholesterols and TAGs. It may also reduce the rate of body weight gain because the MCFAs were metabolized more rapidly in the body compared with soybean oil. SLs of this type are ideal for inclusion into TPN emulsions because of their improved nutritional properties and rapid absorption.

Recently, a long-term study (4 weeks of treatment) was conducted on the efficacy and safety of an SL prepared from coconut oil and soybean oil. Patients receiving the SL were compared with other patients given LCT. This double-blind, randomized, crossover study indicated that the SL was safe and efficient when provided to patients on home parental nutrition on a long-term basis and that it may be associated with possible reduction in liver dysfunction [21].

21.2.1.3 Nutritional Beverages

Several nutritional beverages are commercially available for supplemental use with or between meals, or as a sole source of nutrition. These beverages are manufactured for people on modified diets, at nutrition risk, or with involuntary weight loss, and for patients who are recovering from illness or injury. These products are ideal targets for canola oil replacement with an SL containing MCFAs and LCFAs on the same glycerol backbone to increase the TAG absorption rate and provide essential fatty acids. Beverages formulated with SL have the potential to safely provide consumers with a more rapid energy source than the beverages currently available on the market [21–23]. However, the flavor must not be significantly altered compared with the original beverage formulation, as sensory characteristics of a product greatly influence its performance in the marketplace.

Osborn et al. [24] formulated chocolate-flavored nutritional beverages with canola oil and SL to determine the effect of the SL on the sensory profile of the beverage. The SL was synthesized from canola oil and caprylic acid with an *sn*-1,3 specific lipase obtained from *Rhizomucor miehei*. Differences were determined using a triangle test panel and a trained quantitative descriptive analysis (QDA) panel. Panelists detected a significant difference between samples during triangle testing ($p < .01$). QDA results indicated that sweet flavor was the only prominent attribute significantly influenced when SL was substituted for canola oil in the nutritional beverage. The presence of MCFA on the SL glycerol backbone may be responsible for the increased sweetness, as fatty acids of this length are currently used by the food industry as flavor carriers [25]. Panelists also detected differences in the foaminess of the two beverages, which may be attributed to the changes in fatty acids on the glycerol backbone as well. Studies conducted on wine indicate that increasing the amounts of linolenic acid and palmitic acid results in the formation of more foam [26]. During the acidolysis reaction, a portion of the linolenic and palmitic acids were removed from the canola oil TAG molecule (3.8 and 2.4%, respectively), and may explain the decreased foaminess observed in the beverage formulated with the SL. This study is the first to evaluate the effect of SL on the sensory properties of food emulsions and the results indicate that SL are likely to be suitable for use by beverage manufacturers from a flavor standpoint.

21.2.3 Structured Lipids with Improved Physical Properties

Improvements or changes in the physical characteristics of a TAG can also be achieved when SLs are synthesized. Modifying TAGs to improve melting points is an area of growing interest because replacing hydrogenated oils with SLs results in *trans*-free food products. *Trans* fatty acids are widely believed to have implications in coronary heart disease [27–29].

Margarine, modified butters, and shortenings manufacturers need fats with a steep solid fat content (SFC) curve. Margarine is a w/o emulsion that must contain 80% fat. Manufacturers want their product to be solid in the refrigerator, but spread easily upon removal and melt quickly in the mouth. The spreadability of margarine at refrigerator temperatures is related to its content of solid fats at 2 and 10°C. The solid content at 25°C influences plasticity at room temperature [30]. Desired spreadability occurs within a range of roughly 15 to 35% solids [31]. Interesterification is useful for producing plastic fats and oils suitable for use in margarines and shortenings because chemical properties of the original fat are relatively unaffected and the fatty acids' inherent properties are not changed. Additionally, unsaturation levels stay constant in the fatty acids and there is no *cis–trans* isomerization. When short (SCFAs) chain fatty acids and LCFAs are incorporated, they can produce TAGs with good spreadability and temperature stability.

Marangoni et al. [32] developed a method for enzymatic interesterification of triolein and tripalmitin in a canola lecithin–hexane reverse micelle system that can also be adapted for modification of vegetable oils. Chemical interesterification of butterfat and canola oil was shown to improve the cold-temperature spreadability of butter [33]. Acidolysis, catalyzed by an *sn*-1,3 specific lipase, of stearic acid and triolein produced a *trans*-free SL suitable for use in margarines [34]. This SL was also produced via interesterification with stearic acid methyl ester and a nonspecific lipase. Finally, SLs have been produced successfully through genetic engineering. The high-palmitic soybean oil, described by Stoltzfus et al. [35], has application in the production of *trans*-free margarine-type fats.

Fomuso and Akoh [36] combined two methods of SL production to produce a margarine fat. Stearic acid was enzymatically transesterified with genetically engineered high-laurate canola oil. A stearic acid level of 30% was found to best match the melting characteristics of fat extracted from commercially available stick margarine. This SL was used to prepare nonrefrigerated and refrigerated SL margarines. Slip melting point (SMP), SFC, and hardness index were determined for SL samples and compared with control margarines. Rheological properties of the margarine samples were also studied. The SL was suitable (spreadable with no oil exudation or phase separation) for use in nonrefrigerated margarine, but not spreadable at refrigeration temperatures. The amount of SL in the refrigerated samples was reduced to 60% and the remaining 40% fat was made up with canola oil. The addition of canola oil improved the spreadability at refrigeration temperatures and reduced the hardening effect of the lauric acid in the SL.

The viscoelastic properties of mayonnaise and Italian salad dressing prepared with olive oil or an enzymatically synthesized SL prepared from olive oil and caprylic acid were measured [37]. Olive oil- and SL-based mayonnaise and Italian salad dressings displayed similar viscoelastic properties, which indicates similar structural and textural properties. The SL-based emulsions were significantly more stable to separation than their olive oil-based counterparts.

21.3 OXIDATION OF STRUCTURED LIPID-BASED EMULSIONS

21.3.1 Oxidation Reaction Mechanisms

It is necessary to have a thorough understanding of the mechanisms of lipid oxidation and how these are affected by the molecular environment of the emulsified lipids to effectively design strategies for retarding oxidation in food emulsions. This portion of the chapter will focus exclusively on the oxidation of o/w emulsions because the oxidation properties of oils in this type of emulsion are significantly different than in the bulk phase, whereas, it has been suggested that lipid oxidation in w/o emulsions occurs at a rate similar to that in bulk oils [5].

Food emulsions, prepared with conventional or structured lipids, are susceptible to spoilage through the auto-oxidation of their unsaturated and polyunsaturated oil components [38]. Auto-oxidation reaction mechanisms apply to both systems and entail oxygen from the atmosphere being added to certain fatty acids, creating unstable intermediates that eventually break down to form unpleasant flavor and aroma compounds. Although enzymatic and photogenic oxidation may play a role, the most common and important process by which unsaturated fatty acids and oxygen interact is via a complex, radical chain reaction characterized by three main phases [39]:

$$\text{Initiation: } In^{\bullet} + RH \rightarrow InH + R^{\bullet}$$

$$\text{Propagation: } R^{\bullet} + O_2 \rightarrow ROO^{\bullet}$$

$$ROO^{\bullet} + RH \rightarrow R^{\bullet} + ROOH$$

$$\text{Termination: } ROO^{\bullet} + ROO^{\bullet} \rightarrow O_2 + ROOR$$

$$ROO^{\bullet} + R^{\bullet} \rightarrow ROOR$$

Initiation is frequently attributed in most foods to the reaction of fatty acids with active oxygen species ($In^{\bullet}$). Initiation occurs as hydrogen is abstracted from an unsaturated fatty acid (RH), resulting in a lipid free radical ($R^{\bullet}$), which in turn reacts with molecular oxygen (O_2) to form a lipid peroxyl radical ($ROO^{\bullet}$). The propagation phase of oxidation is fostered by lipid–lipid interactions, whereby the lipid peroxyl radical abstracts hydrogen from an adjacent molecule, resulting in a lipid hydroperoxide (ROOH) and a new lipid free radical. Such interactions continue 10 to 100 times before two radical species ($R^{\bullet}$ or $ROO^{\bullet}$) combine to terminate the process by forming a nonreactive molecule (ROOR) [39]. Most food oils naturally contain enough lipid peroxides to promote lipid oxidation, even if other sources of radical generators are rigorously eliminated [5]. Additional magnification of lipid oxidation in food emulsions occurs through branching or secondary initiation reactions promoted by transition metals or other prooxidants as shown below [40]:

$$Fe^{3+} + ROOH \rightarrow Fe^{2+} + ROO^{\bullet} + H^{+}$$

$$Fe^{2+} + ROOH \rightarrow Fe^{3+} + RO^{\bullet} + OH^{-}$$

Lipid hydroperoxides are not considered harmful to food quality; however, they are further degraded into compounds that are responsible for off-flavors [39]. Formation of the alkoxyl radical ($RO^{\bullet}$) leads to β-scission reactions that result in the generation of a wide variety of different molecules, including aldehydes, ketones, alcohols, and hydrocarbons, which are responsible for the characteristic physicochemical and sensory properties of oxidized oils [40].

The above reaction mechanisms do not illustrate the importance of the physical location of the various reactive species within the emulsion system. Often, hydroperoxides accumulate at the surface of the emulsion droplets, whereas many of the molecular species responsible for accelerating lipid oxidation originate in the aqueous phase (i.e., transition metals or enzymes). Therefore, the inability of prooxidants to come into close contact with lipids at the droplet surface would decrease their ability to accelerate lipid oxidation in food emulsions. Additionally, the rate of lipid oxidation may be limited by the speed that free radicals, hydroperoxides, or lipids can diffuse from one region to another with a droplet [40].

21.3.2 Factors that Influence Lipid Oxidation

21.3.2.1 Chemical Structure of Lipids

The chemical structure of a lipid is the ultimate determinant of its susceptibility to oxidation. The number and location of the double bonds are of particular importance. More unsaturated fatty acids have been observed to oxidize more slowly than less unsaturated fatty acids in o/w emulsions, whereas in bulk oils, the more unsaturated oils oxidize faster [41]. At present the reason for this observation is unknown, although it has been suggested that it is due to differences in the molecular arrangements of the fatty acids within the micelles. The more unsaturated fatty acids may be buried more deeply within the hydrophobic interior of the micelles and therefore may be less susceptible to attack by aqueous-phase prooxidants [40]. SLs may have altered degrees of unsaturation and molecular arrangements compared with their conventional counterparts, so understanding the oxidation properties of SL-based emulsions will require extensive kinetics studies to be completed on a case-by-case basis.

21.3.2.2 Droplet Characteristics

21.3.2.2.1 Oil Concentration

Understanding the kinetics of oxidation in emulsions of varying oil droplet concentrations is critical in the food industry, where emulsified systems range from fruit beverages (<1% oil) to mayonnaise (>80% oil). The effect of oil concentration on lipid oxidation was determined for caprylic acid/canola oil SL-based emulsions [42]. O/w emulsion samples were prepared with 10 or 30% oil and stored at 50°C. Decreasing oil concentrations led to an increase in total oxidation, calculated from the peroxide and anisidine values measured in the emulsion samples. One possible cause of this increase is that the number of radicals generated per droplet increased as the droplet concentration decreased [40]. Additionally, at higher oil concentrations more unsaturated fatty acids may have moved into the interior of the oil droplets, and therefore became less accessible to direct interaction with the prooxidants in the aqueous phase [40].

21.3.2.2.2 Droplet Size

Little is known about the correlation between oil droplet size and oxidation rate in emulsion systems. On one hand, small droplet size signifies a large surface area, implying a high potential for contact between diffusing oxygen, water-soluble free radicals and antioxidants, and the interface. It also implies a high ratio of oxidizable fatty acids located near the interface to fatty acids embedded in the hydrophobic core of the droplets. According to this, decreasing the size of the oil droplets is therefore expected to favor development of oxidation. For emulsions prepared with conventional lipids, data on the influence of droplet size on lipid oxidation are often contradictory. Osborn and Akoh [42] found no effect of particle size on the emulsions prepared with a caprylic acid/canola oil SL. Because limited amounts of hydroperoxides were available in the systems (~0.3 meq/kg oil), they may have all been present at the droplet surface in every o/w emulsion studied, which may explain why changing the droplet size did not affect the oxidation rates [40]. Fomuso et al. [43] recently reported that smaller droplet sizes led to higher oxidation rates because of increased surface area in emulsions prepared with enzymatically synthesized menhaden oil–caprylic acid SL. However, the droplet sizes were similar in most emulsions in this study and it may be deduced that emulsifier concentration, rather than droplet size distribution, caused changes in the oxidation properties.

21.3.2.2.3 Physical State

The physical state of the droplets in an SL-based o/w emulsion would also be expected to influence the rate of lipid oxidation. The droplets in most foods are liquid at room temperature, but may become partially or totally solidified at refrigerated temperatures. Studies on bulk fats indicate that lipid oxidation occurs more slowly when the fat is crystalline than when it is liquid. However, similar studies are needed to determine whether a similar phenomenon occurs in emulsified SLs [40].

21.3.2.3 Interfacial Characteristics

The interfacial region is potentially very important in lipid oxidation since it represents the region where lipid- and water-soluble components interact and it is where surface-active compounds such as peroxides and chain-breaking antioxidants concentrate [44].

21.3.2.3.1 Electrical Charge

The interfacial membrane surrounding the oil droplets may have an electrical charge with a magnitude and sign that is determined by the type and concentration of charged surface-active species present. An electrically charged surface attracts oppositely charged ions (i.e., mineral ions, metal ions, and ionic antioxidants) in the surrounding aqueous phases and subsequently influences the rate of lipid oxidation in emulsions. Several studies have confirmed the importance of droplet charge in controlling the rate of lipid oxidation in o/w emulsions [45,46].

Cationic and anionic emulsifiers are not commonly used by the food industry, but proteins above or below their isoelectric point (pI) could produce negatively or positively charged emulsion droplets. At pH 3.0, whey protein isolate (WPI)-stabilized emulsion droplets have a net positive charge because they are below the pI of ~5.0 for WPI [47]. The positive charge decreases the likelihood of any interaction with positively charged prooxidant ions because of repulsive electrostatic forces. On the other hand, emulsion droplets carry a negative charge at pH 7.0. When the surface charge of dispersed lipids as micelles is negative, metal-catalyzed oxidation rates are much higher than they are at positively charged interfaces [48]. This effect presumably exists because of the electrostatic attraction between the positively charged metal and the negatively charged emulsion droplet membrane [49], and may explain the decreased copper-catalyzed oxidation rates measured in pH 3.0 SL-based emulsions compared with their pH 7.0 counterparts [50].

21.3.2.3.2 Physical and Chemical Barriers

Less is known about how the emulsion droplet interfacial membrane properties impact oxidation rates. Some data suggested that the thickness of the emulsion droplet interfacial membrane could be an important determinant in the ability of lipid peroxides to oxidize fatty acids [49]. The ability of large surfactants to protect lipid hydroperoxides is likely due to the ability of these surfactants to alter the surface activity of lipid hydroperoxides or to provide a protective barrier around the emulsion droplet that would decrease lipid hydroperoxide-continuous phase prooxidant interactions. Similarly, Fomuso et al. [43] found higher oxidation levels in SL-based emulsions prepared with 0.25% emulsifier compared with those containing 1.0%. Higher surfactant concentrations lead to tighter packing of surfactant molecules at the oil–water interface, which resulted in a membrane that acted as an efficient barrier to the diffusion of lipid oxidation initiators into the oil droplets.

Differences in the size of the hydrophobic tail group of emulsifiers may also result in different rates of oxidation in emulsions. Based on hydroperoxide and headspace propanal measurements, greater lipid oxidation occurred in salmon oil emulsion droplets stabilized by Brij-lauryl compared with those stabilized by Brij-stearyl. [51] The ability of increasing hydrophobic tail group length to decrease the oxidation of emulsified salmon oil suggests that the longer tail group size decreases the ability of free radicals originating from hydroperoxides in the aqueous phase to reach the polyunsaturated fatty acids. Studies of this nature have not yet been performed on SL-based emulsions.

Certain types of emulsifier molecules may also be able to act as a chemical barrier to lipid oxidation. Many emulsifier molecules contain either sugar or amino acid moieties that can act as radical scavengers. Adsorbed emulsifiers are likely to be particularly effective at retarding lipid oxidation because of their high local concentration and close proximity to the oxidation substrate [40]. Whey protein isolate has been reported to inhibit lipid oxidation in structured lipid- and salmon oil-based emulsion [42,43,52]. This emulsifier is thought to inhibit oxidation by chelation of iron and copper and by inactivation of peroxyl radicals [53].

Alternatively, some surfactant molecules used by the food industry are sources of hydroperoxides. Tweens and phospholipids have been found to contain 4 to 35 μmol hydroperoxides/g surfactant. Therefore, food manufacturers need to closely monitor the hydroperoxide concentration in their surfactants during storage before adding them to SL-based emulsions.

21.3.2.4 Interactions with Aqueous-Phase Components

Most food emulsions contain a number of ingredients in addition to oil, water, and emulsifiers that may be of importance to the lipid oxidation processes. The hydrophilic components act as either prooxidants or antioxidants depending on their chemical properties, the prevailing environmental conditions, and their interaction with the other molecular species involved in the lipid oxidation reaction.

21.3.2.4.1 Salts, Sugars, Polysaccharides, and Proteins

At pH 3.0, the addition of 0.5 *M* NaCl generally resulted in an increase in secondary oxidation in the SL-based emulsions containing copper [42]. NaCl stimulation of oxidation could be due to the ability of chloride ions to increase the catalytic activity of metals or the NaCl-induced changes in the physical properties of the emulsion droplets, such as reduction in the thickness of the double layer [54]. Because salt is an ubiquitous ingredient in foods, processors must be mindful of its ability to promote lipid oxidation in the presence of metals and incorporate antioxidants into their product formulations accordingly.

Sucrose significantly retarded oxidation in linoleic acid emulsions; however, the effect was not concentration dependent [41]. Addition of sugar to safflower oil emulsions has likewise been found to enhance the oxidative stability of the emulsion [55]. The inhibitory mechanism of sugars has not been fully determined at this point, especially in SL-based emulsions. Sugar's ability to influence lipid oxidation may stem from its ability to increase viscosity and thus reduce the mobility of the reactants and reaction products or other mechanisms including quenching metals and scavenging radicals.

Polysaccharides are often added to o/w food emulsions to enhance the viscosity of the aqueous phase, which imparts desirable textural attributes and stabilizes the droplets against creaming. Polysaccharides may also inhibit oxidation through metal chelation and hydrogen donation [40]. In many food emulsions, there are appreciable quantities of nonadsorbed proteins dispersed in the aqueous phase, which may increase or decrease oxidative stability of emulsions through enzymatic or nonenzymatic mechanisms [40]. The antioxidant capabilities of whey protein in a Tween 20-stabilized salmon oil emulsion were determined by Tong et al. [56]. High-molecular-weight fractions of whey had greater antioxidant activity than low-molecular-weight fractions in this emulsion system. These authors also determined that the sulfhydryl groups were the primary antioxidants in whey proteins. Further research on the effects of polysaccharides and proteins on oxidation of SL-based emulsions is merited.

21.3.2.4.2 Surfactants

After a surfactant concentration is reached that saturates the droplet surface, excess surfactant molecules can form micelles in the continuous phase. It is well-known that micelles have the ability to solubilize certain compounds out of the lipid droplets into the continuous phase of the emulsion [57].

In general, nonpolar antioxidants that are retained in emulsion droplets are more effective inhibitors of lipid oxidation than polar antioxidants that have significant partitioning into the continuous phase of an o/w emulsion [58]. Partitioning of antioxidants in emulsions is mainly influenced by their molecular characteristics. However, surfactants could also play an important role in antioxidant partitioning since they can aid in the solubilization of nonpolar compounds out of the lipid into the water phase. The presence of excess surfactant in the continuous phase of olive oil, salmon oil, and hexadecane emulsions affected the physical location of polar phenolic antioxidants (propyl gallate), but not nonpolar antioxidants such as butylated hydroxyltoluene (BHT). However, the solubilization of antioxidants did not alter the oxidative stability of salmon oil emulsions, suggesting that surfactant micelles influenced oxidation rates by mechanisms other than antioxidant solubilization [59]. Nuchi et al. [57] demonstrated that Brij micelles inhibited lipid oxidation in corn oil emulsions. These authors found that the surfactant micelles solubilized lipid hydroperoxides out of the emulsion droplets, which may better explain the ability of surfactants to inhibit lipid oxidation in o/w emulsions. Similar phenomena are likely to be observed in SL-based emulsions, but studies are needed to confirm.

21.3.2.4.3 Acids, Bases, and Buffers

Acids, bases, and buffers are used to control the pH of food emulsions. The pH of an aqueous phase can impact the oxidative stability of o/w emulsions in a variety of ways and may account for the apparent contradictory results found for the effects of pH on lipid oxidation of emulsions. Recently, increased oxidation rates were measured in SL-based emulsions at pH 7.0 compared with pH 3.0 in the presence of copper, whereas the reverse effect was recorded in the presence of iron [50,60]. The effect of pH on lipid oxidation of SL-based emulsions depends on a variety of factors, including the nature of the oil and surfactant, the oxidation conditions, the absence or presence of added prooxidants (i.e., transition metals), and the analytical methods used to monitor oxidation.

21.3.2.4.4 Transition Metals

Transition metals that can undergo electron redox cycling are a major food prooxidant. These transition metals decrease the oxidative stability of foods through their ability to decompose hydroperoxides into free radicals. Decomposition of hydroperoxides by transition metals may be the most important cause of oxidation in many foods because both metals and hydroperoxides are ubiquitous to lipid-containing foods [40].

Transition metals are common constituents of raw food materials, water, ingredients, and packaging materials. However, not all transition metals in foods are equally reactive. The rate of hydroperoxide decomposition is dependent on the concentration, chemical state, and type of metal present in the emulsion. The most common transition metals in foods capable of promoting lipid oxidation are copper and iron, with the latter generally being in the greater concentration [40]. Although not as abundant in food emulsions, cuprous ions are capable of decomposing hydroperoxides to hydroxyl radicals over 50-fold faster than Fe^{2+}.

The addition of cupric sulfate and ferrous sulfate to model emulsions prepared with a menhaden oil–caprylic acid SL was positively correlated with peroxide values [61]. In emulsions formulated with a canola oil–caprylic acid SL, iron significantly ($p < .05$) increased lipid oxidation rates compared with control emulsions [60]. Lipid oxidation was also significantly increased in canola oil–caprylic acid SL-based emulsions upon addition of cupric sulfate [50].

21.3.2.5 Presence of Antioxidants

One of the most effective means for retarding lipid oxidation in fatty foods is to incorporate antioxidants. This is especially important in SL-based emulsions as enzymatically synthesized SLs require downstream processing to remove free fatty acids, which also results in a loss of natural tocopherols. Akoh and Moussata [62] reported that the total tocopherol content of unmodified canola oil decreased from 45.34 mg/100 g oil to 0.77 mg/100 g oil after the acidolysis reaction with caprylic acid and subsequent purification. The loss of natural antioxidant translated into a lower oxidative stability index for the SL in the bulk phase compared with the unmodified canola oil (3.5 and 9.65 h at 110°C, respectively).

Many "antioxidant" substances retard lipid oxidation under certain conditions, but actually promote it under other conditions [63–65]. Antioxidant activity varies widely depending on the composition of the emulsions system [66]. The efficacy of an antioxidant is affected by many of the colloidal properties discussed earlier for their influence on lipid oxidation of emulsions. Therefore, the term "antioxidant" must be used carefully, and a compound's anti- and prooxidant properties must both be carefully evaluated before it is added to an SL-based emulsion.

There is a growing interest for natural antioxidants as replacements for the synthetic compounds that are currently used in the food industry for food emulsions because of safety concerns and the worldwide trend toward the use of natural additives in foods [67]. Extensive research efforts have been dedicated to the identification of natural antioxidants from various sources.

Tocopherols and ascorbic acid are the most important natural antioxidants in the food industry. The well-understood antioxidant mechanism of α-tocopherol involves the donation of hydrogen to a peroxy radical. In foods, ascorbic acid is a secondary antioxidant with multiple functions, including scavenging oxygen, shifting the redox potential of food systems to the reducing range, acting synergistically with chelators, and regenerating primary antioxidants [68]. Numerous naturally occurring phenolic antioxidants have been identified in plant sources [68]. However, their effectiveness in emulsions is often difficult to predict because there are several different mechanisms by which phenolic compounds influence lipid oxidation rates [69]. Potential natural antioxidants that merit further investigation in SL-based o/w emulsions include quercetin, soybean isoflavones, a wide range of herb/spice extracts, and flavonoids isolated from chrysanthemums, rice, buckwheat, barley, and malt.

Fomuso et al. [61] were the first to investigate the effect of natural antioxidants on SL-based emulsions using a menhaden oil–caprylic acid model system. In their study, citric acid had the highest negative effect on peroxide value for iron-containing emulsions, while tocopherol had the highest negative effects for copper-containing emulsions. Additionally, in the copper-catalyzed emulsions, synergism between citric acid and α-tocopherol was observed.

At pH 3.0, the addition of α-tocopherol, citric acid, or a combination of the two compounds resulted in significantly ($p < .05$) lower peroxide values and anisidine values in canola oil–caprylic acid SL emulsions containing copper. α-Tocopherol and citric acid did not have synergistic effects on hydroperoxide or aldehyde formation in this study [50].

Quercetin and gallic acid exhibited prooxidant effects in SL-based emulsions containing iron at pH 3.0 and 7.0 [60]. Polar antioxidants are able to interact with aqueous-phase iron, and the metal-reducing power of phenolics can increase oxidation reactions. [69] In SL-based emulsions stabilized by sucrose fatty acid esters, α-tocopherol, β-carotene, genistein, and daidzein increased oxidation after 15 d of 50°C storage, compared with control emulsions. However, the same natural compounds were not significant prooxidants in emulsions stabilized with whey protein isolate [70].

The results of the studies discussed above on SL-based emulsions and natural antioxidants demonstrate that many factors affect oxidation of emulsified SLs. More studies are required to fully understand the prooxidant mechanisms behind these natural compounds. In the meantime, care should be exercised when adding α-tocopherol, citric acid, β-carotene, and soy isoflavones to emulsified foods because of their ability to promote lipid oxidation in some SL-based emulsions under certain conditions.

21.6 CONCLUSIONS

The ability of SLs to combine an increased absorption rate for MCFAs and beneficial LCFAs in one triacylglycerol molecule makes them very attractive to the medical community and functional food manufacturers. Lipid oxidation in SL-based o/w emulsions is a highly complex area. Clearly, the molecular environment of SL-based emulsion droplets substantially impacts their oxidative stability. Further studies on the influence of ingredients and antioxidants are needed to expedite the incorporation of SLs into product formulations by the food industry.

REFERENCES

1. Akoh, C.C., Structured lipids, in *Food Lipids — Chemistry, Nutrition, and Biotechnology,* 2nd ed., Akoh, C.C. and Min, D.B., Eds., Marcel Dekker, New York, 2002, p. 877.
2. Osborn, H.T. and Akoh, C.C., Structured lipids — novel fats with medical, nutraceutical, and food applications, *Comp. Rev. Food Sci. Saf.,* 3, 93, 2002.
3. Lee, K.T. and Akoh, C.C., Structured lipids: synthesis and applications, *Food Rev. Int.,* 14, 17, 1998.
4. McClements, D.J., Lipid-based emulsions and emulsifiers, in *Food Lipids — Chemistry, Nutrition, and Biotechnology,* 2nd ed., Akoh, C.C. and Min, D.B., Eds., Marcel Dekker, New York, 2002, p. 63.
5. Coupland, J.N. and McClements, D.J, Lipid oxidation in food emulsions, *Trends Food Sci. Technol.,* 7, 83, 1996.
6. Walstra, P., Dispersed systems: basic considerations, in *Food Chemistry,* 3rd ed., Fennema, O.R., Ed., Marcel Dekker, New York, 1996, p. 95.
7. Decker, E.A., The role of stereospecific saturated fatty acid positions on lipid nutrition, *Nutr. Rev.,* 54, 108, 1996.
8. Willis, W.M., Lencki, R.W., and Marangoni, A.G., Lipid modification strategies in the production of nutritionally functional fats and oils, *Crit. Rev. Food Sci. Nutr.,* 38, 639, 1998.
9. Innis, S.M., Nelson, C.M., Rioux, M.F., and King, D.J., Development of visual acuity in relation to plasma and erythrocyte ω-6 and ω-3 fatty acids in healthy term gestation infants, *Am. J. Clin. Nutr.,* 60, 347, 1994.
10. Lien, E.L., Yuhas, R.J., Boyle, F.G., and Tomarelli, R.M., Corandomization of fats improves absorption in rats, *J. Nutr.,* 123, 1859, 1993.
11. Pai, T. and Yeh, Y.Y., Stearic acid modifies very low-density lipoprotein lipid composition and particle size differently from shorter chain saturated fatty acids in cultured rat hepatocytes, *Lipids,* 32, 143, 1997.
12. Babayan, V.K., Medium chain triglycerides and structured lipids, *Lipids,* 22, 417, 1987.
13. Jensen, G.L. and Jensen, R.G., Specialty lipids for infant nutrition. II. Concerns, new developments, and future applications, *J. Pediatr. Gastroenterol. Nutr.,* 15, 382, 1992.
14. Kennedy, J.P., Structured lipids: fats of the future, *Food Technol.,* 45, 76, 1991.
15. DeMichele, S.J., Karistad, M.D., Babayan, V.K., Istfan, N.W., Blackburn, G.L., and Bistrian, B.R., Enhanced skeletal muscle and liver protein synthesis with structured lipid in enterally fed burned rats, *Metabolism,* 37, 787, 1988.

16. Kruimel, J.W., Naber, A.H., Curfs, J.H., Wenker, M.A., and Jansen, J.B., With medium-chain triglycerides, higher and faster oxygen radical production by stimulated polymorphonuclear leukocytes occur, *J. Parenteral Enteral. Nutr.,* 24, 107, 2000.
17. Chambrier, C., Guiraud, M., Gibault, J.P., Labrosse, H., and Bouletreau, P., Medium- and long-chain triacylglcyerols in postoperative patients: structured lipids versus a physical mixture, *Nutrition,* 15, 274, 1999.
18. Bellantone, R., Bossola, M., Carriero, C., Malerba, M., Nucera, P., Ratto, C., Crucitti, P., Pacelli, F., Doglietto, G.B., and Crucitti, F., Structured versus long-chain triglycerides: a safety, tolerance, and efficacy randomized study in colorectal surgical patients, *J. Parenteral Enteral. Nutr.,* 23, 123, 1999.
19. Kenler, A.S., Swails, W.S., Driscoll, D.S., DeMichele, S.J., Daley, B., Babinead, T.J., Peterson, M.B., and Bistrian, B.R., Early enteral feeding in postsurgical cancer patients: fish oil structured lipid-based polymeric formula versus a standard polymeric formula, *Ann. Surg.,* 223, 316, 1996.
20. Lee, K.T., Akoh, C.C., and Dawe, D.L., Effects of structured lipid containing omega-3 and medium chain fatty acids on serum lipids and immunological variables in mice, *J. Food Biochem.,* 23, 197, 1999.
21. Rubin, M., Moser, A., Vaserberg, N., Greig, F., Levy, Y., Spivak, H., Ziv, Y., and Lelcuk, S., Structured triacylglycerol emulsion, containing both medium- and long-chain fatty acids, in long-term home parenteral nutrition: a double-blind randomized cross-over study, *Nutrition,* 16, 95, 2000.
22. Bell, S.J., Macioli, E.A., Bistrian, B.R., Babayan, V.K., and Blackburn, G.L., Alternative lipid sources for enteral and parenteral nutrition: long- and medium-chain triglycerides, structured triglycerides, and fish oils, *J. Am. Diet. Assoc.,* 91, 74, 1991.
23. Straarup, E.M. and Hoy, C.E., Structured lipids improve fat absorption in normal and malabsorbing rats, *J. Nutr.,* 130, 2802, 2000.
24. Osborn, H.T., Shewfelt, R.L., and Akoh, C.C., Sensory evaluation of a nutritional beverage containing canola oil/caprylic acid structured lipid, *J. Am. Oil Chem. Soc.,* 80, 357, 2003.
25. Stepan Company, Showcase: fats, oils, and emulsifiers — medium-chain triglycerides, *Prepared Foods,* 170, 42, 2001.
26. Pueyo, E., Martin-Alvarez, P.J., and Polo, C., Relationship between foam characteristics and chemical composition in wines and cavas (sparkling wines), *Am. J. Enol. Vitic.,* 46, 518, 1995.
27. Mensink, K.P., Effect of dietary trans fatty acids on high-density lipoprotein and low-density lipoprotein cholesterol levels in healthy subjects, *N. Engl. J. Med.,* 323, 439, 1990.
28. Willett, W.C. and Ascherio, A., Trans fatty acids: are the effects only marginal?, *Am. J. Public Health,* 84, 722, 1994.
29. Lichtenstein, A., *Trans* fatty acids, blood lipids, and cardiovascular risk: where do we stand?, *Nutr. Rev.,* 51, 340, 1993.
30. Brekke, O.L., Soybean oil food products — their preparation and uses, in *Handbook of Soy Oil Processing and Utilization,* Erickson, D.R., Pryde, E.H., Brekke, O.L., Mounts, T.L. and Falb, R.A., Eds., American Soybean Association and American Oil Chemist's Society, St. Louis, MO, 1980, p. 383.
31. de Man, J.M., Fats and oils: chemistry, physics, and applications, in *Encyclopedia of Food Science and Technology,* Hui, Y.H., Ed., John Wiley & Sons, New York, 1992, p. 818.
32. Marangoni, A.G., McCurdy, R.D., and Brown, E.D., Enzymatic interesterification of triolein with tripalmitin in canola lecithin-hexane reverse micelles, *J. Am. Oil Chem. Soc.,* 70, 737, 1993.

33. Rousseau, D., Forestiere, K., Hill, A.R., and Marangoni, A.G., Restructuring butterfat through blending and chemical interesterification: 1. Melting behavior and triacylglycerol modifications, *J. Am. Oil. Chem. Soc.*, 73, 963, 1996.
34. Seriburi, V. and Akoh, C.C., Enzymatic transesterification of triolein and stearic acid and solid fat content of their products, *J. Am. Oil Chem. Soc.*, 75, 511, 1998.
35. Stoltzfus, D.L., Fehr, W.R., Welke, G.A., Hammond, E.G., and Cianzio, S.R. A fap5 allele for elevated palmitate in soybean, *Crop Sci.*, 40, 647, 2000.
36. Fomuso, L.B. and Akoh, C.C., Enzymatic modification of high-laurate canola to produce margarine fat, *J. Agric. Food Chem.*, 49, 4482, 2001.
37. Fomuso, L.B., Corredig, M., and Akoh, C.C., A comparative study of mayonnaise and Italian dressings prepared with lipase-catalyzed transesterified olive oil and caprylic acid, *J. Am. Oil Chem. Soc.*, 78, 771, 2001.
38. Depree, J.A. and Savage, G.P., Physical and flavour stability of mayonnaise, *Trends Food Sci. Technol.*, 12, 157, 2001.
39. Erickson, M.C., Lipid oxidation of muscle foods, in *Food Lipids-Chemistry, Nutrition, and Biotechnology,* 2nd ed., Akoh, C.C. and Min, D.B., Eds., Marcel Dekker, New York, 2002, p. 365.
40. McClements, D.J. and Decker, E.A., Lipid oxidation in oil-in-water emulsions: impact of molecular environment on chemical reactions in heterogeneous food systems, *J. Food Sci.*, 65, 1270, 2000.
41. Ponginebbi, L., Nawar, W.W., and Chinachoti, P., Oxidation of linoleic acid in emulsions: effect of substrate, emulsifier, and sugar concentration, *J. Am. Oil Chem. Soc.*, 76, 131, 1999.
42. Osborn, H.T. and Akoh, C.C., Effect of emulsifier type, droplet size, and oil concentration on lipid oxidation in structured lipid-based oil-in-water emulsions, *Food Chem.*, 84, 451, 2004.
43. Fomuso, L.B., Corredig, M., and Akoh, C.C., Effect of emulsifier on oxidation properties of fish oil-based structured lipid emulsions, *J. Agric. Food Chem.*, 50, 2957, 2002.
44. Mancuso, J.R., McClements, D.J., and Decker, E.A., The effects of surfactant type, pH, and chelators on the oxidation of salmon oil-in-water emulsions, *J. Agric. Food Chem.*, 47, 4112, 1999.
45. Mei, L., McClements, D.J., Wu, J., and Decker, E.A., Iron-catalyzed lipid oxidation in emulsion as affected by surfactant, pH and NaCl, *Food Chem.*, 61, 307, 1998.
46. Mancuso, J.R., McClements, D.J., and Decker, E.A., Iron-accelerated cumene hydroperoxide decomposition in hexadecane and trilaurin emulsions, *J. Agric. Food Chem.*, 48, 213, 2000.
47. Demetriades, K. and McClements, D.J., Influence of sodium dodecyl sulfate on the physicochemical properties of whey protein-stabilized emulsions, *Colloids Surf*, 161, 391, 2000.
48. Decker, E.A., Antioxidant mechanisms, in *Food Lipids— Chemistry, Nutrition, and Biotechnology,* 2nd ed., Akoh, C.C. and Min, D.B., Eds., Marcel Dekker, New York, 2002, p. 517.
49. Silvestre, M.P.C., Chaiyasit, W., Brannon, R.G., McClements, D.J., and Decker, E.A., Ability of surfactant headgroup size to alter lipid and antioxidant oxidation in oil-in-water emulsions, *J. Agric. Food Chem.*, 48, 2057, 2000.
50. Osborn-Barnes, H.T. and Akoh, C.C., Copper-catalyzed oxidation of a structured lipid-based emulsion containing α-tocopherol and citric acid: influence of pH and NaCl, *J. Agric. Food Chem.*, 51, 6851, 2003.
51. Chaiyasit, W., Silvestre, M.P.C., McClements, D.J., and Decker, E.A., Ability of surfactant hydrophobic tail group size to alter lipid oxidation in oil-in-water emulsions, *J. Agric. Food Chem.*, 48, 3077, 2000.

52. Tong, L.M., Sasaki, S., McClements, D.J., and Decker, E.A., Mechanisms of the antioxidant activity of a high molecular weight fraction of whey, *J. Agric. Food Chem.,* 48, 1473, 2000.
53. Donnelly, J.L., Decker, E.A., and McClements, D.J., Iron-catalzyed oxidation of menhaden oil as affected by emulsifiers, *J. Food Sci.,* 63, 997, 1998.
54. Mei, L., Decker, E.A., and McClements, D.J., Evidence of iron association with emulsion droplets and its impact on lipid oxidation, *J. Agric. Food Chem.,* 46, 5072, 1998.
55. Sims, R.J., Fioriti, J., and Trumbetas, J., Effects of sugars and sugar alcohols on autoxidation of safflower oil in emulsions, *J. Am. Oil Chem. Soc.,* 56, 742, 1979.
56. Tong, L.M., Sasaki, S., McClements, D.J., and Decker, E.A., Antioxidant activity of whey in a salmon oil emulsion, *J. Food Sci.,* 65, 1325, 2000.
57. Nuchi, C.D., Hernandez, P., McClements, D.J., and Decker, E.A., Ability of lipid hydroperoxides to partition into surfactant micelles and alter lipid oxidation rates in emulsions, *J. Agric. Food Chem.,* 50, 5445, 2002.
58. Frankel, E.N., Huang, S.W., Kanner, J., and German, J.B., Interfacial phenomena in the evaluation of antioxidants: bulk oils vs. emulsions, *J. Agric. Food Chem.,* 42, 1054, 1994.
59. Richards, M.P., Chaiyasit, W., McClements, D.J., and Decker, E.A., Ability of surfactant micelles to alter the partitioning of phenolic antioxidants in oil-in-water emulsions, *J. Agric. Food Chem.,* 50, 1254, 2002.
60. Osborn, H.T. and Akoh, C.C., Effects of natural antioxidants on iron-catalyzed lipid oxidation of structured lipid-based emulsions, *J. Am. Oil Chem. Soc.,* 80, 847, 2003.
61. Fomuso, L.B., Corredig, M., and Akoh, C.C., Metal-catalyzed oxidation of a structured lipid model emulsion, *J. Agric. Food Chem.,* 50, 7144, 2002.
62. Akoh, C.C. and Moussata, C.O., Characterization and oxidative stability of enzymatically produced fish and canola oil-based structured lipids, *J. Am. Oil Chem. Soc.,* 78, 25, 2001.
63. Fukumoto, L.R. and Mazza, G., Assessing antioxidant and prooxidant activities of phenolic compounds, *J. Agric. Food Chem.,* 48, 3597, 2000.
64. Galati, G., Sabzevari, O., Wilson, J.X., and O'Brian, P.J., Prooxidant activity and cellular effects of the phenoxyl radicals of dietary flavonoids and other polypheolics, *Toxicology,* 177, 91, 2002.
65. Rietjens, I.M.C.M., Boersma, M.G., de Haan, L., Spenkelink, B., Awad, H.M., Cnubben, N.H.P., van Zanden, J.J., van der Woude, H., Alink, G.M., and Koeman, J.H., The pro-oxidant chemistry of the natural antioxidants vitamin C, vitamin E, carotenoids and flavonoids, *Environ. Toxicol. Pharmcol,* 11, 321, 2002.
66. Schwarz, K., Huang, S.W., German, J.B., Tiersch, B., Hartmann, J., and Frankel, E.N., Activities of antioxidants are affected by colloidal properties of oil-in-water and water-in-oil emulsions and bulk oils, *J. Agric. Food Chem.,* 48, 4874, 2000.
67. Sanchez-Moreno, C., Satue-Gracia, M.T., and Frankel, E.N., Antioxidant activity of selected Spanish wines in corn oil emulsions, *J. Agric. Food Chem.,* 48, 5581, 2000.
68. Reische, D.W., Lillard, D.A., and Eitenmiller, R.R., Antioxidants, in *Food Lipids — Chemistry, Nutrition, and Biotechnology,* 2nd ed., Akoh, C.C. and Min, D.B., Eds., Marcel Dekker, New York, 2002, p. 489.
69. Mei, L., McClements, D.J., and Decker, E.A., Lipid oxidation in emulsions as affected by charge status of antioxidants and emulsion droplets, *J. Agric. Food Chem.,* 47, 2267, 1999.
70. Osborn-Barnes, H.T. and Akoh, C.C., Effects of α-tocopherol, β-carotene, and soy isoflavones on lipid oxidation of structured lipid-based emulsions, *J. Agric. Food Chem.,* 51, 6856, 2003.

22 Potential Market for Functional Lipids

Suk-Hoo Yoon

CONTENTS

22.1 PREAMBLE

It is well known that the functional foods market is growing rapidly in most countries and regions of the world and is expected to continue growing in the future until people's concern for better quality of life is met. It is obviously consumers' concern for health maintenance and well-being that drives the growth of the functional foods market. Factors affecting the demand for functional lipids may not be significantly different from those for functional foods. With people's concern for health, socio-economic changes in population age profile, life expectancy, medical conditions, and levels of health support are factors that shape the trends in the consumption of functional lipids [1]. Factors critical to the success of functional lipids include product quality (especially taste), safety, advertising or proof of efficacy, pricing, and market positioning. The functional lipids market, like those of other functional foods, is undoubtedly developing strongly in most countries, and the future of functional lipids will depend, at least in part, on convincing consumers that fats and oils have positive roles in health maintenance.

The functions of food lipids are considered to originate from several characteristics of lipids. The nutritional and biological function of food lipids, which are mainly dependent upon the chain length and degree of unsaturation of fatty acids (e.g., essential fatty acids such as linoleic and linolenic acid) is one of the major functions. Another function would be from the structural properties of lipids, and it is well known that structured lipids like diacylglycerol and olestra have their own nutritional functions. The ordinary edible fats and oils contain various fat-soluble materials like several vitamins and phytosterols, and they also have their own biological functions. Even though a substantial part of the functions of lipids are understood, it is obvious that the complete comprehension of lipid functions needs further research [2].

The scope of functional lipids may be considered to belong to "strict definition" of functional foods rather than "broad definition." Fish oil, for example, contains a substantial amount of omega-3 fatty acid such as eicosapentaenoic acid (EPA), docosahexaenoic acid (DHA), etc. In this chapter, EPA and DHA will be recognized as "functional lipids" whereas fish oil will not be. In the category of functional foods, functional lipids are one of the key items, and this is considered due to the fact that basic materials of lipids like fatty acids are incorporated in human body without further breakage, and thus the functionality of lipids can be almost completely transferred to human body cell.

Sometimes the term "biofoods" is used to elucidate the functional foods, and biofoods can be defined as foods with biofunctionality from biotechnology, and from biosphere [3]. World market of biofoods was forecast at $3.7 billion in 2003, $7.5 billion in 2008, and $12.6 billion in 2013. The annual market rate of increase during 2000 to 2013 was calculated as 28.2% [4]. (Monetary values are U.S. dollars throughout this chapter.)

The size of the functional foods market can be estimated in various ways depending on the definitions of functional foods and calculation methods. By the "strict definition" of functional foods, the global market for functional foods was assumed at $5.8 billion and $33 billion by "broad definition" in 1999. The biggest markets for functional foods by strict definition were achieved in Japan ($2.13 billion), United States ($1.8 billion), and Europe ($1.79 billion) in 1999. The Japanese market increased to $2.4 billion in 2001 [5].

Global market of functional foods by broad definition was estimated at more than $33 billion in 1999 [5] and $62.3 billion in 2001 [6]. It was also estimated at $51.1 billion in 2005 and $66.6 billion in 2010 [7], whereas another analyst forecast that the worldwide market for food and beverages containing added ingredients to improve health or prevent disease will reach $83 billion by 2005 [8]. The U.S. market for functional foods in 2002 was $20.2 billion and predicted at $37.7 billion in 2007 [9]. Based upon data from various sources, a "reasonable" global market for functional foods by broad definition was considered to range from $27.5 billion to $38.5 billion in 2002 [5].

22.2 POLYUNSATURATED FATTY ACIDS

The biological roles of polyunsaturated fatty acids (PUFAs) in the human body have been reported elsewhere, and the consumption of PUFA is known to reduce the risk of coronary heart disease, lower triglyceride levels in blood, prevent irregular heartbeat, decrease high blood pressure, diminish blood clotting potential, improve neurological development in infants, have a mood stabilizing effect, and reduce aggression [10]. PUFAs have been recognized as key items in functional lipids field due to their important biological roles, and, therefore, the market for PUFAs has been a major part of functional lipids. One market analyst estimated the 2001 revenue at $18 million for long-chain PUFAs (LC-PUFAs) and the 2002 revenue at $50 million, and it is expected to reach $300 million annually in the years ahead. The potential global market of LC-PUFAs was estimated at $300 to $400 million [11].

PUFAs are categorized into two major groups, omega-3 (ω-3 or n-3) and omega-6 (ω-6 or n-6) PUFAs, depending upon their molecular structures. Omega-3 PUFAs have different metabolic pathways from those of ω-6 PUFAs, and their metabolisms are influenced by each other [10].

22.2.1 OMEGA-3 PUFA

Omega-3 fatty acids are widely distributed in plants and animal tissues, and the main ω-3 fatty acids in food sources are α-linolenic acid (ALA), EPA, docosapentaenoic acid (DPA), and DHA. Long-chain ω-3 PUFAs are known to increase heart rate variability, decrease the risk of heart stroke, reduce serum triglyceride levels, lower systolic and diastolic blood pressure, reduce insulin resistance, modulate glucose metabolism, have anticancer and anti-inflammatory activity, have a beneficial effect on patients with attention-deficit/hyperactivity disorder and schizophrenia, have a positive effect in managing depression in adults, and decrease plasma/serum total and low-density lipoprotein (LDL) cholesterol levels.

Omega-3 PUFAs have been incorporated into a range of food products, such as bread, milk, yogurt, margarine, mayonnaise, soymilk, ice cream, and cereal bars, as extracted oils but also as powder formulations. Omega-3 PUFAs have been one of the most important items in the functional lipids market due to their critical roles in human health and are, therefore, expected to maintain their status in the market. The practical aspects of introducing ω-3 PUFA in foods should involve the stabilization of ω-3 PUFA in foods, and this is facilitated by addition of antioxidants, blending with more stable vegetable oils, and/or addition of ω-3 PUFA as microencapsulated powder to foods.

ALA has been supplied mostly from dietary plant fats and oils like perilla oil, flaxseed oil, canola oil, soybean oil, and walnut oil. EPA, DHA, and DPA have been obtained traditionally from fish oil and marine animals like the harp seal. DPA is well known to make blood vessels ten times stronger than EPA does, and harp seal oil containing high DPA content is already on the market. It is established that EPA has beneficial effects on cardiovascular disease, coronary heart disease, heart rhythm, atherosclerosis, inhibition of cyclooxygenase activity, ulcerative colitis, and so on.

In 2001, the single cell oil (SCO) blend of DHA and arachidonic acid (ARA, ω-6 PUFA) was given generally recognized as safe (GRAS) status by the U.S. Food and

Drug Administration (FDA), and it has been widely used in infant formula since early 2002 [11]. It is obvious that DHA is able to boost brain power in infants. The global infant formula market is estimated at $7.5 billion, and the U.S. market was estimated at $3.1 billion in 2001. It was reported that U.S. infants on average consume between $1500 and $2000 worth of infant formula during the first year of life [11].

DHA and ARA produced from SCO are sold in more than 20 countries in its formula for term infants, and sales of LC-PUFA-supplemented products in the United States was nearly $10 million in 2001 [12]. The global market for DHA in the food and beverage industry has been estimated at $1 billion [13]. Production of DHA in Japan was calculated as 1000 to 1200 million tons (MT), and the sale price of 22% DHA oil was 1500 yen/kg [14].

It is well known that many factors influence the fatty acid content of breast milk, but diet is the key factor. Because of this fact, DHA-enriched foods are marketed for pregnant and lactating women, besides infant formula and toddler foods. As far as dietary recommendations during pregnancy are concerned, an intake of 1.36 g/d of DHA and 7.9 g/d of ω-6 PUFA are recommended. The minimum intake of DHA for pregnant women is recommended at 250 to 300 mg/d [15,16].

Even though metabolic pathways of ω-3 PUFA such as ALA, EPA, DHA, and DPA are identical, the biological functions are different from each other. It is therefore expected that products targeted for different activities will appear in the future. For example, EPA is expected to be used mostly for atherosclerosis, DHA for brain function and ALA for allergies.

22.2.2 Omega-6 PUFA

Major fatty acids belonging to ω-6 PUFA are linoleic acid, γ-linolenic acid (GLA), and ARA. Until recently, vegetable oils have been the main source of ω-6 PUFA, especially for linoleic acid. Oilseed plants containing high linoleic acid content, however, are being genetically modified to oleic acid–rich plants regardless of its traditional functionality as an essential fatty acid to increase the oxidative stability of oil.

GLA is an important member of the ω-6 family of PUFA and is widely used for nutritional and medicinal purposes, such as treatments of diabetes, atopic eczema, inflammation, stress, cardiovascular disease, breast pain, premenstrual syndrome, and cancer [17]. Major applications of GLA-rich oils (e.g., evening primrose oil, borage oil, and black currant oil) are health food supplements, infant nutrition products, pet foods, and cosmetics such as skin care products (e.g., creams, lotions, and soaps). The total market value of GLA is estimated at about $50 million per year in 2001. China is the largest single producer of evening primrose seeds, at about 10,000 MT, out of a world production of about 12,000 MT. Seeds produced in China are used for oil extraction, and most extracted oil is exported. The market for evening primrose oil has continued to expand and was estimated in 2001 to be in the range of 1300 to 1500 MT. The price of refined evening primrose oil has declined from $45/kg in 1989 to $17.5/kg in 1999 [18].

ARA is found in the body of animals, including humans. ARA is not stored in cells; however, it can be produced and metabolized into mediators very rapidly by one of two enzyme pathways into various prostaglandins (by cyclooxygenase) or leukotrienes (by lipooxygenase). Both prostaglandins and leukotrienes are highly proinflammatory, bronchospastic, and vasodilatory. Ample evidence shows that provision of ARA to growing infants may provide benefit in the neural development either through breast-feeding or through its inclusion in infant formula. The current commercial demand for ARA is predominantly for a single application, infant formula.

Due to its low content in animal organs, ARA is produced by microorganisms, typically by *Mortierella alpina* spp., and microbes-origin ARA (e.g., ARASCO by Martek, United States) is on the market. Companies active in commercial production include Martek, Suntory (Japan), and DSM (the Netherlands), and there are some developments in South Korea and China [19].

Clinical studies show that infant formulas containing ARASCO support normal growth and development in healthy term and preterm infants. A number of government and international scientific bodies have made recommendations with respect to the DHA and ARA contents of preterm and term infant formulas [12].

22.2.3 Current and Future Market of PUFA

The future potential market for LC-PUFA ingredients, primarily LC ω-3 PUFA, is expected to grow up to $540 million, and the food industry is expected to use $55 million in 2006 [20]. The whole current world ω-3 PUFA market was estimated at $500 million in 2002 [21]. Traditional sources of LC-PUFA (i.e., EPA and DHA), for human consumption are fish oils. The pattern for human consumption of fish oil is categorized into four groups: food ingredients, nutraceutical and medical/functional foods, health foods, and pharmaceuticals.

The key problem with the use of microbial oils (e.g., GLA, EPA, DHA, and ARA) is how economically they can be produced. Although the future prospects for PUFA SCO look bright, it should be considered that plant geneticists try to produce PUFA through agricultural routes utilizing microbial genes. It is obvious that competition present between the microbial route and the plant route will be more intense in the coming years, and the results of the competition will provide the production technology of PUFA SCO with more economic benefit [22].

22.3 STRUCTURED LIPIDS

Structured lipids (SLs) have been synthesized to target specific metabolic effects or to improve physical characteristics of fats and oils. Considerable advances have been made over the past 10 years in designing SLs, especially in stereospecific location of fatty acid [23]. There are a number of varieties of SLs such as structured triglycerides (STs), diglycerides (DGs), monoglycerides (MGs), and nonglycerol-based fatty esters. Nonglycerol-based fatty esters include sorbitol fatty esters, fatty alcohol esters of trifunctional carboxylic acid, dialkyl dihexadecyl malonate, esterified propoxylated glycerol, polyglycerol esters, methyl glucose polyester, and polysiloxanes. Commercially available structured lipids like STs and nonglycerol-based fatty esters are aimed as reduced-calorie fats as well as zero energy fat-like substances such as olestra [24].

22.3.1 Structured Triacylglycerols (STs)

One of the approaches to producing reduced-calorie fats is a usage of one or more energy-modulating structural features of triglycerides. These include the reduced energy content of short-chain organic acids, the lower energy content of MCFA, and the reduced gastric absorption of saturated LCFA and partially hydrolyzed triglycerides. Combinations of these structures, of course, have produced several new reduced-calorie lipids [24].

In contrast to LCFA, MCFA has unique biological activities. It is reported elsewhere that MCFA and medium-chain monoglyceride (MCMG) derivatives have beneficial effects on dental caries formation due to their antimicrobial action. Medium-chain triglycerides (MCTs), as opposed to PUFA fats, have no growth-promoting effects in tumor-bearing animals. Data demonstrated that *Helicobacter pylori* is rapidly inactivated by MCMG and lauric acid [25]. Among SLs, STs and DGs have been more emphasized in the market than MGs because of their application in a wide variety of food and nonfood application fields.

There are several commercially available STs on the market. Salatrim is a reduced-calorie fat developed by Nabisco and commercially produced by Nabisco and Danisco Cultor, and is composed of short-chain organic acids (acetic, propionic, and/or butyric acid) and saturated LC fatty acid (mainly stearic acid). Salatrim, having 5 kcal/g of calorie density, can be used for confections, baking, and dairy applications. Caprenin, developed by Procter and Gamble (P&G), is a reduced-calorie saturated triglyceride (5 kcal/g) composed of equimolar amounts of caprylic, capric, and behenic acids. It is useful as a replacement for chocolate and compound coatings. Bohenin, developed by Fuji Oil, is a saturated glycerol ester composed of behenic and oleic acid in 2:1 ratio, and is used for chocolate and compound coating as a calorie-reduced fat (5 kcal/g).

Captrin is the most common form of MCT produced by interesterification of glycerol and caproic, caprylic, capric, and lauric acids derived from coconut oil and palm-kernel oil. It was developed by the Stepan Company, and has been used for sport nutrition drinks, energy bars, infant formula, and cooking oil for over 40 years. Captrin is known to have 8.3 kcal/g by Bomb calorimetry measurement, and 6.8 kcal/g as a net metabolic energy. Consuming MCT at 30 g or less per day does not cause any metabolic problem in humans, but consuming more than 30 g of MCT may cause gastrointestinal problems, and ingestion of large amounts may stimulate ketone body formation. Captrin is on the market under the brand of Neobee by Stepan Company and Akomed by Karlshamns. Healthy Restter is a brand of cooking oil in Japan developed by Nissin Oil through esterification of MCFA and glycerol, and its sales amount in the Japanese market is expected to reach about $1 billion in 2003 [26].

22.3.2 Diacylglycerols

Diacylglycerols (DAGs), known to reduce body fat deposits and to lower the serum triglyceride level, was first marketed in Japan in 1999 by Kao with the brand of Econa. DAG oil can be used as cooking oil for many purposes including frying, salad dressing, spread manufacturing, margarine, snacks, baked foods, curry, and filling oil for canned

tuna fish. Econa, produced primarily from soybean and rapeseed oils by esterification, is known to contain more than 80% DAG with phytosterols. Sales of Econa in Japan in 1999 reached about $66 million, which comprised 8% of the total domestic edible oil market share [27]. It was expected to increase in sales in Japan by more than 25% annually, and thus Econa sales reached more than $150 million in 2002, consisting of approximately 22% of the total edible oil market [28].

Archers Daniels Midland (ADM) initiated commercial scale production of DAG in the United States by a joint venture of ADM and Kao. DAG oil is expected to appear on the market at the beginning of 2004 under the brand of Enova. With approximately ten times more overweight individuals in the United States than in Japan, marketers expect that the sales of DAG oil will show a similar ratio in the United States as shown in Japan [27].

In Korea, DAG oil appeared on the market under the brand of Lowfree in 2002 [29]. The retail price of Lowfree is $5.8 per liter, and is almost four times more expensive than that of normal vegetable cooking oils. The market share of Lowfree in edible oils is not yet substantial in the Korean market, due to the relatively high price of Lowfree. CJ, distributor of Lowfree, expected an increase in sales of Lowfree from $3.1 million in 2002 to $11.5 million in 2003 [26].

It was clearly shown that DAG oil might be useful as an adjunct to the standard diet therapy of fat restriction in the management of diabetics with hypertriglyceridemia [30]. It is expected that the obesity and overweight problems would not be easily overcome in the near future; therefore the sales of DAG oil, based on the continuing trends seen in Japan and Korea, are expected to increase by more than 20% annually up to about 20% of the total edible oil market as a premium cooking oil.

22.3.3 Zero Energy Fat-Like Substances

Besides the developments of reduced-energy lipid products, there has also been an attempt to make a zero energy fatlike materials (i.e., fat mimetics). Among attempts made, P&G successfully developed a sucrose polyester named olestra. The U.S. FDA approved the usage of olestra for a limited range of food in 1996 [31]. P&G has spent over $250 million over the last 25 years for evaluation in a number of animal and clinical studies.

P&G's olestra was first launched in food by the Frito-Lay Division of PepsiCo Inc. in 1998. P&G spent $250 million for olestra plants to supply olestra to Frito-Lay, of which the annual pull-in was about $550 million in manufacturer sales and $900 million in retail sales in 1998. It is reported that consumer attitudes about olestra have mainly been in response to the marketing, education, and media information about olestra. The major potential market of olestra lies in its application for mainly potato chips, and its future is predominantly dependent upon consumer attitudes toward olestra [32]. Glycerol realized from olestra production is not likely to make a significant impact on the glycerol market.

Considerable advances have been made over the past 10 years that are slowly unraveling the differences in designer fat function due to stereospecific fatty acid location SLs particularly in lower calorie fats. To increase the usage of SLs in humans, more research will be needed to expand the scope of application in various

areas such as immune function and metabolism control, and commercial availability in large quantities of specific SL will permit such studies to be conducted more easily and economically.

22.4 PHOSPHOLIPIDS

The total amount of phospholipids (PLs) used in industries was estimated at about 170,000 MT in 2001, most being produced from soybean oil and egg yolk. Usage of PLs in substantial amounts include food (ca. 58%), animal feed (ca. 20%), cosmetic/pharmaceutical purposes (4 to 10%), chemical products such as insecticides, paint, ink, plastics, rubber, leather, textile, and magnetic tapes (ca. 1%), and others [33,34]. Among usages for foods, chocolate and confectionery comprise 21%, baking 15%, margarine 11%, convenience foods 6%, and dietary 4%, respectively. PLs are applicable for various products in various forms such as high-purity PL phosphatidylcholine (PC) content >95%), paste PL (PC content >60%), lysoPL, modified PL, and hydrogenated PL. Dietary lipids contain approximately 10% of PLs in which PC and phosphatidylethanolamine (PE) are the two major components [35]. The intake of dietary PL is estimated to be around 3 to 4 g/d, which amounts to about 5 to 8% of total dietary lipids in many countries, whereas PLs are frequently added to cosmetics at levels between 0.5 and 1% [36].

It was reported that PLs have therapeutic applications for certain neurological disorders and liver cirrhosis [36]. PC is known to affect profiles of serum lipids and lipoproteins, and PE plays a role in altering serum lipoproteins (i.e., lowering the cholesterol level). Feeding phosphatidylinositol (PI) significantly lowered body weight gain, and reduced serum triacylglycerol (TAG) concentration by 50%, cholesterol and PL by 30 to 35%, and liver TAG concentration by 25%. Phosphatidylserine (PS) has been touted to boost memory and concentration in the elderly, and currently PS is commercially available that is derived from soy. Sales of PL, and, in particular, PS, have grown by 100% over the past 18 months. This trend is expected to continue, because the sales of two leading herbal brain products declined rapidly after receiving bad press since 2001 regarding safety concerns [37].

22.5 CONJUGATED LINOLEIC ACID

Since the 1950s, several studies have reported the biological activities of conjugated linoleic acid (CLA) including body fat mass reduction, antimutagenicity against chemical carcinogens, anticancer effects against breast, colon, and prostate cancer, blood glucose control as an insulin sensitizer, inhibition of atherosclerosis, enhancing immune response, and others [38].

Currently, CLA is available on the market almost entirely in supplement form. Future applications of CLA will include the improvement of CLA content of meat, egg, and dairy products, inclusion of CLA in animal tissues and products, incorporation of CLA into a wide range of foods as a nutritional supplement, and clinical applications with purified CLA. Other novel uses for CLA include dietetic foods, skin care products, and cosmetics.

The Lipid Nutrition division of Loders Croklaan of the Netherlands developed a CLA made from natural safflower and sunflower oils, called Clarinol, in oil and powder form. Clarinol is available in free fatty acid form (soft-gel capsule) and triglyceride form for a variety of food and drink applications [39]. There have been many attempts to produce CLA through chemical and biological processes, and conventional hydrogenation process of vegetable oils under specific conditions are also used as a novel production technology [40,41].

The total global sales amount of CLA is considered to continue to increase in the future from $30 to 40 million in 2002 (ca. 1000 MT) to $1.83 to 2 billion in 2014 [42,43]. The sales amount of CLA used for functional foods is calculated at $650 million, $150 million for dietary supplements for diabetes type II, $117 million for immunity-related products, and $90 million for diet foods individually. The animal industry is expected to achieve a market share of $520 million, and the pharmaceutical industry and cosmetics industry are expected to make $260 and $40 million, respectively. The total production of CLA in the United States is estimated at 100 to 150 MT, and its major application is for dietary supplement in mostly capsule form [42,44].

One of the marketing potentials of CLA lies in cheese manufacturing [45]. Specialty cheese sales have grown approximately 4% annually since 1996, an overall growth of 17.5% between 1996 and 2000. Retail sales of gourmet/speciality cheese will continue to grow at the same rate annually to nearly $2.9 billion by 2005. At this moment, however, there are several constraints on the inclusion of high CLA content into cheese since no actual health claims can be made about CLA for humans, and most cheese buyers are not aware of CLA. There are, however, opportunities in usage of CLA in cheese making because the speciality cheese market is increasing as mentioned above, and a trend toward organic/natural products is attractive to makers and consumers, while scientific research on CLA will support the expansion of CLA usage in foods, including cheese.

22.6 OTHERS

22.6.1 TOCOPHEROLS

Tocopherols are well known to have antioxidant and biological activities as vitamin E (VE). Tocopherols have widely been used for food, feed, pharmaceuticals, cosmetics, and resins. In food, tocopherols are used as an antioxidant for frying oil, margarine, fried snacks, and so on.

The production of vitamins is still dominated by chemical synthesis rather than extraction from natural resources. Although the enthusiasm for biotechnology has been attractive, it was not possible to replace chemical processes by fermentation, with the exception of riboflavin — the only success.

The annual worldwide production of synthetic VE, a mixture of eight stereoisomers of α-tocopherol, was approximately 15,000 to 20,000 MT in 1998, whereas natural VE was produced at about 2000 MT from soybean oil distillate (scum) [46]. Natural VE has been primarily used in humans for pharmaceutical purposes, and has been manufactured mainly by Henkel, ADM, Eisai, and Hoffman-La Roche/Cargill. Usages of synthetic VE include feed (71%), pharmaceuticals (24%), cosmetics (3%), and foods (2%), and major producers of synthetic vitamin E are BASF, Rhone-Poulene, Eisai, and Roche.

The worldwide market for vitamins was estimated at $2.65 billion in 1999, and a reduction of selling volume and prices occurred thereafter mainly due to the economic crisis in many regions. The market value of VE fell from $1.1 billion to below $1 billion in the same period. But the market was expected to grow at an average annual growth rate of 0.5% to reach $2.74 billion by 2005, and the global market of oil-soluble vitamins such as A, D, E, and K was estimated to reach $1.49 billion [47].

The Japanese market for (Foods for Specified Health Uses) (FOSHU) is around $3.5 billion [48], and the production of natural VE in Japan is about 450 to 500 MT from 5500 MT soybean oil scum. About 100 MT of natural VE is exported to foreign countries, and domestic usages are for food antioxidants (ca. 60%) and nutrition fortification (ca. 40%). Japan imported about 2000 MT of soybean oil scum for VE production, mainly from the United States and Brazil.

22.6.2 Phytosterols

The term "phytosterols" is used as a collective term for plant sterols and their hydrogenated stanol forms, whether used in the free sterol form or esterified with fatty acids, and commonly used by manufacturers and distributors of these substances. In 2000, the FDA authorized the use of labeling health claims about the role of plant sterol or plant stanol esters in reducing the risk of coronary heart disease (CHD) for foods containing these substances. In February 2003, the FDA expanded the use of phytosterol heart-health claims to a broader range of food products and dietary supplements [49]. Scientific studies showed that 1.3 g of plant sterol esters or 3.4 g of plant stanol esters per day in the diet are needed to show a significant cholesterol-lowering effect through interfering with the absorption of cholesterol in the intestines. To reduce the risk of CHD, a daily dietary intake of phytosterols of 800 mg or more is needed, and the phytosterol mixtures should contain at least 80% β-sitosterol, campesterol, stigmasterol, sitostanol, and campestanol.

Food and pharmaceutical companies are currently active in the study and development of phytosterol-incorporated functional foods and other nutraceuticals, demonstrating that scientific studies and good marketing lead to acceptance by consumers for various types of foods [50].

22.6.3 Cooking Oil

A patented new cooking oil with the brand name of Functional Oil is on the shelves of supermarkets in Canada [51]. It is a blend of tropical oils, olive oil, coconut oil, and flaxseed oil, and contains a high content of MCTs. During the clinical tests of this new oil against conventional cooking oil, the cholesterol level was reduced by as much as 13%.

The vegetable oils fortified with VE have become more popular than before, and the reasons are that VE is beneficial not only in the prevention of oxidation of highly unsaturated vegetable oils but also in the supply of biologically active compounds [48].

22.6.4 Miscellaneous

The first cocoa butter substitutes (CBSs) were produced more than 100 years ago in Denmark, and the basic concept of the product is still same. Unlike cocoa butter, neither CBSs nor cocoa butter replacers (CBRs) require tempering in chocolate production, which makes the production process both simpler and safer for the chocolate manufacturer. All through 2002 cocoa prices increased on the world market, which stimulated the demand for alternative fats. As of August 2003, the chocolate industry may use up to 5% of vegetable fats in products marketed as chocolate within the European Union (EU) according to a new directive [52]. The International Cocoa Organization (ICCO) estimated a total displacement of cocoa butter by cocoa butter alternatives (CBAs) at 74,000 MT, which is equivalent to 184,000 MT of cocoa beans. The global market for chocolate and confectionery products is estimated at 5.6 million MT, with an annual expected growth of 3 to 4%. In 2005/2006, cocoa production would be 50,000 MT lower and prices about 8% lower, with a revenue loss to cocoa producers of $780 million. If there was worldwide adoption of the 5% allowance, the impact could be double that of implementation in EU. In 2005/2006, cocoa production could be 125,000 MT lower with a revenue loss to cocoa producers of more than $1.5 billion.

Compared with other edible oils, milk fat has certain properties that offer a good starting point for developing new milk fat products. The application of milk fat will ultimately be determined by economic factors, and its availability at a reasonable price may be the prerequisite for development of new food applications such as milk with altered fatty acid composition (increased PUFA and CLA content, decreased total *trans* fatty acid content), removed cholesterol, interesterification, hydrogenation, fractionation, and blending [53].

Octacosanol has been reported to lower LDL cholesterol levels while raising high-density lipoprotein (HDL) cholesterol levels, although it does not affect triglyceride levels in serum blood [54,55]. Octacosanol improves hand-to-eye coordination as well as alleviating multiple sclerosis. It regenerates and repairs myelin sheaths, often improves the condition of Parkinson's disease patients, stimulates the production of androgens, and helps to prevent miscarriage during pregnancy. Rice bran wax proved to be the best raw material for 1-octacosanol production, and sugarcane wax, insect waxes, and beeswax are also used for production. The usage of highly pure natural octacosanol in pharmaceuticals is now increasing; and natural octacosanol can be used to formulate sports-type, electrolyte replacement, carbohydrate-loading beverages, and to prepare various types of dietary supplement tablets and/or capsules for the endurance of athletes, the elderly, and those who are concerned about their high blood cholesterol level. Possibilities for consumer products using octacosanol may include sports endurance supplements, weight-loss products, energy-boosting products, stress-reducing supplements, sexual performance products, and cardiovascular support products. Although 1-octacosanol has been researched in Cuba, the United States and China for many years, the marketing of this product is only several years old. However, with the fast development of deep processing products and price increase, a great potential for the market of octacosanol a natural and new material can be envisaged. According to market surveys, many beverages have octacosanol liquid as an additive. It was estimated that the sales volume would be over 10 million cases for octacosanol beverages in the Japanese market in 2002. In the United States there are many series of octacosanol health-care foods, and octacosanol and vitamin formulas on the market. These products have been well received in the U.S. health-care foods market. In Australia, Korea, and Southeast Asian countries, there are various types of natural octacosanol dietary supplements and beverages. In short, in international markets the demand for octacosanol as a material for medicine and health-care food is exceeding supply [56].

22.7 CONCLUSIONS

In normal diets, a specific single fatty acid is not taken as a sole food lipid source. Fats and oils in normal diets contain a wide variety of fatty acids with even nonlipid compounds. No single fat or oil is completely versatile for all purposes such as cooking, processing, preparation of food lipids with nutritional and biological activity, etc. The functions of food lipids in the biological system can be fully understood with extensive and intensive comprehension of the lipid itself and interaction of lipids with other food components, and this understanding can lead us to evolve the optimum conditions to achieve maximum functionality of lipids.

One of the typical nutritional issues concerning fatty acids is the ratio of polyunsaturated and saturated fatty acids content, and the ratio of the ω-3 and ω-6 fatty acids content in foods. It is obviously beneficial for human nutrition when these ratios are properly maintained in regular diets, but the incorporation of lipids containing several functions into foods requires more serious consideration.

A number of functional lipid products with known properties and functions are already on the market, and many research studies on many lipids are in progress to discover and identify new functions of lipids. The important problems to be solved primarily include the improvement of taste and palatability of functional lipid products. Many functional lipids are susceptible to oxidation and therefore prevention of oxidation is also critical for the commercialization of the products.

The progress in genetic rearrangement techniques makes it possible to produce oleaginous materials with specific fatty acid composition. This facilitates the production of natural compounds like special fatty acids and other compounds that are present in low concentrations in nature. The production of those lipid materials in large quantities can lead to practical clinical studies more easily, which are restricted by the limited supply of special fatty compounds.

A bright future for functional foods is commonly indicated by industry, academia, and government, and it is worth emphasizing that fats and oils have a significant role to play in the future growth of functional foods globally. The market size of dietary supplements is estimated at $20 to 25 billion, whereas that of functional foods and medicinal foods is estimated at $5 to 10 billion and $1 to 2 billion, respectively [42]. One of the major challenges that the fats and oils industry faces is to convince consumers that fats and oils have a positive role in health maintenance. With the results of these attempts, functional lipids can be used in many fields such as dietary supplement, functional foods, medicinal foods, the food industry, the pharmaceutical industry, the animal feed industry, and cosmetics/toiletry, without heated opposition from consumers and users of functional lipids.

REFERENCES

1. Young, J., Introduction, in *Lipids for Functional Foods and Nutraceuticals,* Gunstone, F., Ed., The Oily Press, Bridgwater, UK, p. 1, 2003.
2. Himasaki, H., Lipids, in *Materials for Functional Foods II,* Ota, M., Ed., CMC, Tokyo, p. 30, 2001.
3. Yoon, S., Current status and prospect of biofood industry, in *Proc. Bioindustry Symposium,* Bioindustry Assoc. of Korea, Ed., Seoul, 2001.
4. OECD. Data from *Biotechnology and Trade 1997,* 1998.
5. LFRA (Leatherhead Food Research Association), Functional Food Markets, Innovation and Prospect. A Global Analysis, 2000.
6. NBJ (Nutrition Business Journal), *Annual Report of NBJ,* 2001.
7. Nikkei, Data from http://www.nikkei.co.jp, 1999.
8. Grundy, S.M., Consensus statement: role of therapy with "statins" in patients with hypertriglyceridemia, *Am. J. Cardiol.,* 81(4A), 1B–6B, 1998.
9. Anon., Functional foods market growing, *Food Technol.,* 57, 64, 2003.
10. Stam, W., EPA and DHA Supplementation: evidently healthy, in *The International Review of Food Science and Technology,* International Union of Food Science and Technology, p. 51, 2003.
11. Anon., More than just DHA and AA, *INFORM,* 12, 1069, 2001.
12. Anon., DHA in infant formula, *INFORM,* 12, 1073, 2001.
13. Anon., Martek buys Omega Tech; DHA aids postpartum depression?, *INFORM,* 13, 460-461, 2002.
14. Anon., *Food World,* Seoul, August, 76, 2003a.
15. Sattar, N., Berry, C., and Greer, I., Essential fatty acids in relation to pregnancy complications and fetal development, *Br. J. Obstet. Gynaecol.,* 105, 1248, 1998.
16. Simopoulos, A., Leaf, A., and Salem, N., *Annu. Nutr. Metab.,* 43, 127, 1999.
17. Huang, Y. and Ziboh, A., *Recent Advances in Biotechnology and Clinical Applications of GLA,* AOCS Press, Champaign, IL, 2003.
18. Clough, P.M., Specialty vegetable oils containing gamma linoleic acid and stearidonic acid, in *Structured and Modified Lipids,* Gunstone, F., Ed., Marcel Dekker, New York, p. 75, 2001.
19. Streekstra, H., Fungal production of arachidonic acid-containing oil on an industrial scale, *INFORM,* 15, 20, 2004.
20. Anderson, S., *Book of Abstracts,* LFRA (Leatherhead Food Research Association), 1998.
21. NBJ (Nutrition Business Journal), *Annual Report of NBJ,* 2003.
22. Ratledge, C., Microorganisms as sources of polyunsaturated fatty acids, in *Structured and Modified Lipids,* Gunstone, F., Ed., Marcel Dekker, New York, p. 351, 2001.
23. Kim, E. and Yoon, S., Recent progress in enzymatic production of structured lipids, *Food Sci. Biotechnol.,* 12, 721, 2003.
24. Auerbach, M., Klemann, L., and Heydinger, J., Reduced energy lipids, in *Structured and Modified Lipids,* Gunstone, F., Ed., Marcel Dekker, New York, p. 485, 2001.
25. Kabara, J., A health oil for the next millenium, *INFORM,* 11, 123, 2000.
26. Anon., Data from http://www.thinkfood.co.kr, 2003.
27. Anon., Diacylglycerol oil products may be headed for U.S., *INFORM,* 12, 487, 2001.
28. Anon., Diacylglycerol products closer to the U.S. market, *INFORM,* 13, 459, 2002.
29. Anon., September issue, *CJ Family,* CJ, Seoul, 2002.
30. Yamamoto, K. et al., Long-term ingestion of dietary diacylglycerol in type II diabetic patients with hypertriglyceridemia, *J. Nutr.,* 131, 3204, 2001.

31. FDA (Food and Drug Administration), Federal Register (Jan 1996). U.S. Food and Drug Administration. Food additives permitted for direct addition to food for human consumption: olestra; final rule. *Fed. Regist.*, 61, 3118, 1996.
32. Yankah, V. and Akoh, C., Zero energy fat-like substances: olestra, in *Structured and Modified Lipids* Gunstone, F., Ed., Marcel Dekker, New York, p. 511, 2001.
33. Ota, M., Lecithin, in *Development of Functional Foods,* Kamewada, M., Ed., CMC, Tokyo, p. 129, 2001.
34. Gunstone, F., Phospholipids, in *Structured and Modified Lipids,* Gunstone, F., Ed., Marcel Dekker, New York, p. 241, 2001.
35. Pokorny, J., Phospholipids, in *Chemical and Functional Properties of Food Lipids,* Sikorsky, Z. and Kolakowska, A., Eds., CRC Press, Boca Raton, FL, p. 79, 2002.
36. Yanagita, T., Nutritional functions of dietary phosphatidylinositol, *INFORM,* 14, 64, 2003.
37. Anon., Data from http://www.phospholipidonline.com, 2003.
38. Kapoor, R., Westcott, N.D., Reaney, M.J.T., and Jones, S. Conjugated linoleic acid-Seven decades of achievement II: physiological properties, *INFORM,* 14, 482, 2003.
39. Crandall, L., Emerging ingredients, *INFORM,* 14, 62, 2003.
40. Chung, M., Ju, J., Choi, D., Yoon, S., and Jung, M., CLA formation in oils during hydrogenation process as affected by catalyst types, catalyst contents, hydrogen pressure, and oil species, *J. Am. Oil Chem. Soc.,* 79, 501, 2002.
41. Ju, J. et al., Effects of alcohol type and amounts on conjugated linoleic acid formation during catalytic transfer hydrogenation, *J. Food Sci.,* 68, 1915, 2003.
42. Anon., Data from http://www.lipozen.com, 2003.
43. Fernie, C., Conjugated linoleic acid, in *Lipids for Functional Foods and Nutraceuticals,* Gunstone, F., Ed., The Oily Press, Bridgwater, UK, p. 291, 2003.
44. Sugano, M., Conjugated linoleic acid, in *Materials for Functional Food II,* Ota, M., Ed., CMC, Tokyo, p. 133, 2001.
45. Greenburg, L. and Klasna, D., The Marketing Potential of CLA in Cheese, Wisconsin Initiative for Value-Added Development, SARE Program, 2002.
46. Netscher, T., Synthesis and production of vitamin E, in *Lipid Synthesis and Manufacture,* Gunstone, F., Ed., Sheffield Academic Press, Sheffield, UK, p. 251, 1999.
47. BCC (Business Communications Company), Data from http://www.bccresearchcom.com, 2000.
48. Yoshida, S., An overview of the Japanese regulatory framework for foods with health benefit claims and their current market situation. Presented at Natural Products Expo 2003 Asia, 2003.
49. FDA (Food and Drug Administration), Data from http://www.cfsan.fda.gov, 2003.
50. Yankah, V. and Jones, P., Phytosterols and health implications — efficacy and nutritional aspects, *INFORM,* 12, 899, 2001.
51. Ohr, L., Fats for healthy living, *Food Technol.,* 57, 91, 2003.
52. ICCO (International Cocoa Organization), What is the likely impact of using cocoa butter substitutes in the future? Data from http://www.icco.org, 2003.
53. De Greyt, W. and Kellens, M., Improvement of the nutritional and physicochemical properties of milk fat, in *Structured and Modified Lipids,* Gunstone, F., Ed., Marcel Dekker, New York, p. 285, 2001.
54. Castano, G., Canetti, M., and Moreira, M., Efficacy and tolerability of policosanol in elderly patients with type II hypercholesterolemia: A 12-month study, *Curr. Ther. Res.,* 56, 819, 1995.
55. Menendez, R., Arruzazabala. L., and Mas, R., Cholesterol-lowering effect of policosanol on rabbits with hypercholesterolaemia induced by a wheat starch-casein diet, *Br. J. Nutr.,* 77, 923, 1997.
56. Michael, Data from http://forum.agriscape.com, 2003.

23 Structured Lipids Production

Jeung-Hee Lee and Ki-Teak Lee

CONTENTS

23.1 INTRODUCTION

In the past, the food industry's focus was to improve processing and increase the shelf life of the product. Now nutritional quality of food is one of the essential parts for product development. Most natural fats and oils contain highly unsaturated fatty acids at the *sn*-2 position and saturated or monounsaturated fatty acids at the *sn*-1,3 positions of the triacylglycerol (TAG) molecule. Because positional distribution of fatty acids in TAG of conventional fats and oils is not always ideal for human nutrition, the modification of TAG structure has been attempted to improve their physicochemical and nutritional values through biological or technological methods [1].

Structured lipids (SLs), as constituents of functional foods, have been used in formulation for the maintenance of good health as well as for the treatment of disease. A large number of studies on SLs have been conducted using edible oil with specific fatty acids having beneficial properties. The edible oils include palm, coconut, fish, rice bran, perilla, and palm-kernel oils, which are oils produced mostly in Asia. Bohenin (Fuji Oil Company Ltd., Japan), reduced-calorie commercial SLs composed of behenic acid (C22:0) and oleic acid (C18:1), was marketed in Asia and is useful in introducing blooming stability into chocolate and confection fats. In this chapter, SLs produced with edible oils popular in Asia as a functional lipid for pharmaceutical and nutritional purposes, and their application will be discussed.

23.2 EDIBLE OILS IN ASIA

Vegetable oils are part of traditional diets all over the world. They provide energy and are carriers for fat-soluble vitamins and antioxidants. Among edible oils soybean, palm, rapeseed/canola, and sunflower oil, the four major cooking oils, have increased their market share over 40 years. Soybean and sunflower oils have market share at about 22% and 8 to 10% of total vegetable oil production, respectively, while rapeseed/canola oil has about 12 to 13%. The production of palm oil has gradually increased and will exceed that of soybean oil between 2011 and 2020 [2]. During 2000 and 2001, 52.1 million metric tons (MMT) of oils and fats accounting for about 45% of the world total (117.1 MMT) were consumed in Asia. Palm and coconut oils are the major two-commodity oils in Asia, accounting for 20% of global fats and oils. In particular, Asia produces most of the world's palm oil (above 21 MMT). Countries in northern Asia, such as China and Korea, consume more soybean oil than in southeast Asia, where palm oil is prevalent. In the Philippines, the oil of choice is coconut oil, while cottonseed oil is the most commonly consumed oil in Pakistan. Trends in a region vary by economic and cultural characteristics.

23.2.1 Oils Rich in Linoleic Acid

23.2.1.1 Cottonseed Oil

Cottonseed oil typically contains palmitic acid (24%), stearic (3%), oleic (19%), and linoleic acids (53%) (Table 23.1). Minor amounts of cyclopropenoic acids (malvalic and sterculic) are also present, but these are largely removed during refining process. This oil is used to produce a spread since the content of saturated fatty acids is

TABLE 23.1
Fatty Acid Composition of Edible Oils

Oil	8:0	10:0	12:0	14:0	16:0	16:1	18:0	18:1	18:2	18:3	18:4	20:0	20:1	20:4	20:5	22:5	22:6
Canola oil	—	—	—	—	4.7	0.3	1.7	59.3	21.4	9.9	—	0.6	—	—	—	—	—
Coconut oil	8	7	48	16	9	—	2	7	2	—	—	—	—	—	—	—	—
Cottonseed oil	—	—	—	1.0	23.9	0.5	2.9	18.5	52.5	0.3	—	0.4	—	—	—	—	—
Menhaden fish oil	—	—	—	9.1	19.7	1.2	2.9	11.7	1.4	1.4	3.7	0.3	1.7	2.9	16.1	2.5	10.2
Palm oil	—	—	—	1.0	43.8	—	5.0	38.5	10.5	0.3	—	—	—	—	—	—	—
Palm-kernel oil	4.9	4.0	47.5	15.8	7.7	—	2.4	14.3	3.1	—	—	—	—	—	—	—	—
Perilla oil	—	—	—	—	7.1	—	1.9	15.4	14.2	61.4	—	—	—	—	—	—	—
Rice bran oil	—	—	—	—	20.4	0.3	1.9	46.0	31.1	0.3	—	—	—	—	—	—	—
Sesame oil	—	—	—	—	8.7	—	5.5	41.4	44.0	0.3	—	0.1	—	—	—	—	—
Soybean oil	—	—	—	—	11.2	—	4.0	24.4	53.9	7.1	—	—	—	—	—	—	—
Sunflower oil	—	—	—	—	6.5	—	4.3	22.0	68.5	0.4	—	—	0.2	—	—	—	—
Sunflower oil (high oleic)	—	—	—	—	3.0	—	5.9	82.6	7.4	—	—	0.3	—	—	—	—	—

Source: Modified from Hammond, E.W., in *Chromatography for the Analysis of Lipids*, Hammond, E.W., Ed., CRC Press, Boca Raton, FL,1993, p. 169.

higher than in most seed oil. The ratio of C16 to C18 in cottonseed oil is higher, and this chemical characteristic is desirable on its crystallization properties by promoting the β form. Almost one third (32.0%) of the total production was produced in China, followed by India (11.0%). *Oil World* predicts an increase of production in the future, especially in China, India, the former Soviet Union, the United States, and Pakistan.

23.2.1.2 Sunflower Oil

Sunflower oil is composed of linoleic (68%), oleic (20%), palmitic (7%), stearic (4%), and linolenic (<1%) acids (Table 23.1) and also contains usual minor components such as phytosterols and tocopherols (Tables 23.2 and 23.3). High-oleic and midoleic sunflower oils are available by seed breeding techniques and the levels of oleic acid are about 80 and 60 to 65%, respectively.

23.2.1.3 Soybean Oil

Soybean oil, commonly called "vegetable oil," is extracted from whole soybeans; it is the most frequently consumed edible oil in the world. Soybean oil contains low saturated fat (15%) and high unsaturated fat (61% polyunsaturated, 24% monounsaturated) including two essential fatty acids, linoleic (53.9%) and linolenic (7.1%), which are not produced in the human body. Soybean seeds are high in tocopherols, and through oil processing removes over 30% of the tocopherols, the refined soybean oil is still considered to be a good source of tocopherols (α-, γ-, and δ-tocopherols). Soybean oil also contains phytosterols, including β-sistosterol, campesterol, and stigmasterol.

23.2.1.4 Sesame Oil

Sesame oil is obtained from sesame seed (*Sesamum indicum* L.) widely cultivated in India, China, Burma, and east Africa, which are the major producers of sesame seeds, contributing to approximately 60% of its total world production. Sesame is the oldest oilseed and is considered to have nutritional value and medicinal properties. Roasted and unroasted sesame oil is consumed in the eastern Asian countries, especially in Korea, China, and Japan. A considerable amount of tocopherols and a number of ligands, mainly sesamin, sesamolin, and sesamol, were retained in sesame oil, and the presence of these phenolic compounds give this oil a much longer shelf life. Sesamin is the major ligand of sesame and does not have any potential as an antioxidant. However, sesamin along with polyunsaturated fatty acids (PuFAs) have been shown to reduce blood pressure in hypertensive rats [3]. Sesame oil is highly resistant to oxidation compared with other edible oils, although it contains high content of unsaturated fatty acids (85%), including oleic acid (41%) and linoleic acid (LA, 44%) [4]. Sesamolin in sesame oil acts as a precursor of two phenolic antioxidants, sesamol and sesaminol. In refined unroasted sesame oil, the antioxidant activity is mainly due to sesaminol produced from sesamolin during the bleaching process. The strong antioxidant, sesamol, is formed from degradation of sesamolin during roasting process [5]. The roasting process is an important step for producing sesame oil since it produces a distinctive pleasant flavor and enhances the quality of sesame oil.

TABLE 23.2
Tocopherol Composition (mg/kg) of Edible Oils

Oil	α-Tocopherol	α-Tocotrienol	β-Tocopherol	β-Tocotrienol	γ-Tocopherol	γ-Tocotrienol	δ-Tocopherol	δ-Tocotrienol
Canola oil	215	—	56	—	629	—	16	—
Coconut oil	18	3	9	—	2	1	—	—
Cottonseed oil	556	—	44	—	335	—	9	—
Palm oil	240	277	—	3	42	363	2	55
Palm-kernel oil	40	—	22	—	2	2	—	—
Soybean oil	111	—	8	—	1279	—	465	—
Sunflower oil	857	—	23	—	18	—	2	—

Source: Modified from Hammond, E.W., Analytical data, in *Chromatography for the Analysis of Lipids*, Hammond, E.W., Ed., CRC Press, Boca Raton, FL,1993, p. 169.

TABLE 23.3
Sterol Composition (mg/kg) of Edible Oils

Oil	Brassicasterol	Campesterol	Stigmasterol	β-Sistosterol	Δ5-Avenasterol	Δ7-Stigmastenol	Δ7-Avenasterol
Canola oil	612	1530	—	3549	122	306	—
Coconut oil	—	18	296	1322	319	136	—
Cottonseed oil	—	170	42	3961	85	—	—
Palm oil	—	358	204	1894	51	25	—
Palm-kernel oil	—	118	145	924	79	13	—
Rice bran oil	—	3100	1500	6700	—	—	—
Sesame oil	—	600	—	2700	—	—	—
Soybean oil	—	720	720	1908	108	108	36
Sunflower oil	—	313	313	2352	156	588	156

Source: Modified from Hammond, E.W., in *Chromatography for the Analysis of Lipids*, Hammond, E.W., Ed., CRC Press, Boca Raton, FL,1993, p. 169; Cho, E.J. and Lee, K.T., *Korean J. Food Preserv.*, 10, 370, 2003.

23.2.2 Oils Rich in Oleic Acid

23.2.2.1 Rice Bran Oil

Rice bran oil, a by-product of rice milling, is not a popular oil worldwide but it is consumed widely as so-called healthy oil in Asia. Rice bran oil contains mostly oleic acid (46%) and LA (31%) as its unsaturated fatty acids, and palmitic acid (20%) as its saturated fatty acid. Rice bran oil is an excellent salad oil and frying oil with high oxidative stability. In recent years, rice bran oil has received much attention due to its cholesterol lowering, antioxidant, and antiatherogenic activity. This effect may be due to specific components such as γ-oryzanol, tocotrienols, and phytosterols [6,7]. γ-Oryzanol, a natural antioxidant, is a mixture containing ferulate esters of triterpene alcohols and plant sterol and the content varies within the range of 1.1 to 2.6% [8]. Rice bran oil is relatively rich in tocotrienols that inhibit cholesterol synthesis by suppressing Hydroxymethyl-glutaryl-CoA (HMG-CoA) reductase activity [9]. Phytosterols are natural and nonnutritive components. They are poorly absorbed and lower serum cholesterol by inhibiting cholesterol absorption and increasing cholesterol excretion.

23.2.2.2 Rapeseed/Canola Oil

Rapeseed oil is produced from seeds of two *Brassica* species: *B. napus* and *B. campestris*. The nutritional aspects of rapeseed oil were questioned, especially concerning the high content (40 to 55%) of erucic acid, an unsaturated 22-carbon fatty acid suspected to be physiologically harmful, causing heart lesions and an accumulation of fatty material around the heart muscle [10]. In addition, high glucosinolate content in the meal fraction was organoleptically unacceptable as a protein supplement for animal feed formula. Canadian oilseed breeders developed canola by genetically altering rapeseed reducing erucic acid content (<2%) and glucosinolate level (<30 μmol/g). From this improvement, canola oil provided high-quality nutritional properties due to low-saturated and high-unsaturated fatty acid content (94%). It is especially higher in oleic acid, like olive oil, than other common commercial vegetable oils [11]. Canola oil ranks currently the third source of vegetable oil after soybean and palm oil. Canola oil is commonly used for frying due to a high smoke point (224°C) and for shortening, salad dressing, and margarine. It is also used in the manufacture of inks, pharmaceuticals, cosmetics, and other uses.

23.2.3 Oils Rich in Oleic and Palmitic Acid

Palm oil is obtained from the mesocarp of oil palm fruit [12]. The palm is a native of west Africa and was introduced to Malaysia at the start of the 20th century, when palm oil was commercially produced. The two largest producers of palm oil are Malaysia (53.2%) and Indonesia (24.1%). During 2000 and 2001, palm oil was second in world total amounts of produced oil, with a contribution of 20.0%, and palm oil is considered as important as soybean oil (22.8%). Crude palm oil contains approximately 1% of minor components, which are carotenoids, tocopherols, tocotrienols, sterols, phospho- and glycolipids, and terpenic and aliphatic hydrocarbons as well as other trace impurities. Red palm oil is a rich source of β-carotene as well as α-tocopherol and tocotrienols that contribute to the stability and nutritional value of palm oil. The major constituents of tochopherols are generally γ-tocotrienol (44%), α-tocopherol (22%), and δ-tocotrienol (12%). Palm oil contains about 50% of saturated (mainly palmitic acid, 43%), 10% diunsaturated, and 40% monounsaturated fatty acids with oleic acid at the *sn*-2 position in triacylglycerols, making this oil as healthy as olive oil. This unique composition makes it versatile on its application in food manufacturing, in which palm olein and palm stearin are popularly used worldwide in making margarine, shortenings, and confectionery and in frying snack foods. It is also a good raw material for the production of oleochemicals, fatty alcohol, glycerol, and other derivatives for the manufacture of cosmetics, pharmaceutical, household, and industrial products.

23.2.4 Oils Rich in Short- and Medium-Chain Fatty Acids

Lauric oils are coconut and palm-kernel oils, which only grow under tropical conditions. The sources of coconut oil and palm-kernel oil are copra from *Cocos nucifera* and palm kernel from *Elaeis guineensis*, respectively. Palm kernel is present inside the nut of the oil palm fruit and is a coproduct in the extraction of palm oil. Production of lauric oils takes place predominantly in the Philippines (coconut oil), Malaysia (palm-kernel oil), and Indonesia (both oils) [13]. The highly saturated lauric oils differ from most vegetable oils in their high level of lauric acid (45 to 50%) and in a significant proportion of short-chain fatty acids (SCFAs, C2:0–C6:0) and medium-chain fatty acids (MCFAs, C8:0–C14:0). Obviously, coconut oil is preferred as a source of octanoic (caprylic) and decanoic (caproic) acids. The lauric oils are widely used in both food and nonfood products. For food industry, the lauric oils are found in spreads, oils, cocoa butter substitutes, filling creams, ice cream, nondairy creams, and coffee whiteners. For nonfood products they are used in the production of medium-chain triglycerides (MCTs).

23.2.5 Oils Rich in n-3 Fatty Acids

23.2.5.1 Perilla Oil

Perilla is a traditional oilseed grown for its food uses at the household level in some Asian countries, but remains less familiar in Western countries. Perilla oil contains the highest content of α-linolenic acid (ALA), accounting for 61% among vegetable oils. Several studies showed that perilla oil reduced serum cholesterol and triglyceride level in blood, prevented atherosclerosis and cancer, and improved immune function [12,14–16].

23.2.5.2 Fish Oil

Marine fish oils are composed of a wide range of long-chain fatty acids with the number of carbon atoms ranging from 14 to 22. Also, they have a high degree of unsaturation as well as high content of n-3 long-chain polyunsaturated fatty acids (LC-PUFAs), such as eicosapentaenoic acid (EPA) and docosahexaenoic acid (DHA). In Western countries, commercial fish oils are consumed mostly as partially hydrogenated fish oil that is used for making margarines and shortening. Partial hydrogenation can diminish the instability problems from oxidation of PUFAs in the original oils because high PUFA content make them susceptible to oxidation. It is known that diets rich in fish oil can be effective in lowering the TAG levels in human plasma and reducing chronic heart disease [17].

23.3 STRUCTURED LIPID AS FUNCTIONAL LIPIDS

Structured lipids (SLs) are any lipids that have been reconstructed from the native state of natural fats and oils to change the positions of fatty acids, or the fatty acid profile having special functionality or nutritional properties. SLs are produced by either chemical or enzymatic methods. Enzymatic interesterification can create intended structured lipids, resulting in specific placement of the fatty acids on the glycerol backbone. Chemical processing produce randomized SLs. SLs as TAGs contain mixtures of fatty acids (short- and/or medium-chain and long-chain) esterified to the glycerol moiety; thus, the SLs can combine together the unique characteristics of each fatty acid, such as melting behavior, digestion, absorption, and metabolism, and can increase their application in foods, nutrition, and therapeutics. SLs are often referred to as nutraceuticals and functional foods since they can provide medical and health benefits including prevention and/or treatment of disease and management of nutritional deficiency [18].

Enzymes, many of which show useful specificities, can catalyze reversible reactions in either direction. Normally, lipases, associated with lipid hydrolysis, can also effect esterification, transesterification, and acidolysis. Lipases such as those derived from *Aspergillus niger, Mucor javanicus, M. miehei, Rhizopus arrhizus, R. delemar*, and *R. niveus* are particularly useful for enzymic acidolysis. These are used for acyl exchange at the *sn*-1,3 positions while leaving acyl groups at the *sn*-2 position unchanged [19]. Structured TAGs were applied for production of cocoa butter substitutes, human milk fat replacers, and nutraceuticals. Unilever has a patent for upgrading palm midfraction (PMF) as a cocoa butter equivalent (CBE). The PMF is too rich in palmitic acid and has too little stearic acid, but this deficiency can be repaired by enzymic acidolysis with stearic acid. Reaction is confined to the exchange of palmitic acid by stearic acid at the *sn*-1,3 positions with no movement of oleic acid from the *sn*-2 position.

A variety of fatty acids is used in the synthesis of SLs for health benefits, and these fatty acids include SCFAs, MCFAs, saturated LCFAs, monounsaturated fatty acids and PuFAs.

23.3.1 CONJUGATED LINOLEIC ACID (CLA)

CLA is a mixture of positional and geometric isomers of LA (C18:2) containing conjugated double bonds at C10 and C12, or C9 and C11 with possible *cis* and *trans* combination, while LAs contains double bonds between C9 and C10, and C12 and C13. The predominant isomers of eight possible geometric isomers are *cis* 9, *trans* 11 and *trans* 10, *cis*12 CLA, which are found naturally in many animal products and have been shown to have beneficial physiological effects including inhibiting cancer risk, enhancing immune response, reducing atherosclerosis and fat gain, and enhancing growth [20–23].

Dietary CLA has a potential anticancer activity in heterocyclic amine colon cancer of rats via inhibiting extrahepatic enzymes like prostaglandin H synthase [24], and in mammary cancer of rodents via inhibiting arachidonic acid (AA) derived eicosanoid such as prostaglandin E_2 [25, 26]. Apoptosis, in particular, cultured mammary tumor cells and premalignant lesions of rat mammary gland, was induced by CLA, suggesting particular sensitivity of early pathological lesions to CLA [27]. The evidence for CLA in protecting against atherogenesis is still limited. CLA fed in an atherogenic diet reduced the development of early arotic atherosclerosis; however, CLA increased fatty streak formation in an animal model [28–30].

Physiological effects of CLA regarding the body composition are varied. Dietary CLA lowered abdominal white adipose tissue weight in rats through an enhanced fatty acid oxidation with elevated activity of mitochondrial carnitine palmitoyltransferase, and a reduced TAG synthesis with decreased activity of phosphatidate phosphohydrolase [31]. By suppressing an activity of heparin releasable-lipoprotein lipase in 3T3-L1 adipocytes, CLA could reduce body fat storage [32]. In an investigation of rats fed CLA, a reduction of fat pad size was observed by smaller adipocyte size [33]. Dietary CLA enhanced T-cell function in mice and pigs [34,35]. Antioxidant activity of CLA is still an unanswered question. Yang et al. [36] found that CLA as a whole oxidized faster than linoleic acid, since conjugated double bond is more vulnerable to auto-oxidation than a nonconjugated double bond. Of the CLA isomers, four *trans, trans* CLA isomers were relatively stable. Methyl conjugated linoleate was the most susceptible to degradation, followed by methyl linoleate, methyl oleate, and methyl stearate under temperature treatment and illumination [37]. Yu [38] investigated the free-radical scavenging properties of CLA against (2,2-diphenyl-1-picryhydrazyl) DPPH radicals measured by electron spin resonance (ESR) and spectrometry methods. Compared with LA, the kinetics of CLA was very different because CLA provided immediate protection against free radicals while LA quenched the radicals after lag phase.

CLA occurs predominantly in natural products such as meat and dairy products from ruminants, since it is formed by rumen microorganisms involved in LA metabolism [20]. However, even though CLA has beneficial biological properties, the consumption of dietary CLA has decreased during the past 20 years due to replacing milk fat by plant lipid and limiting total energy intake from dietary fat. Therefore, increasing the concentration of CLA in food or improving oxidative stability of CLA has been a recent trend [39–41].

23.3.2 Short-Chain and Medium-Chain Fatty Acids

SCFAs and MCFAs have unique nutritional characteristics. In the stomach these fatty acids are more rapidly absorbed than ordinary LCFAs because of their higher water solubility, smaller molecular size, and shorter chain length. Hydrophilic SCFAs are mainly positioned at the *sn*-3 position of TAG, and they are completely hydrolyzed by pancreatic lipase in the lumen. SCFAs are useful ingredients in the synthesis of low-calorie SLs such as Benefat (Cultor Food Science Inc., New York) since SCFAs are lower in caloric value than MCFAs and LCFAs. The MCFAs are esterified with glycerol to form MCTs. The chemical structure of MCTs affords unique properties such as low viscosity and bland odor and taste. They are very stable to oxidation due to the saturation of the MCFAs and are liquid at room temperature due to much lower melting point (C8:0, 16.7°C; C10:0, 31.3°C) than LCFAs (C16:0, 63.1°C). MCTs deliver fewer caloric values (8.3 kcal/g) than typical LCTs (9 kcal/g). They are easily absorbed and rapidly utilized for energy. In the small intestine, 1 mol of conventional fat/oil is hydrolyzed into 3 mole of LCFAs and 1 mol of glycerol. The LCFAs as a major component of chylomicron are transported into the lymphatic system through the mucosal wall, and then circulate via the lymphatic system to the liver where they are oxidized. The LCFAs that are not oxidized for energy are deposited in fat cells throughout the adipose tissues. Compared to LCTs, the molecular size of MCTs is small. MCTs are readily hydrolyzed as MCFAs and are absorbed primarily into the intestine. After absorption, MCFAs are bound to serum albumin and transported through the portal vein into the liver, where they are oxidized to form ketone body and supply energy quickly [42]. Even MCTs can be absorbed as a form of TAGs instead of free fatty acids in the cases of bile salts or pancreatic lipase deficiency, whereas LCT cannot be absorbed as a triacyglycerol form [43]. Due to their unique characteristics, MCTs have clinical applications in the treatment of patients suffering from fat malabsorption, gallbladder disease, hyperlipidemias, deficiency of the carnitine system, and obesity.

23.3.3 Monounsaturated Fatty Acids

Canola and olive oils are particularly rich dietary source of oleic acid (C18:1), which is synthesized from palmitic and stearic acid through the elongation and saturation. Oleic acid is the most common monounsaturated fatty acid (MUFA) and the precursor for the production of PUFA. As an n-9 family, they do not enter into the prostaglandin cascade like other n-6 and n-3 PUFA. Epidemiologic evidence indicated that Mediterranean diets using olive oil in cooking and in sauces and dressing showed beneficial effects against coronary heart disease. Diets rich in MUFAs may offer a positive means of improving human health since MUFAs favorably affect a number of risk factors for immune [44], diabetes [45], cancer [46], and cardiovascular disease [47] regarding plasma lipids, low-density lipoprotein (LDLs) cholesterol, and LDL oxidation [48–52].

23.3.4 Polyunsaturated Fatty Acids

Usually the important PUFAs can be divided into two families, n-3, n-6, and n-9 fatty acids, depending on the site of the first double bond from the methyl end. The n-3 and n-6 families are PUFAs, whereas n-9 fatty acids are mostly MUFAs. The n-3 fatty acids contained in vegetable oils are ALA and fish oils also have a high content of n-3 fatty acids mostly in the forms of EPA and DHA [53]. So far, a large number of studies have been done on the potential benefits of the n-3 fatty acids in human diseases since a high consumption was associated with protective cardiovascular disease [54–57], renal disorder [58], inflammatory processes, autoimmune disorders

[59], infection [60], cancer [61], and allograft rejection [62]. In 2001, the American Heart Association (AHA) recommended at least 1 g of n-3 PUFA per day for people who have heart disease, and suggested 2 to 4 g/d for people with elevated levels of TAGs [63].

LA is a common n-6 fatty acid found in most vegetable oils. These n-3 and n-6 fatty acids cannot be synthesized by the human body; thus they are considered as essential fatty acids required for keeping physiological health. The ratio of n-3 to n-6 fatty acids in body fat is important since the higher ratio is more protective in the incidence of cancer [64]. Both LA and ALA are the precursors of the n-3 and n-6 families responsible for their active metabolites, AA and EPA/DHA, by desaturation and elongation reactions, respectively (Figure 23.1). Conversion of ALA to EPA and DHA is low in humans and may be further suppressed when intake of dietary LA is high. Under various stimuli, PUFAs such as EPA, DHA, and AA are released from the phospholipid in cell membranes to undergo enzymatic degradation into the eicosanoids, which are all C20 compounds as short-lived lipid derived mediators and, like hormones, have profound physiological effects at extremely low concentration.

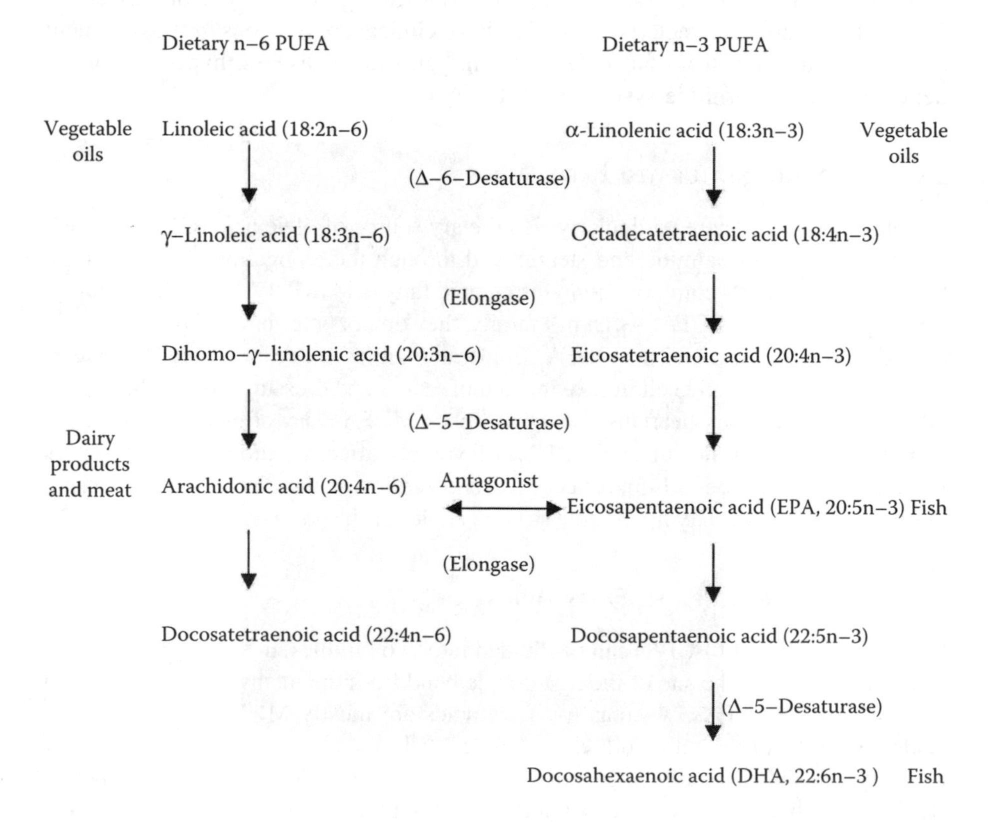

FIGURE 23.1 Pathways of n-6 and n-3 polyunsaturated fatty acid metabolism showing chain elongation and desaturation step.

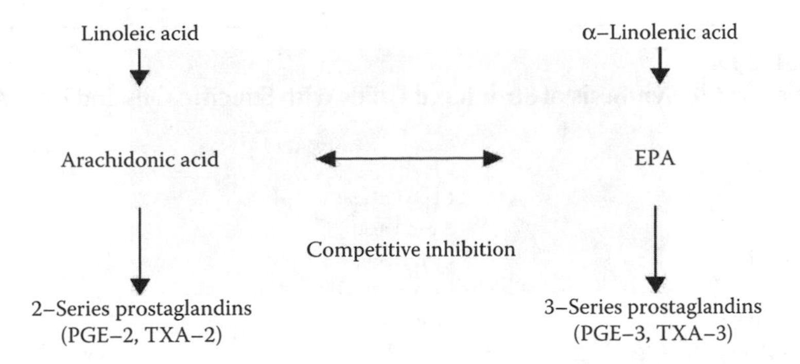

FIGURE 23.2 Pathways of n-6 and n-3 PUFA metabolism showing biosynthesis of eicosanoids.

AA and EPA are the substrates for the formation of two series of eicosanoids including prostagladins, leukotrienes, and thromboxanes (Figure 23.2). AA is metabolized by cyclooxygenase and lipoxygenase to eicosanoids of the 2- and 4-series (prostaglandin E_2 [PGE_2], thromboxane A_2 [TXA_2], and leukotriene B_4 [LTB_4]) that promote tumor growth and aggressiveness in animal tumor models [61]. Metabolism of EPA and DHA produce 3-series prostagladins and 5-series leukotrienes which are noninflammatory and do not promote tumor growth and metastasis. In addition, EPA and DHA are antagonists for the same metabolic pathway of AA resulting in lower production of 2- and 4-series inflammatory eicosanoids. EPA and DHA are incorporated into *sn*-2 position in phospholipids of cell membrane displacing AA by decreasing precursor enzymes that develop prostagladins of E-2 series. Thus, the dietary ratio of n-6 to n-3 is an important nutritional variable and should be considered when searching for the optimum human diet.

23.4 SYNTHESIS OF STRUCTURED LIPIDS

Production of SLs has been conducted with the functional fatty acids in different fats and oils including canola oil, palm oil, soybean oil, rice bran oil, olive oil, fish oil, corn oil, perilla oil, safflower oil, sunflower oil, cottonseed oil, MCT, triolein, tricaprin, and tricaprylin (Table 23.4). Using the functional fatty acids having beneficial effects, lipase-catalyzed acidolysis reaction resulted in the changes in fatty acid composition, triglyceride structure, viscosity, hydrophobicity, minor compound contents, and other physicochemical properties of original substrate oil. According to Fomuso and Akoh [65], SLs from olive oil showed darker color, higher viscosity, and lower melting point than olive oil. It is assumed that the formation of newly synthesized triglycerides altered the melting profiles, and this change can influence the marketing strategy in the fats and oils industry [66].

Furthermore, sensory and nutritional properties of SLs could be comparable or superior to original oil when SLs were applied to beverages or margarines [67,68]. Compared to original oil, however, the oxidative stability of produced SLs was lower than original oil; therefore, addition of antioxidants was needed to protect the SLs from further deterioration [69,70].

TABLE 23.4
Studies on the Synthesis of Structured Lipids with Specific Oils and Fatty Acids

Oil	Fatty Acid	References
Canola oil	Caprylic acid	[67,77]
	Stearic acid	[86]
Coconut oil	Lauric acid	[66]
Corn oil	CLA	[74]
Cottonseed oil	Lauric acid	[66]
Menhaden oil	CLA	[73]
	γ-Linolenic acid	[84]
Olive oil	Caprylic acid	[77]
Palm oil	CLA	[71]
	n-3 Fatty acids	[82]
Palm olein	Lauric acid	[66]
Perilla oil	Caprylic acid	[76]
Rice bran oil	Capric acid	[79]
Soybean oil	CLA	[72]
	Lauric acid	[66]
Sunflower oil	CLA	[72]
Tuna oil	Caprylic acid	[75]
MCT (trilaurin and tricaprylin)	EPA	[81]
	Oil	
Olive oil	Partially hydrogenated palm oil	[68]
Palm oil	Fully hydrogenated soybean oil	[85]

23.4.1 SLs Containing Conjugated Linoleic Acid

Since CLA has been associated with important health benefits, the incorporation of CLA into triglyceride forms of edible oils has been carried out by enzyme-mediated acidolysis to increase the content of CLA in the edible oil. McNeill et al. [71] incorporated CLA into palm oil, which is produced mainly in Malaysia and Indonesia. Lipase from *Rhizomucor miehei* was chosen. After the reaction, the content of CLA was approximately 30% in SL-palm oil, and the CLA was incorporated throughout the triglyceride structures. Other types of SLs-CLA from soybean and sunflower oil were produced, having 24.43 and 23.07 mol% CLA in SL-soybean oil and SL-sunflower oil, respectively. Tocopherol contents in soybean oil and sunflower oil were reduced, and the color and oxidative stability were changed [72]. The production of SLs rich in both CLA and n-3 fatty acids (EPA and DHA) was studied using four commercial lipases (PS from *Pseudomonas cepacia*, G from *Penicillium camemberti*, L2 from *Candida antarctica*, and L9 from *Mucor miehei)*, in which menhaden oil (MO) was modified by lipases at 40°C for 48 h [73]. IM-60 from *Mucor miehei* and Chirazyme L-2 from *Candida antarctica* produced substantial extents of acidolysis from corn oil and CLA at 50°C in 12 h. In their results, hexane, the only organic solvent accepted in food processing by the authorities, was appropriate to maximize the acidolysis reaction rate [74].

23.4.2 SLs Containing SCFAs and MCFAs

SLs with SCFAs or MCFAs have been of much interest. SCFAs and MCFAs are absorbed more easily and thus could help provide proper nourishment for fat malabsorption patients. Pancreatic lipase hydrolyzes ester bonds at *sn*-1,3 positions in triglycerides, and more than 75% of fatty acids at the *sn*-2 position remained in the original position after absorption. SL-triglycerides with SCFAs or MCFAs at *sn*-1,3 positions and other functional fatty acids at the *sn*-2 position could be ideal because SLs having such a composition were rapidly hydrolyzed and efficiently absorbed into the mucosal intestine. SL enriched in caprylic acid (41.9 mol% at *sn*-1,3) and EPA (7.8 mol% at *sn*-1,3; 12.4 mol% at *sn*-2) was produced by acidolysis of tuna oil with caprylic acid (C8:0) [75]. Perilla oil, which contains ALA (about 60%), was acidolyzed with caprylic acid to produce SL enriched in ALA (n-3 fatty acid) and MCFA. Using two lipases, Lipozyme RM IM from *Rhizomucor miehei* and Lipozyme TL IM from *Thermomyces lanuginosus*, caprylic acids were incorporated at 48.5 to 51.4 mol% after 24 h incubation in the presence of *n*-hexane [76].

A packed-bed reactor is frequently used for continuous operation in commercial scale to minimize labor and overhead costs, and to obtain more consistent quality of SL [77]. Acidolysis of caprylic acid and olive oil with immobilized lipase from *Rhizomucor miehei* (IM60) was performed in a bench-scale packed bed bioreactor. SL-canola oil with caprylic acid was synthesized in a packed-bed bioreactor with Lipozyme IM catalyzed acidolysis [67,78]. The SL (2 kg) enriched in caprylic acid (40.1%) and oleic acid (34.7%) was produced under the following optimum conditions: flow rate, 1 mL/min; temperature 60°C; substrate molar ratio, 5:1; water content, 0.2%. Sensory evaluation with quantitative descriptive analysis indicated that a chocolate-flavored nutritional beverage containing the SL-canola oil was found to have enhanced sweet flavor and decreased bubble formation compared with the beverage containing canola oil. The result suggested the possible substitution of the SL-canola oil for canola oil in food applications.

Rice bran oil, consumed widely in Asia, was modified to contain capric acid (C10:0) via lipase (IM 60)-catalyzed acidolysis. The decreases in contents of tocopherol and tocotrienols in SL-rice bran oil compared to rice bran oil were observed during enzymatic modification and further processing [79]. Lauric acid (C12:0) was incorporated into selected vegetable oils such as palm olein, coconut oil, cottonseed oil, rapeseed oil, corn oil, and soybean oil. Mycelium-bound lipase from *Aspergillus flavus* was used for acidolysis, and cottonseed oil showed the highest incorporation rate with lauric acid followed by soybean oil [66].

23.4.3 SLs Containing n-3 Fatty Acid

When n-3 fatty acids are incorporated into vegetable oils, the species of n-3 fatty acids and lipases, and the water content could influence the lipase-catalyzed reaction [80,81]. Lipase IM60 incorporated more EPA into vegetable oils than SP435 from *Candida antarctica.* With IM60, the addition of 5 μl water increased the EPA content by 4.9%; however, the incorporation of EPA by SP435 increased with up to 2 μl water addition. In SL synthesized with medium-chain TAG (trilaurin and tricaprylin) and EPA, the incorporation (mol%) of EPA increased by adding water to the IM 60-catalyzed acidolysis [81]. Incorporation of n-3 fatty acids such as EPA (20.8%) and DHA (15.6%) into palm oil was studied with two lipases, IM60 and QLM [82].

SL containing n-3 PUFA (EPA, 18 mol% and DHA, 15 mol%) and caprylic acid (64 mol.%) was produced from tricaprylin and EPA-rich fish oil, and its dietary effect was compared to soybean oil (oleic acid, 25 mol% and linoleic acid, 66 mol%) in an animal study with mice for 21 d. The experimental diets contained 17% soybean oil or SL and 20% total fat content by weight. The concentrations of total serum cholesterol, LDL cholesterol, and triacylglycerol, and the body weight gain were lowered significantly ($p < .05$) in SL-fed mice than soybean oil-fed mice. The fatty acid composition of liver TAG showed that the mol% of EPA and DHA were much higher in SL-fed mice. The results suggested that this SL could be a therapeutic and medical lipid source and may be useful in enteral and parenteral nutrition [83].

23.4.4 SLs Containing n-6 PUFA

γ-Linolenic acid, n-6 PUFA, is used for dietetic and pharmaceutical purposes. SLs from MO or seal blubber oil (SBO) with γ-linolenic acid were produced by lipase-assisted acidolysis. Under the optimal condition, the incorporation of γ-linolenic acid was 37.1% for SBO and 39.6% for MO, and the ratio of n-3 to n-6 PUFA was 1:3.6 for SBO and 1:2.2 for MO. These SLs enriched with n-3 and n-6 PUFA could be used for clinical and nutritional applications [84].

23.4.5 SLs with Oil Blends

Lipase-catalyzed acidolysis can be an alternative to hydrogenation. SL was produced with olive oil and partially hydrogenated palm oil (weight ratio 3:7) blend with IM60, and the SL was found to have similar physicochemical properties (melting point and solid fat content) and sensory properties (spreadibility and appearances) to soft tub margarines [68]. Cocoa butter is widely used in chocolate and confectionery industries due to its important physical and organoleptic properties. As an alternative, CBEs have been produced from modified vegetable oil and replaced cocoa butter in as much as 10% of the market place. SLs for this purpose were produced with palm oil and fully hydrogenated soybean oil (weight ratio 1.6:1), or high-oleic acid rapeseed oil and stearic acid or methyl stearate. The major triglyceride component, 1(3)-palmitoyl-3(1)-stearoyl-2-monoolein, of cocoa butter was identified in the SLs, and the melting point of SLs was comparable to that of cocoa butter [85,86].

23.5 APPLICATIONS OF STRUCTURED LIPIDS

Several low-calorie oils composed of specific fatty acid have been commercialized, For example, Caprenin (Proctor & Gamble, Cincinnati, Ohio) and Salatrim. SLs having particular fatty acids at specific positions of glycerol are used as CBEs and in an infant formula [1]. LC-PUFAs such as EPA and/or DHA can be introduced into vegetable oils or synthetic glycerides to give products with enhanced nutritional value. Interesterification of a common plant oil, such as sunflower oil, with a MCFA by means of an *sn*-1,3 specific lipase would yield TAGs containing medium-chain acyl moieties at the *sn*-1,3 positions and long-chain acyl moieties at the *sn*-2 position [87]. The commercial structured lipids and their applications worldwide are presented in Table 23.5.

TABLE 23.5
Commercial Structured Lipids and their Food and Medical Applications

Brand Name	Fatty Acid Composition	Application	Sources
Benefat®	C18:0 (high content) and C2:0, C4:0 or C6:0	Incorporation in reduced-calorie baked products, chocolate coating and nutrition bars	Cultor Food Science Inc.
Betapol™	C16:0 (45%)	Infant food formulation	Loders Croklaan
Bohenin	C18:1 and C22:0	Tempering aid and antiblooming agent in the manufacture of chocolate and chocolate coating	Fuji Oil Company Ltd.
Captex®	C8:0/C10:0 (60%) and C12:0 (30%) or C8:0/C10:0 (40%) and C18:2 (40%)	Pharmaceuticals for parenteral and enteric hyperalimentation, and cosmetic industry	Abitec Corp.
Caprenin®	C8:0, C10:0, and C22:0	Confectionary coating fat	Procter & Gamble
Impact®	Interesterification with high-lauric acid oil and high linoleic acid oil	Pharmaceuticals for patients suffering from trauma or surgery, sepsis, or cancer	Novartis Nutrition Corp.
Laurical	C12:0 (40%) and unsaturated fat (C18:1, C18:2, and C18:3)	Medical nutrition products and confectionary coating, coffee whiteners, whipped toppings and filling fats	Calgene Inc.
Neobee®	C8:0, C10:0 and LCFA (n-6 and n-3)	Nutritional or medical beverages	Stepan Corp.
Structolipid	LCT (63%) and MCT (37%) — caprylic (27%), capric (10%), palmitic (7%), oleic (13%), linoleic (33%), and α-linoleic acid (5%)	Intravenous fat emulsion as a rapid source of energy for patients and parenteral nutrition	Fresenius Kabi

23.5.1 Medical Applications

Impact, manufactured by Novartis Nutrition Corp. (Minneapolis, Minnesota), contains randomized structured lipids created by interesterification of high lauric acid oil and high LA oil, and can be used for patients suffering from major trauma or surgery, sepsis, or cancer [88]. Stepan Corp. (Maywood, New Jersey) manufactures Neobee, an SL-based on MCT, that can be a replacement for partially hydrogenated vegetable oil (PHVO) in medical and nutritional applications including salad oils, coating, pastry, breads, and margarine [89]. Abitec Corp. (Columbus, Ohio) developed Captex, which is used in the clinical and cosmetic industry. Captex 350 is an SL based on MCT interesterified with coconut oil providing caprylic/capric acid (60%) and lauric acid (30%), and Captex 810 is prepared from selected high-purity vegetable products and glycerin, producing a high content of LA (40%) and caprylic/capric acid (40%).

23.5.2 Food Applications

SALATRIM (*s*hort *a*nd *l*ong *a*cyl*tri*glyceride *m*olecule) is a family of SLs developed by Nabisco Food Group, and provides approximately half of the calories (5 kcal/g) of normal edible oil (9 kcal/g). Benefat (Cultor Food Science Inc. New York), the brand name of SALATRIM, is composed of at least one LCFA (predominantly C18:0, stearic acid) and at least one short-chain fatty acid (C2, acetic; C3, propionic; or C4, butyric), which can be produced by replacing some of PUFA of hydrogenated canola or soybean oil with specific ratios of SCFA. Benefat is permitted as a food ingredient in the United States, Korea, Japan, and Taiwan, and has been incorporated in a range of reduced-calorie baked products, confectionery, chocolate coating, biscuit filling, and nutrition bars [90]. Caprenin, used as a coating fat for Milky Way II reduced-fat candy bar, is one of the SLs composed of MCFAs (caprylic and capric acids) from coconut and palm-kernel oils, and very LCFAs (C22, behenic acid) from hydrogenated rapeseed oil.

Targeted structured lipids could be created via enzyme-catalyzed reaction to locate specific fatty acids on the glycerol backbone. In Japan, Fuji Oil Company Ltd. (Osaka) marketed Bohenin, which is a triglyceride composed of behenic acid (C22:0) at the *sn*-1 and -3 positions and oleic acid (C18:1) at *sn*-2 position in a 2:1 ratio. The specific structure produced via lipase-catalyzed reaction has received considerable attention. Bohenin is used as an antibloom agent and as a tempering aid in the manufacture of chocolate and chocolate coating, because it has important properties such as high melting temperature and ability to form β2-crystalline polymorphic structures. Human milk contains palmitic acid located predominantly at the *sn*-2 position. Betapol was developed by enzymatic transesterification using a lipase from *Rhizomucor miehei*, in which a palm oil fraction rich in tripalmitin and a mixture of canola and sunflower oils high in oleic acid were used. Such TAGs closely resemble the structure of human milk TAG, and are used in infant food formulation [91]. Calgene Inc. (Davis, California) created Laurical (high-laurate canola) by genetically inserting the lauroyl-APC thioesterase into canola, which does not have lauric acid. This SL contains lauric acid (40%) at *sn*-1 and -3 positions, and unsaturated fat such as oleic, linoleic, or linolenic acid at *sn*-2 positions of the triglycerides, and is used in medical nutrition products and confectionary coating, coffee whiteners, whipped toppings, and filling fats.

ACKNOWLEDGMENT

The authors are grateful for support from the Ministry of Health and Welfare, Republic of Korea (Korea Health 21 R&D Project, 02-PJI-PGI-CH15-001).

REFERENCES

1. Shimada et al., Production of functional lipids containing polyunsaturated fatty acids with lipase, in *Enzyme in Lipid Modification,* Bornscheuer, U.T., Ed., Wiley-Vch, Weinheim, Germany, 2000, chap. 8.
2. Gunstone, F.D., Soybean oil: a primary dietary fat, *INFORM,* 12, 605, 2001.
3. Linda, M.O., Fats for healthy living, *Food Technol.,* 57, 91, 2003.
4. Abou-Gharbia, H.A., Shehata, A.A., and Shahidi, F., Effects of processing on oxidative stability and lipid classes of sesame oil, *Food Res. Int.,* 33, 331, 2000.
5. Mohamed, H.M.A. and Awatif, I.I., The use of sesame oil unsaponifiable matter as a natural antioxidant, *Food Chem.,* 62, 269, 1997.
6. Sugano, M. and Tsuji, E. Rice bran oil and cholesterol metabolism, *J. Nutr.,* 127, 521, 1997.
7. Wilson, T.A. et al., Whole fat rice bran reduces the development of early arotic atherosclerosis in hypercholesterolemic hamsters compared with wheat bran, *Nutr. Res.,* 22, 1319, 2002.
8. Seetharamaish, G.S. and Prabhakar, J.V., Oryzanol content of Indian rice bran oil and its extraction from soap stock, *J. Food. Sci. Technol.,* 23, 270, 1986.
9. Qureshi, A.A. et al., Novel tocotrienols of rice bran modulate cardiovascular disease risk parameters of hypercholesterolemic humans, *J. Nutr. Biochem.,* 8, 290, 1997.
10. Dziezak, J.D., Fats, oils, and fat substitutes, *Food Technol.,* 43, 66, 1989.
11. Carr, R.A., Development of deep-frying fats, *Food Technol.,* 45, 95, 1991.
12. Onogi, N., Okuno, M., and Komaki. C., Suppressing effect of perilla oil on azoxymethane-induced foci of colonic aberrant crypts in rats, *Carcinogenesis,* 17, 1291, 1996.
13. Gunstone, F.D., Lauric oils, *INFORM,* 13:840, 2002.
14. Longvah T., Deosthale, Y.G., and Kumar, P.U., Nutritional and short term toxicological evaluation of perilla seed oil, *Food Chem.,* 70, 13, 2000.
15. Sadi. A.M. et al., Dietary effects of corn oil, oleic acid, perilla oil, and primrose oil on plasma and hepatic lipid level and atherosclerosis in Japanese quail, *Exp. Anim.,* 45, 55, 1996.
16. Jeffery, N.M. et al., The ratio of n-6 to n-3 polyunsaturated fatty acids in the rat diet alters serum lipid levels and lymphocyte functions, *Lipids,* 31, 737, 745, 1996.
17. Burr, M.L. et al., Effects of changes in fat, fish, and fibre intakes on death and myocardial reinfarction: diet and reinfarction trial (dart), *Lancet,* 334, 757, 1989.
18. Kennedy, J.P., Structured lipids: fats of the future. *Food Technol.,* 45, 76, 1991.
19. Gunstone, F.D., Procedure used for lipid modification, in *Structured and Modified Lipids,* Gunstone, F.D., Ed., Marcel Dekker, New York, 2001, chap. 2.

20. Pariza, M.W., Conjugated linoleic acid, a newly recognized nutrient, *Chem. Ind.,* 16, 464, 1997.
21. Pariza, M.W., Park, Y., and Cook, M.E., The biologically active isomers of conjugated linoleic acid, *Prog. Lipid Res.,* 40, 283, 2001.
22. Jaresi, G. et al., Conjugated linoleic acid: physiological effects in animal and man with special regard to body composition, *Eur. J. Lipid Sci. Technol,* 102, 695, 2000.
23. Cook, M.E. and Pariza, M., The role of conjugated linoleic acid (CLA) in health, *Int. Dairy J.,* 8, 459, 1998.
24. Xu, M. and Dashwood, R.H., Chemoprevention studies of heterocyclic amine-induced colon carcinogenesis, *Cancer Lett.,* 143, 179, 1999.
25. Hubbard, N.E. et al., Reduction of murine mammary tumor metastasis by conjugated linoleic acid, *Cancer Lett.,* 150, 92, 2000.
26. Banni, S. et al., Decrease in linoleic acid metabolites as a potential mechanism in cancer risk reduction by conjugated linoleic acid, *Carcinogenesis,* 20, 1019, 1999.
27. Ip, C. et al., Induction of apoptosis by conjugated linoleic acid in cultured mammary tumor cells and premalignant lesions of the rat mammary gland, *Cancer Epidemiol. Biomarkers Prev.,* 9, 689, 2000.
28. Wilson, T.A. et al., Conjugated linoleic acid reduces early aortic atherosclerosis greater than linoleic acid in hypercholesterolemic hamsters, *Nutr. Res.,* 20, 1795, 2000.
29. Munday, J.S., Thompson, K.G., and James, K.A.C., Dietary conjugated linoleic acids promote fatty acid streak formation in the C57BL/6 mouse atherosclerosis model, *Br. J. Nutr.,* 81, 251, 1991.
30. Rudel, L.L., Invited commentary, Atherosclerosis and conjugated linoleic acid, *Br. J. Nutr.,* 81, 177, 1999.
31. Rahman, S.M. et al., Effects of short-term administration of conjugated linoleic acid on lipid metabolism in white and brown adipose tissues of starved/refed Otsuka Long-Evans Tokushima Fatty rats, *Food Res. Int.,* 34, 515, 2001.
32. Lin, Y. et al., Different effects of conjugated linoleic acid isomers on lipoprotein lipase activity in 3T3-L1 adipocytes, *J. Nutr. Biochem.,* 12, 183–189, 2001.
33. Azain, M.J. et al., Dietary conjugated linoleic acid reduces rat adipose tissue cell size rather than cell number, *J. Nutr.,* 130, 1548, 2001.
34. Hayek, M.G. et al., Dietary conjugated linoleic acid influences the immune response of young and old C57BL/6NCrlBR mice, *J. Nutr.,* 129, 32, 1999.
35. Bassaganya-Riera, J. et al., Dietary conjugated linoleic acid modulates phenotype and effector functions of porcine CD8+ lymphocytes, *J. Nutr.,* 131, 2370, 2001.
36. Yang, L. et al., Oxidative stability of conjugated linoleic acid isomers, *J. Agric. Food Chem.,* 48, 3072, 2000.
37. Chen, J.F. et al., Effects of conjugated linoleic acid on the degradation and oxidation stability of model lipids during heating and illumination, *Food Chem.,* 72, 199, 2001.
38. Yu, L., Free radical scavenging properties of conjugated linoleic acids, *J. Agric. Food Chem.,* 49, 3452, 2001.
39. Lin, T.Y., Conjugated linoleic acid concentration as affected by lactic cultures and additives, *Food Chem.,* 69, 27, 2000.
40. Kim, S.J. et al., Improvement of oxidative stability of conjugated linoleic acid (CLA) by microencapsulation in cyclodextrins, *J. Agric. Food Chem.,* 48, 3922, 2000.
41. Kim, Y.J. and Liu, R.H., Increase of conjugated linoleic acid content in milk by fermentation with lactic acid bacteria, *J. Food Sci.,* 67, 1731, 2002.
42. Megremis, C.J., Medium-chain triglycerides: a nonconventional fat, *Food Technol.,* 45, 108, 1991.

43. Bach, A.C. and Babayan V.K., Medium-chain triglycerides: an update, *Am. J. Clin. Nutr.,* 36, 950, 1982.
44. Moussa, M. et al., In vivo effects of olive oil-based lipid emulsion on lymphocyte activation in rats, *Clin. Nutr.,* 19, 49, 2000.
45. Lauszus, F.F. et al., Effect of a high monounsaturated fatty acid diet on blood pressure and glucose metabolism in women with gestational diabetes mellitus, *Eur. J. Clin. Nutr.,* 55, 436, 2001.
46. Kimura, Y., Carp oil or oleic acid, but not linoleic acid or linolenic acid, inhibits tumor growth and metastasis in Lewis lung carcinoma-bearing mice, *J. Nutr.,* 132, 2069, 2002.
47. Jamison, J.R., Cardiovascular health: a case study exploring the feasibility of a diet relatively rich in monounsaturated fats, *J. Nutr. Environ. Med.,* 8, 257, 1998.
48. Archer, W.R. et al., High carbohydrate and high monounsaturated fatty acid diets similarly affect LDL electrophoretic characteristics in men who are losing weight, *J. Nutr.,* 133, 3124, 2003.
49. Pérez-Jiménez, F. et al., Circulating levels of endothelial function are modulated by dietary monosaturated fat, *Atherosclerosis,* 145, 351, 1999.
50. Mata, P. et al., Effect of dietary fat saturation on LDL oxidation and monocyte adhesion to human endothelial cells in vitro, *Arterioscler. Thromb. Vasc. Biol.,* 16, 1347, 1996.
51. Reaven, P., Barbara, G., and Barnett, J., Effect of antioxidants alone and in combination with monounsaturated fatty acid-enriched diets on lipoprotein oxidation, *Arterioscler. Thromb. Vasc. Biol.,* 16, 1465, 1996.
52. Kris-Etherton, P.M., Monounsaturated fatty acids and risk of cardiovascular disease, *Circulation,* 100, 1253, 1999.
53. Alexander, J.W., Immunonutrition: the role of ω-3 fatty acids, *Nutrition,* 14, 627, 1998.
54. Nageswari, K., Banerjee, R., and Menon, V.P., Effect of saturated, ω-3 and ω-6 polyunsaturated fatty acids on myocardial infarction, *J. Nutr. Biochem.,* 10, 338, 1999.
55. Baumann, K.H. et al., Dietary ω-3, ω-6, and ω-9 unsaturated fatty acids and growth factor and cytokine gene expression in unstimulated and stimulated monocytes, *Arterioscler. Thromb. Vasc. Biol.,* 19, 59, 1999.
56. Schacky, C.V., n-3 Fatty acids and the prevention of coronary atherosclerosis, *Am. J. Clin. Nutr.,* 71(suppl), 224S, 2000.
57. Billman, G.E., Kang, J.X., and Leaf, A., Prevention of sudden cardiac death by dietary pure ω-3 polyunsaturated fatty acids in dogs, *Circulation,* 99, 2452, 1999.
58. Donadio, J.V., n-3 fatty acids and their role in nephrologic practice, *Curr. Opin. Nephrol. Hypertens.,*10, 639, 2001.
59. Fernandes, G., Effects of calorie restriction and omega-3 fatty acids on autoimmunity and aging, *Nutr. Rev.,* 53, S72, 1995.
60. Anderson, M. and Fritsche, K.L., (n-3) Fatty acids and infectious disease resistance, *J. Nutr.,* 132, 3566, 2002.
61. Capone, S.L., Bagga, D., and Glaspy, J.A., Relationship between omega-3 and omega-6 fatty acid ratios and breast cancer, *Nutrition,* 13, 822, 1997.
62. Alexander, J.W. et al., Dietary omega-3 and omega-9 fatty acids uniquely enhance allograft survival in cyclosporine-treated and donor-specific transfusion-treated rats, *Transplantation,* 65, 1304, 1998.
63. Kris-Etherton, P.M, Harris, W.S., and Appel, L.J., Fish consumption, fish oil, omega-3 fatty acids and cardiovascular disease, *Circulation,* 106, 2747, 2003.
64. Simonsen, N., Strain, J.J., and van't Veer P., Adipose tissue omega-3 fatty acids and breast cancer in a population of European women, *Am. J. Epidemiol.,* 143, S34, 1996.

65. Fomuso, L.B. and Akoh, C.C., Lipase-catalyzed acidolysis of olive oil and caprylic acid in a bench-scale packed bed bioreactor, *Food Res. Int.*, 35,15, 2002.
66. Long, K. et al., Acidolysis of several vegetable oils by mycelium-bound lipase of *Aspergillus flavus* link, *J. Am. Chem. Soc.*, 74, 1121, 1997.
67. Osborn, H.T., Shewfelt, R.L., and Akoh, C.C., Sensory evaluation of a nutritional beverage containing canola oil/caprylic acid structured lipid, *J. Am. Chem. Soc.*, 80, 357, 2003.
68. Alpaslan, M. and Karaali, A., The interesterification-induced changes in olive and palm oil blends, *Food Chem.*, 61, 301, 1998.
69. Akoh, C.C. and Moussata, C.O., Characterization and oxidative stability of enzymatically produced fish and canola oil-based structured lipids, *J. Am. Chem. Soc.*, 78, 25, 2001.
70. Senanayake, S.P.J.N. and Shahidi, F., Chemical and stability characteristics of structured lipids from borage (*Borago officinalis* L.) and evening primrose (*Orenothera biennis* L.) oils, *J. Food Sci.*, 67, 2038, 2002.
71. McNeill, G.P., Rawlins, C., and Peilow, A.C., Enzymatic enrichment of conjugated linoleic acid isomers and incorporation into triglycerides, *J. Am. Chem. Soc.*, 76, 1265, 1999.
72. Lee, J.H. et al., Characterization of lipase-catalyzed structured lipids from selected vegetable oils with conjugated linoleic acid: their oxidative stability with rosemary extracts, *J. Food Sci.*, 68, 1653, 2003.
73. Torres, C.F., Garcia, H.S., Ries, J.J., and Hill, C.G., Esterification of glycerol with conjugated linoleic acid and long-chain fatty acids from fish oil, *J. Am. Chem. Soc.*, 78, 1093, 2001.
74. Martinez, C.E. et al., Lipase-catalyzed interesterification (acidolysis) of corn oil and conjugated linoleic acid in organic solvents, *Food Biotechnol.*, 13, 183, 1999.
75. Shimada, Y. et al., Production of structured lipids containing docohexaenoic and caprylic acids using immobilized Rhizopus delemar lipase, *J. Ferment. Bioeng.*, 81, 299, 1996.
76. Kim, I.H. et al., Lipase-catalyzed acidolysis of perilla oil with caprylic acid to produce structured lipids, *J. Am. Chem. Soc.*, 79, 363, 2002.
77. Xu, X., Enzyme bioreactors for lipid modifications, *INFORM,* 11:1004–1012, 2000.
78. Xu, X., Fomuso, L.B., and Akoh, C.C., Synthesis of structured triacylgycerols by lipase-catalyzed acidolysis in a packed bed bioreactor, *J. Agric. Food Chem.*, 48, 3, 2000.
79. Jennings, B.H. and Akoh, C.C., Lipase-catalyzed modification of rice bran oil to incorporate capric acid, *J. Agric. Food Chem.*, 48, 4439, 2000.
80. Huang, K.H. and Akoh, C.C, Lipase-catalyzed incorporation of n-3 polyunsaturated fatty acids into vegetable oil, *J. Am. Chem. Soc.*, 71, 1277, 1994.
81. Lee, K.T. and Akoh, C.C., Immobilized lipase-catalyzed production of structured lipids with eicosapentaenoic acid at specific positions, *J. Am. Chem. Soc.*, 73, 611, 1996.
82. Fajardo, A.R., Akoh, C.C., and Lai, O.M., Lipase-catalyzed incorporation of n-3 PUFA into palm oil, *J. Am. Chem. Soc.*, 80, 1197, 2003.
83. Lee, K.T. and Akoh, C.C., Effects of structured lipid containing omega-3 and medium chain fatty acids on serum lipids and immunological variables in mice, *J. Food Biochem.*, 23, 197, 1999.
84. Spurvey, S.A., Senanayake, N., and Shahidi, F., Enzyme-assisted acidolysis of menhaden and seal blubber oils with-linoleic acid, *J. Am. Chem. Soc.*, 78, 1105, 2001.
85. Abigor, R.D. et al., Production of cocoa butter-like fats by the lipase-catalyzed interesterification of palm oil and hydrogenated soybean oil, *J. Am. Chem. Soc.*, 80, 1193, 2003.

86. Gitlesen, T. et al., High-oleic-acid rapeseed oil as starting material for the production of confectionary fats via lipase-catalyzed transesterification, *Ind. Crops Prod.,* 4, 167, 1995.
87. Jandaeck, R.J. et al., The rapid hydrolysis and efficient absorption of triglycerides with octanoic acid in the 1 and 3 positions and long chain fatty acid in the 2 position. *Am. J. Clin. Nutr.,* 45, 940, 1987.
88. Haumann, B.F., Structured lipids allow fat tailoring, *INFORM,* 8, 1004, 1997.
89. Pszczola, D.E., Products and technologies, *Food Technol.,* 57, 54, 2003.
90. Kosmark, R. SALATRIM: Properties and applications, *Food Technol.,* 98, 101, 1996.
91. Quinlan, P.T. and Moore, S.M., Modification of triacylglycerols by lipases: process technology and its application to the production of nutritionally improved fats, *INFORM,* 4, 580, 1993.
92. Hammond, E.W., Analytical data, in *Chromatography for the Analysis of Lipids,* Hammond, E.W., Ed., CRC Press, Boca Raton, FL, 1993, p. 169.
93. Cho, E.J. and Lee, K.T., Analysis of phytosterols and tocopherols, and production of structured lipids from the extracted plant oils, *Korean J. Food Preserv.,* 10, 370, 2003.
94. Wu, G.H., Jarstrand, C., and Nordenström, J., Phagocyte-induced lipid peroxidation of different intravenous fat emulsions and counteractive effect of vitamin E, *Nutrition,* 15, 359, 1999.

Index

B

C

D

E

F

G

H

I

J

K

L

M

N

O

P

R

S

T

Coventry University